ADVANCES IN
GEOPHYSICS

VOLUME 29

Advances in
GEOPHYSICS

VOLUME 29

Anomalous Atmospheric Flows and Blocking

Series Editor

BARRY SALTZMAN

Department of Geology and Geophysics
Yale University
New Haven, Connecticut

Guest Editors

ROBERTO BENZI

IBM
Rome, Italy

AKSEL C. WIIN-NIELSEN

Danish Meteorological Institute
Copenhagen, Denmark

1986

ACADEMIC PRESS, INC.
Harcourt Brace Jovanovich, Publishers

Orlando   San Diego   New York   Austin
Boston   London   Sydney   Tokyo   Toronto

COPYRIGHT © 1986 BY ACADEMIC PRESS, INC.
ALL RIGHTS RESERVED.
NO PART OF THIS PUBLICATION MAY BE REPRODUCED OR
TRANSMITTED IN ANY FORM OR BY ANY MEANS, ELECTRONIC
OR MECHANICAL, INCLUDING PHOTOCOPY, RECORDING, OR
ANY INFORMATION STORAGE AND RETRIEVAL SYSTEM, WITHOUT
PERMISSION IN WRITING FROM THE PUBLISHER.

ACADEMIC PRESS, INC.
Orlando, Florida 32887

United Kingdom Edition published by
ACADEMIC PRESS INC. (LONDON) LTD.
24–28 Oval Road, London NW1 7DX

LIBRARY OF CONGRESS CATALOG CARD NUMBER: 52-12266

ISBN 0–12–018829–5

PRINTED IN THE UNITED STATES OF AMERICA

86 87 88 89   9 8 7 6 5 4 3 2 1

CONTENTS

FOREWORD xi
PREFACE xiii

Part I. Historical Introduction

Global Scale Circulations — A Review
A. C. WIIN-NIELSEN

1. Introduction 3
2. The Distant Past 3
3. The Near Past 8
4. Blocking as a Structural Entity 15
5. Some General Considerations 24
6. A Concluding Remark 25
   References 26

Part II. Observations

The Life Cycles of Persistent Anomalies and Blocking over the North Pacific
RANDALL M. DOLE

1. Introduction 31
2. Data 32
3. Procedure 33
4. Development of 500-mbar Height Anomaly Patterns 34
5. Vertical and Thermal Evolutions during Development 50
6. Synoptic Characteristics of Development 54
7. Breakdown 59
8. Discussion 64
9. Conclusions 66
   References 68

On Atmospheric Blocking Types and Blocking Numbers
HEINZ-DIETER SCHILLING

1. Introduction 71
2. Energy Parameters and Data 72
3. Blocking Numbers 77
4. Kinetic Energy Budget and Blocking Numbers 80

5. Relevant Energy Fluxes	85
6. Types of Blocking	89
7. Conclusions	94
Appendix A	96
Appendix B	97
Appendix C	98
References	99

Observational Characteristics of Atmospheric Planetary Waves with Bimodal Amplitude Distributions

ANTHONY R. HANSEN

1. Introduction	101
2. Frequency Distribution of the Planetary-Wave Amplitude Indicator	103
3. Wave Structure of the Two Modes	105
4. Energetics and Enstrophy Budgets for the Two Modes	113
5. Role of Cyclone Waves in Blocking	127
6. Conclusion	131
References	132

A Case Study of Eddy Forcing during an Atlantic Blocking Episode

G. J. SHUTTS

1. Introduction	135
2. The Eddy Straining Mechanism	136
3. Data Manipulation and the Synoptic Situation	139
4. E Vectors and the Sense of Momentum Forcing	142
5. Eddy Vorticity Flux Divergence Patterns	145
6. Ertel Potential Vorticity Analysis	147
7. Summary and Discussion	158
References	161

Part III. Theory

The Effect of Local Baroclinic Instability on Zonal Inhomogeneities of Vorticity and Temperature

R. T. PIERREHUMBERT

1. Introduction	165
2. Eddy Fluxes in the Two-Layer Model	167
3. Fluxes and Tendencies Associated with Local Baroclinic Instability	169

4. Conclusions	180
References	182

Forcing of Planetary-Scale Blocking Anticyclones by Synoptic-Scale Eddies

J. Egger, W. Metz, and G. Müller

1. Introduction	183
2. Stochastically Forced Planetary Modes	184
3. Results	186
References	197

Deterministic and Statistical Properties of Northern Hemisphere, Middle Latitude Circulation: Minimal Theoretical Models

A. Speranza

1. Introduction	199
2. A Reexamination of CDV	200
3. Modification of the CDV Wave Equation	209
4. Baroclinic Energetics	215
5. Resonance Bending in a Baroclinic Model Atmosphere	219
6. Summary and Conclusions	223
References	224

Probability Density Distribution of Large-Scale Atmospheric Flow

Alfonso Sutera

1. Introduction	227
2. Theoretical Background	228
3. The Data	230
4. Nonparametric Probability Density Estimation	231
5. Results	233
6. Connection with Patterns of the 500-mbar Geopotential Height	235
7. Discussion	243
8. The Zonal Wind	246
9. Conclusions	247
References	248

Stationary Planetary Waves, Blocking, and Interannual Variability

R. S. Lindzen

1. Introduction	251
2. How Persistent Are Anomalies?	253

3. Multiple Equilibria?	255
4. Teleconnections—The Tropical Connection?	259
5. Linearized Response to Stationary Forcing	262
6. Free Rossby Waves and the Meaning of Persistence	269
7. Concluding Remarks	271
References	272

Part IV. Numerical Experiments

Instability Theory and Nonlinear Evolution of Blocks and Mature Anomalies

J. S. FREDERIKSEN

1. Introduction	277
2. Three-Dimensional Instability Theory	278
3. Time Evolution of Observed Mature Anomalies	289
4. Nonlinear Simulation	291
References	301

Numerical Prediction: Some Results from Operational Forecasting at ECMWF

A. J. SIMMONS

1. Introduction	305
2. The ECMWF Forecasting System	306
3. Methods of Assessment	308
4. The Accuracy of Forecasts in the Medium Range	310
5. The Prediction of Blocking and Cutoff Lows	318
6. Developments in Predictive Skill	323
7. The Representation of Monthly-Mean Anomalies	328
8. Systematic Model Errors	333
9. Concluding Remarks	335
References	336

Envelope Orography and Maintenance of the Quasi-Stationary Circulation in the ECMWF Global Models

STEFANO TIBALDI

1. Introduction	339
2. Orographic Forcing and the Systematic Error of the ECMWF Gridpoint Model: The Envelope Orography	340
3. The January 1981 Set of Experiments	343
4. The Experiments with the Blended Orographies	355

5. Mountain Torque and Zonal Flow	359
6. The Effects of the Envelope Orography in the Tropical Regions	363
7. Summary and Conclusions	370
References	372

Numerical Forecasts of Tropospheric and Stratospheric Events during the Winter of 1979: Sensitivity to the Model's Horizontal Resolution and Vertical Extent

CARLOS R. MECHOSO, MAX J. SUAREZ, KOJI YAMAZAKI, AKIO KITOH, AND AKIO ARAKAWA

1. Introduction	375
2. Selected Features of the Atmospheric Circulation during the Northern Hemisphere Winter of 1979	377
3. Description of the Model	380
4. Tropospheric Forecasts	381
5. Stratospheric Forecasts	392
6. Impact of the Upper Boundary on Tropospheric Forecasts	398
7. Conclusions	411
References	413

Mechanistic Experiments to Determine the Origin of Short-Scale Southern Hemisphere Stationary Rossby Waves

EUGENIA KALNAY AND KINGTSE C. MO

1. Introduction	415
2. Analysis and Control Experiments	417
3. "No Andes" Experiment	427
4. "Reduced Tropical Heating" Experiment	429
5. "Suppressed Regional Heating" Experiments	435
6. "Easterly Deceleration" Experiment	437
7. Summary and Conclusions	439
References	441

SST Anomalies and Blocking

J. SHUKLA

1. Introduction	443
2. Influence of Tropical SST Anomalies on Extratropical Circulation	445
3. Influence of Extratropical SST Anomalies on Extratropical Circulation	449
References	451

INDEX	453

FOREWORD

It has been our custom occasionally to devote a single volume of *Advances in Geophysics* to a special topic of timely interest. In the present volume we consider the rapidly growing subject of low-frequency weather variability. This variability is exemplified by the persistence of certain large-scale "anomalous" features of the atmospheric flow that are often related to the so-called blocking weather patterns. It is widely acknowledged that if an extension of the range of reliable weather prediction is to be achieved it will depend on an understanding of the behavior of these persistent anomalous flows. A notable feature of this collection of reviews and original articles is its broad coverage of the observational, theoretical, and numerical–experimental aspects of this subject by many of the leading investigators in this field. Preliminary versions of these articles were presented at a workshop meeting held in Rome in the summer of 1984 under the auspices of IBM.

The articles give an excellent overview of the subject, clearly elucidating the full scope of the problems encountered in trying to understand low-frequency weather phenomena. Perhaps the most significant new results to emerge are the observational confirmations, reported here for the first time, of the existence of a bimodality in large-scale wave structure of the atmospheric flow. The inference from this bimodality is that an important long-wave instability must be present that is probably associated with the nonhomogeneous topographic and/or thermal properties of the lower boundary. In a sense, these findings represent a triumph of a rapidly emerging new paradigm of dynamical meteorology, pioneered by E. N. Lorenz, based on examining the finite-amplitude nonlinear properties of the atmosphere from the viewpoint of low-order dynamical systems analysis. Although the first (barotropic) theoretical scenarios proposed along this line do not seem applicable to the particular form of bimodality revealed by the data, some of the theoretical ideas proposed in this volume offer much promise that a successful and illuminating low-order theory is in the offing. In spite of some continuing reservations concerning these ideas, a few even voiced in this volume, it appears that a major new opening has been achieved that, one hopes, will lead to measurable improvement in longer range atmospheric predictability.

Barry Saltzman

PREFACE

In the last few years a good deal of scientific attention and work has been devoted to trying to understand the physical causes and mechanism of anomalous circulation in the atmosphere. As often happens in meteorology, most of the effort has been addressed to posing the right questions to be answered. At the end of August 1984, a workshop was organized in Rome, bringing together most of the leading experts in the field, with the purpose of summarizing and clarifying the scientific results obtained thus far. The central question for the workshop was as follows: does there exist a possible physically sound definition of anomalous circulation or blocking, and if so which processes are connected to the development and decay of anomalous flows so defined?

Looking back at the scientific literature it is easy to recognize that the starting point of recent interest in this subject was the winter of 1976, when extreme cold weather conditions were experienced in North America. Since that year a number of papers were devoted to describing or simulating blocking events in the atmosphere. A challenging problem was also posed concerning the degree to which medium-range weather forecasting in general could improve qualitatively and quantitatively, by improving numerical forecasting of blocking events.

One of the outcomes of the Rome meeting has been to restate the problem in terms of a number of well-defined scientific questions to be answered. From this point of view, this book represents the contribution by all the speakers of the workshop with the common intention of underlining the scientific strategy for the future. Here we shall briefly synthesize the major results.

It has been known since the work of Charney (1947) and Eady (1949) that the driving engine for the atmospheric motion is the baroclinic instability acting on a scale L of about 4000 km. The discovery of this process had made possible the development of operational numerical weather forecasts in which cyclogenetic processes could be predicted. The characteristic range of validity of numerical forecasts at present is estimated to be of the order of 4–5 days. This characteristic time corresponds quite well to the time scale L/U where U is the zonal wind speed typically of the order of 10 m sec^{-1}. Simmons' article offers a systematic review of the ECMWF operational model used for the last few years. It is pointed out that the loss of predictability in this time range is mainly due to inability to predict ultralong waves of about 10,000 km. If we assume that a different physical process is acting on the very large scale of atmospheric circulation, we could guess that the range of numerical weather prediction, at least for the slow component of the system, can be extended to time scale of the order of 10,000 km/U, namely, 10–15 days. This speculation, underlying most of the meeting discussion, is directly linked with the observational studies presented by many authors (Dole, Schilling, Sutera, Hansen). In particular Dole and Schilling, analyzing Pacific and Atlantic blocking events, found that the characteristic time scale for such anomalous flows is of the order of 10 days. Moreover, the results presented by Egger *et al.* and Sutera show that the probability distribution of the duration in time of anomalies is with good accuracy a Poisson distribution with average time of about 10 days. During the workshop discussion it was clarified that a Poisson distribution is not the consequence of an absence of a characteristic time scale but, on the contrary, it is indicative that an instability mechanism is acting on the system. In agreement with this are the results of Frederiksen, who studied the set of normal mode instabilities acting on the climatological state. Stationary waves dominated the growing unstable field after 5 days, with a pattern resembling North Pacific and Atlantic anomalies. Assuming that an instability different from ordinary baroclinic instability is acting on the ultralong planetary waves in middle latitude circulation, we are faced with the question whether this process is triggered by changes in the boundary conditions or in the forcing terms (such as tropical forcing or sea surface temperature anomalies) or, alternatively, is mainly due to the internal variability of the atmospheric system. Kalnay and Mo's article addresses directly this question for the January 1979 Southern Hemisphere anomaly using the general circulation model (GCM) of GLAS. In that case tropical forcing seems to be particularly important to explain both the duration and the location of the anomaly. Shukla discusses explicitly the Northern Hemisphere response to an "El Niño" year by GCM simulations. The correlations between sea surface temperature (SST) anomalies and blocking events is indicative of a nonnegligible response of Northern Hemisphere circulation to tropical forcing. It is in general very difficult to use GCMs to distinguish between internal and external variability of atmospheric circulations. In the article by Mechoso *et al.* the effects of high and low spatial resolutions (both vertical and horizontal) on the simulation of Northern Hemisphere anomalies using the UCLA GCM model are discussed. Special attention is focused on the ability to forecast stratospheric sudden warming. Resolution problems for the stationary forcing due to orography are also discussed in Tibaldi's article. The numerical experiments performed by Tibaldi show that systematic corrections of the zonal flow probably have the most important effect in improving the predicted behavior of the ultralong planetary waves. By this brief discussion of GCM simulations we conclude that the representation of the anomalous behavior in numerical models is not yet perfected; much more work is therefore needed. Nevertheless, as GCM simulations are the analogs in meteorology of laboratory experiments in physics, we can expect a systematic use of GCMs in the future to test different theories proposed to explain anomaly events.

The question of the relative importance of the external forcing and the internal variability for the atmospheric circulation is discussed theoretically in several articles in this book. Three main different aspects are studied, namely, the theory of multiple equilibria, the role of synoptic forcing connected with the regional character of blocking events, and the possible effects of ultralong two-dimensional traveling Rossby waves. The introductory article by Wiin-Nielsen gives an extensive review of the present state of the art in both theoretical and observational studies. The theory of multiple equilibria was independently proposed in 1979 by Charney and DeVore and Wiin-Nielsen. The basic idea of the theory is to study the coarse-grained structure of the phase space for the atmospheric "slow" variables, namely, the ultralong planetary waves. In its first formulation, the theory raised many doubts because of its apparent inconsistency with the observed atmospheric flows. However, the effect of nonlinear dynamics and a reinterpretation of the form drag equation are able to improve significantly the quality of the theory, as discussed in Speranza's contribution. A systematic check of multiple equilibria theory is presented by Sutera, who estimates the probability distribution of the amplitude of ultralong planetary waves (wave numbers 2, 3, 4) in the Northern Hemisphere winter circulation. An extensive analysis of a 20-year record gives strong evidence of a bimodal distribution as predicted by the theory of multiple equilibria. The persistence in time around each equilibrium has a Poisson distribution in agreement with the above-mentioned observations. The energetics of the two modes is presented in Hansen's article, in which the relevance of the stationary waves in both modes and the baroclinic source of energy in maintaining the high-amplitude mode are discussed.

The kind of instability assumed in the theory of multiple equilibria has a global nature, typically involving wave number 3 in the atmosphere. Superimposed on this scenario there is the possibility that some of the anomalous flows have a local character. It is implied that synoptic scale forcing plays a leading role in this case. The article by Egger *et al.* discusses such a role of synoptic scale forcing, as computed from observational data, on planetary waves using the barotropic vorticity equation. A good resemblance is found between computed and observed blocking events. Shutts describes the role of eddy vorticity forcing by which some Atlantic blockings are initiated and maintained. Finally Pierrehumbert discusses the concept of local multiple equilibria arising from absolute instability in both barotropic and baroclinic flows. It is worthwhile to remark that local theories of blocking are neither inconsistent nor at variance with global theories even if a unified approach has never been undertaken.

At variance with the interpretation of anomalous events as being the results of global or local dynamical instabilities, Lindzen proposed a theory depending on the behavior of two-dimensional traveling neutral Rossby waves. The theory is supported by observational analysis of FGGE data over the whole hemisphere, using Hough's modes. The characteristic period of such ultralong waves is about 20 days, and adjustment processes are supposed to be the triggering mechanism of these waves. It is pointed out that interference between waves of different scale length can produce patterns which synoptically are similar to anomaly flows.

We are not able at this stage to distinguish between the different interpretations discussed at the meeting, and future work is needed to solve some of the more puzzling problems. We feel confident, however, that we are pointing in the right direction and making fundamental strides toward improved understanding of large-scale anomalous states of the atmosphere with the hope that the range of weather forecasting can be extended.

ROBERTO BENZI
AKSEL C. WIIN-NIELSEN

Part I

HISTORICAL INTRODUCTION

GLOBAL SCALE CIRCULATIONS — A REVIEW

A. C. WIIN-NIELSEN

Danish Meteorological Institute
DK-2100 Copenhagen
Denmark

1. Introduction

It has been a pleasure for me that the organizers of the workshop on Global Scale Anomalous Circulation in the Atmosphere and Blocking (Rome, Italy, 1984) asked me to serve as a chairman of that interesting event. At the same time they were kind in inviting me to present an introductory lecture at the opening of the workshop. I selected to prepare a review of our understanding of the global circulation, and it was of course the most recent events which were the most interesting. In this regard I faced a dilemma. It was my privilege to know some of the studies which would be presented later on in the program, but I tried as far as possible to avoid stealing thunder from subsequent presentations.

The IBM Scientific Center in Rome, Italy, deserves great praise for its support of the workshop. On behalf of all the participants I should like to express appreciation to IBM for inviting us to the eternal city of Rome to discuss some of the most exciting meteorological problems facing our community. It is most welcome that a private corporation shows such an interest in our meteorological field. We want naturally to include our colleague, Dr. Roberto Benzi, who took all of the trouble in taking care of the scientific organization of the workshop.

My main task was to try to set the stage with the introductory lecture and otherwise to help create a constructive discussion of the problems which had been selected as main topics for the workshop. What follows herein is a version of the lecture presented at the workshop.

2. The Distant Past

When I was a young student at the newly created International Meteorological Institute in Stockholm and worked under Professor Carl-Gustaf Rossby, and we had the problem of building a small library, it was the general story that Rossby should have said that you can safely disregard all literature published before 1940. I do not know if this story is true, but it is nevertheless a fact that for our present problems at the workshop there is little to gain in the very old literature.

Let me try to give an eye-witness account of these early developments. It follows that such an account necessarily will be somewhat personal, but I shall naturally try to cover the main events. We may conveniently take our starting point in the upper-air network which was created during and immediately after World War II. It was considered by some forward-looking meteorologists, primarily Bjerknes and Mintz at the University of California, Los Angeles and V. Starr and his colleagues at the Massachusetts Institute of Technology, that this network was good enough to study the general aspects of the atmospheric circulation. The first attack was on the primary quantities such as meridional cross-sections of the zonally averaged wind and temperature fields, but at the same time there were calculations of the meridional transport of momentum and specific heat. The main question at that time, approximately 1950, was to determine the relative role of the mean meridional circulation and the transport of the eddies. Particularly, the MIT school under the leadership of V. Starr came out very strongly in favor of the important role of the transient eddies in accomplishing the necessary transport of both momentum and heat.

A rather strong reaction to their somewhat overstated case came particularly from E. Palmen of Finland, who argued convincingly that the mean meridional circulation was still of great importance in the tropics as compared to the eddies. A lively discussion developed both at scientific conferences and in the literature, but it was difficult to settle the disagreements because the network was not good enough to calculate the mean meridional circulations with sufficient accuracy.

A very important theoretical development took place with Lorenz's (1955) formulation of the concept of available potential energy and his description of the atmospheric energy cycle, resulting in the famous four-box energy diagram in which both available potential energy and kinetic energy were divided into the energy contained in the zonally averaged fields and that contained in the eddies. Many followers of Lorenz, including the speaker, used many investigations to calculate the various generations, conversions, and dissipations.

It has of course been known for many years that the radiation from the sun drives the atmosphere, and that the kinetic energy ultimately is dissipated, i.e., turned back to heat, by the forces of friction in the atmosphere. Lorenz's (1955) major contribution concerning atmospheric energetics was to show that the sum of potential and internal energy is at a minimum when the isentropic and the isobaric surfaces coincide with the horizontal surfaces.

Internal and potential energy which are proportional in an integrated sense in a hydrostatic atmosphere are *generated,* according to Lorenz's theory, when on average the warm air masses are heated and the cold air masses are cooled. On the other hand, the kinetic energy in the atmosphere is maintained by an internal process whereby internal and potential energy are *converted* into kinetic energy. Such a *conversion* will take place when on average warm air masses are rising and cold air masses are sinking. When such a process takes place it is seen that the center of gravity for the atmosphere is lowered, thus giving less potential energy. Finally, the kinetic energy is turned back into heat by the work of the frictional force, i.e., kinetic energy is dissipated.

The "easy" quantities were the conversions from zonal to eddy energies for both available potential energy and kinetic energy, because the main part of these could be computed from the temperature field and the horizontal wind field, and because the already mentioned meridional transports of sensible heat and momentum by the eddies were important elements of these calculations.

Much more difficult were the calculations of the generations of zonal and eddy available potential energies because they depend on a knowledge of the so-called diabatic heating of the atmosphere. Equally difficult were the conversions between available potential energies and kinetic energies in both zonal and eddy form because they require a knowledge of the vertical velocity.

We (Wiin-Nielsen and Brown, 1960) attempted to calculate the generations of available potential energy by first calculating the heating field as a residual from the thermodynamic equation every 12 hr, then proceeding to the calculations of the generations. This work was later carried further by Brown (1964).

Our work became immediately quite controversial because while the total generation $G(A)$ and the zonal generation $G(A_Z)$ were positive, it turned out that the eddy generation $G(A_E)$ was negative. We could of course argue that the negative generation indicated that on the average the cold air masses were warmed and the warm air masses were cooled, and that this indeed was the case in the radiation and in the exchanges between the atmosphere and the underlying surface whether it was continents, ice or snow fields, or the oceans. On the other hand, when it came to the heat from condensation and evaporation processes, all we could say was that the former processes gave a larger contribution than the latter.

In these years there was also an important contribution by Barry Saltzman (1957), who formulated the energy relations in the spectral domain. His formulation was still based on a Fourier transformation along individual latitude circles, but the normal field representation in the meridional and vertical directions. We had not yet moved completely into the spectral domain in 1957. Both Saltzman and I were independently engaged in the special aspects of the energy conversions from eddy available potential energy to eddy kinetic energy. I used the vertical velocities as they were computed from a two-level, quasi-geostrophic model at use at the Joint Numerical Weather Prediction where I was employed, and Saltzman and Fleisher used a similar although not identical procedure.

The main result was a spectrum with a marked maximum corresponding to the scale of baroclinic instability, and another maximum for the very long waves. It was difficult to explain the second maximum, but it was demonstrated that mountains and heating could reduce it if they had been incorporated in the model used at the time.

The time series used for these early calculations were pitifully short, and the results could not be considered as more than snapshots of the general circulation. It became quite a task during the coming years to extend all the calculations into the spectral domain and to cover a time period of at least a year. I had the opportunity to do this during my time at the National Center for Atmospheric Research (NCAR) and later at the University of Michigan. The calculations of the energetics were extended by distinguishing between the baroclinic and barotropic components of the kinetic energy, corresponding to the kinetic energy of the deviations from the vertical mean flow and the kinetic energy of the mean flow itself. I had the opportunity to develop this concept from a theoretical as well as an observational point of view in some early papers (Wiin-Nielsen, 1962). For the concept itself I was certainly inspired by J. Smagorinsky (1963), who later used it in his analysis of the so-called basic experiment with his general circulation model.

I am now approaching the subject of this workshop, i.e., the anomalous circulation in the atmosphere. My co-workers, Margaret Drake and John A. Brown, Jr., and I (1964) had decided that we would use the year 1962 as a representative year for calculations of the energetics in the spectral domain. We started from January 1962 and did the calculations day by day, making monthly averages as we went along. The results looked absolutely normal for the larger part of 1962 in the sense that the directions of the energy generations, conversions, and dissipations agreed with the now classical picture, i.e., $C(A_Z, A_E)$ and $C(K_E, K_Z)$ were positive. As we were about to finish the year and looked at the average results for December 1962, we were astonished to find that $C(K_E, K_Z)$ was negative, i.e., the eddies received kinetic energy from the zonal flow. Our first reaction was that something had gone wrong with the data, obtained from the National Meteorological Center (NMC) on magnetic tapes, or some silly mistake had come into the program used for the calculations. Checking and rechecking confirmed that everything was in order and we had to conclude that we were dealing with an anomalous situation.

We decided then out of curiosity that it would be interesting to continue the calculations into 1963, and we completed the first few months of that year with the results that $C(K_E, K_Z)$ stayed negative well into March 1963.

It was thus discovered from energy calculations that it is entirely possible for the atmosphere to work in at least two quite different modes. Also this result was, of course, controversial, and the reaction was almost immediate from MIT. Professor V. Starr questioned our results due to the fact that our formulations were different from his, because $C(K_E, K_Z)$ can be formulated in two different ways, depending on an integration by parts. The question came up because we used a less than hemispheric domain, and it was therefore a question of boundary conditions. John A. Brown, Jr., who was a graduate of MIT, was put on the problem, and he was able to show that although the two formulations gave different results because of the boundary conditions, it was definitely not enough to reverse the sign of $C(K_E, K_Z)$, which stayed negative.

Our calculations were of course not done in real time, but we were not far behind as you can see from the publication dates. When it was possible somewhat later to look at the real circulation during these anomalous winter months, it became clear that the winter 1962/1963 was indeed very far from normal. The upper air maps show rather typical blocking situations. All kinds of weather records were broken during the period, depending on where a given locality was situated with respect to the very longest waves in the upper air flow. Both the spectra of available potential energy and of kinetic energy showed that the greater portion of the energies were in the very longest waves.

Our conclusion must therefore be that it is possible for the atmosphere to work in at least two separate regimes. One of these dominates most of the time. It is characterized by a typical baroclinic regime in which the scale, characteristic of baroclinic instability, is at work more or less continuously giving rise to positive values of $C(A_Z, A_E)$ and $C(K_E, K_Z)$. The other regime is in a sense more barotropic because $C(K_E, K_Z)$ is negative as it would be during instability of the barotropic kind. In any case, the larger scales dominate, containing most of the energy, the flow is quite anomalous, and it is of blocking nature.

My own engagement in observational studies of atmospheric energetics came more or less to an end in 1967, when I wrote a summary paper on the annual variation and spectral distribution of energy in the atmosphere. I discovered the third power law for the distribution of kinetic energy in the atmosphere for relatively high wave numbers. Since then this work has been continued by F. Baer, L. Steinberg, V. Barros, and above all by M. Chen, who still continues to contribute to the study of atmospheric energetics.

3. The Near Past

Most meteorologists agree, probably, that the present intense interest in anomalous circulations, including blocking situations, stems from recent theoretical papers written some 5 years ago. It is of course true that the forecasters, especially those engaged in the longer ranges, have been aware of very anomalous circulations for a long time. In this regard, we may recall the investigations of blocking carried out by Rex from a synoptic and statistical point of view. The more recent studies by Austin (1980) in the United Kingdom, along the same lines and forming a natural extension of Rex's work (1950), are a welcome addition. I understand that further observational studies will be presented at this workshop, and I am looking forward to them.

The theoretical studies started with Charney and DeVore (1979) and Wiin-Nielsen (1979), who independently showed that a low-order, nonlinear barotropic system, with forcing and dissipation for certain parameter values, may possess multiple steady states, of which several may be stable. The bifurcation depends on the strength of the forcing, and both investigations suggest that the blocking phenomenon may indeed be one such stable state. In more detail, one finds typically two stable steady states, of which one has a rather strong zonal current (high index) and relatively small amplitudes in the eddies, while the other has a much weaker zonal current (low index) and a larger amplitude of the waves (the blocking situation).

My own interest in such problems started in the early 1970s because of the catastrophe theories developed by the French mathematician Thom, and because I got interested in certain aspects of population dynamics as carried out by ecologists who have developed relatively simple, but still nonlinear models of natural or laboratory situations of two or more competing populations. In my writings on the subject you will find some examples selected from these fields.

The most recent simple example from the meteorological field which displays the properties of multiple steady states is probably the spectral form of the advective equation with or without forcing and dissipation:

$$\partial u/\partial t + u(\partial u/\partial x) = -\epsilon u + B(u_E - u) \qquad (1)$$

in which u is the velocity in the x-direction, t is time, and u_E is a forcing.

Using an expansion of u and u_E in trigonometric series and reducing the spectral equations to just two components, we find, with the proper scaling,

$$dx/dt = xy - x + x_E$$
$$dy/dt = -x^2 - y + y_E \qquad (2)$$

Considering the case $x_E = 0$ we find that the steady states have either $x = 0$ or $y = 1$. The steady states are therefore

$$(0, y_E); \quad ((y_E - 1)^{1/2}, 1); \quad (-(y_E - 1)^{1/2}, 1)$$

We note that only one steady state exists for $y_E < 1$, while three exist for $y_E > 1$. The perturbation equations for a given steady state are obtained in the usual way. They are (with primes indicating perturbation quantities)

$$dx'/dt = xy' + yx' - x'$$
$$dy'/dt = -2xx' - y' \qquad (3)$$

from which we seek solutions of the form $(x', y') = (x'_0, y'_0)x\, e^{\sigma t}$. The frequency equation is:

$$(\sigma + 1)^2 - y(\sigma + 1) + 2x^2 = 0 \qquad (4)$$

Considering first the steady state $(0, y_E)$, we find that $\sigma = -1$ and $\sigma = y_E - 1$, indicating that this steady state is stable when it exists alone ($y_E < 1$), while it becomes unstable when the other two steady states exist. For these steady states $(\pm(y_E - 1)^{1/2}, 1)$ we find that $\sigma = 1/2(-1 \pm (9 - 8y_E)^{1/2})$ when $y_E > 1$. The solution turns out to indicate stability for all permissible values of y_E. The only difference is that for $1 < y_E < 9/8$ we have negative values of σ, while for $y_E > 9/8$, σ is complex with a negative real value.

We may thus conclude from this analysis that the nature of the steady state depends entirely on the magnitude of the forcing. For $y_E < 1$ there is only one steady state which is stable. For $y_E > 1$ there are three steady states of which the two new steady states are stable, while the former single steady state has become unstable. It may also be noted that the single steady state which exists for $y_E < 1$ has small amplitudes, while the two steady states for $y_E > 1$ may have very large amplitudes for large values of y_E. Finally, the selected steady state depends on the initial condition.

The analysis described above was carried out in 1976 and is published in the proceeding of the first European Centre for Medium Range Weather Forecasts (ECMWF) seminar (Wiin-Nielsen, 1976). Further extensions of the analysis to more complicated systems of the same kind have been made by Källen and Wiin-Nielsen (1980), and Wiin-Nielsen (1982, 1983).

We shall now proceed to other systems. Most investigations so far have been of low-order, nonlinear systems based on the barotropic vorticity equation with forcing and dissipation. This is true also for the pioneering studies of Charney and DeVore (1979) and Wiin-Nielsen (1979). These studies normally include a zonal component and one or more wave components. The forcing may be classified as vorticity forcing and/or mountain forcing, where the vorticity forcing, in my opinion, is a simulation of heating. A discussion has developed on whether or not both kinds of forcing are necessary to obtain multiple steady states. It is, however, quite clear by now that they supplement each other, but multiple steady states may exist if only one kind of forcing is present.

A much more important question is related to how realistic these investigations are. For the time being we shall discuss the solutions and leave the question of how realistic low-order models are to be answered later. Several serious questions can be raised with respect to the realism of the solutions. We note, for example, that (1) to bifurcate, i.e., to create multiple steady states when a zonal component is included, rather large values of the amplitude of the mountain and heating components are needed in the models mentioned above; (2) the energy levels of the solutions produced by most investigators are much higher than those obtained in observational studies; (3) correspondingly, generations, conversions, and dissipations are higher than observational studies show that they should be.

I decided about a year ago (Wiin-Nielsen, 1984) to investigate these questions in detail for the simple reason that my own studies are suspect in this regard. For this purpose I used a model with the spherical components $(0, 1)$ $(0, n)$ $(1, n_1)$, and $(1, n_2)$. As a general result it turned out that the energy levels are quite sensitive to the meridional scales involved, i.e., n, n_1, and n_2, in the sense that the smaller the scale, the smaller the energy amounts.

Concerning the energies, it turns out that the steady state energies can be expressed in u_0 and z, which are the amplitudes of the components $(0, 1)$ and $(0, n)$. Figure 1 shows the zonal kinetic energy K_Z as a function of z for various values of u_0. It is seen that only small values of z are permissible if K_Z shall have a reasonable value. Figure 2 shows, on the other hand, the eddy kinetic energy K_E as a function of the same parameters. For $z > 0$ it is seen that K_E is an increasing function of z, while distinct minima exist for $z < 0$. If the energy values of K_E shall be reasonable, it is a question to have z around the minimum for the given value of u_0. These considerations indicate the ranges for which reasonable solutions can be found. As can be seen, it is a tight squeeze without much margin. It is nevertheless possible to find triple solutions. The energetics of such stable solutions are shown in Fig. 3.

Even if it is possible as shown above to obtain reasonable answers, it is nevertheless a fact that the "window" in parameter space for proper solutions is quite narrow. One may therefore question if these models, which are so strongly dependent on the zonal current, describe the true mechanism in the atmosphere. Considering the synoptic development leading to blocking

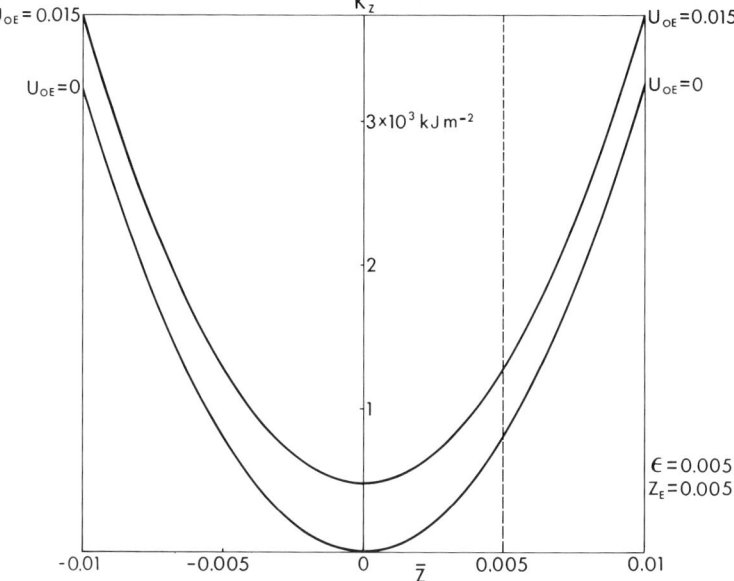

FIG. 1. The kinetic energy of the zonal flow, K_Z, as a function of z for various values of u_0.

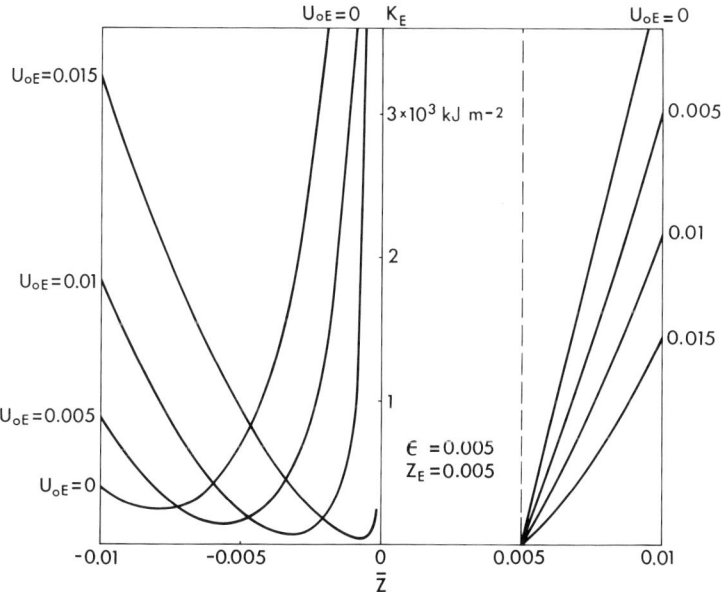

FIG. 2. The eddy kinetic energy as a function of z for various values of u_0.

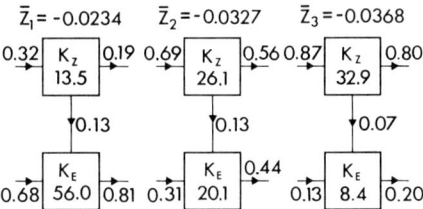

Fig. 3. The energy diagram for stable solutions.

it is rather unlikely that a critical value of the zonal current is the decisive factor. Rather, the impression is that the blocking configuration is the dying ground of a set of traveling disturbances, and that it is maintained by them.

In discussions with A. Sutera some months ago it occurred to us that the block as a nonlinear yet global phenomenon perhaps depended on a wave–wave interaction rather than interaction between waves and the zonal current. As a preliminary contribution to the solution of this problem I have investigated the nonlinear aspects of wave–wave interaction with a view toward multiple steady states and bifurcation. The starting point is also in this case the barotropic vorticity equation with mountain and vorticity forcing.

Using the scaling $a^2\Omega$ for the streamfunction (a = radius of the earth, Ω = angular velocity), Ω^{-1} as the time scaling, and $H_0 = RT_0/g$ as the mountain scaling, we get the following nondimensional equation:

$$\partial\nabla^2\psi/\partial t = J(\nabla^2\psi, \psi) + A_0 J(h, \psi) - 2(\partial\psi/\partial\lambda) - \Gamma(\psi - \psi^*) \quad (5)$$

Following Platzman (1962) and using his notation we have the following spectral equation:

$$\frac{d\psi_\gamma}{dt} = \frac{1}{2}i\sum_\beta\sum_\alpha \frac{c_\beta - c_\alpha}{c_\gamma} K(\gamma, \beta, \alpha)\psi_\beta\psi_\alpha$$

$$- \frac{1}{2}i\frac{A_0}{c_\gamma}\sum_\beta\sum_\alpha K(\gamma, \beta, \alpha)(h_\beta\psi_\alpha - h_\alpha\psi_\beta)$$

$$+ i\frac{2l_\gamma}{c_\gamma}\psi_\gamma - \Gamma(\psi_\gamma - \psi_\gamma^*) \quad (6)$$

where $K(\gamma, \beta, \alpha)$ is the interaction coefficient and $c = n(n + 1)$.

It is now possible to use the selection rules to reduce Eq. (6) to three complex equations for a triplet. If we further neglect the β term we may get a closed system of three equations by assuming that only the imaginary parts of the amplitudes are nonzero. The system then becomes

$$\frac{dy_1}{dt} = \frac{1}{2}g_1 y_2 y_3 + \frac{1}{2}\delta_1(m_3 y_2 - m_2 y_3) - \Gamma(y_1 - y_1^*)$$

$$\frac{dy_2}{dt} = \frac{1}{2}g_2 y_1 y_3 + \frac{1}{2}\delta_2(m_1 y_3 - m_3 y_1) - \Gamma(y_2 - y_2^*) \quad (7)$$

$$\frac{dy_3}{dt} = \frac{1}{2}g_3 y_1 y_2 + \frac{1}{2}\delta_3(m_2 y_1 - m_1 y_2) - \Gamma(y_3 - y_3^*)$$

where

$$g_1 = \frac{c_2 - c_3}{c_1} K; \quad g_2 = \frac{c_3 - c_1}{c_2} K; \quad g_3 = \frac{c_1 - c_2}{c_3} K$$

$$\delta_1 = \frac{A_0}{c_1} K; \quad \delta_2 = \frac{A_0}{c_2} K; \quad \delta_3 = \frac{A_0}{c_3} K$$

The simplifications leading to Eq. (7) are quite drastic, and the results should be viewed with some caution. At a later time I hope to report on the results obtained without the neglect of the β effect and considering the complete waves.

The energy relations for this system are easy to obtain. They are given schematically in Fig. 4 in which we notice that they consist of

(i) generations of the form: $c_i y_i y_i^*$
(ii) dissipation of the form: $c_i y_i^2$
(iii) nonlinear interactions: $(c_i - c_j) K y_1 y_2 y_3$
(iv) mountain effects: $A_0 K(m_i y_j y_k - m_k y_j y_i)$

We notice in this regard that the sums of the nonlinear interaction terms and the mountain effects each are zero, indicating that they do not change the total energy, but are distributive terms.

For purposes of illustration we shall look at an example. The following components are selected: (1, 3), (2, 2), (3, 4). The energy diagram is shown in Fig. 5 in which the amounts of energy are given in the units kJ m^{-2}, while generations, conversions, and dissipations are shown in 10^2 W m^{-2}. Three steady states have been found, and it can be shown that (a) and (c) are stable, while (b) is unstable. Comparing (a) and (c) we notice first of all that they differ greatly in the energy distribution, because (a) has most of the energy on K component (1) while (c) has a maximum on component (2). In addition, component (1) has the largest generation in case (a), while component (2) takes over this role in case (c). It should also be mentioned that the mountain heights used in this example were taken as those components which were

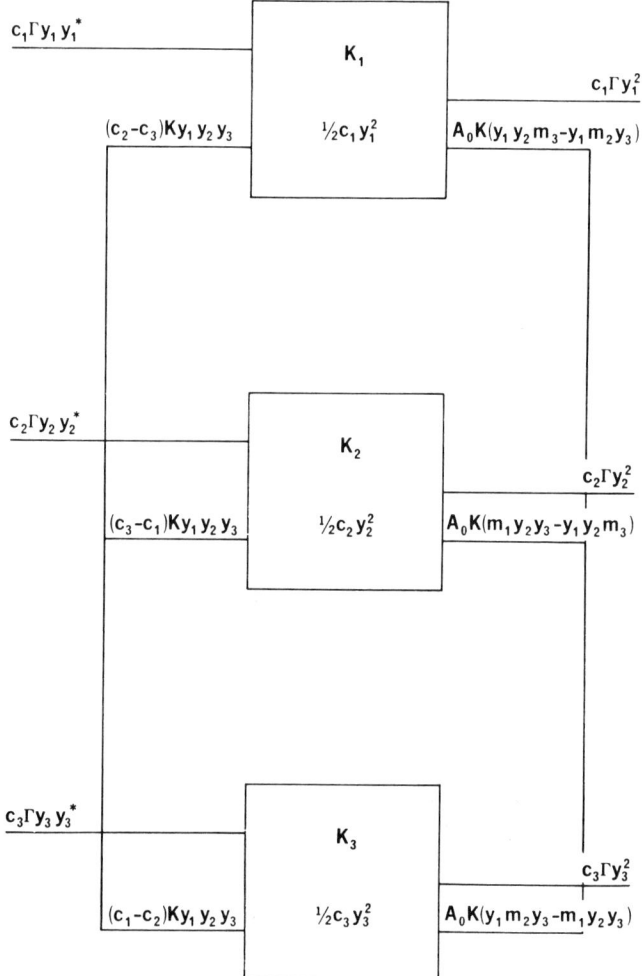

FIG. 4. Energy relations for a wave–wave interaction barotropic model.

obtained by a spherical harmonic decomposition of the real mountain height; m_1, m_2, and m_3 are therefore realistic.

It may on the other hand be interesting to look at the results if the mountains were excluded. This situation is depicted in Fig. 6. It is seen that multiple steady states can still exist without the mountains, that the transformations maintain the same directions as before although the energy amounts are changed, and that the relative distributions of energy on the components are maintained. It would thus appear that the mountain effects

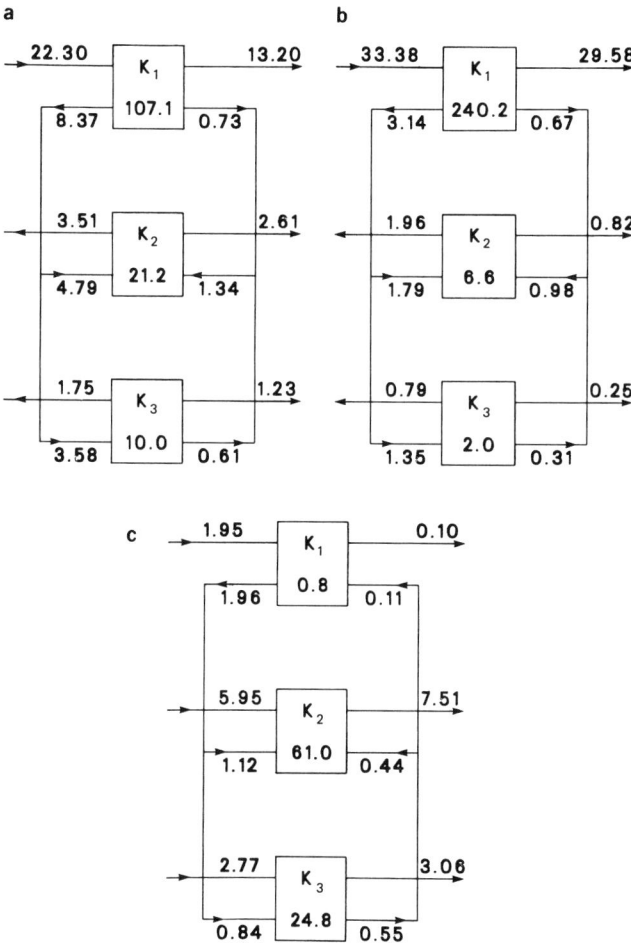

FIG. 5. A specific energy diagram for the model mentioned in Fig. 4. Energy amounts in the unit in kJ m^{-2}, others in 10^2 W m^{-2}.

are not vital for bifurcation, but have modifying influence. Studies of this kind are of course never conclusive, but the present study indicates that wave–wave interactions provide a better mechanism for multiple steady states because the "window" in parameter space is more open.

4. Blocking as a Structural Entity

Time-averaged maps of the atmosphere at various pressure levels, the so-called normal maps, where the averaging time is many years, show certain

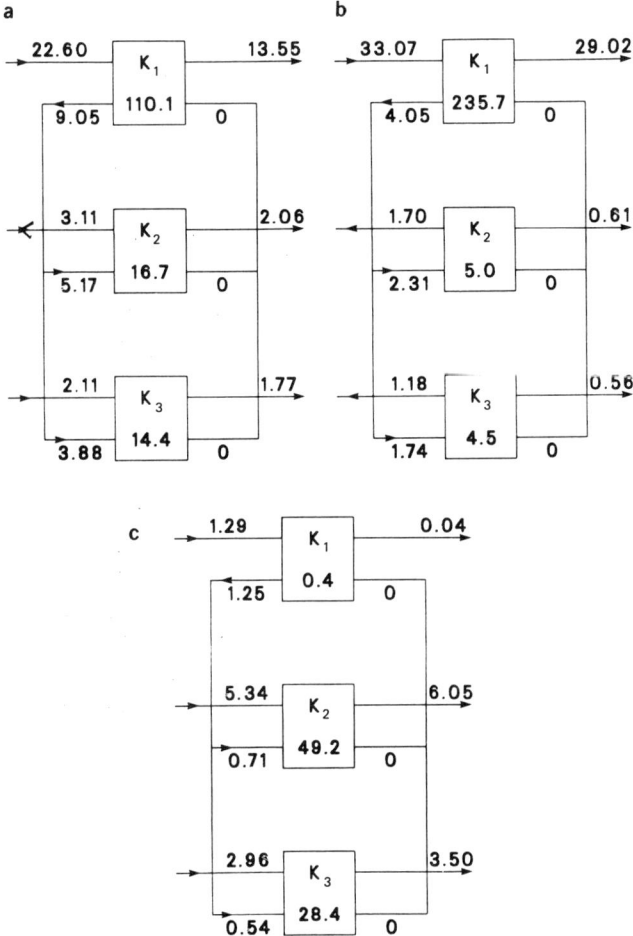

FIG. 6. The same situation as in Fig. 5 but without the mountain effect.

characteristic features such as well-defined troughs and ridges. These features are presumably of a stationary nature since otherwise they would have been eliminated in the averaging process. Indeed, it is these maps which are used to represent the stationary flow (zonal flow and waves) of the atmosphere.

Synoptic experience shows that the atmosphere additionally has other structural entities which are of shorter duration, but which nevertheless are of a quasi-stationary nature, and has a typical lifetime far exceeding that of a traveling disturbance. Some of these quasi-stationary atmospheric structures are seasonally dependent such as the continental wintertime anticyclone or the monsoon circulation. These structures will normally be seen on time-averaged maps for the proper individual month or season because their creation and maintenance and disappearance are strongly connected with the seasonal cycle more than the general circulation in the annual sense.

On a still shorter time scale it is observed that the atmosphere from time to time creates features which have a typical structure and exist for a certain period in an essentially unchanged form. The blocking situation is an example of such a structure. If the phenomenon of blocking as an atmospheric structure is to be discussed from a theoretical or practical point of view it would appear desirable to have a well-formulated definition of it. While such definitions exist, it is nevertheless true that there is no general agreement on them.

One of the first to take a theoretical interest in blocking was Rossby (1950) based on synoptic investigations by Berggren *et al.* (1949), Elliott and Smith (1949), and Namias and Clapp (1949). A more extensive "climatology" of blocking was carried out by Rex (1950), who also provided a detailed description of some blocking cases from a synoptic point of view as well as an analysis of the climatological aspects of blocking. The definition of blocking action provided by Rex (1950) is that the blocking case must exhibit the following characteristics:

1. The basic westerly current must split into two branches.
2. Each branch must transport an appreciable mass.
3. The double jet system must extend over at least 45° of latitude.
4. A sharp transition from zonal-type flow upstream to meridional-type downstream must be observed across the current split.
5. The pattern must persist with recognizable continuity for at least 10 days.

On the basis of this definition Rex (1950) was able to make a statistical analysis of blocking using the years 1933–1949 (inclusive). A main result was that blocking seems to have preferred locations in the eastern Pacific and eastern Atlantic regions with mean values around 150°W and 10°W, respectively. While Atlantic blocking cases may exist at all times of the year there is a minimum around August and a maximum in April. Pacific blocking cases are very rare in August and September, but show a broad maximum in the period March–May. From the data it is also possible to compute the percentage of blocked days. For the Atlantic cases one gets 27.6% and for the Pacific case 12.2%. Considering that the Atlantic cases are 68% of all cases, one gets as an average that 22.7% of all days are blocked days. This result is not inconsistent with a much later result obtained by A. Sutera (see this volume), who, using a different criterion, found that about 30% of all days were blocked days, but his results were obtained using a condition that the block should exist for only 4 days. Concerning the persistence of Atlantic blocking cases it is found by Rex (1950) that the average duration is about 17 days while the highest frequency is found for blocking cases which last 14 days. These studies show also the typical anomalies in temperature and precipitation which make the blocking action very important from a predictive point of view.

Austin (1980), in a study of blocking, is less stringent in the definition of the phenomenon. A distinction is made between the synoptician's point of view and that of the theoretical meteorologist. The first view is simply that blocking is a spatially isolated phenomenon consisting of a stationary high-pressure cell, warm in the troposphere and cold in the stratosphere, persisting in a region where westerly winds are normally found. The requirement is that the persistence is substantially longer than 3 days. On the other hand, it is also stated that, to those who are engaged in spectral studies, blocking is any stationary long wave of large amplitude, meaning that blocking is seen as a global-scale phenomenon.

Although the split in the jetstream is not made part of the definition itself it is nevertheless mentioned that it occurs, particularly in those cases where a high-pressure cell at about 60°N is located to the north of a low-pressure cell at about 40°N.

It will be noted that both of these definitions emphasize the connection between the split of the jet upstream and the location of the blocking high-pressure cell downstream. The definition as such does not imply any causal relationship between the split of the jetstream and the blocking anticyclone. On the other hand, the description of the irregular variation between the low- and high-index zonal circulation and blocking, as, for example, described by Namias and Clapp (1951), could lead to the impression that the creation of the blocking action follows the split in the jetstream in time. It is, indeed, this idea which forms the basis for the theoretical considerations put forward by Rossby (1950), who investigated the possibility that the blocking phenomenon is analogous to a hydraulic jump. This theory is far from satisfactory.

Another attempt at understanding the behavior of the zonally averaged flow and especially its interaction with the large-scale eddies was made by Thompson (1957) who developed a heuristic theory of the long-period variations in the zonal flow, i.e., a prediction equation for the time variation as influenced by the statistics of the eddy behavior. The theory is for two-dimensional barotropic flow, and it should be stressed that it is entirely internal in the sense that frictional and topographical effects are excluded. It is obvious that in forming the prediction equation for the zonally averaged flow it is a necessity to parameterize the influence of the eddies on this flow. Such a parameterization is possible only under a general hypothesis, which in Thompson's case is as follows:

1. The most effective agencies of momentum transport are eddies of very large scale, i.e., eddies with a meridional scale comparable with the width of the channel.

2. The statistical measures of the scale and intensity of turbulence at any time may be identified with the statistics of a v-field whose amplitude is a constant multiple of the amplitude of the initial v-field, and whose scale in the x-direction is a constant multiple of the scale of the same initial field.

The general assumptions are equivalent to a closure, and they permit the formulation of the prediction equation for the zonally averaged flow. This equation has, after some simplification, the form

$$\partial^2 U/\partial t^2 = -k^2 v_0^2 [\beta - \partial^2 U/\partial y^2 - (2/\lambda_0^2)U] \qquad (8)$$

where U is the zonally averaged wind, v_0^2 the initial variance of the meridional velocity, β the Rossby parameter, λ_0^2 a scale parameter, and k a constant.

Thompson (1957) shows by numerical integrations of this and similar, but more complicated, equations that a well-defined sharp jet in middle latitudes will split into two jetstreams. The positions of the jet maxima travel to the north and the south relative to the single original maximum. The typical time scale of the phenomenon is several days, and good agreement is obtained between a theoretical calculation of the speed of the traveling maxima and a determination of this speed from observational studies. The theory permits also a calculation of the changes in the variance of the eddy vorticity determined from the predicted changes in the zonal wind. The main conclusion from the study is that "a mechanism for the initiation and development of blocking is contained in the theory of barotropic flow."

The mechanism for the splitting of the simple jetstream into two separate jets was tested by Wiin-Nielsen (1961) in a low-order, barotropic system containing six components. The time scale as calculated by Thompson for typical cases was confirmed by these experiments, which did not use any other closure mechanism than the severe truncation in wavenumber space inherent in a low-order model.

The studies which have been summarized very briefly above are in sharp contrast to the blocking mechanism proposed by Charney and DeVore (1979) which was mentioned in an earlier section of this paper. We should note that the zonal wind in the latter study is a measure of the total angular momentum which in the model is influenced by the topography and the drag effect from the planetary boundary layer. The main point is, however, that the zonal wind in this model does not vary with latitude and, consequently, any description of a split in the jetstream is thus excluded from the beginning. This is not the case for the studies by Wiin-Nielsen (1979) and Källen (1981), which at least contain the interaction between the eddies and the zonal flow albeit in a low-order model.

Most of the examples given by Thompson (1957) are related to a sharp symmetrical jet. For these examples it can be shown that the general equation with good approximation can be reduced to Eq. (8). The numerical integrations show that the sharp jet in general will split in two branches, one traveling north and the other south. It may be of some interest to make a further analysis of Eq. (8). In the following discussion we shall consider a case which assumes that the *initial* convergence of the momentum transport vanishes, which in two-dimensional barotropic flow is equivalent to $\partial U/\partial t = 0$ at $t = 0$. In addition, in agreement with the previous study we shall also assume that $v^2 = v_0^2 = \text{const.}$, where v_0^2 is the initial value at the center of the channel. If we under these assumptions determine $U_s(y)$ in such a way that the right side of Eq. (8) vanishes, we know that $\partial^2 U/\partial t^2 = 0$. It is then easy to show that the third and all other higher derivatives vanish or, in other words, $U_s(y)$ is a steady state. To obtain this steady state in a convenient form we nondimensionalize the independent variables as follows:

$$\tau = t/d, \quad \eta = y/W, \quad d = 1 \text{ day} = 0.864 \times 10^5 \text{ sec} \quad (9)$$

and the steady state equation is

$$\partial^2 U_s/\partial \eta^2 + (2W^2/\lambda_0^2)U_s = \beta W^2 \quad (10)$$

Equation (10) is solved with the boundary condition

$$U_s = U_0 = \bar{u}_0 - \tilde{u}_0, \quad \eta = 0; \quad U_s = U_1 = \bar{u}_1 - \tilde{u}_0, \quad \eta = 1 \quad (11)$$

where $\tilde{u}_0 = \int_0^1 \bar{u} \, d\eta$ is constant and can be calculated from the initial distribution $\bar{u}_0(\eta)$. The solution to Eq. (10) is then

$$U_s = \frac{1}{2}\beta\lambda_0^2 + \left(U_1 - \frac{1}{2}\beta\lambda_0^2\right)\frac{\sin[\sqrt{2}(W/\lambda_0)\eta]}{\sin[\sqrt{2}(W/\lambda_0)]} + \left(U_0 - \frac{1}{2}\beta\lambda_0^2\right)\frac{\sin[\sqrt{2}(W/\lambda_0)/(1-\eta)]}{W} \quad (12)$$

Using Thompson's theory we may adopt the following values for the various parameters:

$$W^2/\lambda_0^2 = \pi^2(1 + W^2/L^2)$$

$$\beta\lambda_0^2 = \beta\frac{W^2}{\pi^2(1 + W^2/L^2)}$$

$$\beta = 16 \times 10^{-12} \text{ m}^{-1} \text{ sec}^{-1}$$

$$W = 5 \times 10^6 \text{ m}$$

$$L = 28 \times 10^6 \text{ m}$$

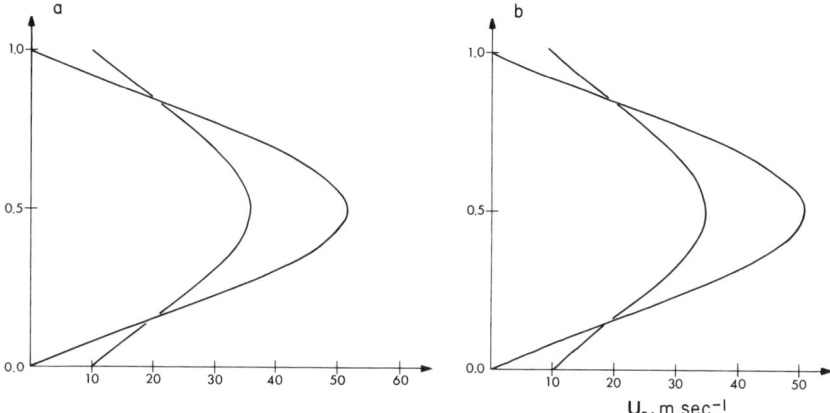

FIG. 7. Steady-state wind profiles for various boundary conditions. (a) Obtained by solving the finite-difference equations; (b) based on the analytical solution.

Figure 7 shows $U_s(\eta)$ for two values of U_0 and U_1. If $U_0 = U_1$ it is seen that $U_s(\eta)$ is symmetric around the center of the channel. We return next to Eq. (8). From this equation and the equation

$$0 = -\overline{v_0^2}(d^2/W^2)[\beta W^2 - \partial^2 U_s/\partial\eta^2 - (2W^2/\lambda_0^2)U_s] \qquad (13)$$

we obtain by subtraction and denoting

$$u_1 = U - U_s \qquad (14)$$

$$\frac{\partial^2 u_1}{\partial \tau^2} = \overline{v_0^2}(d^2/W^2)[\partial^2 u_1/\partial\eta^2 + (2W^2/\lambda_0^2)u_1] \qquad (15)$$

The solutions of Eq. (15) are obtained by integration of the corresponding finite-difference equation. For this purpose we introduce a grid size of $\Delta\eta$ and a time step of $\Delta\tau$. Counting the time steps by n and the space variable by m we have

$$u_1(n+1, m) = 2u_1(n, m) - u_1(n-1, m) + \overline{v_0^2}\frac{d^2}{W^2}\left(\frac{\Delta\tau}{\Delta\eta}\right)^2$$
$$\times \left[u_1(n, m-1) + u_1(n, m+1)\right.$$
$$\left. - \left(2 - \frac{2W^2}{\lambda_0^2}\Delta\eta^2\right)u_1(n, m)\right] \qquad (16)$$

As mentioned earlier we shall be particularly interested in the possibility that the initial jetstream may split in two branches. We may get a first idea of this possibility by investigating the initial tendency in the center of the channel if we restrict this part of the investigation to a jet which is symmetrical around the center of the channel. Setting $n = 1$ and denoting $m = M$ for the center we note that

$$u_1(1, M) = u_1(0, M) \tag{17}$$

and we shall consider the quantity $u_1(2, M) - u_1(1, M)$, which, according to Eq. (16) and recalling the symmetry, may be written as follows:

$$u_1(2, M) - u_1(1, M) = \overline{v_0^2} \frac{d^2}{W^2} \left(\frac{\Delta \tau}{\Delta \eta}\right)^2$$
$$\times \left[2u_1(1, M+1) - \left(2 - \frac{2W^2}{\lambda_0^2} \Delta \eta^2\right) u_1(1, M)\right] \tag{18}$$

The jet will have a tendency to split if the left-hand side of Eq. (18) is negative or if the bracket in the equation is negative. From this condition we get

$$u_1(1, M+1)/u_1(1, M) \leq 1 - (W^2/\lambda_0^2)(\Delta \eta)^2 \tag{19}$$

Equation (19) says that the tendency for the splitting of the jet will occur if the jet is sufficiently sharp. Using the same parameter values as before and $\Delta \eta = 0.1$ we find that

$$u_1(1, M+1)/u_1(1, M) \leq 0.9 \tag{20}$$

Equation (19) is an initial tendency only and does not necessarily result in a division of the jet in two branches. It may, however, be used as a guide for numerical experiments. For these experiments we have to select an initial wind profile $u_1 = u_1(\eta)$. The following profile was selected:

$$u_1 = u_0 \exp\{-[(\eta - a)/b]^2\} \tag{21}$$

Since we consider a jet which is symmetrical around the middle we select $a = 0.5$ and continue to use $\Delta \eta = 0.1$. The center value is u_0, while the value one grid-size away is

$$u_0 \exp[-(1/10b)^2] \tag{22}$$

The value b corresponding to Eq. (20) is then $b = 0.3$. If b is smaller than this value there is an initial tendency for a decrease of the center value. Such a case is shown in Fig. 8 calculated for $b = 0.25$. After 1 day we observe that the maximum value has decreased and that the jet has become wider. This process goes on further and after 1.5 days it is observed that the two branches start to appear. A case with $b = 0.35$ is shown in Fig. 9. In this case the jet continues to increase and become broader. The cases illustrated in Figs. 8 and 9 are close to the dividing case where the equality sign applies in Eq.

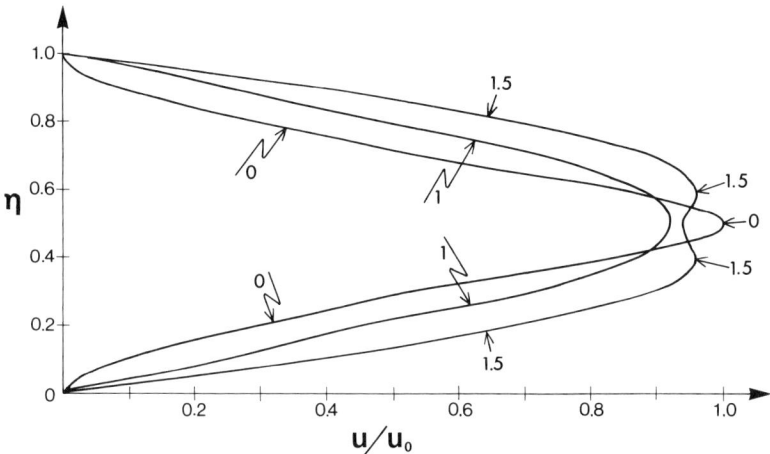

Fig. 8. A time integration leading to a division of the jet after about 1.5 days. Parameters: $a = 0.5$, $b = 0.25$.

(19). More extreme cases of the type considered by Thompson (1957) are easily computed. Figure 10 illustrates a case of a very sharp and narrow jet which after 1 day has developed two branches. The two maxima move away from the center.

The brief review of Thompson's theory made above shows that in its simplified form it has a steady state determined by the β-effect and the boundary condition (Fig. 7). The equation for the time development of the zonally averaged flow contains the mechanism for the splitting of the jet if it is sufficiently sharp, as shown by the examples (Figs. 8 and 9). In extreme

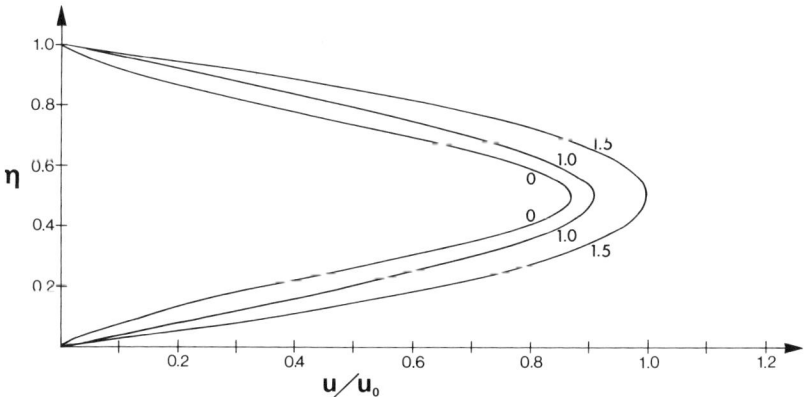

Fig. 9. A time integration leading to an increase and a broadening of the jet. Parameters: $a = 0.5$, $b = 0.35$.

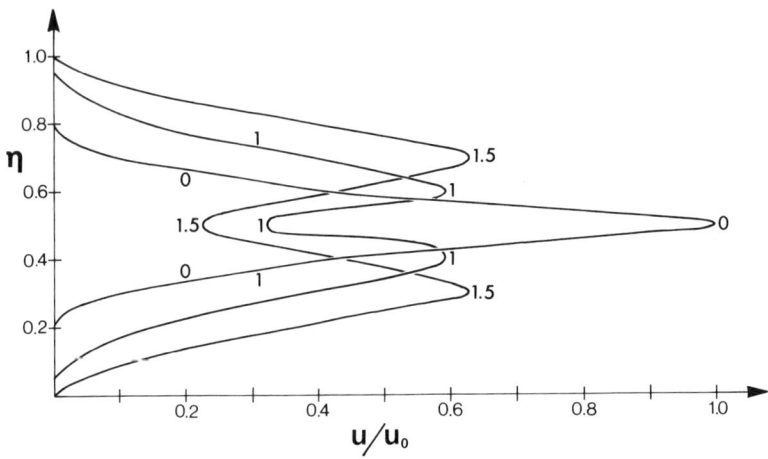

FIG. 10. A time integration of an initially very sharp and narrow jet leading to two branches after about 1 day.

cases (Fig. 10) it is possible to reproduce the northward and southward movement of the two maxima. Under the assumption that the splitting of the jetstream appears first and the development of the blocking pattern is a secondary effect, this theory is relevant to the phenomenon of blocking.

5. Some General Considerations

It is my hope that I succeeded in giving a reasonable introductory lecture to the workshop on anomalous circulations including blocking. The participants undoubtedly provided many more original contributions. Before the end of the lecture it seemed appropriate to discuss the state of affairs in this important, but specialized field:

1. It is an important field of research because anomalous circulations are connected with equally anomalous weather situations. The duration of these anomalies is also such that the time scale reaches into the lower end of the climatic time scale, which deals with the problems of long-range predictions and interannual variation. It is perhaps here that the anomalous circulations are particularly important from a practical point of view.

2. The present studies of anomalous circulations are concerned with the causes for their creation, maintenance, and dissipation. The physical factors being investigated at the moment are the interactions between the atmosphere and the oceans and continents on the one hand and the mountain effect on the other. The distinct possibility exists that the long-lasting anoma-

lies cannot be understood without a coupled ocean–atmosphere model. Further progress may depend on how successful we are in creating the proper models for a two-way interaction with the oceans.

3. Even if the coupled models may not be necessary for the understanding of the anomalies of shorter duration, it is nevertheless of importance to know how dependent simulations or predictions of anomalous circulations are on the specifications of the forcing.

4. As you will have seen from the workshop lecture, there has been considerable emphasis on the use of simple, although nonlinear, models in many studies, with emphasis on multiple steady states as a possible explanation of the existence of anomalous circulations. Several questions arise in this regard. A weakness in the theory is that it does not provide an answer to the question of how the models (or the atmosphere) select a given steady stable state. Is this an initial value problem? One may think so, because the selected stable steady state could be determined by whether or not the initial state is within the attractor basin of the given asymptotic state. On the other hand, how does the atmosphere move from one attractor basin to another?

5. A property of the low-order, nonlinear systems is the existence of multiple steady states, of which at least one corresponds to a stable anomalous circulation. Is this picture, which is so clear for the simple models, a good background for a study of the real atmosphere? Does the real atmosphere possess such multiple, quasi-stationary states? — or are we dealing with illusions or "barking up the wrong tree"?

6. There seems to be no doubt that the atmosphere from time to time can operate in rather different regimes, as we can see from blocking situations or a comparison between very anomalous and more normal winters and summers. An area of research for the future is to use our global, high-resolution models to simulate such anomalous situations, while the object of the GCM studies in the past seems to have been to simulate the normal or long-term average situation as well as possible.

6. A Concluding Remark

This paper is written in a personal style, and it is an attempt to review only some aspects of anomalous circulations. It has not been the purpose to write a complete review and, consequently, many important contributions from the distant and near past remain unmentioned. With the recent greatly increased interest in anomalous circulations it is certainly too early to attempt a final account of the truly important contributions or those which, in retrospect, looked interesting and important, but turned out to be a sidetrack or a move in a wrong direction.

Acknowledgment

I want to thank Ms. Charlotte Wiin-Nielsen, who contributed to the paper and its calculations and, furthermore, brought it into final form.

References

Austin, J. F. (1980). The blocking of middle latitude westerly winds by planetary waves. *Q. J. R. Meteorol. Soc.* **106**, 327–350.

Berggren, R., Bolin, B., and Rossby, C.-G. (1949). An aerological study of zonal motion, its perturbations and breakdown. *Tellus* **1**, 14–37.

Brown, J. A. (1964). A diagnostic study of tropospheric heating and the generation of available potential energy. *Tellus* **16**, 371–388.

Charney, J. G., and Devore, J. G. (1979). Multiple flow equilibria in the atmosphere and blocking. *J. Atmos. Sci.* **36**, 1205–1216.

Elliott, R. D., and Smith, T. B. (1949). A study of the effects of large blocking highs on the general circulation in the Northern Hemisphere westerlies. *J. Meteorol.* **6**, 67–85.

Källen, E. (1981). The nonlinear effects of orographic and momentum forcing in a low-order barotropic model. *J. Atmos. Sci.* **38**, 2150–2163.

Källen, E., and Wiin-Nielsen, A. (1980). Non-linear, low-order interactions. *Tellus* **32**, 393–409.

Lorenz, E. N. (1955). Available potential energy and the maintenance of the general circulation. *Tellus* **7**, 157–167.

Namias, J., and Clapp, P. F. (1949). Confluence theory of the high tropospheric jet stream, *J. Meteorol.* **6**, 330–336.

Namias, J., and Clapp, P. F. (1951). Observational studies of the general circulation pattern. *Comp. Meteorol. Am. Meteorol. Soc.* 551–567.

Rex, D. (1950). Blocking action in the middle troposphere and its effect on regional climate. *Tellus* **2**, 196–211, 275–301.

Rossby, C.-G. (1950). On the dynamics of certain types of blocking waves, *J. Chin. Geophys. Soc.* **2**, 1–13.

Saltzman, B. (1957). Equations governing the energetics of the larger scales of atmospheric turbulence in the domain of wave number. *J. Meteorol.* **14**, 513–523.

Smagorinsky, J. (1963). General circulation experiments with the primitive equations I: The basic experiment. *Mon. Weather Rev.* **91**, 99–164.

Thompson, P. D. (1957). A heuristic theory of large-scale turbulence and long-period velocity variations in barotropic flow. *Tellus* **9**, 69–91.

Wiin-Nielsen, A. (1961). On short- and long-term variations in quasi-barotropic flow. *Mon. Weather Rev.* **89**, 461–476.

Wiin-Nielsen, A. (1962). On transformation of kinetic energy between the vertical shear flow and the vertical mean flow in the atmosphere. *Mon. Weather Rev.* **90**, 311–323.

Wiin-Nielsen, A. (1976). Predictability and climate variation illustrated by a low-order system. *ECMWF Semin.*

Wiin-Nielsen, A. (1979). Steady states and stability properties of a low-order, barotropic system with forcing and dissipation. *Tellus* **31**, 375–386.

Wiin-Nielsen, A. (1982). On simple stochastic-dynamic systems with forcing. *Bulg. Geophys. J. Bulg. Acad. Sci.* **8**, 3–16.

Wiin-Nielsen, A. (1983). On low-order non-linear stochastic-dynamic systems. *Tellus* **35A**, 1-16.

Wiin-Nielsen, A. (1984). Low- and high-index steady states in a low-order model with vorticity forcing. *Contrib. Phys. Free Atmos.* **57**, 291-306.

Wiin-Nielsen, A., and Brown, J. A. (1960). On diagnostic computations of atmospheric heat sources and sinks and the generation of available potential energy. *Proc. Int. Symp. Numer. Weather Predict., Tokyo* pp. 593-613.

Wiin-Nielsen, A., Brown, J., and Drake, M. (1964). Further studies of the energy exchange between the zonal flow and the eddies. *Tellus* **16**, 168-180.

Part II

OBSERVATIONS

THE LIFE CYCLES OF PERSISTENT ANOMALIES AND BLOCKING OVER THE NORTH PACIFIC

RANDALL M. DOLE

Center for Meteorology and Physical Oceanography
Massachusetts Institute of Technology
Cambridge, Massachusetts 02139-4301

1. INTRODUCTION

Recently, Dole and Gordon (1983) studied the geographical variations in the persistence characteristics of wintertime Northern Hemisphere height anomalies. They focused particular attention on the behavior of anomalies that persisted at a given location beyond the durations associated with synoptic-scale variability ("persistent anomalies"), without regard to the character of the concurrent large-scale flow patterns. They identified three key regions characterized by relatively high numbers of persistent anomalies (durations of 10 days or longer): the North Pacific to the south of the Aleutians (PAC), the North Atlantic to the southeast of Greenland (ATL), and from the northern Soviet Union northeastward to over the Arctic Ocean (NSU).

Subsequently, Dole (1986) described the typical time-average structures of flow patterns occurring with persistent anomalies in these regions. He found that, for each region, the majority of persistent anomaly cases were associated with the amplification of a single basic anomaly pattern, with one phase of the pattern frequently associated with blocking. For the PAC and ATL regions, the primary patterns also resembled certain prominent teleconnection patterns described by Wallace and Gutzler (1981) [the "Pacific–North American" (PNA) pattern, and the "Eastern Atlantic" (EA) pattern, respectively]. Analyses of variance and lag autocorrelation characteristics suggested that most of the temporal variability of the primary patterns was contributed by low-frequency, intraseasonal fluctuations (predominantly by periods between 10 and 90 days).

Although the previous studies described certain statistical aspects of the temporal variability of persistent anomaly patterns, they did not examine specific temporal relationships associated with the development and decay of the patterns. In the present investigation, we will extend the previous studies by attempting to identify systematic aspects of the temporal evolution of persistent anomalies, concentrating on periods around the development and decay of the anomaly patterns. For brevity, we will focus here mainly on a description of the evolution of the anomaly and flow patterns associated with persistent height anomalies over the central North Pacific, but we will also attempt to summarize systematic aspects of the evolution of persistent anomalies in the other regions.

2. DATA

The data base for the 500-mbar height analyses is that employed previously by Dole and Gordon (1983) and Dole (1986). It consists of twice-daily (0000 and 1200 GMT) National Meteorological Center (NMC) final analyses of the Northern Hemisphere 500-mbar heights for the 14 120-day winter seasons (beginning 15 November) from 1963/1964 through 1976/1977. The data base for the vertical structure analyses consists of the twice-daily NMC final analyses of geopotential heights at five levels (1000, 700, 500, 300, and 100 mbar) for the 11 120-day winter seasons (beginning 15 November) from 1965/1966 through 1975/1976. Details of the NMC analyses are described in Dole and Gordon (1983). Missing analyses were linearly interpolated in time. All analyses were performed on a 5° by 5° latitude–longitude grid. Data were spatially interpolated from the NMC grid to the latitude–longitude grid by the procedure described in Dole and Gordon (1983).

Raw anomalies are defined as the departures of the analyzed values from the corresponding long-term seasonal trend values. For the 500-mbar height anomaly analyses, the seasonal trend time series at a point is determined by a least-squares quadratic fit to the 14-winter mean time series for that point (the first value of the 14-winter mean time series is the average of the 14 different 0000 GMT 15 November values, the second value is the average of the 1200 GMT 15 November values, etc.). Anomalies in the vertical structure analyses are similarly defined as departures from the corresponding long term (11-winter) seasonal trend values.

In addition, for the 500-mbar analyses, the raw height anomalies h' have been further normalized (as in Dole and Gordon) by a scale factor which varies inversely with the sine of latitude:

$$z'_\theta = [(\sin 45°)/\sin \theta)]h'$$

This scaling is motivated by a recent study on atmospheric energy dispersion (Hoskins *et al.*, 1977) showing that, as a result of the latitudinal variation of the Coriolis parameter, height field analyses provide a poor indication of the meridional component of energy propagation. Hoskins *et al.* suggest that quantities such as streamfunction or vorticity provide better indicators of horizontal energy propagation. Note that this normalization is similar to that used in obtaining a geostrophic streamfunction from height data.

3. Procedure

Most of the results to be described subsequently are determined from composite time evolution analyses, where the compositing is performed relative to objectively defined case onset (or termination) times. In the present study, we will focus on the evolutions of persistent anomaly cases that occurred in the PAC key region; characteristics of the evolutions for the other two regions are described elsewhere (Dole, 1982). Our procedure for identifying persistent anomaly cases follows that of Dole and Gordon (1983) and so will only be briefly summarized here. As in Dole and Gordon, we define the occurrence of a persistent positive (negative) anomaly at a point if an anomaly at that point remains equal to or greater (less) than a specified magnitude threshold M for at least T days. The total duration D of the case is defined as the time from which the anomaly first crosses M (onset) to the next crossing of M (termination). In this study, we will focus on results obtained for one set of criteria values [(+ 100 m, 10 days) and (− 100 m, 10 days)] that appear broadly representative of a large range of values.

For all analyses, case onset and termination dates are determined by applying the selection criteria to 500-mbar height anomaly time series obtained at certain "key" points that were previously identified (Dole and Gordon) as the locations of the regional maxima in the total numbers of (positive and negative) persistent anomaly cases. In order to remove possible effects of brief transient fluctuations associated with passing synoptic scale disturbances on the starting (and ending) times, the time series were first lightly low-pass filtered before applying the selection criteria. The specific filter used removes periods of less than approximately 10 days; its characteristics are described in detail by Blackmon (1976). The "key point" identified for the PAC cases, 45°N 170°W, had 15 positive and 13 negative cases, about double the numbers obtained from the unfiltered data (Dole and Gordon, 1983). Dole (1986) provides a list of the case dates and durations along with a description of the composite time-average three-dimensional structures of these cases.

Composite time evolution fields for cases of a given region and sign (for example, the PAC positive cases) are obtained by averaging over all of the cases relative to the time when the low-pass filtered anomaly crosses the threshold value at the key point (relative to the onset time for the development analyses and termination time for the breakdown analyses); this threshold-crossing time is defined as day 0. For the development analyses, composites were constructed for each day for the 30-day period from 10 days prior to onset (day −10) to 19 days after onset (day +19). For the breakdown analyses, composites were constructed from 10 days prior to termination (day −10) to 10 days following termination (day +10).

4. Development of 500-mbar Height Anomaly Patterns

We will first study the "slow" time evolution of the 500-mbar anomaly patterns by constructing composites from low-pass filtered data. This provides an overview of the developments and will enable us to identify certain subtle features that are partially obscured by high-frequency fluctuations. We will then examine corresponding composites constructed from unfiltered data around certain key times when the patterns rapidly evolve. Following this, we will present additional simple analyses to provide a more complete view of the developments. In the following discussion, we will primarily focus on selected times around day 0.

4.1. Low-Pass Filtered Height Anomaly Patterns

Figure 1 presents composite anomaly maps and the corresponding confidence levels of the composite anomalies (estimated by a two-sided t-test with a null hypothesis of zero mean) for 15 PAC positive anomaly cases at 2-day intervals from 4 days before onset (day −4) to 6 days after onset (day +6). There is little evidence in these analyses of a precursor until a few days prior to onset; indeed, at day −4, only a limited region of the northern Soviet Union to the northwest of the Tibetan plateau has t-values exceeding the 99% confidence level. Note that, through day −2, the mean anomalies over the entire North Pacific north of 20°N are not significantly different from zero. The structure of the anomaly patterns prior to (and to a lesser extent following) onset suggests that the associated wind anomalies are primarily in the zonal component.

A single major positive center becomes established over the key region at day 0. Subsequent to this time, anomaly centers form to the south of, and in sequence downstream from, the main center. Intensification of these centers occurs with little evidence of phase propagation. By day +4 the PAC positive pattern is established. The gross features of the development downstream from the main center are strongly reminiscent of the behavior seen in simple models of energy dispersion on a sphere away from a localized, transient

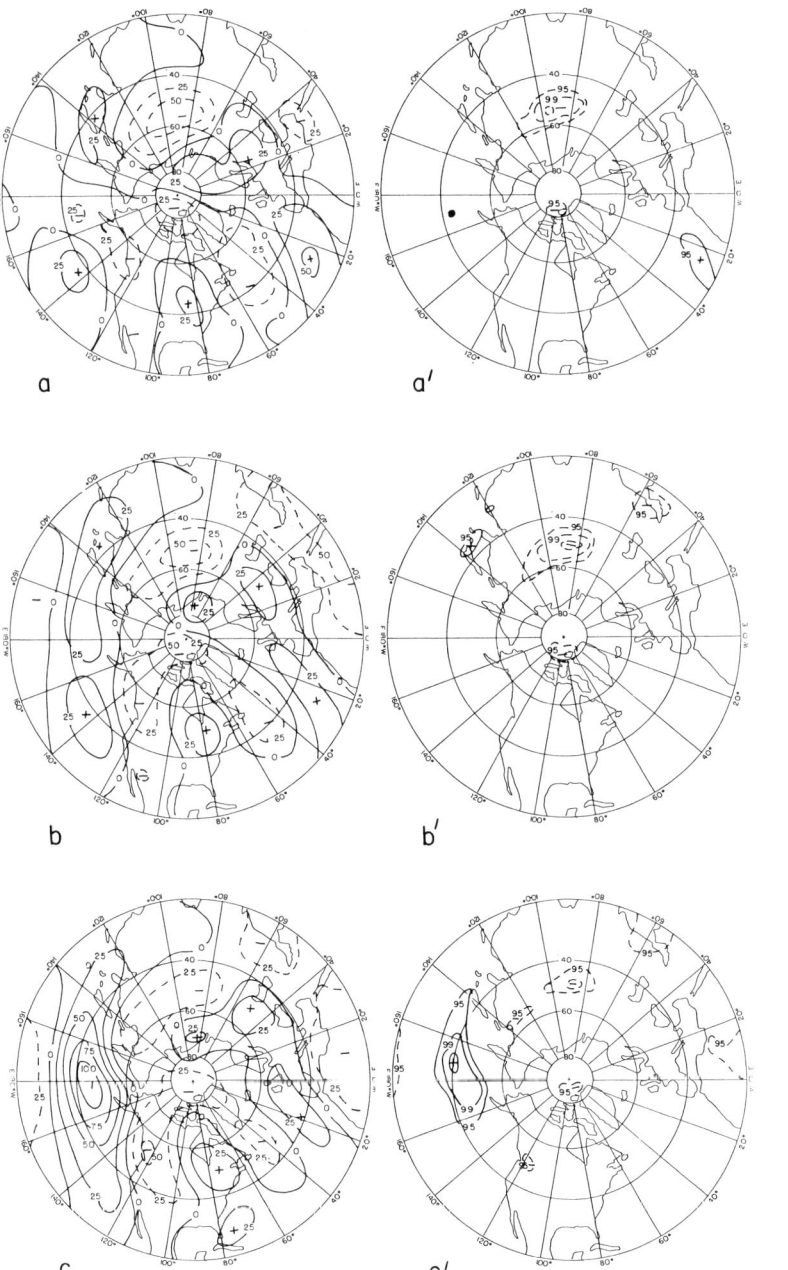

FIG. 1. Composite mean time evolution of low-pass filtered 500-mbar height anomalies (in meters) obtained from 15 PAC positive anomaly cases for days (a) −4, (b) −2, (c) 0, (d) +2, (e) +4, and (f) +6, relative to the onset time. Areas where the composite mean anomalies are greater or less than zero at varying confidence levels are shown for the same times in a′−f′. Dashed lines indicate negative anomaly values.

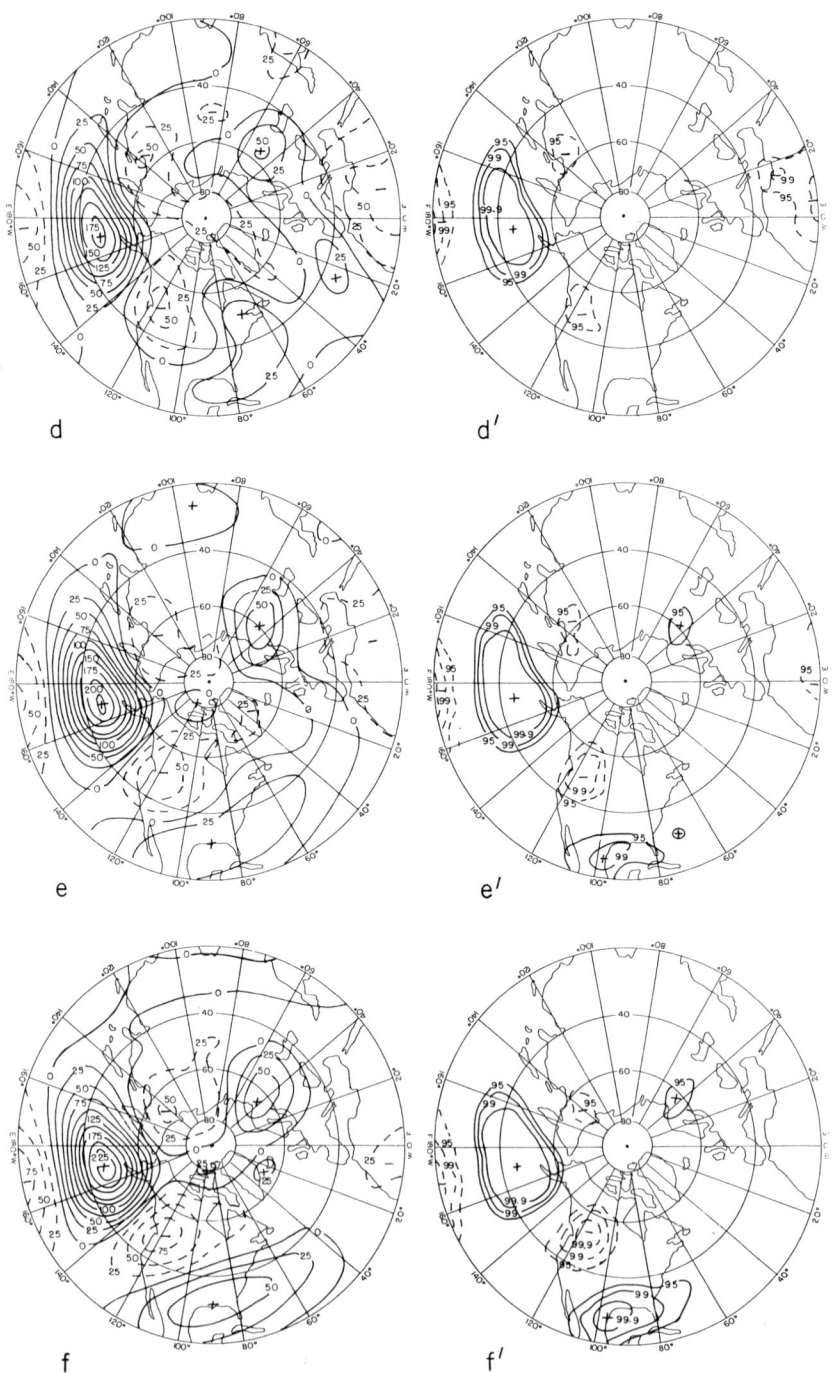

FIG. 1d–f'. See legend on p. 35.

(e.g., switch-on) source of vorticity (Hoskins *et al.,* 1977). Over the Pacific, however, the near simultaneity of development, the almost north–south orientation of the centers, and the absence of tilts in the anomaly axes make determination of the meridional component of energy propagation (if any) difficult. Indeed, over this region, the pattern somewhat resembles a growing standing wave.

Similar maps of the PAC negative anomaly composites and corresponding confidence levels are presented in Fig. 2. We see that the pattern of evolution displays considerable similarity to the sequence seen previously for the positive cases. At day -4, the largest area of significant anomalies is again located upstream over the Asian continent, with values exceeding the 99% confidence level over a large region extending from the Tibetan plateau eastward to Japan. In contrast to the positive cases, however, the major area of significant anomalies is now located principally over and to the south of the Himalayas. Through day -2, the significant anomalies are mainly confined to this region; as for the positive cases, the associated wind anomalies at this time appear mainly in the zonal flow. Through this time, height anomalies over the central North Pacific are not significantly different from zero. A single major center becomes established over the key region at day 0 and, subsequently, centers deepen in sequence downstream. The largest discrepancies between the positive and negative patterns appear from day 4 onward, over the North Atlantic, where the intense positive anomaly center evident in the negative cases has no counterpart in the positive cases.

The symmetry in development (with sign reversal) can be evaluated more readily by constructing difference maps of the evolutions of the positive and negative cases. Figure 3 displays the composite anomaly differences (positive – negative) and associated confidence levels for the differences between means (null hypothesis of equal means) at 2-day intervals from day -4 to day $+6$. As suggested by the previous analyses, prior to the development of the main center, the principal differences between the positive and negative cases appear upstream, primarily over eastern Asia. At day -2, the southern portion of high differences extends eastward into the western North Pacific; note, however, that over the North Pacific north of 20° N and east of 170° E, differences between the mean anomalies for the positive cases and negative cases are not statistically significant until day 0. These analyses suggest that, in 500-mbar height data, the most prominent features preceding the development of the North Pacific anomaly pattern are located upstream, and are primarily associated with the zonal flow over eastern Asia and the extreme western North Pacific. In contrast, knowledge of the sign of the anomalies over the central North Pacific immediately preceding development is likely to be of little value in distinguishing whether a positive case or negative case will subsequently occur.

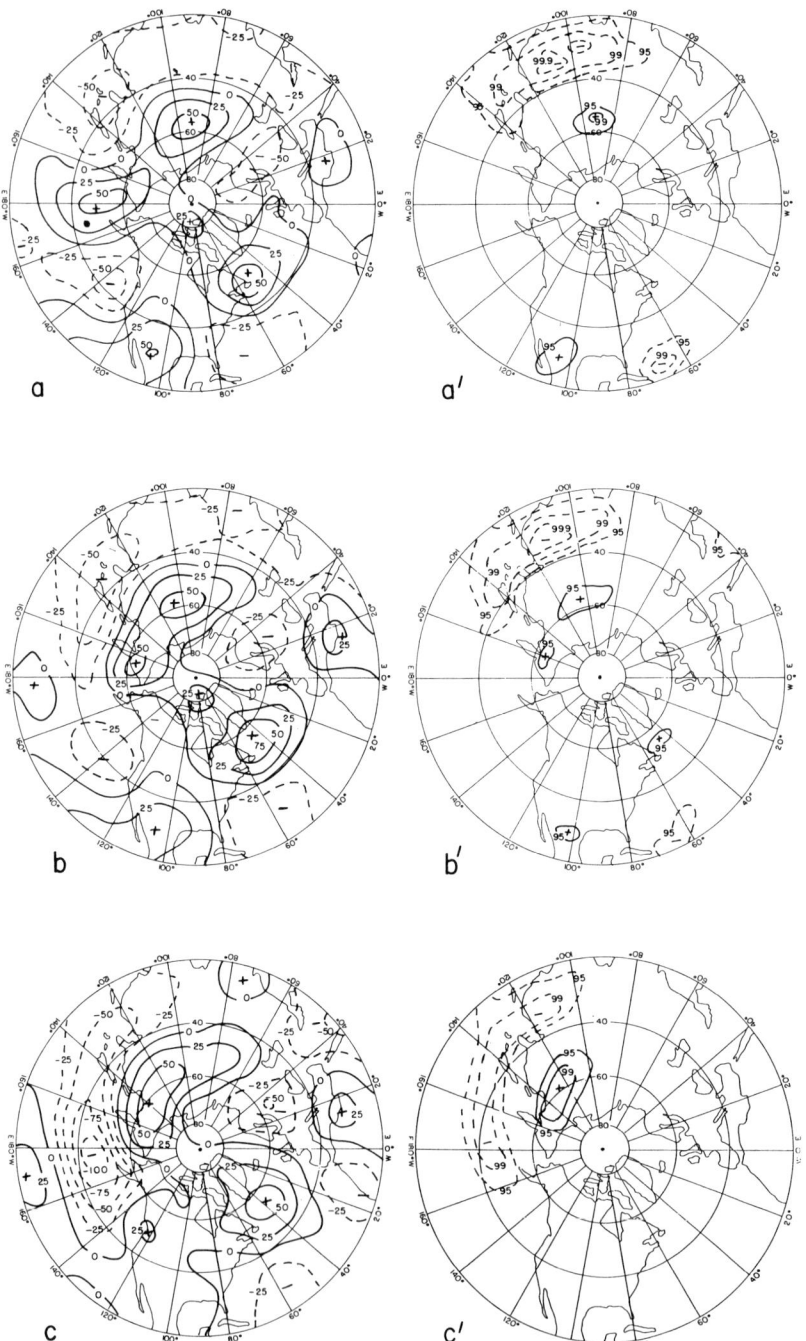

FIG. 2. As in Fig. 1, for the composite time evolution of low-pass filtered 500-mbar height anomalies (in meters) obtained from 13 PAC negative anomaly cases for days (a) −4, (b) −2, (c) 0, (d) +2, (e) +4, and (f) +6, relative to the onset time.

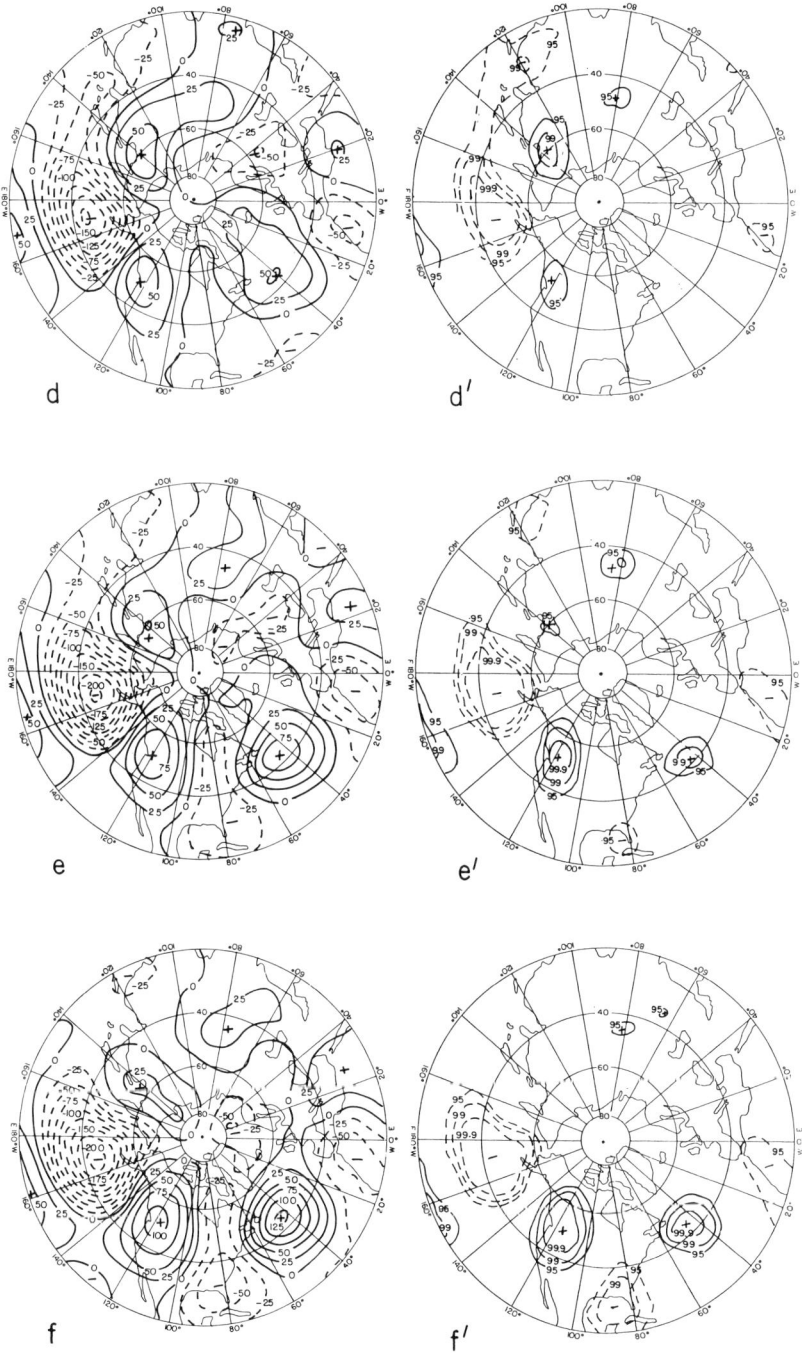

FIG. 2d–f'.

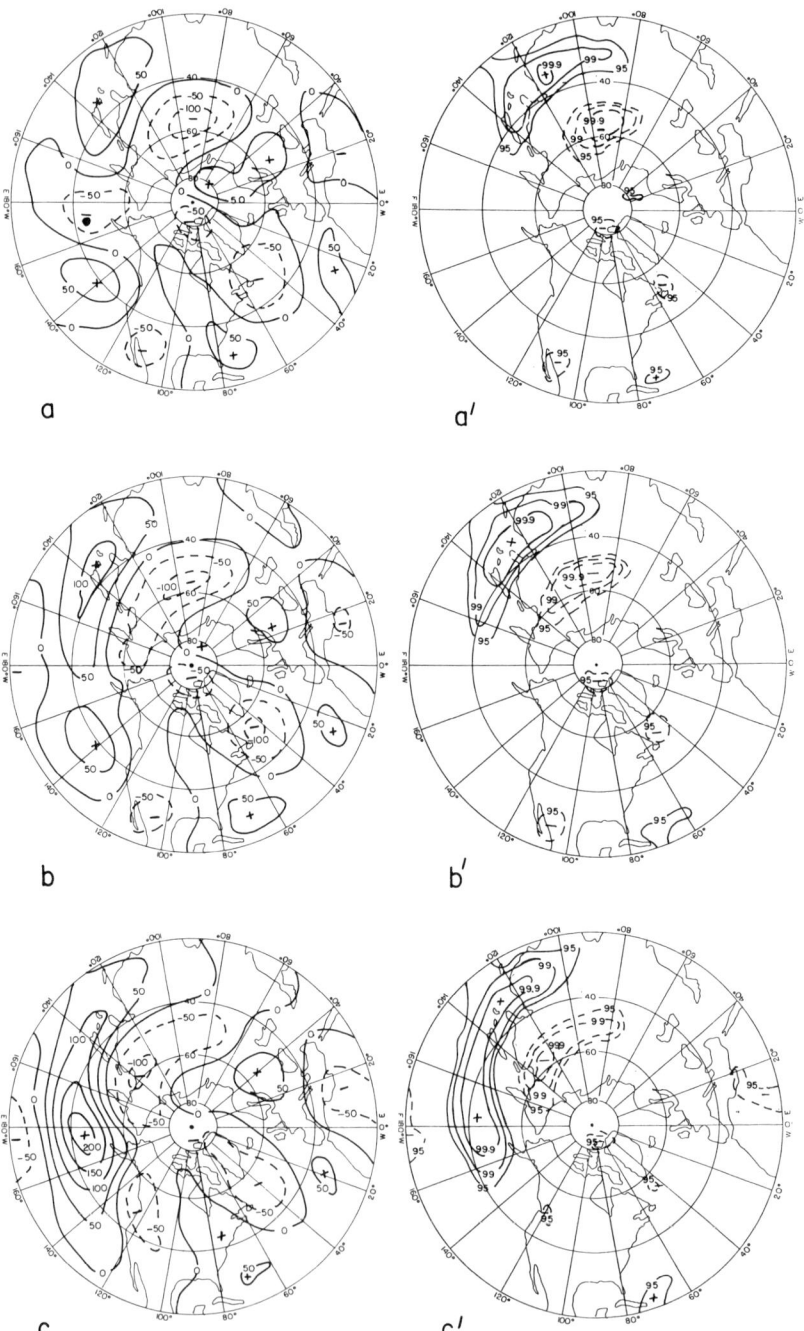

FIG. 3. As in Fig. 1, for the differences in the composite mean time evolutions of low-pass filtered 500-mbar height anomalies (in meters) between the 15 PAC positive and 13 PAC negative anomaly cases for days (a) −4, (b) −2, (c) 0, (d) +2, (e) +4, and (f) +6, relative to the onset time. Areas where the differences between positive and negative means are significantly different from zero are shown at varying confidence levels for the same times in a′–f′. Dashed lines indicate negative anomaly differences.

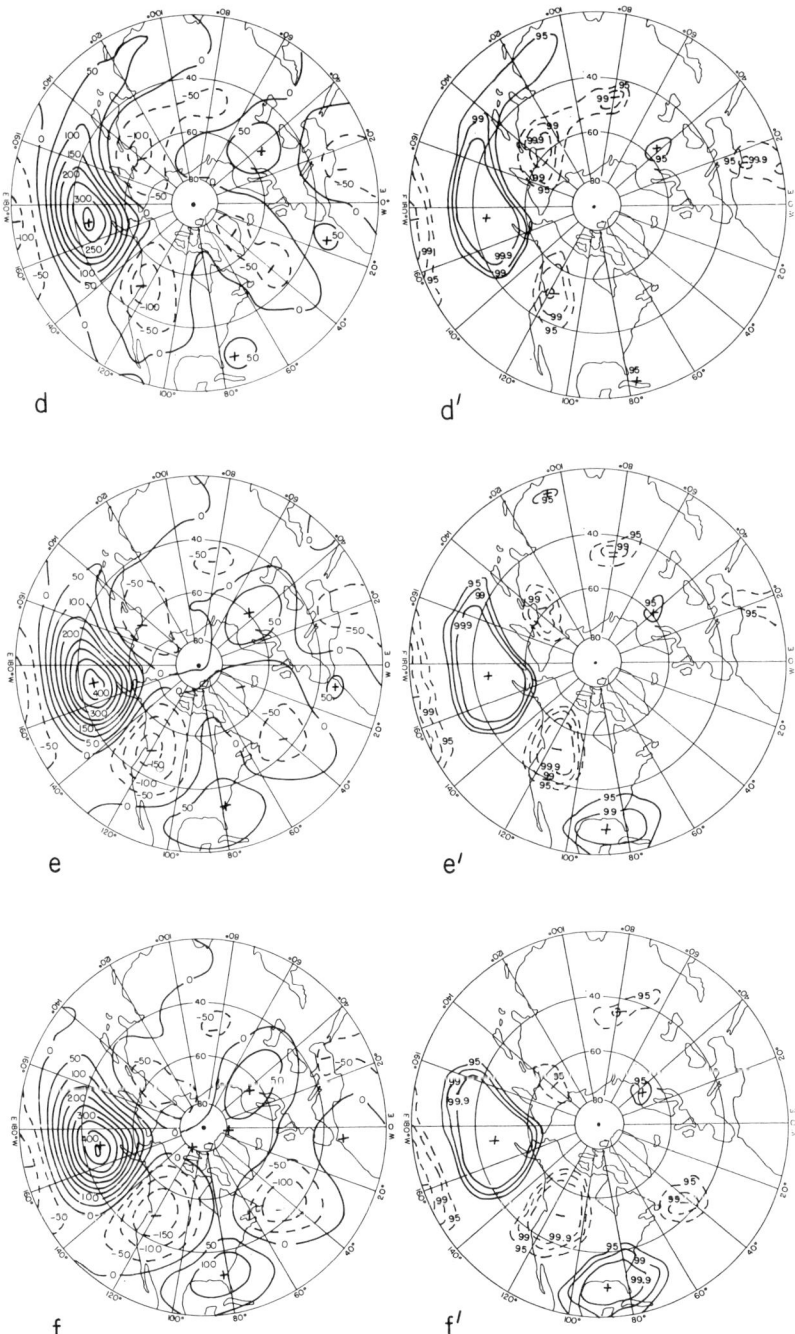

FIG. 3d–f'.

42   RANDALL M. DOLE

4.2. Unfiltered Anomaly Analyses

Although data used in the above analyses were low-pass filtered, the main centers appeared to develop quite rapidly. We now examine similar analyses conducted on unfiltered data for further clues to the character of this rapid development. Note that the starting dates are *identical,* so that the only difference from the previous analyses is the filtering process.

Figure 4 displays the unfiltered anomaly composite evolution and the

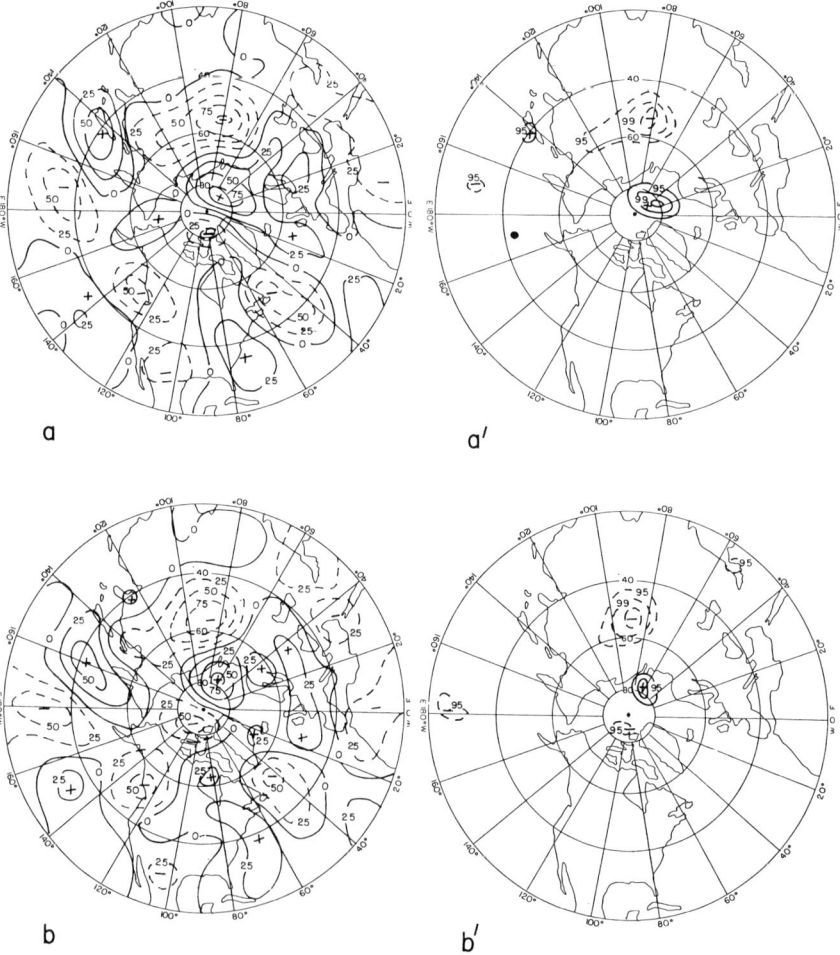

FIG. 4. Composite time evolution of unfiltered 500-mbar height anomalies (in meters) obtained from the same PAC positive anomaly cases as in Fig. 1, for days (a) −3, (b) −2, (c) −1, and (d) 0. Corresponding confidence levels are shown in a′–d′.

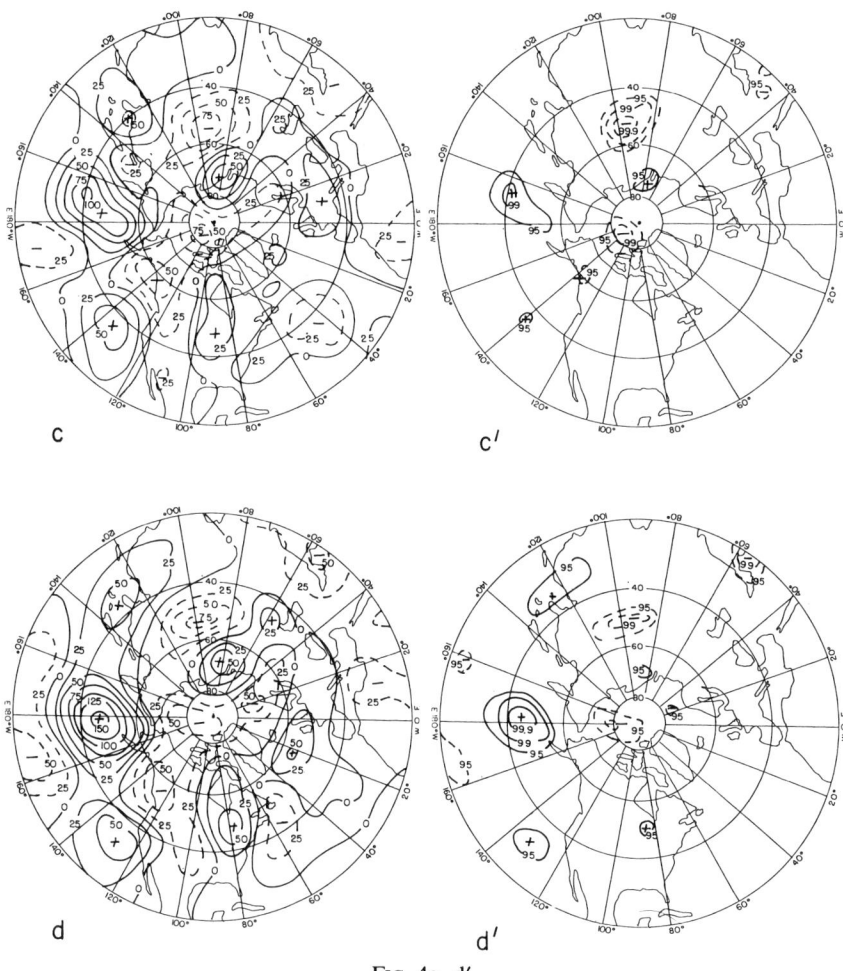

Fig. 4c–d'.

corresponding confidence levels for the positive cases at 1-day intervals from day −3 to day 0. The major differences in evolution from the filtered analyses are mainly associated with a positive anomaly center located to the east of Japan at day −3. This center propagates eastward and intensifies through the period, reaching the key region at day 0. In advance of this feature, a negative center moves southeastward to the subtropical central North Pacific by day 0. This sequence of development suggests that the initial rapid growth of the main center is partly associated with the propagating, intensifying disturbance which originates in midlatitudes near Japan. This disturbance slows up but continues to intensify as it approaches the key region. Similarly, the intensification of the main center over the subtropical North Pacific appears related in part to a disturbance originating in midlatitudes. Concurrent with these developments, a larger scale pattern of height rises over the midlatitude North Pacific, with falls to the south, is again evident. Maps following day 0 (not shown) indicate that subsequent developments are qualitatively similar to those displayed in the low-pass analyses.

As for the filtered analyses, the unfiltered PAC negative anomaly evolution (Fig. 5) in many respects resembles the corresponding positive se-

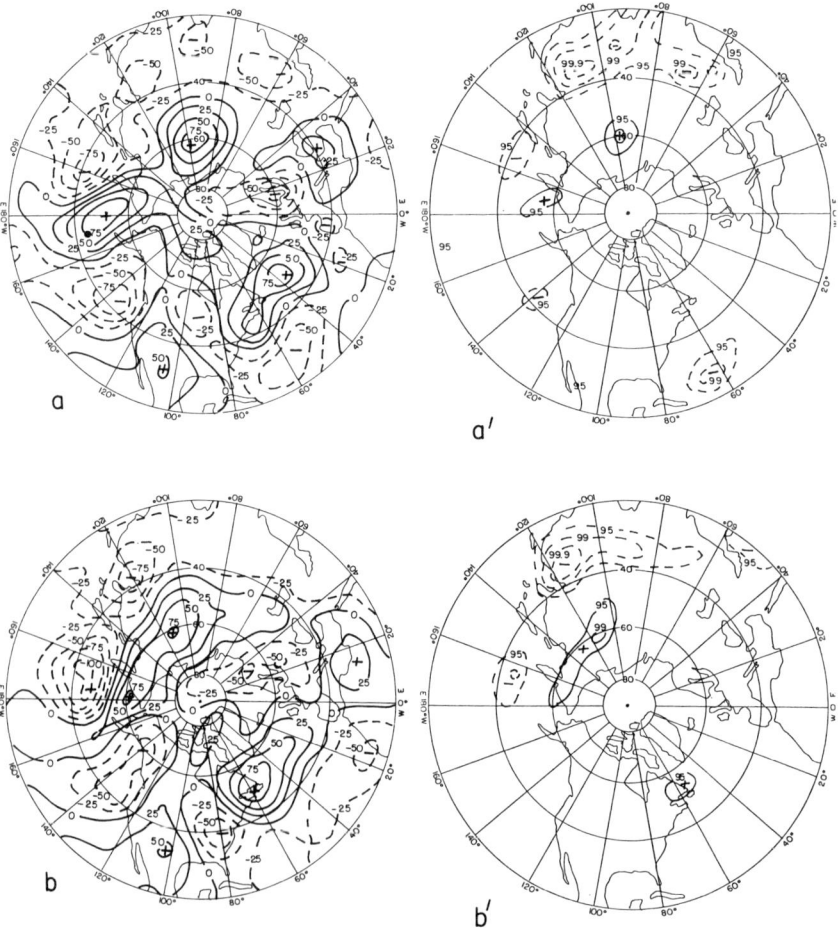

FIG. 5. Composite time evolution of unfiltered 500-mbar height anomalies (in meters) obtained from the same PAC negative anomaly cases as in Fig. 2, for days (a) −3, (b) −2, (c) −1, and (d) 0. Corresponding confidence levels are shown in a′–d′.

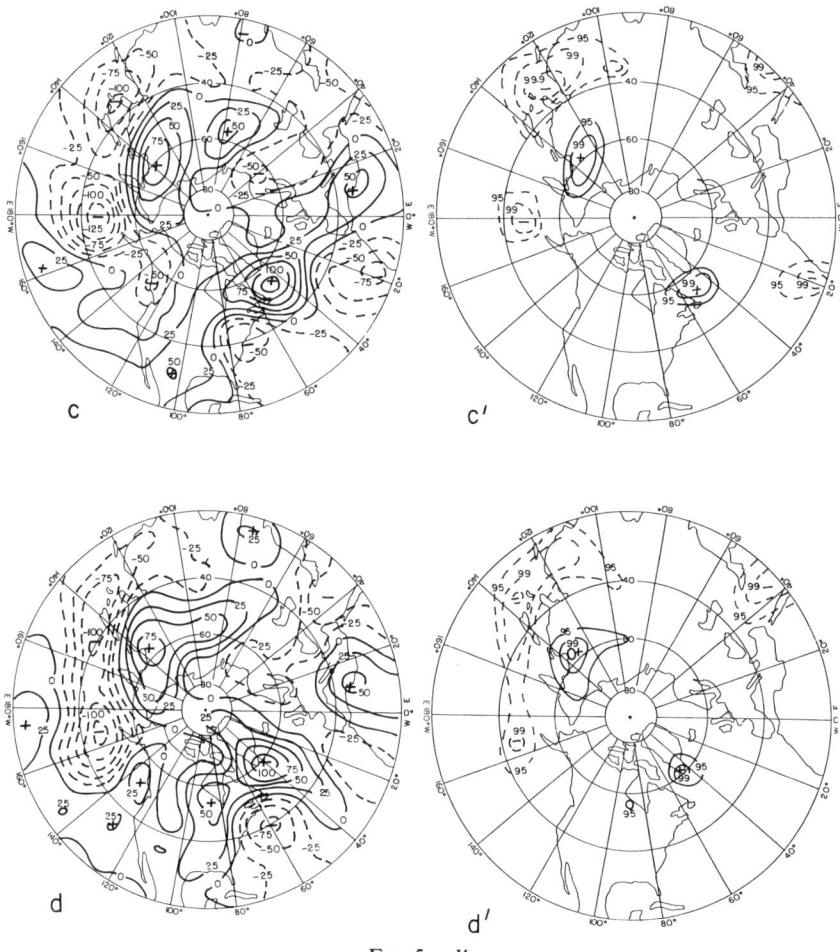

FIG. 5c–d'.

quence. The negative center to the east of Japan on day -3 in the negative analyses, however, is considerably more intense than its positive counterpart. This negative center intensifies and moves eastward to the key region by day 0; the second negative anomaly center over the East China Sea on day -2 follows a similar course, eventually merging with the main center after day 0 (not shown). The difference and corresponding t-test results (Fig. 6) confirm our expectations based on the separate positive and negative analyses: immediately in advance of development, the major additional differences between positive and negative cases not identified in the low-pass filtered analyses are associated with the propagating, intensifying midlatitude disturbances initially near Japan.

4.3. Relationship to the Zonal Flow in the Pacific Jet Region

The variations in the zonal flow over eastern Asia and the western Pacific in advance of the developments over the central Pacific show an interesting

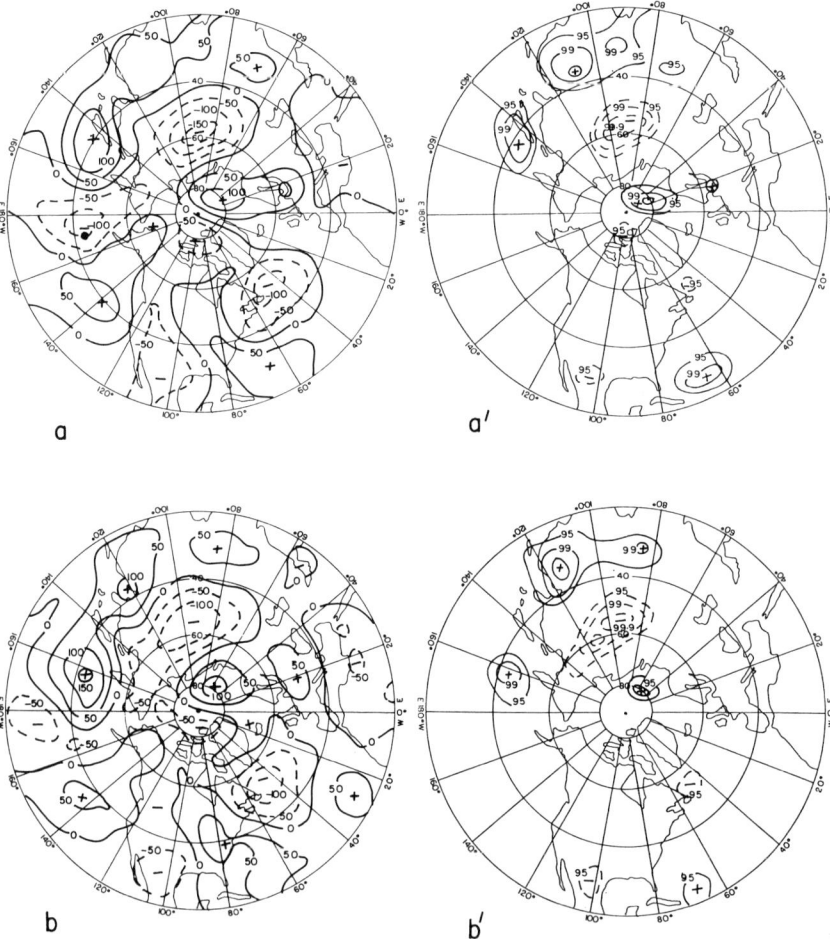

FIG. 6. Composite time evolution of unfiltered 500-mbar height anomaly differences (in meters) between the positive and negative anomaly cases for days (a) −3, (b) −2, (c) −1, and (d) 0. Corresponding confidence levels for the differences between means are shown in a′–d′.

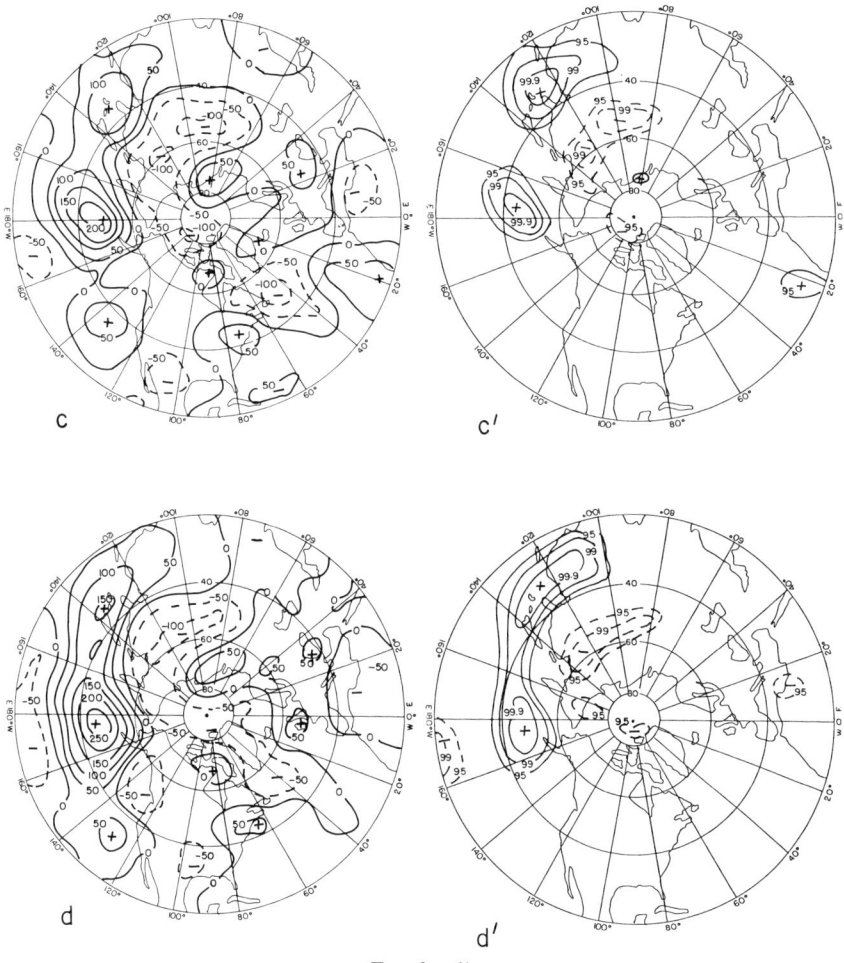

FIG. 6c–d'.

relationship to the climatological mean jet structure, with a tendency for the largest meridional gradients in the height anomalies to occur over the mean jet axis. This suggests that the evolving flow anomalies are associated with major changes in the jet intensity and structure over this region. This relationship can be seen more clearly in Fig. 7, which displays the time evolution of the geostrophic wind anomalies for the PAC negative cases evaluated along the wintertime climatological mean jet axis from central Asia to the eastern North Pacific. We see that well prior to development there is a modest intensification of the jet well upstream of the key region, with a weaker than normal jet to the east near the jet exit region. Beginning at about

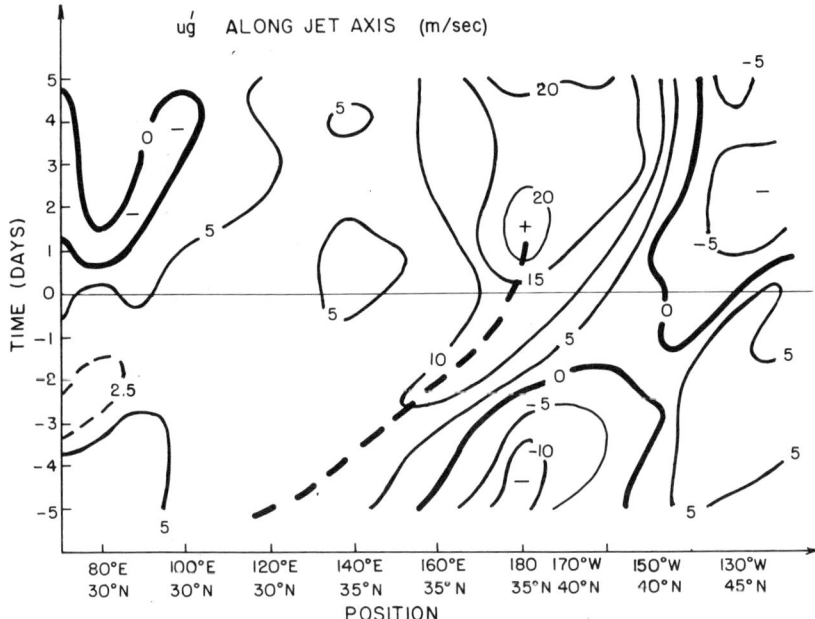

FIG. 7. Time evolution of the composite geostrophic wind anomalies (in m sec^{-1}) evaluated along the climatological mean 500-mbar jet axis from eastern Asian to western North America for the period from 5 days before to 5 days following onset of the Pacific negative anomaly cases.

day -3, the zonal wind anomalies in the western Pacific strengthen and begin to propagate eastward, reflecting an intensification and eastward shift of the jet maxima. Following development, the major flow anomalies remain nearly quasi-stationary over the Pacific, reflecting the intensified and eastward-elongated jet structure. Parallel analyses on the positive cases (not shown) display a weakening of the jet upstream immediately prior to development, with pronounced weakening of the jet in the jet exit region over the central North Pacific during development. Analyses conducted on the Atlantic cases (not shown) display similar variations in the jet intensity and structure, with a tendency for a weakening of the upstream jet over eastern North American and the western Atlantic prior to the development of the positive cases (which are associated with blocking over the eastern Atlantic) and an intensification of the jet prior to the negative cases.

4.4. Representation of Development in Zonal Fourier Harmonics

Investigators have suggested that blocking over the oceans is often characterized by the amplification of certain zonal wavenumbers of the 500-mbar

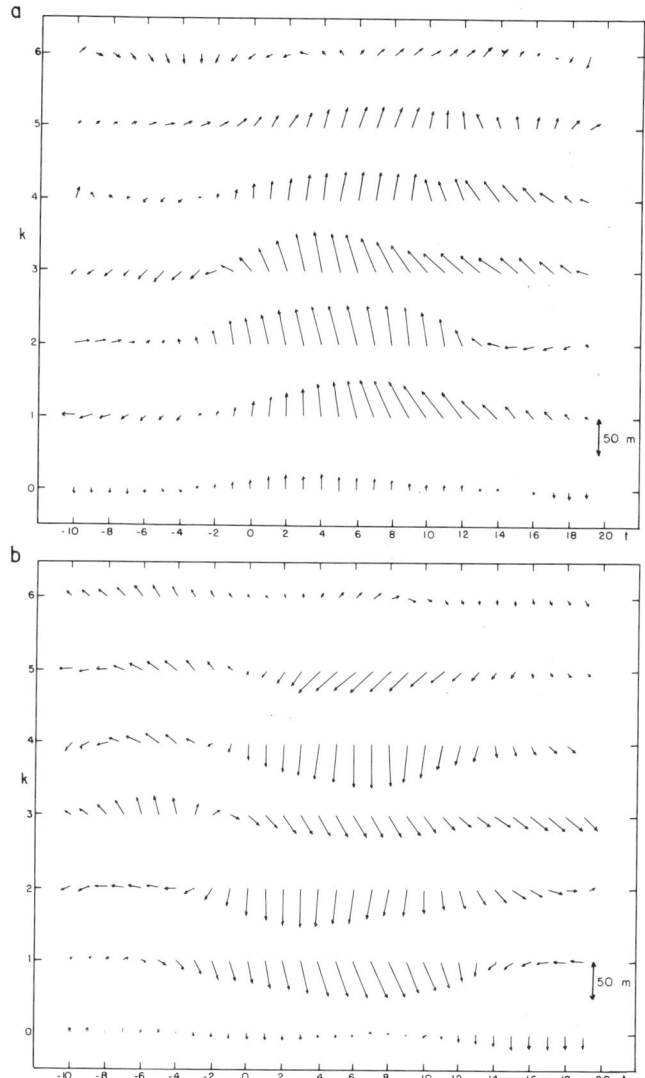

FIG. 8. Composite vector time series of zonal Fourier components $k = 0-6$ of the low-pass anomalies at latitude 45°N for the PAC (a) positive and (b) negative anomaly cases. The amplitude of a component is proportional to the length of the vector. The phase is given by the angle measured in the clockwise direction between the positive vertical axis, through the origin of the vector, and the vector itself (positive angles eastward); thus clockwise (counterclockwise) rotation with increasing time indicates eastward (westward) propagation. The reference longitude is chosen as 170°W, so that the vertical component of a vector is proportional to the contribution by that wavenumber to the observed anomaly at that longitude and time.

height fields (Austin, 1980; Colucci *et al.,* 1981). We have examined this possibility for our cases by constructing composite vector time series of zonal Fourier components, $k = 0-6$, of the anomalies at the corresponding "key" latitudes (Fig. 8). Consistent with the time mean composites presented in Dole (1985), we see that important contributions to the observed anomalies at the key point are provided by a relatively broad band of components, primarily wavenumbers $k = 0-4$. Thus, these analyses do not suggest a strong dominance by a particular wavenumber in the development of the anomaly patterns.

These analyses also provide no clear indication of a systematic precursor in the wavenumber domain: prior to onset, the amplitudes are small, typically under 10 m, and the vectors tend to have a haphazard orientation with respect to the vertical axis, indicating a lack of wave coherence locally. Before the major amplification, most components show some rotation, indicating propagation, with a tendency for the lowest wavenumbers ($k = 1-2$) to be retrogressive and the higher wavenumbers ($k > 5$) to be progressive. During the period of maximum amplification, however, little propagation is evident for wavenumbers $k < 5$, consistent with an interpretation of this stage of the development in terms of waves having phase speeds near zero.

5. Vertical and Thermal Evolutions during Development

A further indication of the character of the developments is obtained by examining how the vertical structures evolve in time. For this purpose, parallel development composites for the 1000-, 700-, 500-, 300-, and 100-mbar height anomalies were prepared from unfiltered data following the procedures described previously. Data at all levels were available only for the 11 winters from 1965/1966 through 1975/1976, so that the composites in this section were formed from all of the previous cases [listed in Dole (1986)] that fell within this period. For brevity, only the PAC evolutions are discussed in detail.

Figure 9 presents longitude–pressure cross-sections at 45°N and 20°N of the unfiltered PAC positive composite anomalies at 1-day intervals from day -3 to day 0. We see that, consistent with the previous 500-mbar analyses, the rapid development of the main center appears to be partly associated with an amplifying, eastward-propagating midlatitude disturbance. This feature has pronounced westward tilts with height during this period, suggesting that a substantial baroclinic contribution is involved in its amplification. There are two maxima in the vertical structure, with peaks at 1000 mbar and near 300 mbar. A similar structure is observed in numerical studies of nonlinear effects on growing baroclinic waves of long synoptic scales (Gall, 1976;

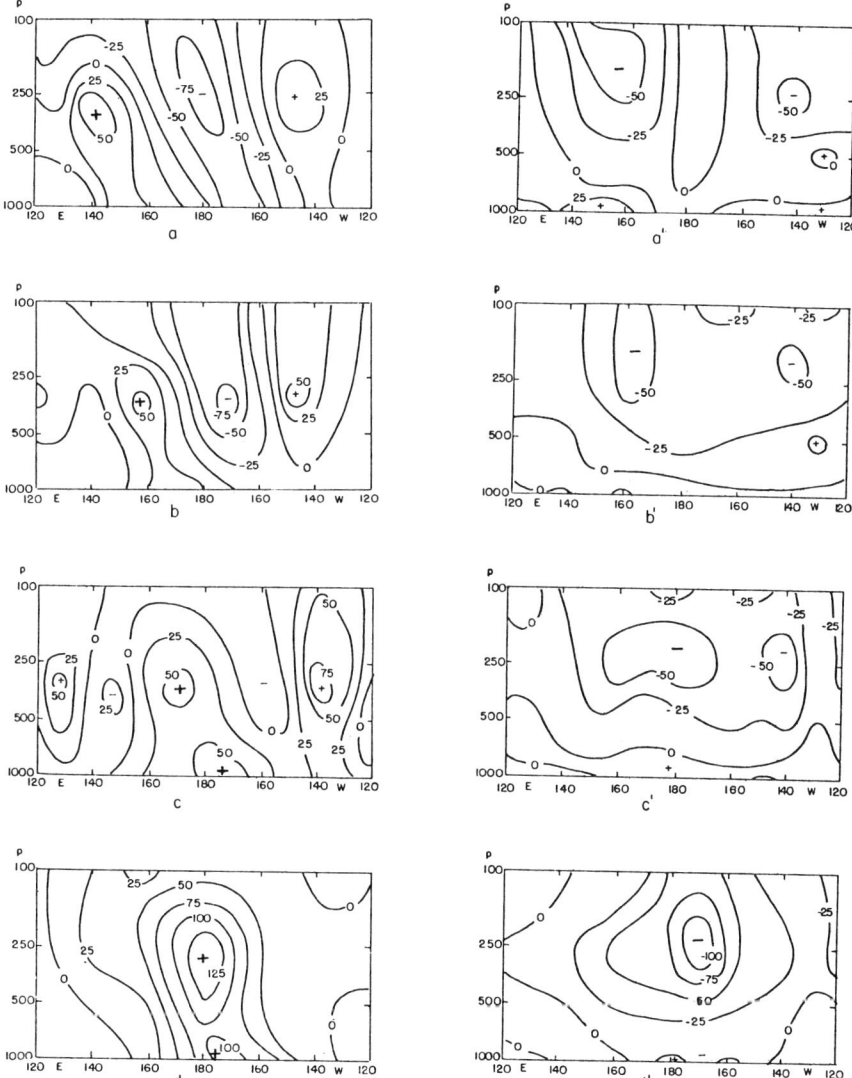

FIG. 9. Composite longitude–pressure (p, millibars) cross sections of unfiltered anomalies (in meters) for the PAC positive anomaly cases at 45°N for days (a) −3, (b) −2, (c) −1, and (d) 0. Corresponding cross sections constructed instead at 20°N are presented in a′–d′.

Simmons and Hoskins, 1978). Such a structure, however, may also be partly an artifact of data coverage because, over this region, data are more plentiful at these levels. Nevertheless, comparison of the relative positions of the centers at the data-rich levels suggests that there are pronounced westward tilts with height throughout the troposphere, and that, at an early stage in the development, the maximum anomalies are realized in the upper troposphere.

The corresponding development at 20°N indicates that, in parallel with its midlatitude positive counterpart, the main negative center progresses eastward across the Pacific through day 0. Subtropical negative anomalies are initially confined primarily to the upper troposphere. Following day 0 (not shown), however, 1000-mb heights continue to fall over the central Pacific, leading to the establishment of a cold-core negative center with little or no evidence of tilts throughout the troposphere.

Figure 10 shows similar analyses for the PAC negative cases. In many respects, the vertical evolution parallels that of the positive cases. The main center propagates eastward and intensifies through the period. The associated trough axis initially tilts strongly westward with height, but becomes nearly vertical by day 0. Double maxima are also evident in the vertical structure early in the evolution, giving way to a single major center in the upper troposphere by day 0. The subtropical patterns are initially rather ill-defined, but a positive center is evident over the subtropical mid-Pacific from day -2 onward. By day 0, positive anomalies associated with this feature extend from the surface to above 100 mbar. Vertical tilts are relatively small, although there is some indication of eastward tilts with increasing height up to the tropopause level, suggesting the possibility of downward energy propagation from this level.

For both positive and negative cases, then, the strongest anomalies occur in the upper troposphere throughout development. At midlatitudes, there are also substantial low-level anomalies that are initially displaced eastward relative to the upper level centers; this relative displacement generally decreases with time as the anomalies approach maximum amplitudes. In contrast, the subtropics display little evidence of substantial surface anomalies prior to development; vertical tilts, if any, appear to be predominantly eastward with increasing height. These structures are quite distinct from the vertical structures of disturbances forced by local anomalous heat sources as obtained in simple models of the stationary wave response to thermal forcing (Hoskins and Karoly, 1981), suggesting that local diabatic heating anomalies are unlikely to be the proximate source for their development.

The evolving jet structure located over eastern Asia and the western Pacific prior to development also appears to be associated with pronounced thermal anomalies. This is illustrated for the negative cases in Fig. 11, which

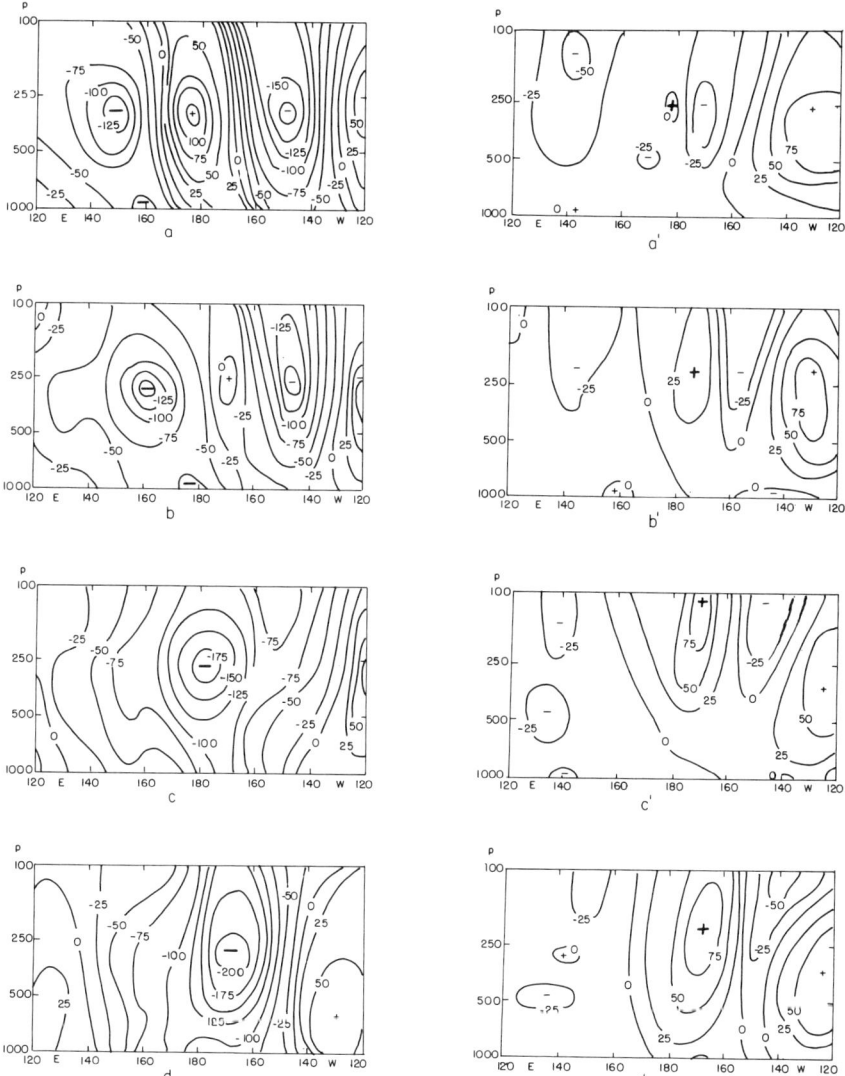

FIG. 10. Composite longitude–pressure (p, millibars) cross sections of unfiltered anomalies (in meters) for the PAC negative anomaly cases at 45°N for days (a) −3, (b) −2, (c) −1, and (d) 0. Corresponding cross sections constructed instead at 20°N are presented in a′–d′.

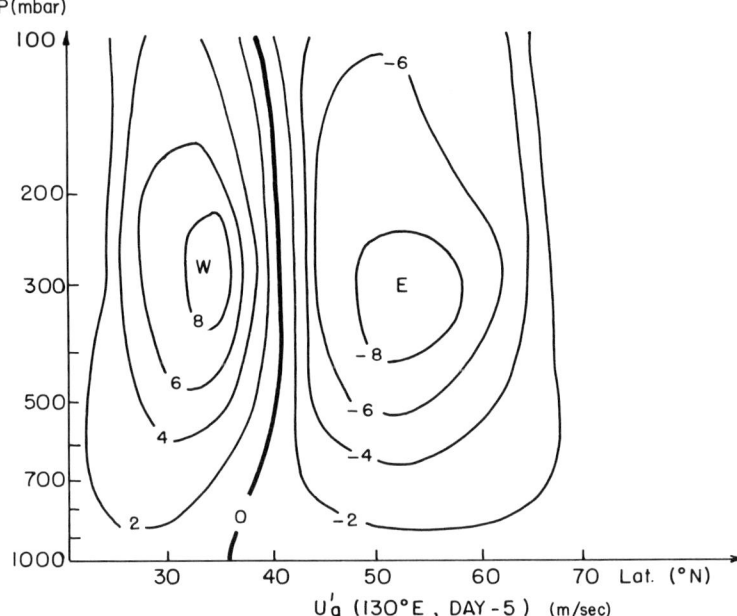

FIG. 11. Cross section of the geostrophic wind anomalies (in meters) at longitude 130°E at day −5.

displays the vertical structure of the geostrophic zonal wind anomalies over this region 5 days before development. The anomalous vertical shear along this cross-section suggests that, through most of the troposphere, there are pronounced cold temperature anomalies to the south of about 40°N, with relatively warm anomalies further north. The structure is also indicative of an upward increase in positive vorticity, with maximum positive vorticity anomalies located near the tropopause at about 40°N. In the presence of mean westerly winds, we anticipate through quasi-geostrophic theory (e.g., Holton, 1979) that this structure should also be associated with anomalous ascent and low-level convergence downstream of the vorticity maxima, with an enhanced likelihood for strong cyclogenesis in this region.

6. Synoptic Characteristics of Development

The analyses so far have illustrated the evolution of the height anomaly fields. We will now briefly examine the synoptic characteristics of the evolution of the height and thermal patterns associated with the persistent anomaly development. For brevity, we will focus mainly on the development of the Pacific negative cases, which are often associated with blocking over the eastern Pacific and western North America. All of the analyses to be discussed subsequently are derived from the unfiltered series from the 11 winter seasons.

Figure 12 displays the evolution of the composite 500-mbar height patterns for the PAC negative cases at 2-day intervals from 5 days before onset until 5 days following onset. Several aspects of this evolution are familiar to synoptic meteorologists. At day -5, the Aleutian low is abnormally weak, with a center displaced far to the west of its climatological mean position. At this time, a weak ridge is centered over the central North Pacific. As will be illustrated later, this pattern resembles the final stages of the PAC positive pattern.

Between day -3 and day -1, the jet initially confined to the far western Pacific intensifies and begins to extend eastward over the western Pacific, while the ridge over the central Pacific weakens, with the remnants of the anticyclonic center drifting northwestward toward Siberia. From day $+1$ onward, a major upper level low center becomes established over the Aleutians, with downstream amplification of both the ridge near the west coast of North America and, subsequently, the trough over eastern North America clearly evident. By day $+5$, the highly amplified 500-mbar height pattern characteristic of the time-average structure of the PAC negative cases (Dole, 1986) is well established.

Figure 13 displays the corresponding evolution of the 1000-mbar height fields. We see that several days before development, there are two distinct lows with centers over the western and eastern North Pacific, with a weak high-pressure ridge over the Aleutians. This tendency for a split of the Aleutian low center into two centers along the continental margins is typical of the Pacific positive pattern (Dole, 1986), and reflects a tendency for the Aleutian low to vary between a single intense center located over the central North Pacific in the negative cases and two centers located in the northwest and northeast Pacific (Dole, 1986) for the positive cases. Similarly, persistent anomaly cases in the Atlantic are typically associated with a tendency for either a single abnormally intense Icelandic low centered in the central North Atlantic in the negative anomaly cases or a split into two relatively weak centers in the northwest and northeast Atlantic with a ridge over the central Atlantic in the positive cases (Dole, 1986).

Between day -3 and day -1, the low centered to the northeast of Japan propagates eastward and intensifies, so that, by day $+1$, a single major center is located over the central North Pacific. This center continues to intensify while remaining nearly stationary through day $+3$. The circulation around the low at this time essentially spans the entire North Pacific north of 35°N.

The relationship between the surface low and the evolving thermal struc-

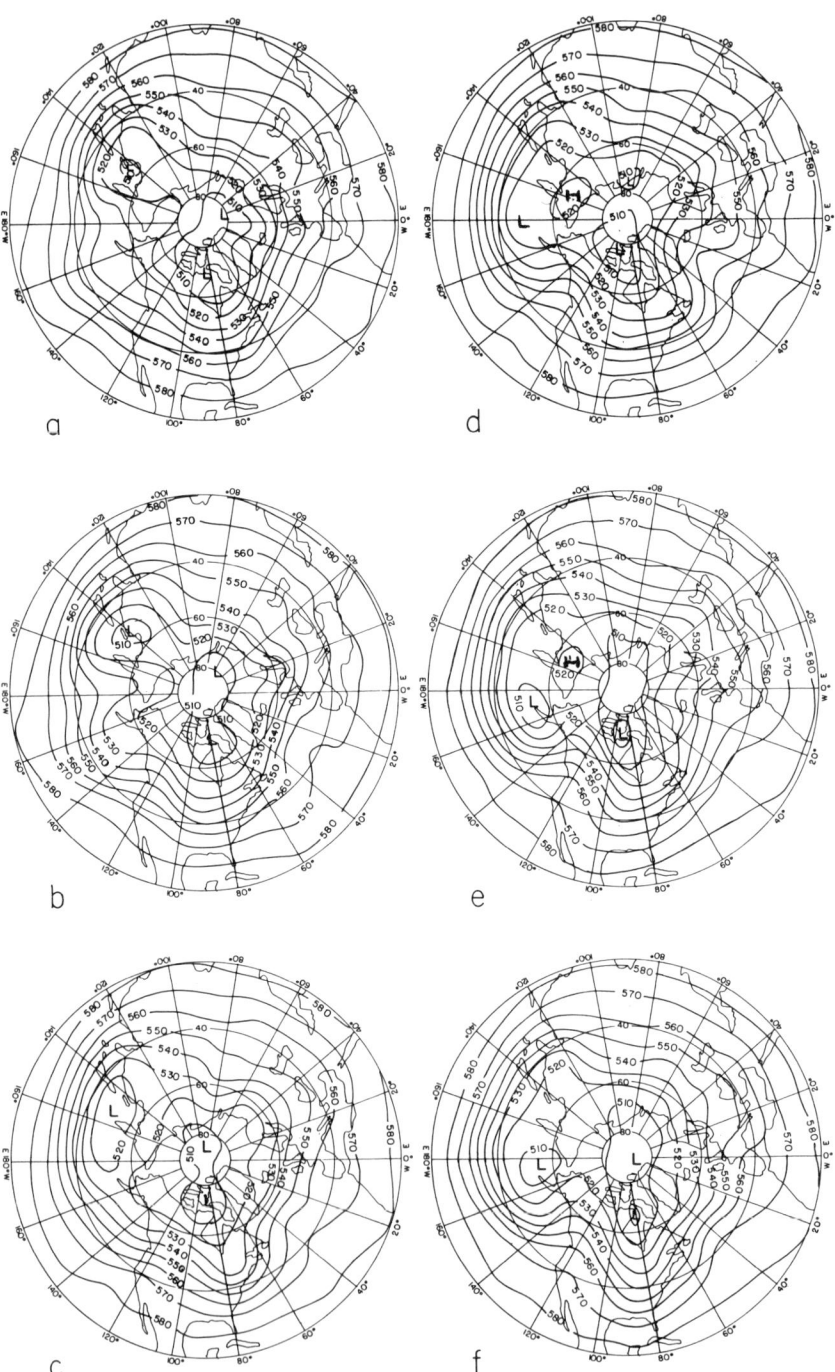

FIG. 12. Composite time evolution of the unfiltered 500-mbar height fields (in meters) for the PAC negative anomaly cases for days (a) −5, (b) −3, (c) −1, (d) +1, (e) +3, and (f) +5.

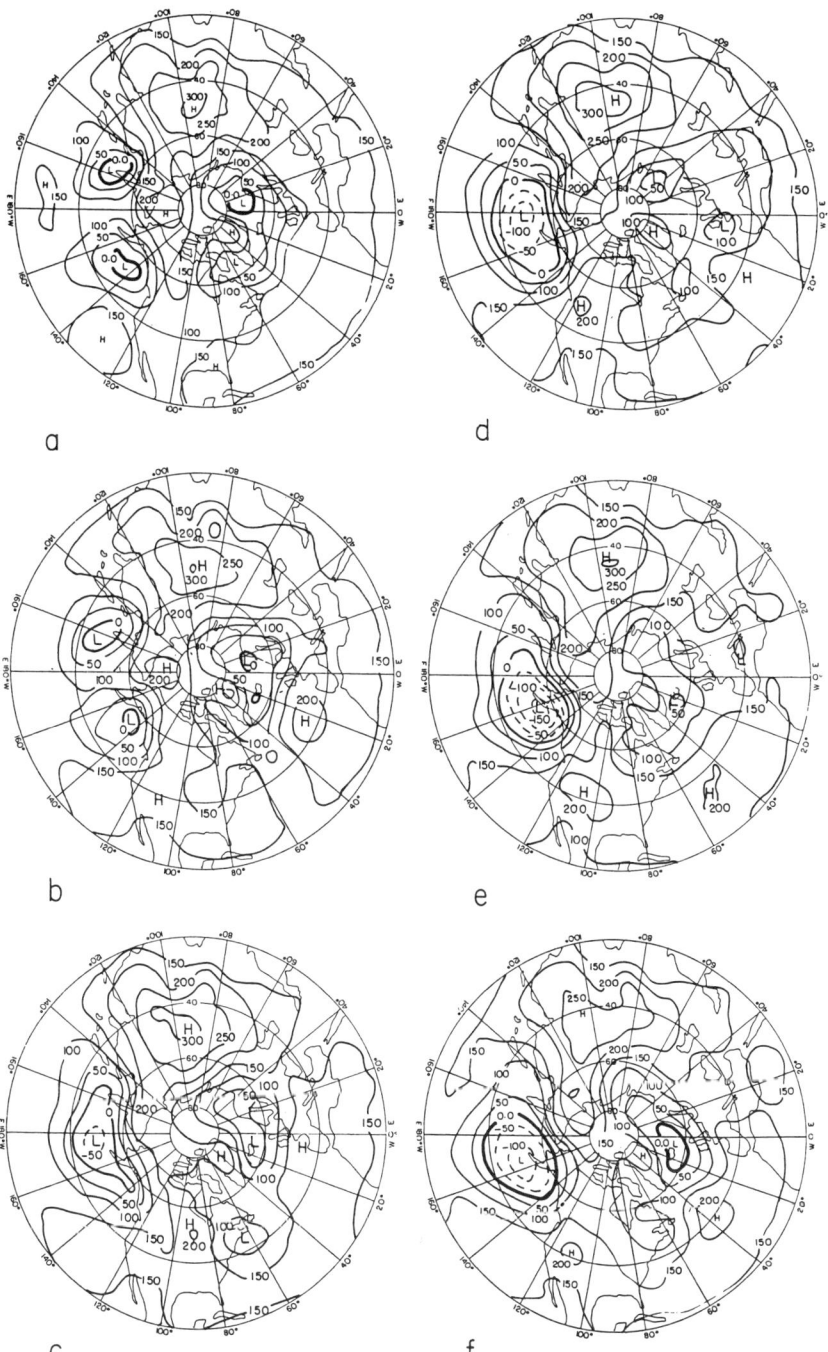

FIG. 13. Composite time evolution of the unfiltered 1000-mbar height fields (in meters) for the PAC negative anomaly cases for days (a) −5, (b) −3, (c) −1, (d) +1, (e) +3, and (f) +5.

ture (as reflected in the 1000- to 300-mbar mean thickness fields) during the period of most rapid intensification is displayed in Fig. 14. In many respects, the gross aspects of the evolution resemble that of an amplifying baroclinic wave, although on a scale that appears considerably larger than for typical baroclinic developments. The low to the northeast of Japan on day -3 is developing on the cyclonic shear side of the jet in a region of pronounced baroclinicity. On its western flank, there is enhanced cold advection off the Asian continent over the western North Pacific on the north side of the jet

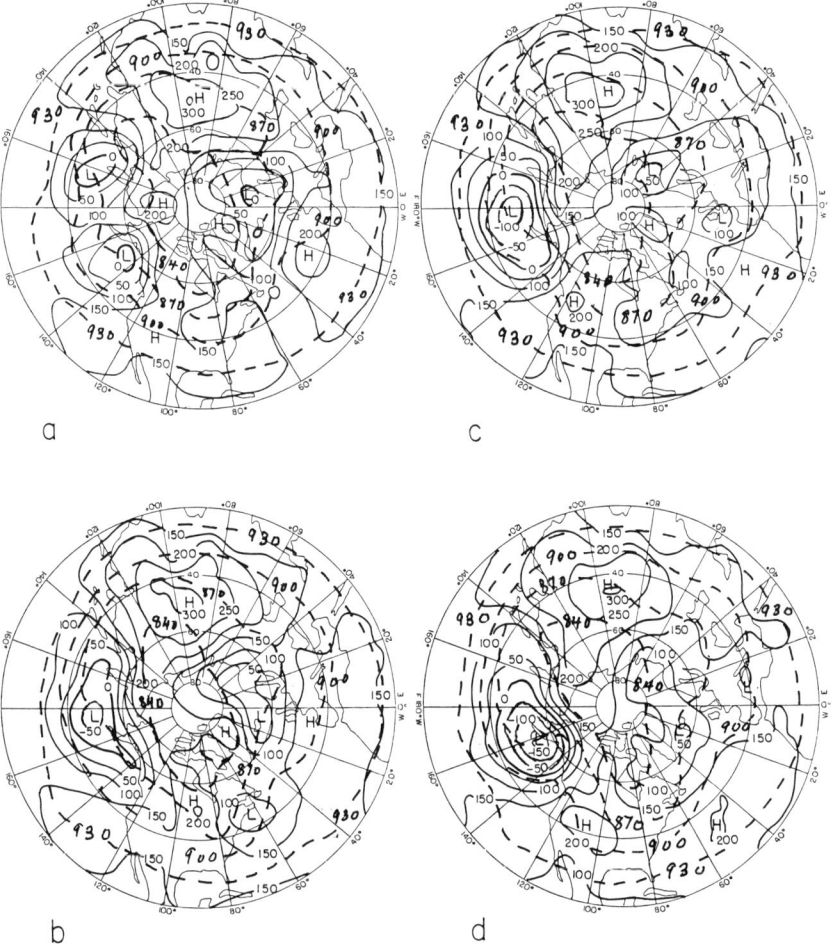

FIG. 14. Composite 1000-mbar heights (solid lines, meters) and 1000- to 300-mbar thickness (dashed lines, dekameters) for the PAC negative anomaly for days (a) -3, (b) -1, (c) $+1$, and (d) $+3$.

axis, somewhat reminiscent of an eastern Asian "cold surge" (e.g., Joung and Hitchman, 1982); the associated deformation field tends to support increasing temperature gradients over this region near the latitude of the intensifying jet.

As the developing low propagates eastward, the region of cold advection to the north of the jet axis extends eastward to the central Pacific, again tending to increase temperature gradients over this region centered at latitudes near 40°N. By day +1, the developing southerly flow to the east of the major low center produces strong warm advection over the northeastern North Pacific, northwestern North America, and Alaska under and to the west of the developing upper level ridge. The thermal advections at day +1 and day +3 tend to bring relatively cold air southward and eastward from the Asian continent and relatively warm air northward and westward over the northeast Pacific and Alaska; we might anticipate that heat flux statistics obtained through this period would show considerable correlations in the zonal ($u'T'$) as well as meridional ($v'T'$) fluxes. Following day +3 (not shown), there are reduced advections as the low assumes a roughly cold-core structure with little tilt with height.

7. Breakdown

The previous analyses suggest that persistent anomaly patterns often develop rapidly, with corresponding positive and negative cases displaying several parallel features. We now briefly examine the subsequent breakdowns of the patterns for further evidence of systematic behaviors. Composites for this section are constructed relative to the time when the anomaly first falls below the threshold value at the key point. The unfiltered analyses provided no obvious indication that small-scale, mobile disturbances were systematically involved in the breakdowns; for this reason, only low-pass composites are presented.

Figures 15 and 16 present, respectively, composite low-pass analyses for the PAC positive and negative cases at 2-day intervals from 4 days prior to breakdown (day −4) to 6 days following breakdown (day +6). We see that during breakdown the evolutions also display a number of striking similarities. Up until day −2, the patterns strongly resemble the PAC composite patterns described previously. Breakdown then proceeds rapidly. By day 0, the main centers have moved northwestward to the Bering Sea and have weakened considerably. These features then remain nearly quasi-stationary and continue to decay. Anomalies over the key region are not significantly different from zero beyond day +2.

Differences between the composite maps immediately preceding break-

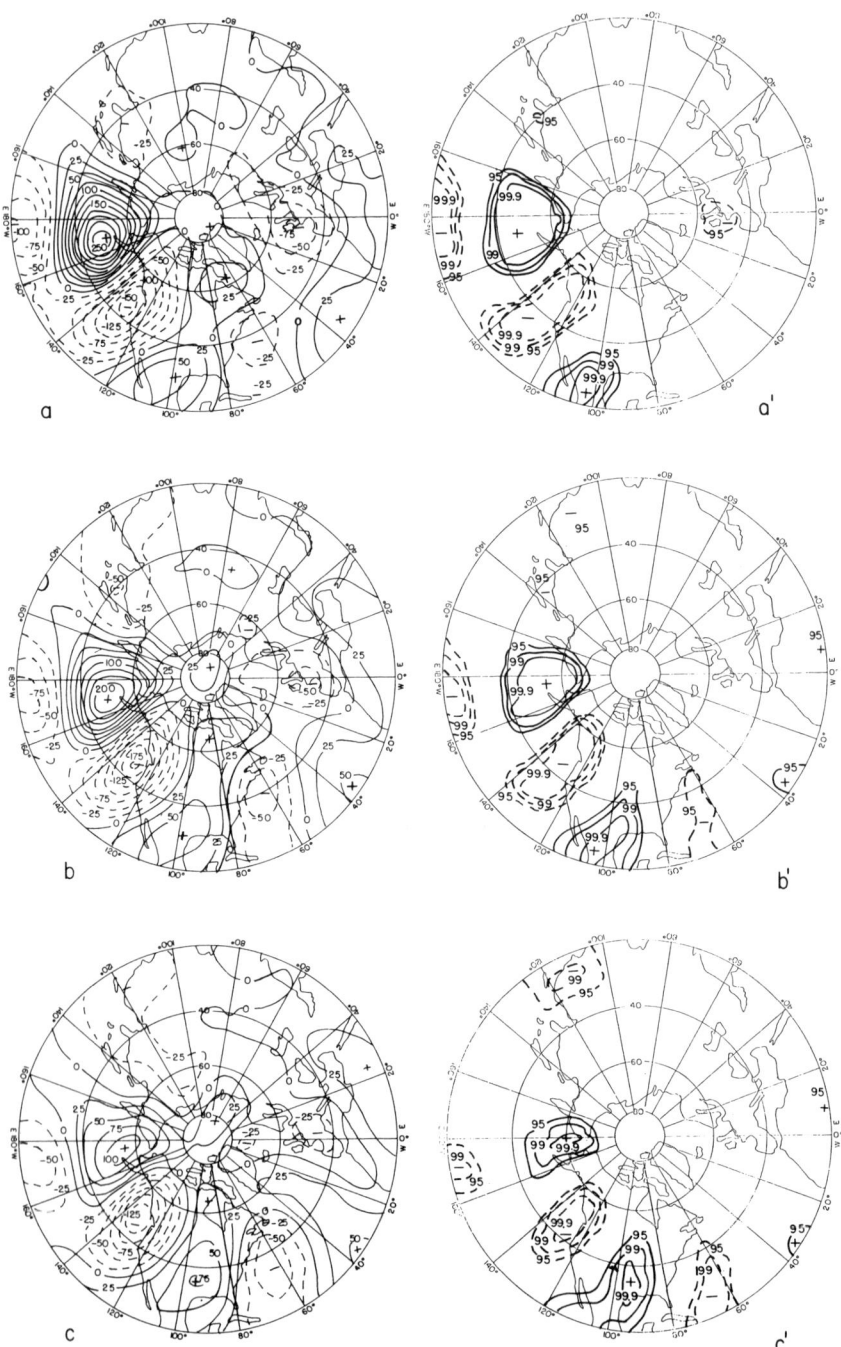

FIG. 15. Composite mean time evolution of low-pass filtered 500-mbar height anomalies (in meters) obtained from 15 PAC positive anomaly cases for days (a) −4, (b) −2, (c) 0, (d) +2, (e) +4, and (f) +6, relative to the breakdown time. Areas where the composite mean anomalies are greater or less than zero at varying confidence levels are shown for the same times in a′–f′. Dashed lines indicate negative anomaly values.

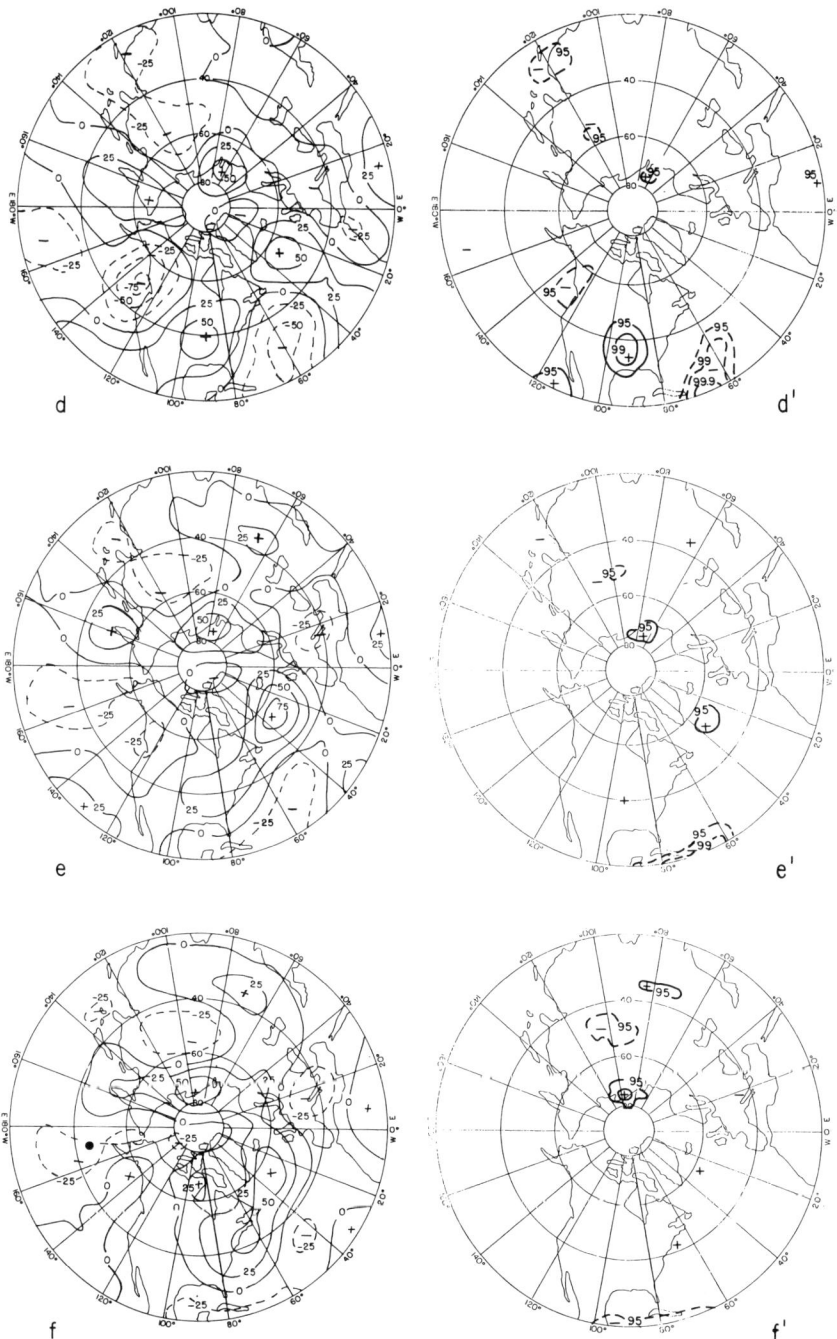

Fig. 15d–f'.

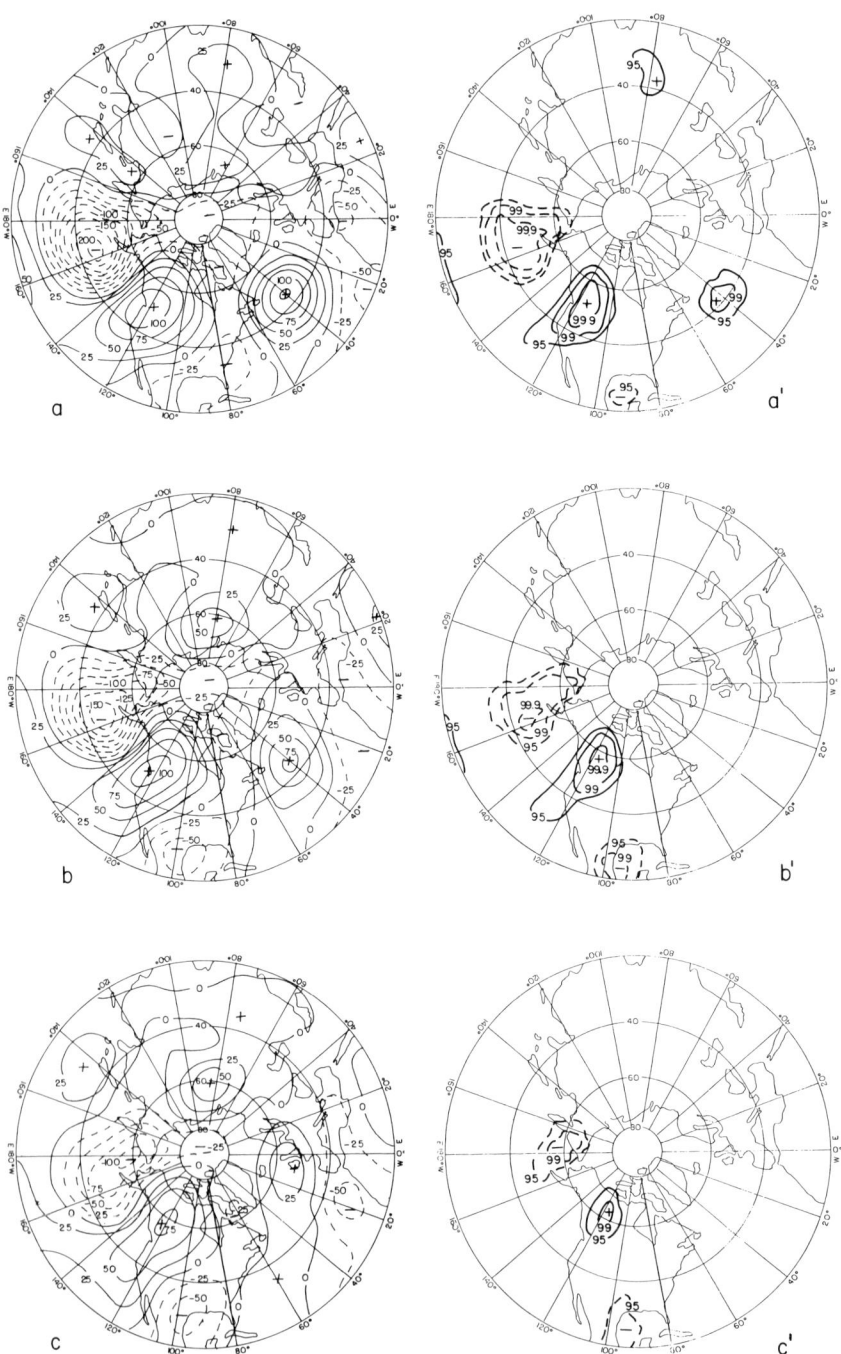

FIG. 16. As in Fig. 15, for the composite mean time evolution of low-pass filtered 500-mbar height anomalies (in meters) obtained from 13 PAC negative anomaly cases for days (a) −4, (b) −2, (c) 0, (d) +2, (e) +4, and (f) +6, relative to the breakdown time.

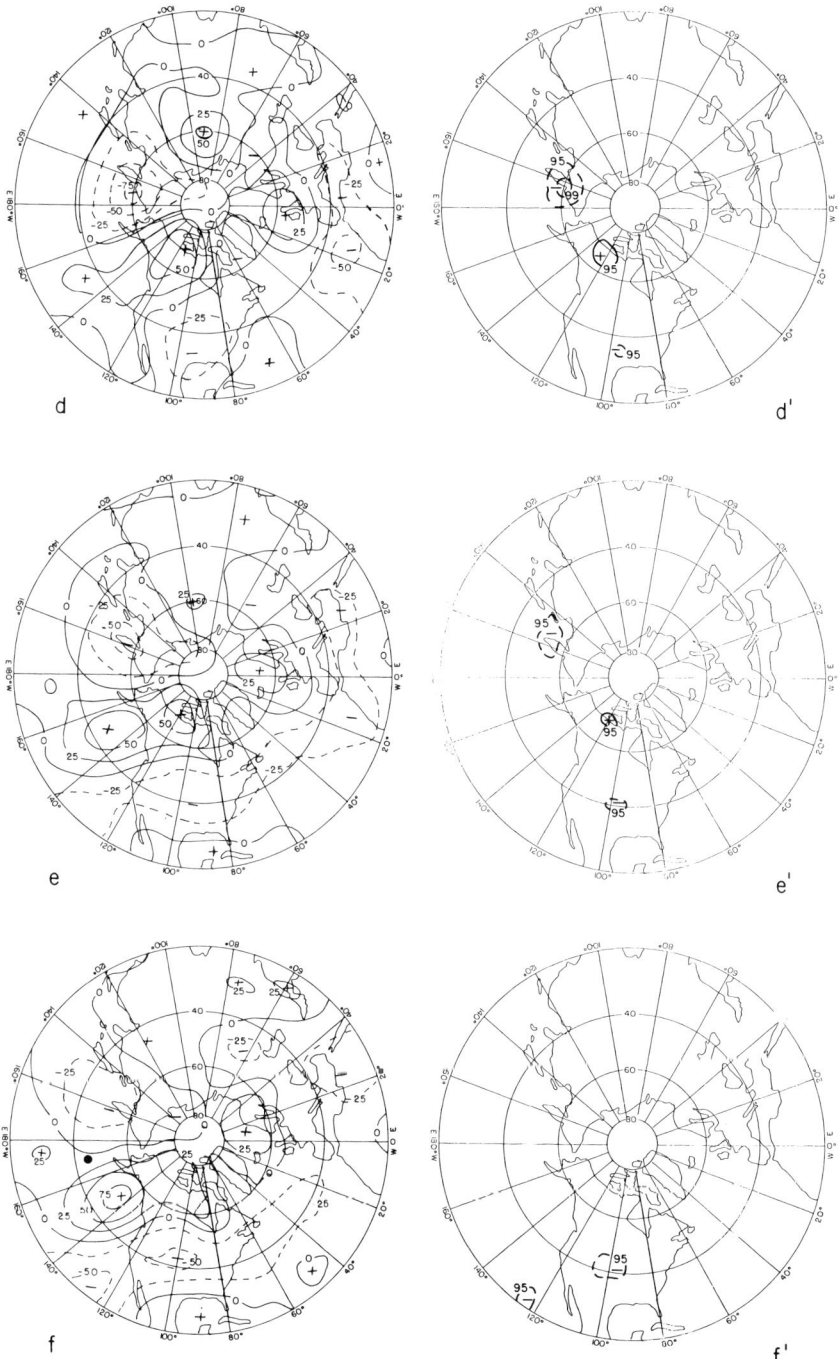

Fig. 16d–f'.

down and the corresponding maps following development may suggest clues to the cause of the breakdown. Comparison of the maps 2 days before breakdown with (for example) the maps 6 days after onset, however, reveals few striking changes. For the positive cases, the most noticeable differences are the reduction in the apparent wavelength downstream from the main center and the intensification of the negative anomaly center in the northeast Pacific. We suspect that the latter development, which is also reflected in an asymmetry between the positive and negative case mean composites discussed in Dole (1986), is at least partly the result of orographic effects: in particular, northerly flow over the Brooks Range associated with an anomalously strong ridge over the Aleutians (as is characteristic of the positive cases) is often followed by intense cyclogenesis in the northeast Pacific (Winston, 1955). The negative cases, in contrast, show little change in the location or intensity of the downstream center.

Another difference evident in both the positive and negative cases when comparing postdevelopment with prebreakdown maps is the change in sign of the anomaly centers near Japan. Whereas at and immediately following development anomalies in this region have the same sign as in the key region, immediately preceding breakdown anomalies in this area and in the key region are of opposite sign; however, the anomalies are of sufficiently weak intensity that not much confidence can presently be attached to this result.

8. Discussion

The time scales for the development and decay of persistent anomalies appear to be short compared to the time scales associated with major changes in the boundary conditions (e.g., sea surface temperature anomalies). In addition, the sequence of development for the PAC cases suggests that the initial rapid growth of the main center may be partly associated with a propagating, intensifying, synoptic-scale disturbance which apparently originates in midlatitudes over eastern Asia. These observations suggest that the patterns may often, and perhaps primarily, grow and decay while the external forcing remains nearly fixed. General circulation modeling experiments (Lau, 1981) provide further evidence that anomalies in the external forcing are not required to produce patterns similar to those described here.

Nevertheless, Horel and Wallace (1981) have recently presented convincing observational evidence of a modest relationship between interannual variability in tropical Pacific sea surface temperature anomalies (related to the El Niño–Southern Oscillation phenomenon) and the Pacific–North American teleconnection pattern (closely resembling our PAC anomaly pattern). They have interpreted the extratropical circulation anomalies as essentially a forced stationary Rossby wave response to tropical heating anomalies centered over the central equatorial Pacific. Our results, however, provide no clear indication for Rossby wave trains propagating northward from this region immediately prior to development. If a link exists between tropical sea surface temperature anomalies and the Pacific pattern, then it may be more subtle than indicated by this picture.

An intriguing clue toward a possible link is provided by the significant pattern located upstream over Asia and the extreme western Pacific preceding the development of the PAC cases. The structure of this pattern suggests that the associated wind anomalies are primarily in the zonal flow over both the Himalayas and the southwestern North Pacific. The anomalous flow patterns we observe in this region may be a reflection in part of changes in tropical forcing over the far southwest Pacific and Indonesian regions, areas where interannual precipitation anomalies are known to be strongly linked to different phases of the Southern Oscillation (e.g., Horel and Wallace, 1981). In addition, we note that recent modeling studies (Simmons, 1982; Branstator, 1983; Simmons *et al.*, 1983) have displayed a marked sensitivity in the extratropical response centered over the central North Pacific to forcing changes located in southwestern Pacific.

Some aspects of the evolution observed in the 500-mbar analyses are reminiscent of the evolutions seen in a study of the barotropic instability of the Northern Hemisphere 300-mbar time-mean flow (Simmons *et al.*, 1983; cf. their Figs. 11 and 19). In particular, that study indicates that zonally elongated eddies (having u'^2 greater than v'^2) located in regions where the basic-state zonal flow is decreasing downstream are capable of growing barotropically by extracting kinetic energy from the basic flow. The growing disturbances that we have described have this general structure and are primarily located in the jet exit region over the central North Pacific, suggesting that this mechanism may contribute positively to their growth. Barotropic instability of the zonally varying time-mean flow therefore provides one possible source for the initial developments in the PAC region. In addition, the evolution of the 500-mbar anomaly pattern downstream from the central North Pacific appears qualitatively similar to that seen in simple linear barotopic models of horizontal energy dispersion on a sphere away from a quasi-stationary localized source of vorticity (Hoskins *et al.*, 1977), suggesting that this process may account for important aspects of the downstream development.

The vertical structures of the growing disturbances over the central North Pacific, however, display considerable tilts with height as they amplify, indicating that their developments may be significantly influenced by baroclinic processes or by the possibility of vertical energy propagation. Indeed, in several aspects the structures resemble those of amplifying baroclinic waves, although their spatial scales are considerably larger and propagation speeds considerably smaller than for typical disturbances. In addition, for the negative anomaly cases, the associated deformation fields produce horizontal temperature advections that are highly favorable for concentrating temperature gradients along the evolving jet axis. These observations raise the distinct possibility that barotropic or equivalent barotropic models will be inadequate for modeling important aspects of the initial developments.

Few attempts have been made to analyze the possible influences of baroclinicity on the development of persistent anomalies. White and Clark (1975) have suggested that baroclinic instability modified by sensible heat exchange is responsible for the development of blocking over the central North Pacific. For theoretical support they apply the results of Haltiner's (1967) analysis of the effects of diabatic heating on baroclinic instability in a two-layer quasi-geostrophic model. Geisler (1977), however, criticizes the use of Haltiner's model, since Geisler and Garcia (1977) find that in a continuously stratified model Newtonian cooling acts to reduce the growth rates at all wavelengths, with the greatest reductions at long wavelengths. Further, all unstable modes propagate eastward at a rate slightly larger than the basic-state surface wind, and therefore would not remain geographically stationary in the presence of a westerly surface flow.

More recently, Frederiksen (1983) has attempted to analyze the instability characteristics of the three-dimensional Northern Hemisphere wintertime mean flow using a two-layer spherical quasi-geostrophic model. With this model, he identifies certain unstable modes whose structures and evolutions resemble in many gross aspects the features that we have described. Frederiksen interprets the development of the PAC pattern as essentially a two-stage process involving both baroclinic and barotropic processes (where the stages are essentially determined by variations in the basic-state stability). During the stage in the development where baroclinic conversions dominate, the unstable mode is eastward propagating; the larger scale, quasi-stationary mode is associated mainly with barotropic conversions. Further investigation will be required, however, to ascertain the extent to which Frederiksen's model agrees with observed developments and, in particular, whether it can provide any predictive information for the onset and evolution of the Pacific patterns.

9. Conclusions

Our results suggest a number of typical characteristics in the evolution of persistent anomaly patterns:

1. Development rates are often rapid (full establishment in less than a week).

2. Over the key region, there is little evidence of an atmospheric precursor until just prior to onset.

3. Following onset, anomaly centers develop and intensify in sequence, forming a quasi-stationary wave train downstream from the main center. Intensification occurs with little indication of phase propagation. This leads to the establishment of the persistent anomaly pattern.

4. Vertical cross-sections indicate that the PAC anomaly centers have substantial tilts with height during development. Corresponding 1000-mbar height and 1000- to 300-mbar thickness analyses for the PAC negative cases provide further evidence that the developments may have a markedly baroclinic character, with horizontal temperature advections tending to increase the temperature gradients to the south and southeast of the intensifying surface low near the latitude of the jet maximum. The blocking ridge developing downstream over the eastern Pacific also amplifies in a region of strong thermal advection and is characterized by a westward tilt with height during development.

5. Breakdown rates are also often rapid. Until immediately prior to breakdown, the patterns resemble the corresponding patterns obtained following development.

6. From development through decay, corresponding positive and negative patterns display striking similarities in their evolutions.

The evolution of the 500-mbar anomaly pattern downstream from the central North Pacific appears qualitatively similar to that seen in simple time-dependent models of horizontal energy dispersion on a sphere away from a quasi-stationary localized source of vorticity, suggesting that this process may account for important aspects of the downstream development. For the PAC cases, the most systematic precursors are related to variations in the jet intensity and structure over eastern Asia and the southwestern North Pacific and to an eastward-propagating, intensifying anomaly center that appears to originate at midlatitudes.

Some aspects of the initial development are reminiscent of the evolutions seen in studies of the barotropic instability of the time-mean flow; however, the growing disturbance over the Pacific is associated with pronounced temperature advections and, in the negative anomaly cases, a deformation field favorable for concentrating temperature gradients along the jet axis. The structure of this disturbance bears considerable resemblance to that of a classic amplifying baroclinic wave, although its spatial scale is considerably larger and its propagation speed considerably smaller than for typical baroclinic disturbances. Thus, the observations cast some doubt on whether barotropic or equivalent barotropic models will be sufficient for modeling important aspects of the developments, or whether models that take into account the full three-dimensional structure of the flow will be required to adequately describe the development of many persistent anomaly cases.

Acknowledgments

I thank Ms. Isabelle Kole for her help in drafting several of the figures. Some of the calculations were performed on the Goddard Space Flight Center computer system located at Greenbelt, Maryland; the remainder were performed on the National Center for Atmospheric Research computer system. NCAR is supported by the National Science Foundation. Support for the research was provided by NASA Grant NASA-g NAGw-525.

References

Austin, J. F. (1980). The blocking of middle latitude westerly winds by planetary scale waves. *Q. J. R. Meteorol. Soc.* **106**, 327–350.

Blackmon, M. L. (1976). A climatological spectral study of the geopotential height of the Northern Hemisphere. *J. Atmos. Sci.* **33**, 1607–1623.

Branstator, G. (1983). Horizontal energy propagation in a barotropic atmosphere with meridional and zonal structure. *J. Atmos. Sci.* **40**, 1689–1708.

Colucci, S. J., Loesch, A., and Bosart, L. (1981). Spectral evolution of a blocking episode and a comparison with wave interaction theory. *J. Atmos. Sci.* **38**, 2092–2111.

Dole, R. M. (1982). Persistent anomalies of the extratropical Northern Hemisphere wintertime circulation. Ph.D. thesis, Massachusetts Institute of Technology, Cambridge, Mass. 02139. Thesis available from the author on request.

Dole, R. M. (1986). Persistent anomalies of the extratropical Northern Hemisphere wintertime circulation: Structure. *Mon. Weather Rev.* **114**, 178–207.

Dole, R. M., and Gordon, N. D. (1983). Persistent anomalies of the extratropical Northern Hemisphere wintertime circulation: Geographical distribution and regional persistence characteristics. *Mon. Weather Rev.* **111**, 1567–1586.

Frederiksen, J. S. (1983). A unified three-dimensional instability theory of the onset of blocking and cyclogenesis. II. Teleconnection patterns. *J. Atmos. Sci.* **40**, 2593–2609.

Gall, R. (1976). Structural changes in growing baroclinic waves. *J. Atmos. Sci.* **33**, 374–390.

Geisler, J. E. (1977). On the application of baroclinic instability and sensible heat exchange to explain blocking ridge development. *J. Atmos. Sci.* **34**, 311–321.

Geisler, J. E., and Garcia, R. P. (1977). Baroclinic instability at long wavelengths on a beta plane. *J. Atmos. Sci.* **34**, 311–321.

Haltiner, G. J. (1967). The effects of sensible heat exchange on the dynamics of baroclinic waves. *Tellus* **19**, 183–198.

Holton, J. R. (1979). "An Introduction to Dynamic Meteorology." Academic Press, New York.

Horel, J. D., and Wallace, J. M. (1981). Planetary scale atmospheric phenomena associated with the Southern Oscillation. *Mon. Weather Rev.* **109**, 813–829.

Hoskins, B. J., and Karoly, D. (1981). The steady linear response of a spherical atmosphere to thermal and orographic forcing. *J. Atmos. Sci.* **38**, 1179–1196.

Hoskins, B. J., Simmons, A. J., and Andrews, D. G. (1977). Energy dispersion in a barotropic atmosphere. *Q. J. R. Meteorol. Soc.* **103**, 553–567.

Joung, C. H., and Hitchman, M. H. (1982). On the role of successive downstream development in East Asia cold air outbreaks. *Mon. Weather Rev.* **110**, 1224–1237.

Lau, N. C. (1981). A diagnostic study of recurrent meteorological anomalies appearing in a 15-year simulation with a GFDL general circulation model. *Mon. Weather Rev.* **109**, 2287–2311.

Reinhold, B. B., and Pierrehumbert, R. T. (1982). Dynamics of weather regimes: Quasi-stationary waves and blocking. *Mon. Weather Rev.* **110**, 1105–1145.

Simmons, A. J. (1982). The forcing of stationary wave motion by tropical diabatic heating. *Q. J. R. Meteorol. Soc.* **108**, 503–534.

Simmons, A. J., and Hoskins, B. J. (1978). The life-cycles of some nonlinear baroclinic waves. *J. Atmos. Sci.* **35**, 414–432.

Simmons, A. J., Wallace, J., and Branstator, G. (1983). Barotropic wave propagation and instability and atmospheric teleconnection patterns. *J. Atmos. Sci.* **40**, 1363–1392.

Wallace, J. M., and Gutzler, D. S. (1981). Teleconnections in the geopotential height field during the Northern Hemisphere winter. *Mon. Weather Rev.* **109**, 784–812.

White, W. B., and Clark, N. E. (1975). On the development of blocking ridge activity over the central North Pacific. *J. Atmos. Sci.* **32**, 489–502.

Winston, J. S. (1955). Physical aspects of rapid cyclogenesis in the Gulf of Alaska. *Tellus* **7**, 481–500.

ON ATMOSPHERIC BLOCKING TYPES AND BLOCKING NUMBERS

HEINZ-DIETER SCHILLING

Meteorology Institute
University of Munich
D-8000 Munich 2
Federal Republic of Germany

1. INTRODUCTION

During the past few years work on blocking in the westerlies has grown rapidly. Many studies have suggested a variety of possible blocking mechanisms. Whereas each proposal is by itself interesting and suggestive, it must be proved to be an essential feature of the formation or maintenance of at least some real blocking cases. On the other hand, the study of blocking cases with either observational or model data raises certain difficulties. This article addresses two of them.

The first problem is whether the study should be made circumpolar or not. Using local methods the consequences of the blocking event on other regions are not taken into account (see double blocking structure and the monthly 700-mbar anomaly charts published by the *Monthly Weather Review* for months with strong, long-lasting blocking; e.g., Stark, 1965; Dickson, 1967, 1969; Taubensee, 1975). Using the circumpolar method we could possibly mask some special features of the blocking case by averaging with other far downstream or upstream events. Therefore it is desirable to develop a criterion to characterize the longitudinal extent of the influence of the blocked zone. Our proposed solution is to introduce a blocking number defined in terms of circumpolar parameters which therefore responds to circumpolar characteristics of the streamfield. Moreover, this single number is given at any instant. In this way, it is possible to study the response of distant regions to blocking activity in certain other locations. This is done here by computing the temporal correlation between the 500-mbar topography and the blocking number. This response, in turn, helps to answer the question as to whether or not the anomaly studies of the type of Charney *et al.* (1981) are representative for the circumpolar type of blocking.

The second problem in data studies is that the dynamics and energetics of blocking events seem to vary widely from case to case. As a first step, case studies can contribute to the understanding of different mechanisms. However, we believe, a statistical approach is the next step to obtain insight into the cooperation of the main proposed mechanisms of blocking formation.

This raises the question as to how we can group blocking events together to perform sample statistics. Since there are many proposed or observed blocking mechanisms, the definition of types of blocking is by itself useful to bring order into the somewhat confusing collection of previous results. Clearly the question as it stands is very complicated and needs some simplification.

In the following treatment one simplification will be to concentrate on dominant waves (or wave groups) and on the energy fluxes feeding them. We expect that statistics based on this approach will not necessarily weaken signals from other dynamical processes that trigger the block formation or stabilize the formed block rather than feed the blocking waves. If their influence on blocking is systematic it will show up more clearly in such sample statistics, if calculated for each type of block separately.

As to the blocking number, we note another potential advantage of it. Being an averaged quantity, the blocking number should have an extended predictability. Therefore, correctly predicting this parameter's tendency some days in advance will be the equivalent to predicting the atmosphere's tendency toward blocking states. In practice, however, the predicted absolute value of the blocking number is not as important as the increase above a certain threshold value. That critical value should be chosen so that it indicates high possibility of occurrence of blocking states.

2. Energy Parameters and Data

According to present knowledge, there are too many aspects of observed blocking for a proper treatment by only one method. However, we decided to concentrate on circumpolar energetical aspects of the transformation of the flow under blocking conditions. A considerable amount of evidence has been given, indicating that even simple circumpolar energy parameters of the blocked latitude belt show a distinct behavior during blocking (Schilling, 1982; Chen and Shukla, 1983; Fischer, 1984; Hansen and Chen, 1982; Speth and Meyer, 1984). We choose here the following parameters (for defining relations see Appendix A):

K_E, the kinetic energy of pertubations of zonal-mean flow
K_z, the kinetic energy of the vertical mean of zonal-averaged flow
K_T, the kinetic energy of the vertical shear of zonal-mean flow, roughly representing A_z, the available potential energy (APE) of the zonal mean

These and other energetic parameters were computed from once-daily height field data at three levels (300, 500, and 1000 mbar) for the period

1967–1976. These data were prepared by the German Weather Service (DWD).

Figure 1 shows how these parameters develop during 20 cases of Atlantic/European blocking. The cases are taken from the catalog of Schilling (1981). This catalog was recently revised on the basis of the following definition (after Schilling, 1982, and Egger, 1978): A blocking is defined as (a) a 500-

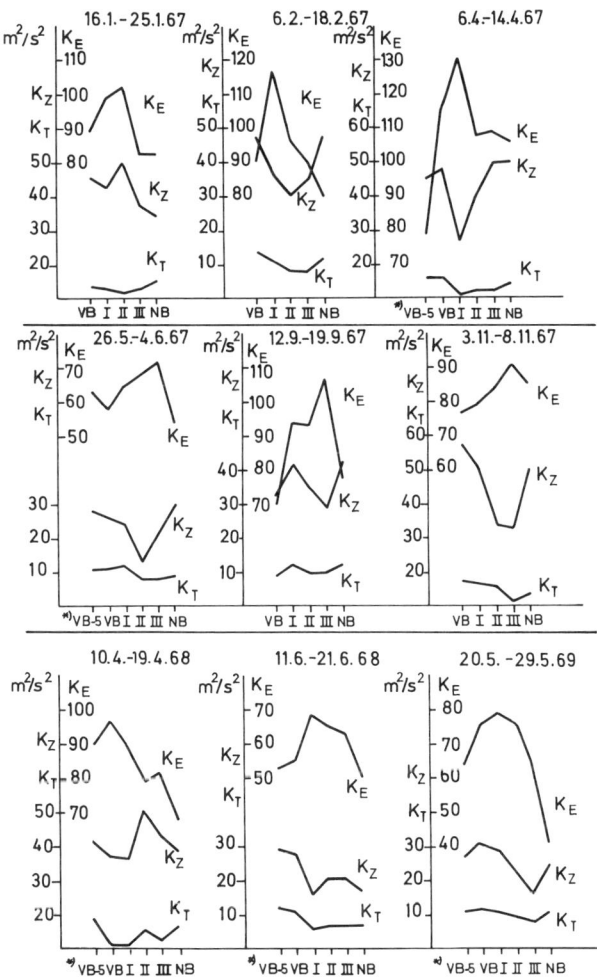

FIG. 1. K_E, K_z, and K_T averaged over several stages of Atlantic blocking (20 cases). For comparison, two nonblocking cases are provided. See text for definitions. VB − 5, Average over 5 days before VB; NB + 5, average over 5 days after NB; I − 5, III + 5 analogous to definitions above.

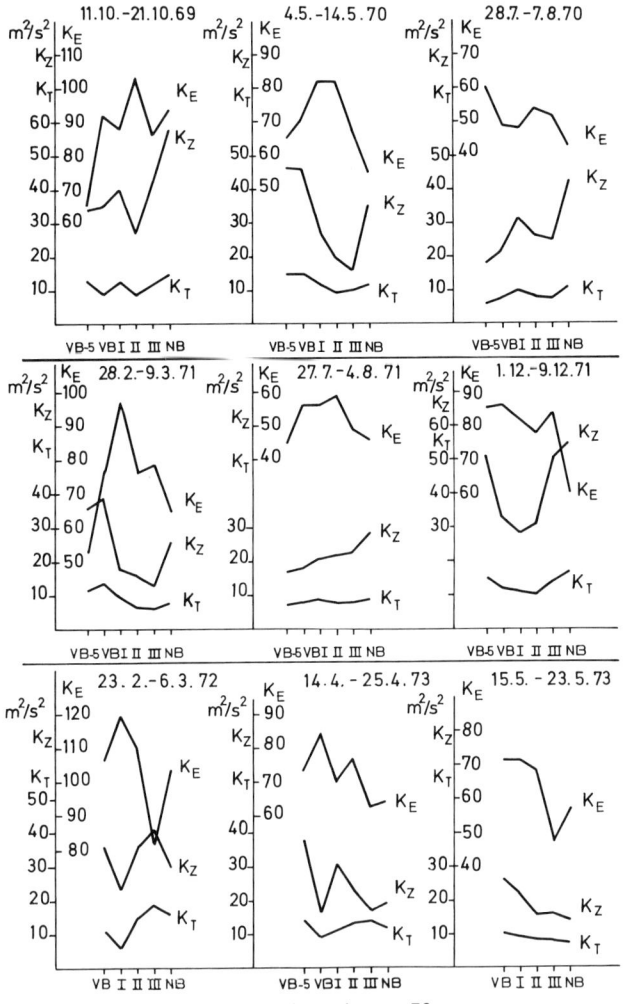

FIG. 1. See legend on p. 73.

mbar high with at least one closed contour (in a representation with 4-gpdm increments) which (b) is persistent for at least 5 days, and (c) lies in the region $70°W \leq \lambda \leq 70°E$ and $50° \leq \varphi \leq 80°$. Additionally (d) the high cell is embedded in a split jet with branches of about equal masses. (e) If criteria (a–c) are fulfilled at least for 5 consecutive days and there exists a further day, fulfilling (a), (c), and (d) but separated from the other blocking days by a day with arbitrary 500-mbar flow, this entire period is taken as one blocking case.

Out of this revised Atlantic/European blocking catalog selected cases are

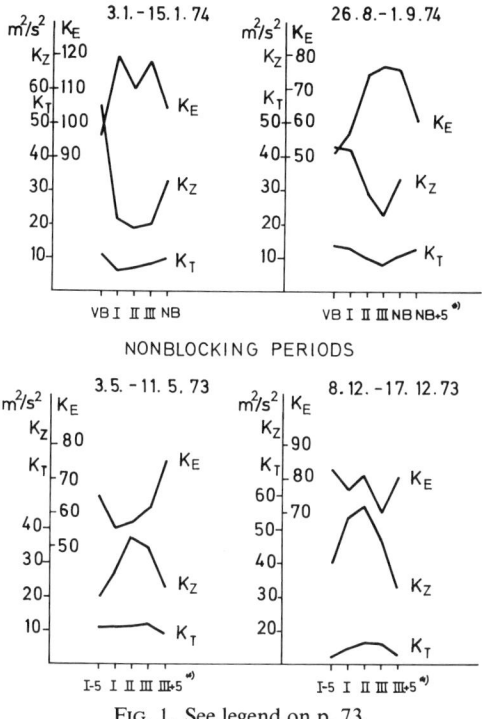

FIG. 1. See legend on p. 73.

given in Appendix C. The complete catalog is available from the author upon request.

In this section we are interested in several stages of evolution: VB, the 5-day period before onset; I, II, and III, which constitute three equal subdivisions of the block; and NB, the 5-day period following the block. For K_E, K_T, patterns as displayed in Fig. 1 are similar in almost every case, but for K_z the similarities are not as striking (see also Lejenäs, 1977). Therefore we exclude the latter from further consideration.

We introduce parameters describing the observed jet-splitting tendency and the dominance of larger scales in the zonal Fourier spectrum of perturbations. This is done by kinetic energy weighted averages of the respective wavenumbers of appropriate height field decompositions (e.g., zonal Fourier components m with $m \neq 0$ or order numbers n of normal modes of the Laplacian ∇^2 on the sphere, having eigenvalues κ_n^2 (see Appendix A).

$$m_c^2 = \sum_{m=1}^{M} m^2 K_m / \sum_{m=1}^{M} K_m; \quad M = 8 \tag{1a}$$

$$L_c = 2\pi a \cos \varphi_c/m_c; \quad \varphi_c = 60°\text{N} \tag{1b}$$

$$\kappa_c^2 = \sum_{n=1}^{N} \kappa_n^2 \bar{u}_n^2 / \sum_{n=1}^{N} \bar{u}_n^2; \quad N = 15 \tag{2a}$$

$$B_c = \pi/\kappa_c \tag{2b}$$

The kinetic energy of zonal wavenumber m is denoted by K_m; κ_n and $\bar{u}_n$ are the wavenumber and amplitude of the meridional decomposition of $\bar{u}$, the zonal-mean flow. Other symbols are defined in Appendix A if not explained in the text. These parameters show reasonably good response to blocking conditions. The tendency for jet splitting is chararacterized by smaller than normal values of B_c and the dominance of ultralong waves by larger than normal values of L_c. This can be seen for the 20 examples in Fig. 2 (averaged over the 20 cases).

The time evolution of the measure of static stability

$$L_{\text{crit}}^2 = \sigma p_{1000}^2/f^2(\varphi_c); \quad \sigma = [-1/\rho)(\partial \ln\theta/\partial p)]_{20°\text{N}}^{85°\text{N}} \tag{3}$$

is shown also in Fig. 2. It exhibits almost no variations during blocks and seems to be quite unaffected by the transformation of flow.

It cannot be expected that each of the components of the set $C = (K_E, K_T, L_c, B_c)$ will respond with constant reliability. Sometimes only a subset of C is well developed and the rest is masked.

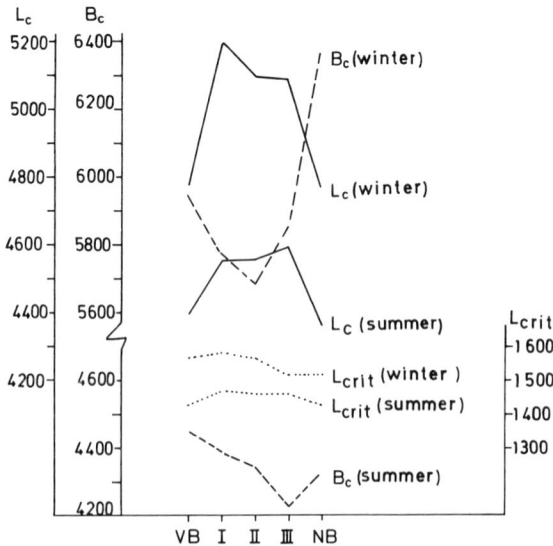

FIG. 2. L_c, B_c, and L_{crit} averaged over several stages of Atlantic blocking (20 cases). See text for definitions. Winter, 1 Oct to 31 Mar; summer, 1 Apr to 30 Sep.

3. Blocking Numbers

It is quite attractive to form a single blocking number out of the foregoing parameters. By doing so, it is hoped that weak development of one or two parameters will be compensated by other strongly developing ones.

We decided to arrange the parameters in a such way as to obtain the largest values of the number during the mature stage of blocking. However, there are different possibilities for the definition:

$$Bl_1 = K_E L_c^2 / (K_T B_c^2) \qquad (4)$$

$$Bl = K_E L_c^2 / (K_T L_{crit}^2) \qquad (5)$$

The latter was motivated by the fact that B_c sometimes exhibits an erratic behavior. Moreover, it seems that the above-mentioned set C of characteristics is redundant. This is because B_c tends to decrease whenever L_c increases, as can be shown in simple models (e.g., barotropic ones conserving energy and enstrophy). Figure 3 justifies Eq. (5), showing that this criterion performs as well as Eq. (4) on average. However, Bl is smoother than Bl_1 in single cases.

In view of A. Hansen's (see this volume) amplitude index $[Z_{2-4}]$, which is restricted to zonal wavenumbers $m = 2, 3$, and 4, a more selective version of Bl was tested:

$$Bl_{2-4} = Bl \cdot K_{2-4}/K_E \qquad (6)$$

Here, K_{2-4} means the kinetic energy of zonal wavenumbers $m = 2, 3$, and 4, averaged over the same latitudinal belt as before ($52.5°N \leq \varphi \leq 82.5°N$).

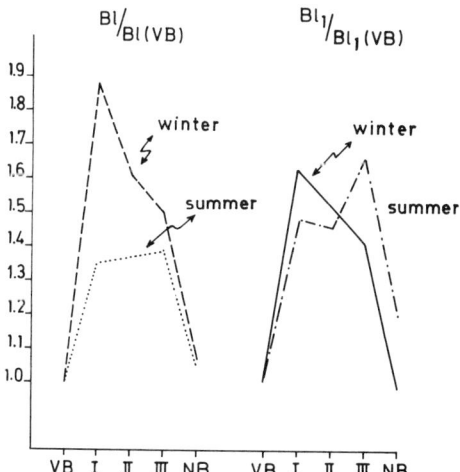

FIG. 3. Same as Fig. 2, except for Bl_1 and Bl as normalized with the value of period VB.

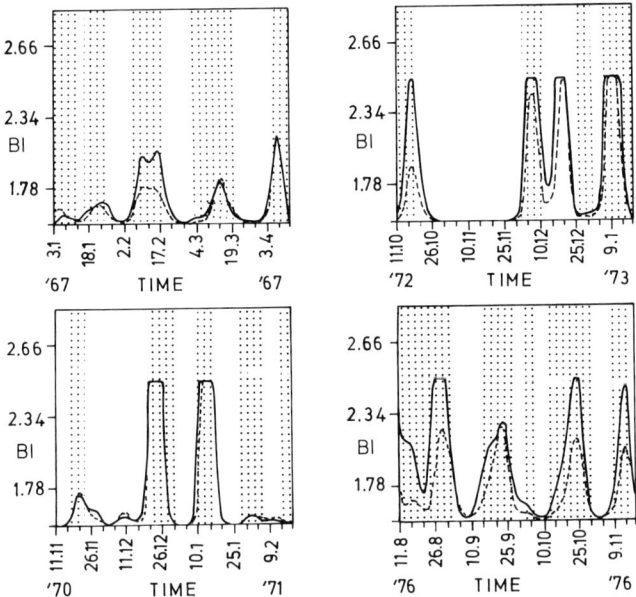

FIG. 4. Bl curves for selected periods during 1967–1976. Nonlinear ordinate scale. Blocking cases are indicated by dotted areas. Solid lines, $Bl_{2-4}/\langle Bl_{2-4}\rangle_{mon}$; dashed lines, $Bl/\langle Bl\rangle_{mon}$. Values above 2.51 were cut off.

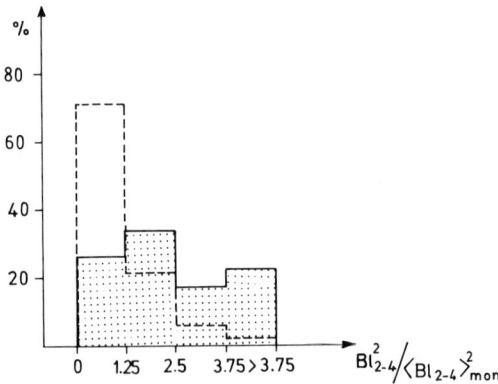

FIG. 5. Frequency distribution of $Bl_{2-4}^2/\langle Bl_{2-4}\rangle_{mon}^2$ being averaged over one blocking period except for periods > 15 days. Such long cases were split into two halves, each one an entry. The distribution for all Atlantic/European as well as Pacific cases during 1967–1976 is indicated by dotted area. The distribution for nonblocking periods longer than 4 days is given (dashed lines), with averaging analogous to block cases.

To eliminate the seasonal variation of those numbers, they were normalized by their long-term monthly means $\langle Bl \rangle_{\text{mon}}$ or $\langle Bl_{2-4} \rangle_{\text{mon}}$, taken only over nonblocking days. Some examples are given in Fig. 4 to show how the normalized Bl and Bl_{2-4} behave during selected periods in 1967–1976.

Blocking periods (dotted areas in Fig. 4) are taken from the author's catalog of Atlantic/European cases (see Appendix C), or in the case of Pacific blocks from Treidl et al. (1981) (some Pacific cases are also given in Appendix C). The overall response of Bl to Atlantic/European cases is not significantly different from that to Pacific cases (not shown).

Many events are marked by a well-developed Bl or Bl_{2-4} number. Those clearly have circumpolar characteristics strong enough to justify the use of circumpolar energetics in their study. According to Fig. 5, about 30% of the blocks show relatively small Bl_{2-4} averages, comparable to those recorded for the majority (70%) of nonblocking periods. Those blocks definitely require a local description.

Since strong development of Bl generally coincides with strong development of Bl_{2-4} (see Fig. 4) we can expect that many of the well-developed circumpolar cases are dominated by the zonal waves $m = 2, 3$, and 4. This finding is confirmed by Fig. 6a and b. The 500-mbar stream field, consisting of $m = 1-5$ (and defined on a regular grid), was correlated with blocking numbers for a 2-yr period. In both (a) and (b) it can be seen first that $m = 2$ is dominating the response with relatively fixed phase, and second that

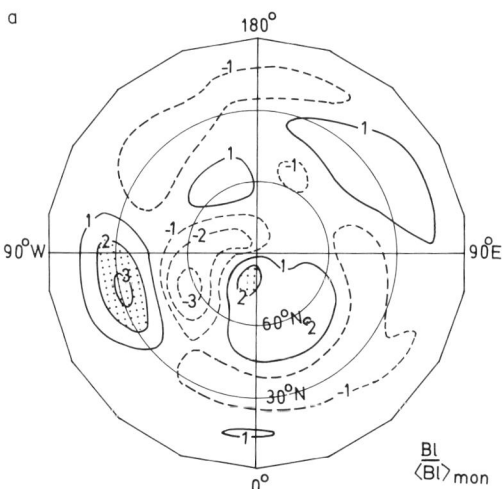

FIG. 6. (a) Pointwise correlation of ψ' in 500 mbar (consisting of waves $m = 1, \ldots, 5$) with $Bl/\langle Bl \rangle_{\text{mon}}$ for the period mid-1969 to mid-1971 (720 days). (b) Same as (a) except for $Bl_{2-4}/\langle Bl_{2-4} \rangle_{\text{mon}}$.

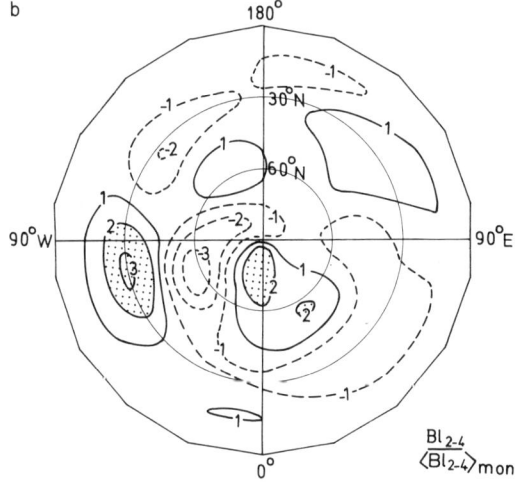

FIG. 6b. See legend on p. 79.

"blocks" (given by peaking Bl or Bl_{2-4} curves) influence remote regions (compare with dotted 95% significance areas).

We are interested now in how the dynamics varies according to the blocking numbers. This question is answered in the next section. Also, what can be concluded from the increase of L_c during blocking events? At least in the case of relatively long-lived blockings the longer waves should have enhanced baroclinic activity. This is necessary since the ever-present differential heating has to be balanced by eddy heat fluxes in a long-term sense. Whenever the "synoptic" wave regime's baroclinic activity is reduced during strong blocking (this reduction is documented in Fig. 7d for $m = 5-8$), the longer waves have to perform this task, at least partially.

4. KINETIC ENERGY BUDGET AND BLOCKING NUMBERS

We now examine the kinetic energy budget in the blocked latitude belt using statistical methods. Our starting point is the quasi-geostrophic version of the vorticity equation with linear damping. The spectral kinetic energy tendency consistent with this equation reads

$$dK_m/dt = -k_m K_m + \text{BTP}_m + \widetilde{\text{BCL}}_m + \text{ORO}_m \\ + \text{NLLS}_m + \text{NLLL}_m + \text{BOUND}_m \qquad (7)$$

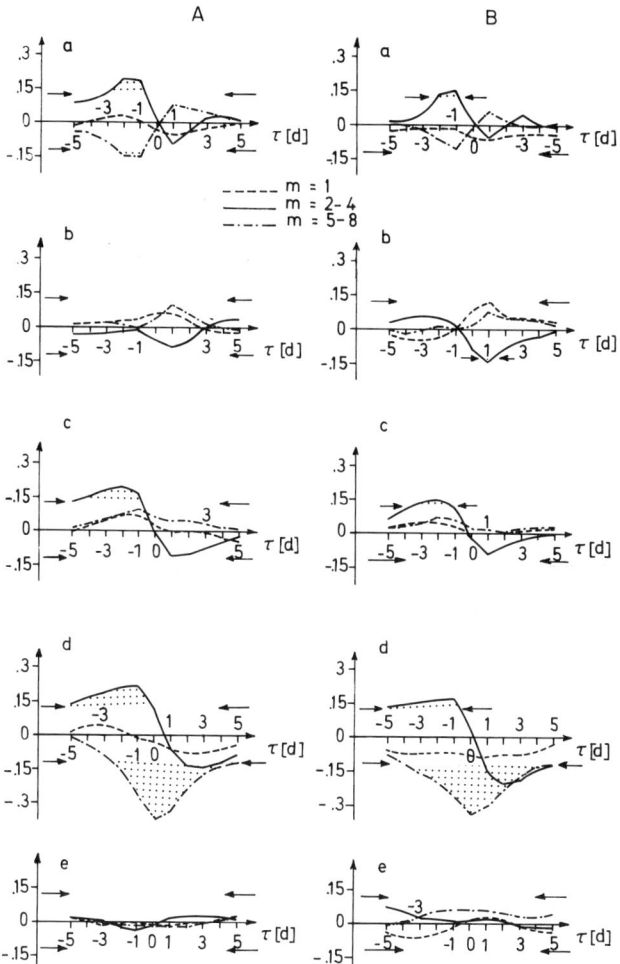

FIG. 7. Time-lagged correlations of $Bl_{2-4}/\langle Bl_{2-4}\rangle_{mon}$ with energy fluxes defined in the text. Solid lines, $m = 2-4$; dashed lines, $m = 1$; dash-dotted lines, $m = 5-8$. Arrows indicate the 95% significance levels; $\tau < 0$, Bl lags the fluxes; $\tau > 0$, Bl leads the fluxes. A, Summer (1 Apr to 30 Sep); B, winter. a, NLLS; b, NLLL; c, BTP; d, BCL; e, TOP.

with

$$K_m = [(u_m'^2 + v_m'^2)/2]_{52.5°N}^{82.5°N} \qquad (8)$$

$$\text{BTP}_m = -\left[\bar{u}\frac{1}{a^3\cos^2\varphi}\frac{\partial}{\partial\varphi}\left\{\cos\varphi\frac{\partial\psi_m'}{\partial\lambda}\frac{\partial\psi_m'}{\partial\varphi}\right\}\right]_{52.5°N}^{82.5°N} \qquad (9)$$

$$\widetilde{BCL}_m = [f_0 \omega'_m \partial \psi'_m / \partial p]_{52.5°N}^{82.5°N} \tag{10}$$

$$ORO_m = -\frac{f_0}{a}\left[\bar{\rho}_B \bar{u}_B h'_m \frac{1}{\cos\varphi}\frac{\partial}{\partial\lambda}\psi'_{1000,m}\right]_{52.5°N}^{82.5°N} \tag{11}$$

$$NLLS_m = [\psi'_m J(\psi'_{1-5}, \nabla^2 \psi'_{6-12})|_m + \psi'_m J(\psi'_{6-12}, \nabla^2 \psi'_{m \neq m})|_m]_{45°N}^{80°N} \tag{12}$$

$$NLLL_m = [\psi'_m J(\psi'_{1-5}, \nabla^2 \psi'_{1-5})|_m]_{45°N}^{80°N} \tag{13}$$

$$BOUND_m = \frac{1}{a^2}\left\{\left[\cos\varphi_2\left(\psi'_m \frac{\partial^2 \psi'_m}{\partial t \partial \varphi}\right)\right]_{85°N}^{85°N} - \left[\cos\varphi_1\left(\psi'_m \frac{\partial^2 \psi'_m}{\partial t \partial \varphi}\right)\right]_{45°N}^{45°N}\right\}; \quad \varphi_1 = 45°N, \varphi_2 = 85°N \tag{14}$$

The term Eq. (11) is introduced by the vertical integration of the divergence term in the vorticity equation and the lower boundary condition involving orography h. In the derivation of this term the lower boundary condition for the eddies was linearized (see Appendix B). The upper boundary condition was simply $\omega = 0$. Moreover, it was assumed that the zonal-mean flow at the lower boundary can be computed geostrophically by an extrapolation formula for $\bar{\phi}_B(\varphi)$.

All terms, Eqs. (9)–(14), are possible candidates for providing suitable energy input during blocking episodes. Indeed, there have been many suggestions for blocking mechanisms based on one or more types of energy conversion; for example (a) barotropic instability BTP_m (Thompson, 1957); (b) baroclinic instability BCL_m (Chen and Shukla, 1983; Hansen and Chen, 1982); (c) baroclinic instability BCL_m plus nonlinear interaction $NLLL_m$ between longer waves (Schilling, 1982); (d) mountain torque induced conversion ORO_m (Egger, 1978; Charney and Devore, 1979); (e) nonlinear interaction of longer with shorter scales $NLLS_m$ (Fischer, 1984; Hansen and Chen, 1982).

A first step in studying the relevance of those inputs to blocking dynamics is to correlate the above-mentioned energy fluxes with blocking numbers. To this end we should remove the nonstationarity associated with the trivial seasonal trend. Therefore fluxes and energies are normalized to give zero mean and variance 1:

$$\tilde{Z}(t_{j,l}) = \frac{Z(t_{j,l}) - \langle Z \rangle_l}{\{\langle Z^2 \rangle_l - \langle Z \rangle_l^2\}^{1/2}} \tag{15}$$

$t_{j,l}$ is the jth day of the lth month continuously counted during the 10 years 1967–1976; $\langle \ \rangle_l$ is a 10 yr monthly mean related to the lth month.

Figure 7a–c shows the seasonally stratified response of some of the energy fluxes to variations of Bl_{2-4} (and vice versa). Interestingly, we find a significant signal for the nonlinear interactions with synoptic scales $NLLS_{2-4}$ and the barotropic instability BTP_{2-4}.

On the other hand, topographically induced energy input seems to have nothing to do with the circumpolar type of blocking (Fig. 7e). The same is true for $BOUND_m$ (not shown).

The most prominent result for seasonally stratified and time-lagged Bl_{2-4} correlations (Fig. 7d) is that with the flux BCL_m, a substitute of $\widetilde{BCL}_m$, given by

$$BCL_m = -\frac{1}{\sigma}\left[\frac{\partial^2 \overline{\phi}}{\partial p \partial \varphi} \frac{1}{a^2 \cos \varphi}\left(\frac{\partial \psi'_m}{\partial \lambda}\frac{\partial \phi'_m}{\partial p}\right)\right]_{52.5°N}^{82.5°N} \quad (16)$$

(σ as before). It was found that BCL_m and $\widetilde{BCL}_m$ are well correlated for $m \gtrsim 2$, especially during episodes of large energies K_m (Schilling, 1986).

Figure 7d shows the strong response of blocking numbers to the baroclinic activity of ultralong waves. Here $\tau < 0$ means a leading flux. Above normal baroclinic activity of $m = 2-4$ results in above normal Bl_{2-4} 1 or 2 days later. On the other hand above normal Bl_{2-4} results generally in below-normal baroclinic fluxes 1 or 2 days later. The wave group $m = 5-8$ tends to suppress baroclinic activity during episodes of large Bl (compare with related results of A. Hansen, this volume)!

We discuss now these results in more detail by scaling arguments related to the baroclinic part BCL.

If $A_z = (1/2\sigma)[(\partial\overline{\phi}/\partial p - \partial\overline{\phi}/\partial p)^2]$ is the available potential energy of the zonal flow we note that

$$(d/dt)A_z = -\sum_m BCL_m + \text{other terms}$$

We introduce some scale quantities ($\langle \ \rangle$ is a time average) for the zonal stream and the eddies separately:

Zonal stream: for velocity $\langle K_T \rangle^{1/2}$, geopotential $(\partial\overline{\phi}/\partial p)p_{1000} \sim f_0 B_c \langle K_T \rangle^{1/2}$, resulting in $A_z \sim B_c^2 \langle K_T \rangle / L_{crit}^2$, and length scale B_c.

Eddies of wave group m: for velocity $\langle K_m \rangle^{1/2}$, geopotential $\phi'_m \sim f_0 L_c \langle K_m \rangle^{1/2}$, and length scale L_m.

Since the APE of the eddies A_m is found to be of the same order as K_m, we obtain for $(\partial\phi'_m/\partial p)p_{1000} \sim \sqrt{\langle K_m \rangle} f_0 L_{crit}$.

We can estimate now the order of magnitude for BCL_m:

$$BCL_m \sim \frac{\langle K_m \rangle \langle K_T \rangle^{1/2}}{L_{crit}}$$

Since longer waves are characterized by longer time scales, $T_0 = (L_c B_c / L_{crit}) \langle K_E \rangle^{-1/2}$ is a characteristic time of the system. The magnitude of BCL relative to the zonal-mean flow APE tendency ($\sim A_z/T_0$) is

$$M_z \sim \langle K_m \rangle / \langle K_E \rangle \sqrt{Bl_1}$$

The magnitude of BCL relative to the eddy APE tendency ($\sim A_m/T_0$) is

$$M_e \sim (L_c^2/L_{\text{crit}}^2)(1/\sqrt{Bl_1})$$

It can be shown that $\widetilde{\text{BCL}}$ is of the same relative magnitude as BCL if ω'_m is scaled after Green (1970).

Hence, Bl_1, and also Bl, measure the relative strength of meridional heat flux convergence. This in turn is a reflection of zonal-mean flow baroclinic instability to long wave perturbations. From M_z we see that increased Bl_1 enhances the decrease of A_z. Too small values of Bl_1 would decrease the energy support from this reservoir. Likewise, strong baroclinic input favoring the eddies would enhance $Bl_1 \sim K_E/K_T$ since K_E would increase and K_T would decrease. On the other hand, the measure M_e for eddy baroclinic activity is proportional to $Bl_1^{-1/2}$. If Bl_1 is very large (e.g., during the mature stage) the importance of BCL for the eddies is small. Since L_c counteracts this effect, longer waves would be affected only if Bl_1 is very large, but dominant shorter waves even in the case of moderate Bl_1. These conclusions can be drawn for Bl too. Hence, the baroclinic energy exchange between A_z and K_E works like a negative feedback, controlled by Bl or Bl_1. From this point of view linear instability theories are not able to predict the evolution of blocks with strong peaked Bl. Are those linear theories able at least to model the sharp increase of Bl or Bl_1 as shown in Fig. 3? Only theories with standing waves as part of the mean flow ought to be considered, since otherwise $Bl \sim K_E/K_T$ is set to be very small according to the linear assumption.

Because of the standing waves as part of the basic flow we have $K_E \gtrsim K_T$ and $L_c \gtrsim 4L_{\text{crit}}$ from the beginning. According to Fig. 3 those theories have to explain the doubling of Bl during the buildup period of block (about 4 days long).

We can write

$$Bl = \epsilon(1 + a_0)^2/(1 - a_0)^2 \sim K_E/K_T \quad \text{for} \quad t = t_0$$

and

$$Bl(t_0 + 4 \text{ days}) = \epsilon(1 + \gamma a_0)^2/(1 - \gamma a_0)^2$$

with ϵ being a constant which cancels out. This ϵ contains the number of wave modes considered in the unstable wave packet, the number of perturbations of the zonal-mean flow, a characteristic wavenumber, and the largest standing wave amplitude. The amplification rate of the fastest growing mode in the ultralong wave range is denoted by $\gamma = \exp(4 \text{ days}/T_e)$. Of course, L_c/L_{crit} is nearly constant for that wave packet. The initial amplitude a_0, normalized with the basic flow amplitude, is assumed to be the same both for the eddies and the zonal-mean flow perturbations. To yield $Bl(t_0 + 4 \text{ days})/Bl \simeq 2$ as in Fig. 3, the amplification rate γ must be at least 2.55, 1.74, and 1.46 for $a_0 = 0.1, 0.2,$ and 0.3, respectively. In other words, the e-folding times T_e should be 4.2 days for $a_0 = 0.1$, 7.2 days for $a_0 = 0.2$, and 10.6 days for $a_0 = 0.3$. These e-folding times correspond to the lower end of theoretical values for ultralong waves (e.g., about 10 days; after Sasamori and Youngblut, 1981). Therefore, in principle, this type of linear theory is able to explain the evolution of Bl if the initial perturbation is in the upper range of linearly consistent amplitudes. In fact, a_0 should be even larger since the initial perturbation of the zonal-mean shear wind is $a_0/2$ at best, as can be computed from the results reported by Sasamori and Youngblut. Given this we have to have an e-folding time of 6.5 days for $a_0 = 0.3$ to explain a doubling of Bl.

The application of linear baroclinic instability theories is therefore a little unsatisfactory in the case of strong blocking. Moreover, those theories rely strongly on the forcing (or standing wave). Therefore, a more natural approach would be a nonlinear baroclinic instability theory (e.g., Schilling, 1984), which permits amplification rates $Bl(t_0 + 4 \text{ days})/Bl(t_0) \gtrsim 2$ simply by increasing a_0 above the limit demanded for linear theories. Moreover, there is no more necessity for external forcing if not required.

5. Relevant Energy Fluxes

5.1. Correlation Analysis

Synoptically defined blocking cases correspond not always to periods with sharp peaking blocking numbers. Hence, it is useful to analyze the impact of the energy conversions on the kinetic energy for the synoptic blocking situations. The seasonally stratified response of K_m to various energy inputs is given in Fig. 8a–e. Again, the fluxes and energies are standardized to zero mean and variance 1 [see Eq. (15)].

We begin with BTP_m (Fig. 8c), a measure of barotropic instability. Its impact on $m = 2-4$ for all blocking days (Pacific plus Atlantic regions) during 1967–1976 is significant in summer and winter (on a 95% level), but somewhat weaker in winter. Indeed, BTP_{2-4} can be related to blocking wave dynamics since $\tau < 0$ means that BTP_m precedes K_m. The significance for $m = 1$ and $m = 5-8$ is not as clear.

As to the measure of baroclinic instability BCL_m the correlation (Fig. 8d) with K_m during blocking days shows considerable impact on all scales, especially on the wavegroup $m = 2-4$. Hansen (this volume) has chosen another

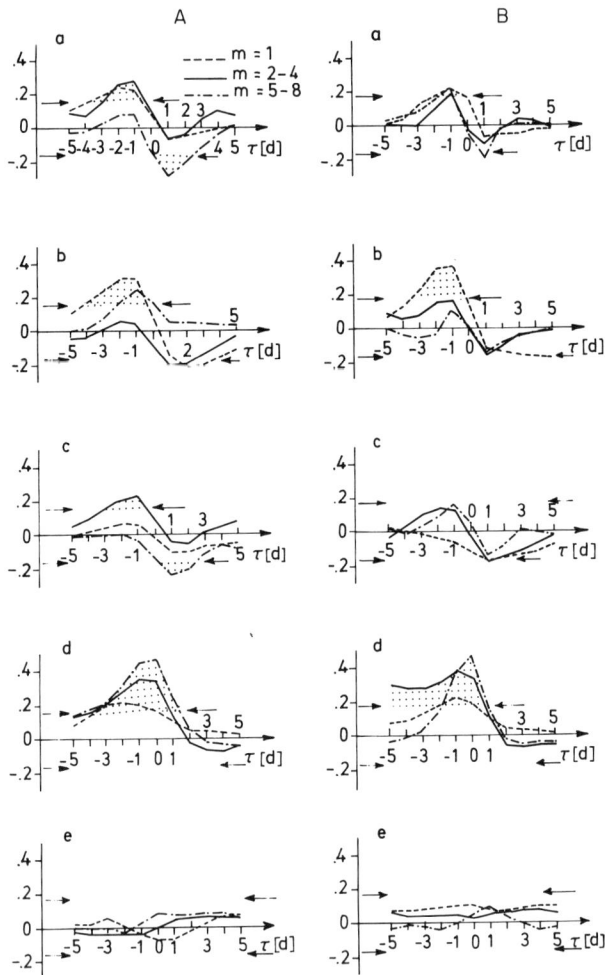

FIG. 8. Time-lagged correlations of K_m with various fluxes in the same format as in Fig. 7. Only blocking days were considered. A, Summer; B, winter. a, NLLS; b, NLLL; c, BTP; d, BCL; e, TOP.

way to investigate the impact of various energy transformations on the amplified ultralong waves. Nevertheless, he found similar results especially with regard to the baroclinic interaction with the zonal-mean flow.

The topographic influence ORO_m on wave dynamics is given in Fig. 8e. Irrespective of the season, this input does not affect the wave's energetics. Moreover, the values of ORO_m are quite small compared to those of BCL_m, BTP_m (Table I). Therefore we neglect this mechanism in the following sec-

TABLE I. SEASONAL AVERAGES OF SEVERAL ENERGY FLUXES[a]

Season	BTP_{1-4}	BTP_{5-8}	BCL_{1-4}	BCL_{5-8}	ORO_{1-4}	ORO_{5-8}	$NLLS_{1-4}$	$NLLS_{5-8}$
Dec–Feb	−0.331	−0.104	1.861	0.417	0.021	0.007	0.177	−0.206
Mar–May	−0.083	−0.102	1.244	0.576	0.007	0.013	0.156	−0.246
Jun–Aug	−0.119	−0.186	0.606	0.459	0.001	0.009	0.040	−0.083
Sep–Nov	−0.336	−0.315	1.712	0.758	0.021	0.006	0.090	−0.187

[a] Averages are for the ultralong wavegroup (index 1–4) and the "synoptic" wavegroup (index 5–8). For definitions of the fluxes see text. Units are W m^{-2}.

tions. This is not to say that it may be completely irrelevant, e.g., it can provide for a triggering mechanism.

The next effect, the input from smaller scales by nonlinear interaction $NLLS_m$ (Fig. 8a) shows a significant signal for the blocking wavegroup $m = 2-4$ in summer and winter.

The last flux $NLLL_m$, measuring the interaction within the long wave packet, is shown in Fig. 8b. It seems to have little effect on $m = 2-4$ in summer, but a stronger effect in winter. Except for $m = 1$, the impact of this term on the wave dynamics in a long-term sense is not quite clear.

5.2. Regression Analysis

There is another way to demonstrate the ability of these energy conversions to build up the long waves kinetic energy. To this end we use a regression model derived from the tendency equation of K_m [Eq. (7)] in the relevant latitude belt. Using an idea of Y. Hayashi (personal communication, 1985), we derive the model from the original equation in two steps. First we fit the (artificial) damping constants k_m to the reduced energy tendency after

$$(d/dt)K_m - BTP_m - BCL_m - NLLS_m - NLLL_m - ORO_m = Y_0 \quad (17)$$

$$Y_0 = -k_m K_m + \text{residue}$$

This is done for $m = 1$ and for the wavegroups $m = 2-4$ and $m = 5-8$. The lag-1-day autocorrelation of $BOUND_m$ is about 0.12 and is low compared to the same correlation for BCL_m (about $0.5 \rightarrow 0.8$) or for K_m (about 0.7). This justifies our assumption that the boundary term $BOUND_m$ is noisy and can be included in the residue. The (artificial) damping k_m is in the order of $\frac{1}{3}$ day. Secondly, having the maximum likelihood estimate of k_m from Eq.

(17), we turn over to the model

$$(d/dt)K_m + k_m K_m = \alpha_1 \text{BTP}_m + \alpha_2 \text{BCL}_m + \alpha_3 \text{NLLS}_m$$
$$\alpha_4 \text{NLLL}_m + \alpha_5 \text{ORO}_m + \text{residue}$$
$$= Y + \text{residue} \tag{18}$$

Retaining only some regression coefficients α_i and setting the other to zero gives a hierarchy of nested more and more complicated models Y. The "skill" SKILL^2 of those truncated models is given by

$$\text{SKILL}^2 = \frac{\text{VAR}(Y)}{\text{VAR}(dK_m/dt + K_m K_m)} \tag{19}$$

where VAR denotes variance. One hierarchy of model skills is presented in Fig. 9 for each wavegroup. These hindcast skills show the relative importance of the various terms in predicting the amplification during blocking cases (the regressions and skills are restricted to block periods). For $m = 2-4$ the baroclinic flux BCL plays the most prominent role, followed by NLLL$_{2-4}$. The term NLLS$_{2-4}$ increases the skill by a smaller but nevertheless reasonable amount. Again, ORO seems to be meaningless for the evolution of K_m in

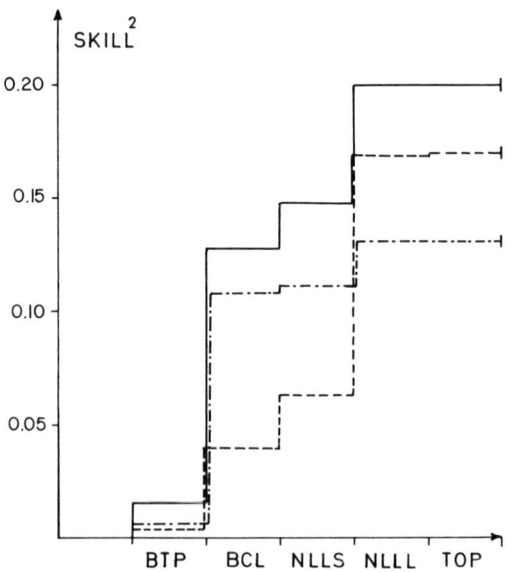

FIG. 9. Skills of various regression models of $dK_m dt + k_m K_m$. The hierarchy of nested models is BTP only (left), BCL added (next row), etc. The next added flux is given at the abscissa. The step curves give skill levels for each model. Solid lines, $m = 2-4$; dashed lines, $m = 1$; dash-dotted lines, $m = 5-8$.

all cases. Similar results hold for $m = 5-8$, but the skill of the most complicated model Y is somewhat smaller in these cases. For $m = 1$ the skill contribution of $NLLL_1$ is much larger than that of BCL_1 in contrast to the other wavegroups. Since these regression models were intended to show previous results from another point of view, there is no need to analyze their significance in this case.

We end this section with the conclusion that ORO can be discarded in the following analysis. Moreover, to make things simpler we deal with the combined flux BTP+ from now on. In this term BTP and NLLL are added for each wavegroup before it is normalized after Eq. (15). Therefore BTP+ is comparable to the other fluxes.

6. Types of Blocking

It is of value to separate blocking cases into various types. We are aware of the fact that the energy conversions can cooperate or compete and are interested in whether certain patterns of cooperation are preferred by the fluxes or not. Here we show that the cooperation patterns of the energy fluxes associated with the dominant wave are helpful for our goal of rational classification of cases.

As a simplification we establish the dominant wave (group) for each blocking case. To this end we use the following approach. First we average the normalized fluxes $BTP+_m$, BCL_m, and $NLLS_m$ for each case separately. Since often energetics has inhomogenous characteristics during longer periods, we have split up blocks with a duration longer than 15 days into two halves. Those segments are treated as separate cases in this context. The triplet of averaged fluxes together with the averaged kinetic energy stand for the energetics of a wave/wavegroup in each case.

Second, we establish a regression model for dK_m/dt similar to Eq. (18) but with BTP+ instead of BTP and NLLL and neglecting ORO. With this model we obtain an estimate for K_m by an Euler backward approximation ($\Delta t = 1$ day):

$$K_m(t) = \exp(-k_m \Delta t) \cdot K_m(t - \Delta t) + \Delta t \cdot \exp(-k_m \Delta t/2) \\ \cdot [\alpha_1 BTP+_m(t) + \alpha_2 BCL_m(t) + \alpha_3 NLLS_m(t)] \quad (20)$$

This formula can be regarded as being a score with weighted inputs K_m, BCL_m, $BTP+_m$, and $NLLS_m$. Since its form is conserved if averaged over a case, we simply consider the wave/wavegroup with maximum averaged score K_m to be dominant.

Accordingly, we can split up the total of 222 cases into three subsets with dominant $m = 1$ (30% of cases), $m = 2-4$ (48%), and $m = 5-8$ (22%). The dominant wave/wavegroup is an additional entry of the catalog of Appendix C.

The cooperation of the energy inputs is studied for each subset separately by means of an empirical orthogonal function (EOF) analysis. The EOFs are shown in Fig. 10. They are all significant to at least the 95% level as revealed by a Monte Carlo experiment with 3000 outcomes. In this test the computation of eigenvalues was repeated after interchanging fluxes of the same kind randomly between cases, thereby destroying the cooperation patterns. The test variable was the sum of absolute eigenvalue deviations from the trivial ones (= autocovariances of the fluxes).

The EOF1 explains about half of the variance and has approximately the

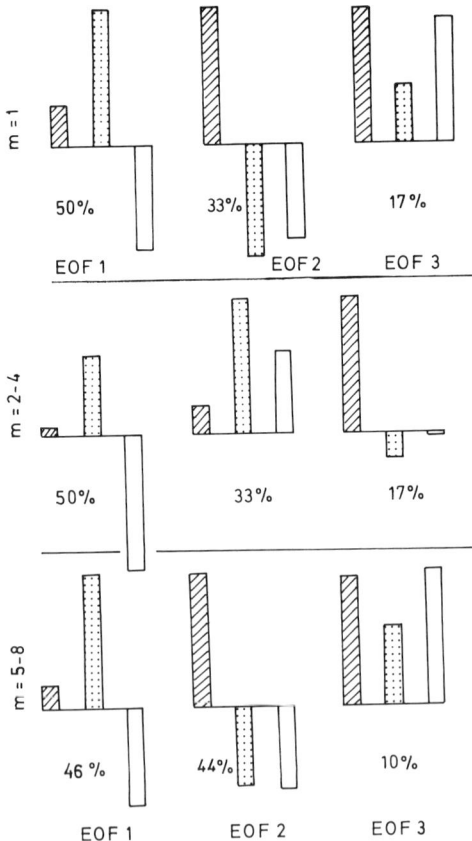

FIG. 10. EOFs of the triplet ($NLLS_m$, BCL_m, $BTP+_m$); subscript m stands for three wavegroup subsets: $m = 1$, $m = 2-4$, and $m = 5-8$.

same form for each subset. If taken positive, it describes a combined input by BCL and NLLS, counteracted by BTP+. This EOF represents one of the basic mechanisms for feeding in energy to the leading blocking wave. It bears close resemblance to the process described by Reinhold and Pierrehumbert (1982) (in their model, topography seems to be needed for fixing the long wave's phase only). The negative EOF1 consists completely of barotropic inputs, namely the interaction with $m = 1-5$ and the zonal flow. The term BCL is then below normal (note that all these fluxes are standardized). The negative EOF1 is comparable to barotropic instability theory.

The next EOF2 explains about 33% of the variance and differs from subset to subset with respect to the role of NLLS. Basically, if taken negatively, it describes the cooperation of BCL and BTP+ as inputs. NLLS acts against it or is at least small. This second main pattern was reported earlier by Hansen and Chen (1982) and Chen and Shukla (1983) in cases of dominant $m = 2-4$. If nonlinear interaction within the long wave packet is the leading term in BTP+, this EOF fits the model results of Schilling (1982) too. The same pattern taken positive would fit barotropic models with the synoptic scale input NLLS only (J. Egger, W. Metz, and G. Müller, this volume).

The third EOF explains about 20% of the variance and differs considerably from subset to subset. The positive pattern describes the leading role of NLLS alone ($m = 2-4$) or together with BTP+ ($m = 1$, $m = 5-8$). Fischer (1984) reported that a similar blocking mechanism prevailed in his GCM blocking cases. If taken negatively the EOF3 would lead to a dissipation of blocking waves, except for $m = 2-4$ where a small baroclinic input could serve as an energy support.

Note that other blocking models like that by Charney and DeVore (1979) or Legras and Ghil (1985) fail completely in finding an EOF counterpart. This is because those models are mainly based on a strong input by ORO which generally is absent in real cases.

Unfortunately many of our 222 cases can not be projected onto only one pattern. Nature is much more complex. Nevertheless, we can find out the pattern or combination of patterns which optimizes the projection of flux vector **a** onto a subspace for each case. If all combinations with n patterns failed to project at least 75% of the length of vector **a**, a next try was made with $n + 1$ patterns. The procedure was terminated when this criterion was first met. The resulting n patterns were set to be the governing pattern of the case. Since this procedure yields about 10 different combinations or more it was decided to collect similar governing patterns in types. Additionally, all cases with no single flux being larger than 0.25 were collected into a separate class named "no-type."

Besides this no-type category we found another six types corresponding to six certain EOF patterns, respectively. For example, type "BCL" is the set of all cases characterized by a positive EOF1, or its combination with another EOF which conserves the positive EOF1s main features.

The resulting types and their characteristics are given in Table II. The dominating ensemble- and type-averaged fluxes are relatively large. All values except those in parentheses are significantly different from normal (= zero) at a 95% level.

Types with NLLS input are more important for $m = 1$ than for other dominant wavegroups: their relative frequency is 29% ($m = 1$), 22% ($m = 2-4$), and 10% for $m = 5-8$. Types with positive BCL were found in 43% ($m = 1$), 43% ($m = 2-4$), and 56% ($m = 5-8$) of the cases. Positive BTP+ was recorded in 37.5% ($m = 1$), 38% ($m = 2-4$), and 44% ($m = 5-8$) of the cases.

The type-averaged blocking number Bl is also shown in Table II. Circumpolar types are found mainly in the $m = 2-4$ subset. As expected, Bl is considerably smaller for the $m = 5-8$ subset in general. Indeed, dominating waves with $m = 5-8$ should emphasize local structures. Therefore we should define the lower bound of Bl related to circumpolar behavior in such a way that most cases of the $m = 5-8$ subset have smaller Bl values.

There is no general relation between Bl and the dominant energy input. On the other hand Bl is somewhat smaller in the cases where BCL is the only input. This is especially true for $m = 5-8$ cases, a result that may be a consequence of the Bl dependence of baroclinic activity as discussed in Section 4.

Such classifications of cases may be helpful in organizing case studies involving historical data. Therefore we add the type for each block to the catalog of Appendix C. The cases in that catalog are selected to be characterized by $Bl \geq 2$. Therefore the circumpolar averaged fluxes are especially appropriate for our classification. We are aware of the fact that boundaries between blocking cases will be always a bit artificial due to the complexity of nature. Classification will be more successfull the more the characterizing parameter points of cases tend to build up separate clusters in parameter space. We show now that this tendency is observable in the space spanned by the fluxes BCL, BTP+, and NLLS. Each case is given by the point (BCL, BTP+, NLLS) of averaged fluxes. Figure 11a and b show BCL-BTP+ projections of this space for two different ranges of NLLS. All cases are taken into account and we observe a relative large spread of points over the entire plane. However, there are three clusters separated by relatively low point density regions. Moreover, those clusters correspond roughly to different types given above. Since the EOFs behind these types are significant at the 95% level, it is presumed that these clusters are not artifacts of our sample. As to this point we made tests with flux averages over periods shifted forward 1 or 2 days relative to the blocking period. Thereby we found that the clusters

TABLE II. CHARACTERISTICS OF TYPES

Type	NLLS			BCL			BTP+			Bl			Relative frequency (%)		
	$m=1$	$m=2-4$	$m=5-8$	$m=1$	$m=2-4$	$m=5-8$	$m=1$	$m=2-4$	$m=5-8$	$m=1$	$m=2-4$	$m=5-8$	$m=1$	$m=2-4$	$m=5-8$
None	−0.28	−0.09	(−0.02)	(−0.15)	−0.18	(−0.03)	(−0.08)	−0.24	(−0.11)	1.22	1.67	1.18	21	23	21
BTP+	−0.47	(−0.02)	(−0.17)	−0.78	−0.32	−0.40	0.68	0.60	0.62	1.56	1.90	1.25	11	13	15
BCL/BTP+	−0.43	(0.02)	−0.40	0.59	0.54	0.62	0.60	0.45	0.41	1.47	1.93	1.11	14	23	29
NLLS/BTP+	0.64	0.46	—	(0.03)	(−0.46)	—	0.50	0.57	—	1.52	1.72	—	12.5	2	—
NLLS	0.67	0.57	0.95	(−0.20)	(0.06)	−0.17	0.29	(0.09)	(−0.49)	1.24	1.91	1.45	12.5	19	8
BCL	(0.12)	(0.07)	(−0.06)	0.64	0.72	0.90	−0.52	−0.57	−0.34	1.33	1.41	0.92	25	19	25
BCL/NLLS	0.50	0.57	1.50	1.02	0.43	0.59	(0.26)	−0.31	−0.97	1.83	1.30	1.09	4	1	2

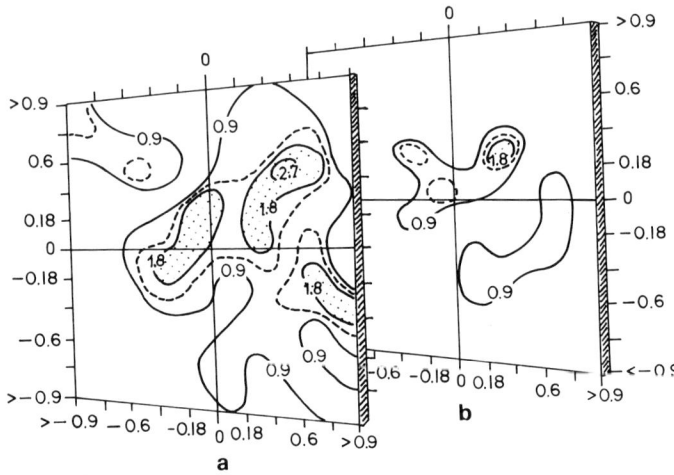

FIG. 11. (a) BCL_m-$BTP+_m$ projection of flux space distribution of all blocking points 1967–1976 with $NLLS_m \leq 0.37$; subscript m stands for the dominant wavegroup of the case. (b) Same as (a), except for blocking points with $NLLS_m > 0.37$. (Units in percentage.)

are unaffected by changing details of our procedure. The clusters, however, are not so pronounced that the boundaries between types can be considered as very well defined.

7. Conclusions

Our goal has been to characterize blocking cases by a small number of circumpolar energetic indices. We have found that as a first step, five or six numbers would be sufficient. These are a blocking number, characterizing the longitudinal extent of the influence of blocked zones, the dominant zonal wavenumber(s), and three or four circumpolar averaged energy fluxes related to the dominant wavegroup. It was found also that the baroclinic instability of the ultralong waves plays an important role in the energy budget of the blocking waves. The explanation of this fact by current linear baroclinic instability theories is only possible if some steady forcing is considered to provide for a standing wave response as part of the mean flow. It has been demonstrated that initial perturbation amplitudes needed for observed rapid amplifications are no less than the upper bound of values compatible with the linear assumption. That nonlinear baroclinic instability prevails is demonstrated by the observed cooperation with nonlinear wave–wave interaction as well.

There are many cases with dominant energy inputs from barotropic interactions accompanied by a below-normal baroclinic activity. Therefore, it seems that barotropic models can give considerable information on the formation of certain kinds of blocks. However, models which crucially depend on large mountain torques should be used with caution for explaining blocking activity. There is no counterpart for this mechanism in blocking energetics as long as the blocked latitude belt is considered. Nevertheless, strong signals from the orographic terms can be extracted out of model data (Fischer, 1985) as well as from observational data (Metz, 1985) taken from south of the blocked zone, although they can not serve as direct input to the blocking waves.

Local blocking numbers have been defined previously in the literature (Elliott and Smith, 1949; Namias, 1964), but Bl and Bl_{2-4} are the first circumpolar ones. They perform well in many cases and serve as measure of the circumpolar character of the blocked regime. Bl, and even more so Bl_{2-4}, respond most to the amplifying wavegroup $m = 2-4$, a characteristic they have in common with the amplitude index $[Z_{2-4}]$ of A. Hansen (this volume).

In general, $m = 2-4$ is the dominant wavegroup (about 50%). However, $m = 1$ or $m = 5-8$ cases are not seldom. On the other hand about 75% of the $m = 5-8$ cases are characterized by $Bl < 1.32$, whereas such small Bl numbers occur in 52% of the $m = 1$ cases and 33% of the $m = 2-4$ cases. Accordingly the dominant wavenumber(s) may indicate the circumpolar extent of blocking activity, but it is not quite reliable in doing that. This is so because dominance of one wavegroup does not necessarily imply the unimportance of the other ones. The Bl number is not redundant in giving the desired information.

Last but not least we tried to classify blocking cases by means of energy inputs. We found a tendency toward clustering in flux space which is based on cooperation patterns. Many of these types can be related to an observation or model result as reported in the literature. However, the structure of blocking energetics is too complex to give sharp boundaries between types. On the other hand, the energetics of blocks is not dispensable in the study of events, since in many cases the energy inputs are much larger than normal. Even ensemble and time averages could show relative abnormal values and therefore strong signals.

It is hoped that this study will help to initiate the discussion on appropriate classification criteria for blocking events. A broad agreement on this issue will certainly be the basis for better structuring further studies on blocking dynamics.

Acknowledgments

I would like to thank Dr. A. R. Hansen for his helpful comments on the text. This work was partly supported by the Fraunhofer Gesellschaft, München, Federal Republic of Germany.

APPENDIX A

φ, λ	Latitude, longitude
$\overline{X}$	Zonal mean
X'	$X - \overline{X}$
X'_m	X' restricted to zonal wave m (or a wave group specified in the text)
u, v, ω	Zonal and meridional components of the geostrophic wind, $\omega = dp/dt$
ψ'_{1000}	ψ' at 1000-mbar level
X_B	X at lower boundary
ϕ	Geopotential
$[X]_{\varphi_1}^{\varphi_2}$	Average over the latitude band $\varphi_1 \leq \varphi \leq \varphi_2$ and tropospheric depth 1000 mbar $\geq p \geq$ 300 mbar $= g^{-1} \sum_{j=1}^{J} \Delta pj \{\sum_{k=0}^{K} \cos \varphi_k \overline{X}(\varphi_k, pj)\}/\{\sum_{k=0}^{K} \cos \varphi_k\}$; $J = 2$ for σ, BCL, $\overline{BCL}$; $J = 3$ for BTP, NLLS, NLLL, K_m, A_m, BOUND
$\langle X \rangle_{\text{mon}}$	Time average over 10 years for each month separately
$\langle\ \rangle$	Time average
$\bar{u}_n$	Amplitude of an eigenmode of the discrete Laplacian $\nabla^2 = \cos \varphi^{-1} \partial_\varphi \{\cos \varphi \partial_\varphi\}\vert_{\varphi=\varphi_k}$, replacing $\cos \varphi^{-1} (\partial/\partial\varphi)(\cos \varphi (\partial/\partial\varphi))$. Eigenmodes are vectors $\mathbf{Y}$ related to an equidistant grid and obeying $\nabla^2 \mathbf{Y}_n = -\kappa_n^2 \mathbf{Y}_n$; ∇^2, matrix of entries ∇^2. The scalar $\bar{u}$ is given on the same gridpoints ($\varphi_k = \varphi_1 + k\Delta\varphi \leq \varphi_K$ with $k = 0, \ldots, K$; $\Delta\varphi = 5°$ and $\varphi_1 = 10°$N, $\varphi_K = 90°$N) by $[\bar{u}(\varphi_1), \ldots, \bar{u}(\varphi_K)]^T = \sum_n \bar{u}_n \mathbf{Y}_n$. To compute $\mathbf{Y}_n$ it was assumed $\partial_\varphi \mathbf{Y}_n = 0$ for φ_1, φ_K with ∂_φ as a discrete φ derivation
a	Radius of earth
$J(X, Y)\vert m$	$a^{-2} \cos \varphi^{-1} [(\partial X/\partial\lambda)(\partial Y/\partial\varphi) - (\partial Y/\partial\lambda)(\partial X/\partial\varphi)]\vert_m =$ projection of J on to wave m
$\text{VAR}(X)$	$\{\langle X^2 \rangle - \langle X \rangle^2\}$
$\text{COR}(X, Y)$	$\{\langle XY \rangle - \langle X \rangle\langle Y \rangle\}/\{\langle X^2 \rangle - \langle X \rangle^2\}^{1/2}/\{\langle Y^2 \rangle - \langle Y \rangle^2\}^{1/2}$
φ_c	Latitude characteristic for blocked zone, $\varphi_c = 60°$, say
f_0	Coriolis parameter f at φ_c
ρ, p, T	Density, pressure, temperature
θ	$T(p_{1000}/p)^{R/c_p}$
p_{1000}	1000 mbar
β	$a^{-1} \partial f/\partial\varphi$
g	Earth gravity constant
K_m	Kinetic energy of wave m [see Eq. (8)]
K_E	$\sum_{m=1}^{8} K_m$
K_z	$(\bar{u}^2/2)_{52.5°N}^{82.5°N}$
K_T	$\frac{1}{2}[(\partial\bar{u}/\partial p)^2 \Delta p^2]_{52.5°N}^{82.5°N}$; $\Delta p = 200$ or 500 mbar

Appendix B

The mountain-induced energy flux stems from the vertical averaging of the baroclinic term in the kinetic eddy energy equation

$$-[f_0\psi'\partial\omega'/\partial p] = -[f_0(\partial/\partial p)(\psi'\omega')] + [f_0\omega'\partial\psi'/\partial p]$$

The last term denotes the transformation $\widetilde{BCL}_m$. With $\omega' = 0$ at the upper boundary, the second term is transformed into

$$-[f_0(\partial/\partial p)(\psi'\omega')]_{\varphi_1}^{\varphi_2} = (f_0/g)\overline{\psi'_B\omega'_B}^{\lambda\varphi} \quad \text{(average over } \lambda, \varphi)$$

with $\varphi'_B = \varphi'(p = p_B)$; $\omega'_B = \omega'(p = p_B)$. Let $\omega_B = -\rho_B g J(\psi, h)$ with orography $h(\lambda, \varphi)$, and $\bar{h} = 0$.

A linear approximation of the boundary condition gives

$$-[f_0(\partial/\partial p)(\psi'\omega')]_{\varphi_1}^{\varphi_2} \approx f_0[\bar{\rho}_B\bar{u}_B(\psi'_B/a\cos\varphi)(\partial h'/\partial \lambda)]_{\varphi_1}^{\varphi_2} \quad (B1)$$

Since we can establish a quadratic interpolation formula for $\phi(p)$ and three levels, we can do this for $\psi(p)$ too:

$$\psi = \psi_{500} + \frac{p_{500}^2(\psi_{500} - \psi_{300}) + p_{200}^2(\psi_{1000} - \psi_{500})}{p_{200}p_{700}p_{500}}(p - p_{500})$$
$$+ \frac{p_{200}(\psi_{1000} - \psi_{500}) + p_{500}(\psi_{300} - \psi_{500})}{p_{200}p_{700}p_{500}}(p - p_{500})^2 \quad (B2)$$

(with $\psi_j = \psi(p = p_j)$; $p_{200} = 200$ mbar, etc). Using Eq. (B2) we are able to compute $\bar{\rho}_B = -(\partial\bar{\phi}/\partial p)_B^{-1}$, $\bar{u}_B$, ψ'_B for $p = p_B$. The linearized hydrostatic relation gives (with $\bar{h} = 0$)

$$p'_B = -g\bar{\rho}_B h' \quad (B3)$$

We have therefore

$$\psi'_B = \psi'_{1000} - p'_B\bar{\psi}_{500}\frac{p_{700}^2}{p_{200}p_{700}p_{500}} + p'_B\bar{\psi}_{300}\frac{p_{500}}{p_{200}p_{700}}$$
$$+ p'_B\bar{\psi}_{1000}\frac{p_{1200}}{p_{700}p_{500}} = \psi'_{1000} + h' \cdot \text{function}(\bar{\psi}_{500}, \bar{\psi}_{300}, \bar{\psi}_{1000}) \quad (B4)$$

and for the zonal mean

$$\overline{\psi'_B(\partial h'/\partial \lambda)}^\lambda = \overline{\psi'_{1000}(\partial h/\partial \lambda)}^\lambda = \overline{\psi'_{1000}(\partial h/\partial \lambda)} \quad (B5)$$

after Eqs. (B4) and (B3). Using this relation and computing $\bar{u}_B$, $\bar{\rho}_B$ after Eq. (B2) for $p = p_B$ gives the mountain-induced flux Eq. (11).

Appendix C

Atlantic/European Blockings

Period (from–to)[a]	Fig. 1	$\langle Bl \rangle$	Dominant wavegroup	Overlapping with Pacific block?	Type
12.09.–21.09.67	Yes	2.15	2–4		BCL/BTP+
18.11–22.11.67		2.11	2–4		BCL
03.04.–08.04.68		2.09	2–4		BCL
03.08.–12.08.68		2.74	2–4	Partly	NLLS
15.01.–21.01.69		2.43	1	Partly	BTP+
06.02.–04.03.69		2.16	5–8	Partly	BCL/BTP+
		1.24	1		BCL
20.05.–29.05.69	Yes	2.08	2–4		NLLS
22.09.–26.09.69		2.37	1		NLLS/BCL
25.12.–05.01.69		2.15	1		NLLS/BTP+
13.01.–31.01.70		1.87	2–4		None
		2.01	2–4		None
21.02.–27.02.70		3.01	2–4		NLLS
09.09.–16.09.70		2.91	1		NLLS/BTP+
22.12.–31.12.70		2.74	1		BCL/BTP+
11.01.–15.01.71		4.03	2–4	Yes	BCL/BTP+
24.04.–28.04.71		2.22	5–8		None
25.01.–07.02.72		4.57	2–4		NLLS
19.02.–07.03.72		3.03	2–4		BCL/BTP+
	Yes	1.32	2–4	Yes	BTP+
05.01.–23.01.73		2.46	2–4		BCL/BTP+
		1.10	2–4		BTP+
03.01.–18.01.74		5.13	2–4	Yes	BCL/BTP+
	Yes	3.46	2–4		None
27.01.–09.02.75		2.77	2–4	Yes	None
14.09.–29.09.76		1.73	2–4		NLLS
		2.24	2–4		BTP+
12.10.–29.10.76		1.41	1		BTP+
		2.13	2–4	Yes	BTP+
12.12.–31.12.76		2.32	2–4		NLLS/BTP+

[a] Periods are given as day, month, year; e.g., 12.09.–21.09.67 is 12 September to 21 September 1967.

Pacific Blockings[a]

Period (from–to)[b]	$\langle Bl \rangle$	Dominant wave or wavegroup	Overlapping with Atlantic blocking?	Type
07.08.–16.08.68	3.36	2–4	Partly	NLLS
19.01.–28.01.69	2.15	1	Partly	None
12.10.–17.10.69	2.24	2–4	Yes	BCL/BTP+
09.01.–16.01.71	3.14	2–4	Yes	BCL/BTP+
21.02.–29.02.72	3.11	2–4	Yes	BTP+
02.12.–12.12.72	2.03	2–4		BCL
02.11.–09.11.73	2.52	1		BTP+
28.12.–09.01.73	3.46	2–4		BCL/BTP+
12.01.–19.01.74	2.77	2–4	Yes	None
04.02.–09.02.75	3.07	2–4	Yes	BTP+
	3.24	2–4		None

[a] From Treidl et al., 1981.
[b] Periods are given as day, month, year; e.g., 12.09.–21.09.67 is 12 September to 21 September 1967.

References

Charney, J. G., and DeVore, J. G. (1979). *J. Atmos. Sci.* **36**, 1205–1216.
Charney, J. G., Shukla, J., and Mo, K. C. (1981). *J. Atmos. Sci.* **38**, 762–779.
Chen, T.-C., and Shukla, J. (1983). *Mon. Weather Rev.* **111**, 3–22.
Dickson, R. R. (1967). *Mon. Weather Rev.* **95**, 143–152.
Dickson, R. R. (1969). *Mon. Weather Rev.* **97**, 830–834.
Egger, J. (1978). *J. Atmos. Sci.* **35**, 1788–1801.
Elliott, R. D., and Smith, T. B. (1949). *J. Meteorol.* **6**, 67–85.
Fischer, G. (1984). *Contrib. Phys. Atmos.* **57**, 183–200.
Fischer, G. (1985). *Contrib. Phys. Atmos.* **58**, 291–303.
Green, J. S. A. (1970). *Q. J. R. Meteorol. Soc.* **96**, 157–185.
Hansen, A. R., and Chen, T.-C. (1982). *Mon. Weather Rev.* **110**, 1146–1165.
Legras, B., and Ghil, M. (1985). *J. Atmos. Sci.* **42**, 433–471.
Lejenäs, H. (1977). *Atmosphere-Ocean* **15**, 89–113.
Metz, W. (1985). *J. Atmos. Sci.* **42**, 1880–1892.
Namias, J. (1964). *Tellus* **16**, 394–407.
Reinhold, B. B., and Pierrehumbert, R. T. (1982). *Mon. Weather Rev.* **110**, 1105–1145.
Sasamori, T., and Youngblut, C. E. (1981). *J. Atmos. Sci.* **38**, 87–96.
Schilling, H.-D. (1981). Dynamik und Energetik blockierender Wellen in der Atmosphäre. Dissertation, Universität Hamburg, FRG. Hamburger Geophys. Einzelschr. 50
Schilling, H.-D. (1982). *J. Atmos. Sci.* **39**, 998–1017.
Schilling, H.-D. (1984). *Contrib. Phys. Atmos.* **57**, 150–168.
Schilling, H.-D. (1986). *Mon. Weather Rev.*, submitted.
Speth P., and Meyer, R. (1984). *Contrib. Phys. Atmos.* **57**, 463–476.
Stark, L. P. (1965). *Mon. Weather Rev.* **93**, 337–342.
Taubensee, R. E. (1975). *Mon. Weather Rev.* **103**, 562–566.
Thompson, P. D. (1957). *Tellus* **9**, 69–91.
Treidl, R. A., Birch, E. C., and Sajecki, P. (1981). *Atmosphere-Ocean* **19**, 1–23.

OBSERVATIONAL CHARACTERISTICS OF ATMOSPHERIC PLANETARY WAVES WITH BIMODAL AMPLITUDE DISTRIBUTIONS

Anthony R. Hansen

Meteorology Research Center
Control Data Corporation
Minneapolis, Minnesota 55420

1. Introduction

Persistent, large-scale circulation anomalies and "blocking" have been topics of great interest in recent years. The connection between quasi-stationary blocking ridges and stationary forcing due to the earth's topography and surface thermal contrasts, and the question of whether blocking is a regional or global phenomenon, are central issues in the investigation. It has long been known that the long-term mean waves found in Northern Hemisphere midlatitudes owe their existence to the zonally asymmetric distribution of the earth's topography and land–sea thermal contrasts (Charney and Eliassen, 1949; Smagorinsky, 1953). Recently, much interest has been generated by the work of Charney and DeVore (1979) and Hart (1979), which suggested for given forcing parameters, multiple stable steady states of the zonal-mean wind and planetary wave amplitude were possible due to orographically induced instability in a simple barotropic model. Other work illustrated similar possibilities in baroclinic models (e.g., Charney and Strauss, 1980; Pedlosky, 1981). In these models, one stable steady state corresponds to a relatively strong zonal flow while another corresponds to a state with a weak zonal flow and a wave pattern reminiscent of blocking. Bimodality in the probability density distribution of the zonal wind is predicted by these theories (e.g., Benzi *et al.*, 1984).

An extension of the Charney–DeVore idea to a barotropic model including nonlinear wave–wave interaction by Benzi *et al.* (1986) shows that the presence of the wave–wave interaction causes the orographically forced wave amplitude resonance in Charney and DeVore's model to become a "folded" resonance in which, for a given value of the zonal wind, more than one stable planetary-scale wave amplitude is possible. This folding of the resonance is due simply to the presence of the anharmonic terms in the governing equations of the model. Sutera (1986) has shown that the probability density distribution of the amplitude of the planetary-scale wave packet formed by zonal harmonic wavenumbers 2–4 is bimodal. This wavenumber band was chosen because it represents waves with wavelengths near the resonant wavelength for the earth's topography and observed values of the zonal wind. Conversely, the zonal-mean wind exhibits a unimodal distribution.

In the present observational study, we examine stationary wave structure and mean spectral energy and enstrophy budgets to establish whether consistent, systematic differences appear between these statistics computed for the two modes. We wish to provide further confidence in the significance of the observed planetary-wave bimodality based on physical consistency rather than simply on statistical tests. An analysis of this kind cannot by itself provide definitive explanation of the dynamics of these circulation regimes. However, the information provided by this approach can indicate suitable directions to pursue in formulating theoretical understanding of the dynamics of these important components of the general circulation.

Our results highlight a particular class of persistent quasi-stationary wave phenomenon in which stationary baroclinic forcing plays an apparently central role and for which the attendant wave pattern appears more nearly hemispheric rather than regional in character. The synoptic patterns associated with the large-amplitude wave mode generally correspond to what Rex (1950) termed "amplified waves." We will contrast this pattern with what appears to be another distinct blocking phenomenon for which the wave pattern appears more regional and for which the stationary forcing appears secondary compared to interactions between this pattern and transient cyclone-scale waves.

Our study differs from many previous studies of circulation anomalies because we use a global measure of wave amplitude rather than the local departure of the 500-mbar height from climatology. The studies of Dole and Gordon (1983) and Charney et al. (1981), for example, used local, gridpoint departures of the 500-mbar height from climatology or the average departure along meridians, whereas we use the global measure of planetary-wave amplitude. However, we should note that the two modes found in the wave amplitude probability density estimation are not "anomalous" circulations. Rather, they apparently were fundamental aspects of the general circulation during the four winters considered.

Our analysis is based on data archived at the European Center for Medium Range Weather Forecasts (ECMWF) from the winters of 1980/1981, 1981/1982, 1982/1983, and 1983/1984, where winter is defined by the period from 1 December through 28 February. We will first consider the frequency distribution of the planetary-wave amplitude in a manner similar to that used by Sutera (1986). Based on the bimodal wave amplitude distribution, we then examine the stationary wave structure and construct ensemble mean spectral energy and enstrophy budgets of the two wave modes to illustrate the important physical processes occurring during each. Finally, we consider another class of blocking phenomenon with quite different budget characteristics from those in the large-amplitude wave mode.

2. Frequency Distribution of the Planetary-Wave Amplitude Indicator

First consider the frequency distribution of an indicator of the planetary-wave amplitude. We will proceed in a manner very similar to that used by Sutera (1986). The data used to construct the histograms and probability density estimates are daily 500-mbar geopotential heights compiled at ECMWF on a 5.625° latitude–longitude grid from 1 March 1980 through 31 May 1984. For each day, we compute a wave amplitude indicator by averaging the gridpoint 500-mbar heights, $z(\lambda,\phi)$ with respect to latitude, ϕ, from 22.5°N, to 78.5°N, and Fourier decompose the result in the zonal direction. Then, we compute the wave amplitude indicator for the planetary waves synthesized from wavenumbers 2–4 as follows:

$$[Z_{2-4}] = \left(\sum_{m=2}^{4} 2Z_m^2 \right)^{1/2}$$

To be precise, $[Z_{2-4}]$ is the square root of the wave 2–4 height variance, not the amplitude. The actual amplitude is given by multiplying $[Z_{2-4}]$ by $\sqrt{2}$. In this equation, Z_m denotes the Fourier coefficient for wave m. The wavenumber band $m = 2-4$ is chosen because this range represents the topographically resonant wavelengths for the observed values of the zonal wind based upon Rossby's two-dimensional formula. This procedure will select planetary waves of broad latitudinal extent over those with more complicated structure in the meridional direction.

The resulting time series is then filtered to remove high-frequency variability with periods of 5 days and less, and low-frequency variability associated with the seasonal cycle and interannual variability. It is especially important to remove the interannual variability. This was done by performing a Fourier transform of the time series of $[Z_{2-4}]$, zeroing the unwanted frequencies, and resynthesizing the filtered time series. Periods longer than 170 days were removed to eliminate the interannual variability as well as annual through semiannual periods. The histogram in Fig. 1 was constructed from the winter data of the filtered $[Z_{2-4}]$ time series. A similar procedure was used to obtain a histogram of the zonal-mean wind $[u_z]$ [see Sutera (1986) for these results].

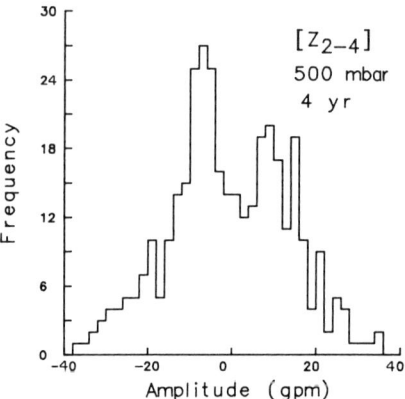

FIG. 1. The frequency distribution histogram of the amplitude of the wave packet formed from zonal wavenumbers 2–4, $[Z_{2-4}]$. See text for details.

Alternatively, an area-averaged wave amplitude indicator was also calculated as follows:

$$\overline{[Z_{2-4}]} = \int_{\phi_1}^{\phi_2} \left(\sum_{m=2}^{4} 2Z_m^2(\phi) \right)^{1/2} \cos \phi \, d\phi \bigg/ \int_{\phi_1}^{\phi_2} \cos \phi \, d\phi$$

where the latitudinal variation of the 500-mbar height is not averaged prior to the Fourier analysis. Applying the same filtering discussed above results in a time series that also yields a bimodal frequency distribution. Generally, differences between the two methods of calculating the wave amplitude are minor, but the area-average approach systematically places more days in the large-amplitude mode than does the procedure outlined earlier. However, the particular method of computing the wave amplitude does not affect the budget results to be presented in any substantial way. We will employ the former method in what follows.

As shown by Sutera (1986), the striking aspect of the histograms is the bimodality in the $[Z_{2-4}]$ histogram. The zonal-mean wind exhibits a unimodal distribution (Sutera, 1986). The obvious question of the statistical significance of the minimum in the $[Z_{2-4}]$ histogram has been discussed in detail by Sutera (1986). Nonparametric estimates of the probability density using the maximum penalized likelihood method (Good and Gaskins, 1980), based on these data and using mesh spacings comparable to our bin widths, yield probability density distributions that look quite similar to our histograms. Probability density distributions of $[Z_{2-4}]$ for the individual winters each exhibit similar bimodal characteristics. A bimodal probability density estimation based on the sample histogram provides much greater statistical confidence in the fit than does a Gaussian estimate of the distribution. These

TABLE I. FREQUENCY OF VARIATION IN $[Z_{2-4}]$

	Mode 1 days[a]	Mode 2 days
1980/1981	47 (52%)	43 (48%)
1981/1982	48 (53%)	42 (47%)
1982/1983	48 (53%)	42 (47%)
1983/1984	52 (58%)	38 (42%)
Total	195 (54%)	165 (46%)

[a] The number of days from each winter belonging to each ensemble. The number in parentheses is the percentage of days in each year or in the total of the 4 years belonging to each ensemble.

results are discussed in greater detail by Sutera. The main purpose of the present study is to illustrate that by stratifying the data based upon the two modes in the $[Z_{2-4}]$ probability distribution, we find consistent, systematic differences in various diagnostic quantities. These results add confidence in the physical significance of the bimodality and provide insight into the dynamics involved.

Henceforth, days for which the value of $[Z_{2-4}]$ is less than the minimum in Fig. 1 will be referred to as Mode 1 days, and those for which $[Z_{2-4}]$ is greater than the minimum will be called Mode 2 days. We get 195 Mode 1 days (54% of the total) and 165 Mode 2 days (46% of the total). The breakdown for the individual years is given in Table I. The average duration of both Mode 1 and Mode 2 events is roughly 11 days, with transitions between modes taking 4 days on the average. Individual events persist anywhere from a few days to over 3 weeks.

3. WAVE STRUCTURE OF THE TWO MODES

The large-scale structure of the wave pattern for the two modes is illustrated by constructing maps of the mean 500-mbar heights for each mode (Fig. 2a and b). The mean 500-mbar height for Mode 1 has a very zonal appearance, while the Mode 2 pattern exhibits a prominent ridge over the eastern Pacific and deep troughs over the western Pacific and eastern North America. The climatological mean stationary 500-mbar pattern is a weighted average of the patterns associated with the two modes and depends upon the asymmetry in the statistics of the two modes.

By synthesizing the 500-mbar pattern represented by zonal wavenumbers 2–4, denoted Z_{2-4} (Fig. 3a and b), a clearer picture is presented of the differences between the two modes. Comparing the resulting 500-mbar

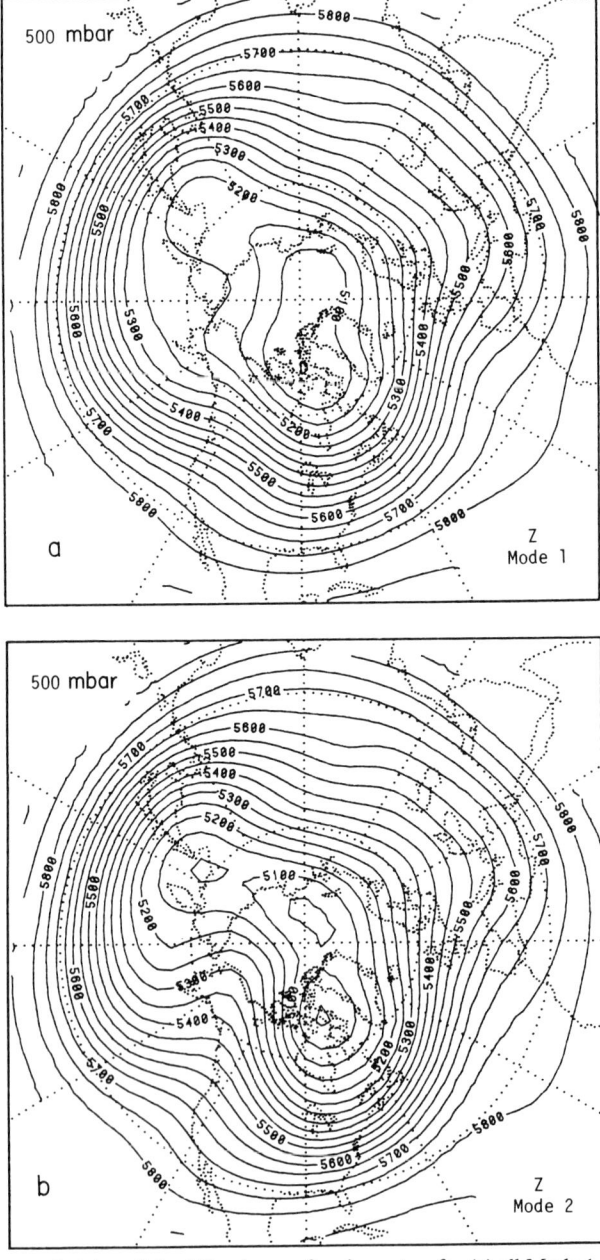

FIG. 2. Average heights of the 500-mbar surface in meters for (a) all Mode 1 days from the four winters and (b) all Mode 2 days. The contour interval is 50 m.

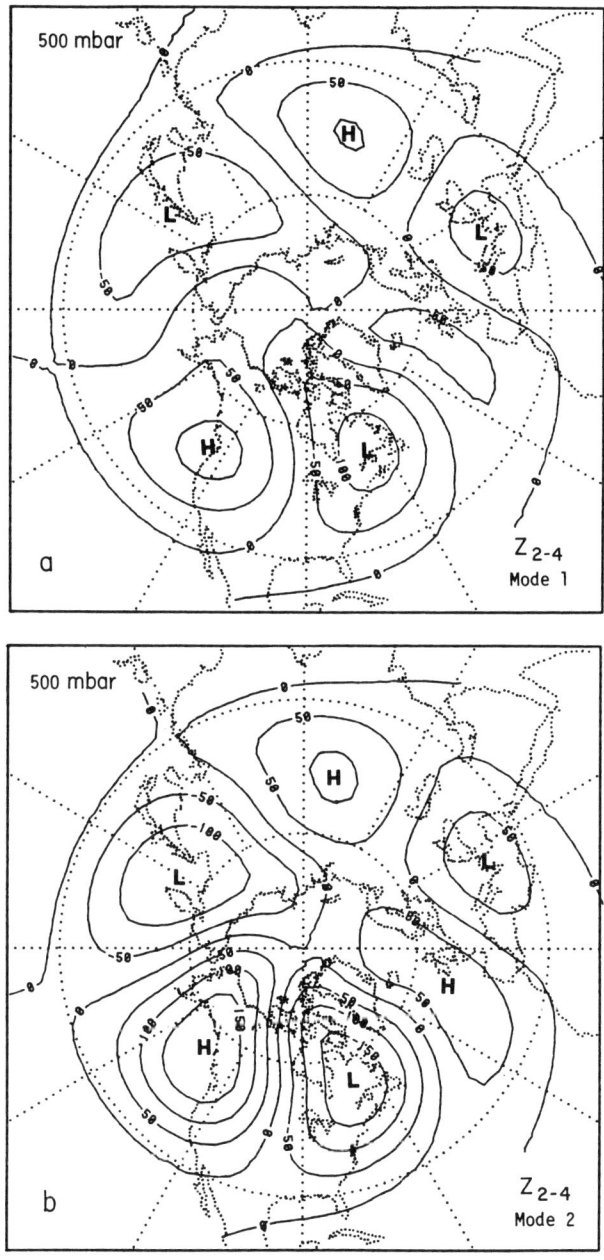

FIG. 3. Mean 500-mbar heights synthesized from zonal wavenumbers 2–4, Z_{2-4}, for (a) Mode 1, (b) Mode 2, and (c) ΔZ_{2-4}, the difference between the Mode 2 and Mode 1 Z_{2-4} maps. The contour interval is 50 m.

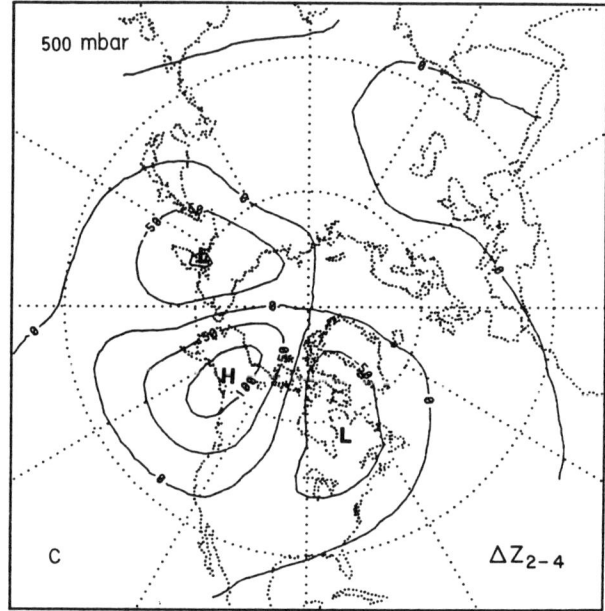

FIG. 3c. See legend on p. 107.

height fields indicates that for Mode 2 the western Pacific trough, the eastern Pacific ridge, and the Hudson Bay trough are sharply amplified compared to their Mode 1 counterparts. The difference between these two fields (Fig. 3c) highlights the major differences in 500-mbar heights that extend eastward from eastern Asia to the east coast of North America. The difference in the total height field between the two modes is represented largely by the differences in wavenumbers 2–4 (not shown).

Inspection of maps of the difference in Z_{2-4} for the two modes (Mode 2 minus Mode 1) for each of the individual years shows that the differences between the modes extend farther downstream in the form of a ridge over the Atlantic Ocean or Europe. This feature is not prominent in the 4-yr mean field, however, because the phase of this ridge, relative to the surface of the earth is variable from year to year. For example, in 1980/1981 (Fig. 4a) the Mode 2 Atlantic ridge axis is along 20°W, while in 1981/1982 (Fig. 4b) it is along 10°E. These two winters exhibit gridpoint differences in Z_{2-4} at 500 mbar of over 100 m for the eastern Pacific ridge, the western Pacific trough, and the North American trough. The 1982/1983 winter (Fig. 4c) exhibited lower amplitude planetary waves and thus smaller differences in the synthesized 500-mbar heights, Z_{2-4}, between the two modes. The overall pattern in 1982/1983 was similar to the other years except the western Pacific low did

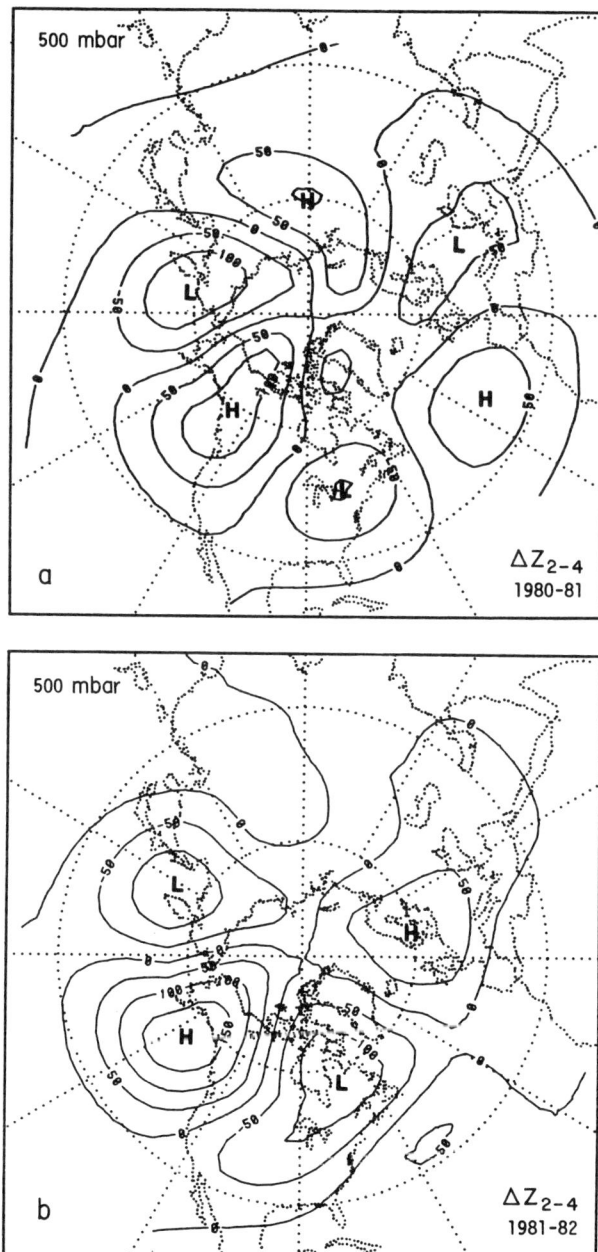

FIG. 4. The difference between the Mode 2 and Mode 1 maps of Z_{2-4} for each of the individual winters: (a) 1980/1981, (b) 1981/1982, (c) 1982/1983, and (d) 1983/1984.

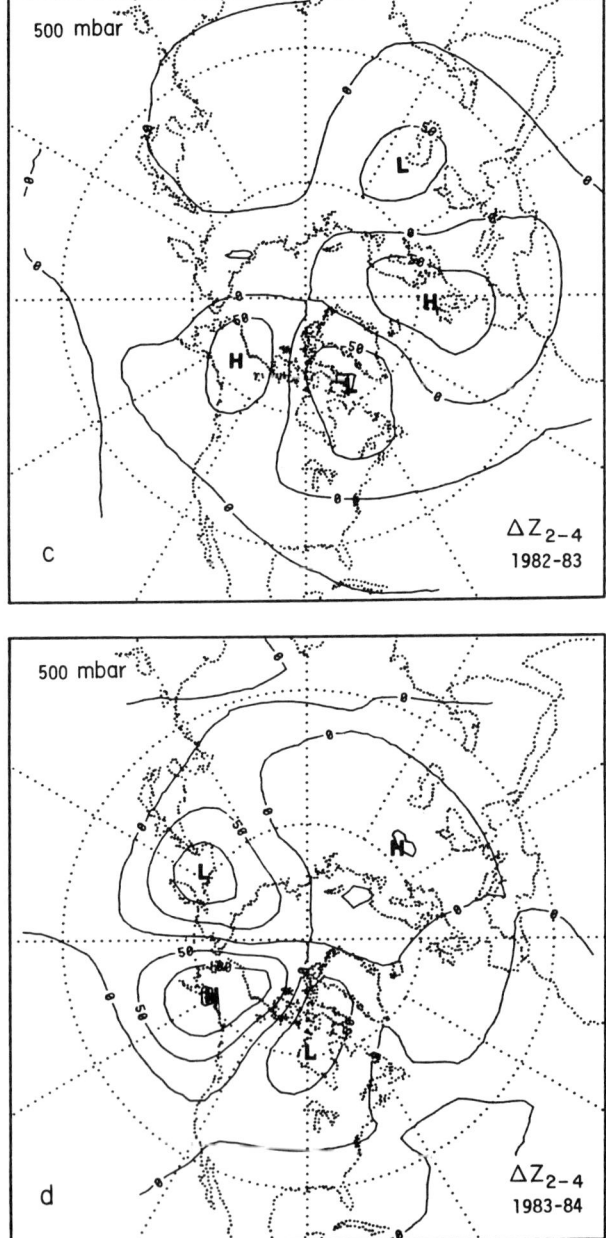

FIG. 4c and d. See legend on p. 109.

not experience the same deepening for Mode 2 that it did in the other years. The generally lower amplitude of the midlatitude quasi-stationary waves during this winter may have been connected with the El Niño event in progress at the time. In 1983/1984, the western Pacific low and the eastern Pacific ridge experienced sharp amplification for Mode 2 (greater than 100 m) with a lesser deepening of the North American low (Fig. 4d). Height increases for Mode 2 also existed along 40°E, but these are not as well defined as the analogous feature in the other 3 years. These results give a general impression of the interannual variability in the wave patterns associated with the bimodality. It is also evident that the differences between the two modes extend over a large portion of the Northern Hemisphere midlatitudes.

By superposing the wavenumber 1 field on top of the Z_{2-4} patterns (Fig. 5a and b), the overall mean planetary-wave fields for each mode are illustrated. Since the amplitude and phase of wavenumber 1 is nearly identical in each case, the greatest contrast in patterns lies over western North America and the eastern Pacific, with the pattern elsewhere being similar in phase for the two modes but of greater amplitude for Mode 2.

To see the temporal variation of the 500-mbar pattern for the wavenumber 2–4 ensemble as it intermittently switches between Mode 1 and Mode 2 behavior, consider the longitude–time section of the synthesized Z_{2-4} height field averaged between 20° and 80°N for 1981/1982 as an example (Fig. 6). The time periods designated as belonging to Mode 2 are indicated by the brackets on the right-hand side of the figure. Note that for the three major Mode 2 events during this winter, a fairly consistent pattern characterized by a western Pacific trough, eastern Pacific ridge, eastern North American trough, and Atlantic–European ridge is present.

Synoptically, the Mode 2 events manifest themselves as what Rex (1950) would call "amplified wave" patterns. Rex drew a distinction between this type of pattern and what is now called "Rex blocking," in which a split westerly jet, with an easterly flow between and the familiar high–low doublet pattern, exists. The Mode 2 events in general do *not* include Rex-type blocking events. Inspection of daily synoptic charts indicates that Mode 2 episodes can be identified quite well without knowledge of the wave amplitude indicator by searching for patterns characterized by hemispheric-scale amplified waves. On the other hand, regionally isolated Rex-type blocking patterns are uniformly *not* included in the Mode 2 category. These events have quite different characteristics than do the Mode 2 events. This aspect will be discussed further in Section 5.

The overall impression is left that rapid transitions are occurring between large and small wave amplitudes for fixed-phase waves. Next consider the mean spectral energy and enstrophy budgets of the two modes to determine

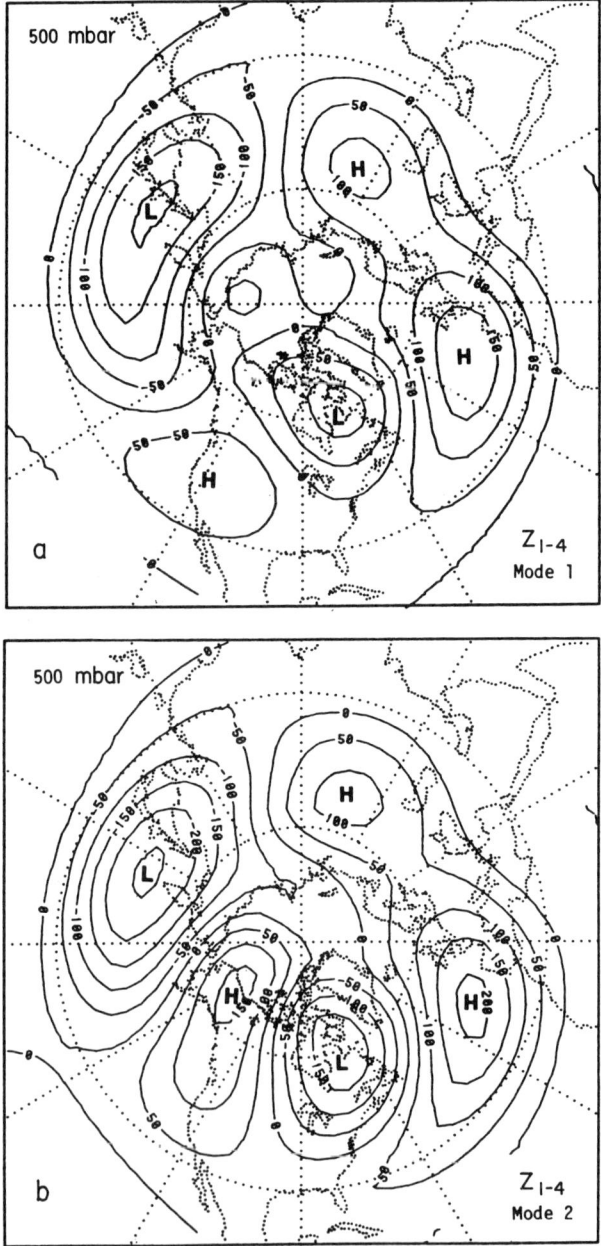

FIG. 5. Mean 500-mbar heights synthesized from zonal wavenumbers 1–4, Z_{1-4}, for (a) Mode 1 and (b) Mode 2.

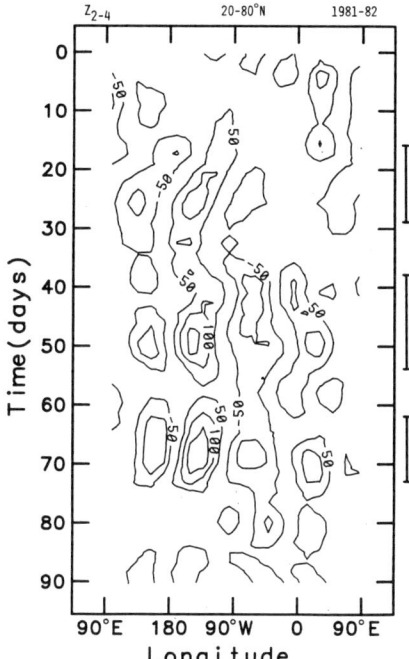

FIG. 6. The longitude-time section of Z_{2-4} averaged between 20°N and 80°N for 1981/1982. The brackets along the right-hand side of the figure indicate the time periods belonging to Mode 2. The contour interval is 50 m. The zero contour has been removed for clarity.

what processes supply energy and enstrophy toward the maintenance of the large-amplitude waves.

4. Energetics and Enstrophy Budgets for the Two Modes

4.1. Data

Before proceeding with the analysis, let us briefly discuss the data sets being employed. The data used in this study are the initialized fields for the operational model of the European Center for Medium Range Weather Forecasts. We obtained the data for the horizontal wind, vertical velocity, geopotential height, and temperature at 10 levels in the troposphere (1000, 850, 700, 500, 400, 300, 250, 200, 150, and 100 mbar) interpolated to a 2.5° by 2.5° latitude-longitude grid. The fields for 0000 GMT of 1 December through 28 February of the years 1980/1981, 1981/1982, 1982/1983, and

1983/1984 were used. These data were objectively analyzed from observations with a multivariate optimum interpolation scheme (Lorenc, 1981) and were balanced with a normal mode initialization technique (Temperton and Williamson, 1981; Williamson and Temperton, 1981). Further discussion of the data handling procedures used at ECMWF as well as the potential shortcomings of the data can be found, for example, in Hoskins *et al.* (1983). The major problem with the data is the attenuation of the divergent wind in the tropics caused by difficulties in incorporating diabatic heating in the initialization procedure. In the present study, we are concerned with midlatitude phenomena so this will not be a problem. In general, it is widely believed that the ECMWF analyses have the highest accuracy of any currently available global dataset (e.g., Hoskins *et al.*, 1983). As the balanced initial fields of an operational model, they are necessarily model dependent, but the fact that they include assimilated data from numerous sources in addition to conventional radiosondes (e.g., satellite soundings and aircraft reports over the oceans) and that they are a three-dimensionally balanced dataset largely compensates the disadvantage of model dependency. In addition, in the present study we will be primarily interested in differences between ensembles of fields within the dataset so we do not have serious reservations about the data source.

4.2. Energetics and Enstrophy Budget

4.2.1. Formulation. The technique of computing the energetics of the atmosphere using a zonal harmonic representation of the basic-state variables was developed by Saltzman (1957). The strength of this technique lies in decomposition of the longitudinal variance of these variables in midlatitudes into contributions from different wavelength features. As such, it can yield useful information about the energy cycle for the various zonal harmonic waves and has been used in innumerable diagnostic studies. Its drawback is that it yields only mass-averaged results and does not explicitly account for variance in the meridional direction. As such, regionally localized processes may be masked in the spectral results, although if a localized process is sufficiently intense, it will leave a marked signature in the averaged results (Hansen and Chen, 1982). Since we are investigating a phenomenon whose spatial amplitude variations are primarily oriented in the east-west direction, the zonal harmonic representation of the data is the most appropriate for our purposes.

The formulation of the spectral energetics equations is given by Saltzman (1970) and Tomatsu (1979). In symbolic form, the rates of change of wavenumber m kinetic energy (K_m) and available potential energy (A_m) are given by

$$(\delta/\delta t)K_m = C(K_z, K_m) + C(A_m, K_m) + C_K(m|n, l)$$
$$+ F(K_m) + F(\Phi_m) + D(K_m)$$
$$(\delta/\delta t)A_m = C(A_z, A_m) - C(A_m, K_m) + C_A(m|n, l)$$
$$+ G(A_m) + F(A_m)$$

Here, a z subscript denotes a zonal mean quantity

$$(\ )_z = \frac{1}{2\pi} \int_0^{2\pi} (\ ) \, d\lambda \quad \text{where} \quad \lambda = \text{longitude}$$

and an m subscript denotes a departure from this mean for wavenumber m. The notation $C(A, B)$ represents a conversion of energy from reservoir A to reservoir B. The term $C_X(m|n, l)$ denotes the rate of increase of the quantity X (where $X = K$ or A) at wavenumber m due to nonlinear triad interactions with all possible combinations of wavenumbers n and l.

Terms $F(K_m)$ and $F(A_m)$ denote the combined horizontal and vertical boundary fluxes of K_m and A_m, respectively, and $F(\Phi_m)$ is the boundary flux of geopotential energy. The kinetic energy dissipation $D(K_m)$ will be estimated as a residual. The generation of available potential energy, $G(A_m)$, was not computed but it will generally make a negative contribution (Brown, 1964). The $G(A_m)$ could be estimated as a residual also.

Similarly, the rate of change of wavenumber m enstrophy (E_m) can be written as

$$(\partial/\partial t)E_m = C(E_z, E_m) + C_E(m|n, l) + \beta_m + G(E_m) + T(E_m)$$
$$+ F(E_m) + D(E_m)$$

The enstrophy equation is derived from the equation for the vertical component of the atmosphere's vorticity. The notation follows that in the energy equations except that β_m represents the β-effect ($-\mathbf{v} \cdot \nabla f$) from the vorticity equation. The term $C(E_z, E_m)$ represents the wave-mean flow interaction, $C_E(m|n, l)$ represents the wave–wave interactions, and $F(E_m)$ and $D(E_m)$ are, respectively, the boundary fluxes and the dissipation of enstrophy. Enstrophy generation, $G(E_m)$, is due to the vortex stretching mechanism that results from divergent motions. The $T(E_m)$ represents the effect on the enstrophy budget of the twisting and tilting term in the vorticity equation. It is a transfer of vorticity (and therefore enstrophy) from the horizontal components to the vertical component. The $G(E_m)$ represents the major source of enstrophy in midlatitudes. The contribution due to $T(E_m)$ is generally small and was not computed in the present study. The formulation of the spectral enstrophy equation can be found in Chen (1985).

The terms in these equations were integrated from 20°N to 80°N and from 1000 to 100 mb. The zonal harmonic representation of the basic state variables was truncated at $m = 20$.

Each of the terms in these budget equations was evaluated for each day of our four-winter dataset. Using the wave amplitude probability density distribution as a guide (Sutera, 1986), each observation was assigned to the Mode 1 or Mode 2 sample depending upon the value of $[Z_{2-4}]$ on that day. The mean energetics presented below represent an arithmetic average over all of the days in each mode. In addition, the ensemble mean fields for each of the two modes were computed for each winter from which the "stationary" contribution to the energetics of each mode was computed.

A weighted average of the individual year's mean stationary contributions was then computed as follows:

$$(\ )_s^{(j)} = \sum_{i=1}^{4} w_i^{(j)} (\ )_{s_i}^{(j)}$$

where $w_i^{(j)}$ is the number of days in mode j during the ith winter (Table I) divided by the total number of mode j days in the four winters, $(\ )_{s_i}^{(j)}$ is the stationary part of a given budget term for the jth mode in the ith winter, and $(\ )_s^{(j)}$ is the 4-yr mean value of the given stationary term for mode j. This approach was used to reduce the influence of interannual variability on the computation of the stationary contribution. If the stationary contributions are computed simply by averaging the basic state variables over the four winters for each mode and then computing the energetics, the stationary contributions are uniformly smaller (20–50% smaller for Mode 1 and 20% smaller for Mode 2) because the variance associated with the interannual variability is then represented in the transient components of the budget terms.

4.2.2. Energetics for Each Mode. First consider the kinetic energy and available potential energy spectra for the two modes (Fig. 7). (In this and subsequent figures, stationary contributions are denoted by the shading.)

Notice that the total K_m for wavenumbers 2–4 increases markedly for Mode 2, with doubling of the stationary contributions to waves 2 and 3 (see summary in Table II). This results in a more steeply sloped K_m spectrum at its low-wavenumber end. The A_m spectrum also exhibits large increases in the total and stationary A_m for wavenumbers 2 and 3. The overall increase in K_{2-4} ($= \Sigma_{m=2}^{4} K_m$) for Mode 2 compared to Mode 1 is 30%, with 92% of this due to the stationary component, while A_{2-4} increases 32% for Mode 2, with 92% of the increase contributed by the stationary part.

TABLE II. ENERGY SUMMARY OF TWO MODES FOR WAVENUMBERS 2-4

Energy budget[a]

	K_{2-4} ($\times 10^5$ J m^{-2})	$C(K_z, K_{2-4})$ (W m^{-2})	$C(A_{2-4}, K_{2-4})$ (W m^{-2})	$C_K(2-4)$ (W m^{-2})	F_{2-4} (W m^{-2})	Residual (W m^{-2})
Mode 1	4.0	−0.01	1.33	−0.11	−0.10	−1.11
	(1.2)	(−0.04)	(0.49)	(−0.15)	(0.03)	(−0.33)
Mode 2	5.2	0.03	1.47	0.00	−0.03	−1.47
	(2.3)	(−0.02)	(0.60)	(−0.12)	(0.02)	(−0.48)
	A_{2-4} ($\times 10^5$ J m^{-2})	$C(A_z, A_{2-4})$ (W m^{-2})	$C_A(2-4)$ (W m^{-2})	$F(A_{2-4})$ (W m^{-2})	Residual (W m^{-2})	
Mode 1	4.1	1.30	−0.05	0.11	0.03	
	(2.0)	(0.65)	(0.05)	(0.32)	(−0.53)	
Mode 2	5.4	1.96	−0.32	−0.24	0.07	
	(3.2)	(1.32)	(0.13)	(0.27)	(−1.12)	

Zonal-mean energy levels[b]

	K_z ($\times 10^6$ J m^{-2})	A_z ($\times 10^6$ J m^{-2})
Mode 1	1.24	3.95
	(1.20)	(3.91)
Mode 2	1.26	3.81
	(1.23)	(3.78)

[a] The stationary contribution to each term is in parentheses. $C_K(2-4)$ denotes the gain in $K_{2-4} = \Sigma_{m=2}^4 K_m$ due to wave–wave interactions. F_{2-4} denotes the contribution due to all boundary fluxes in the K_{2-4} budget.

[b] The stationary contribution is in parentheses. Differences between terms in the budgets for K_z and A_z are small.

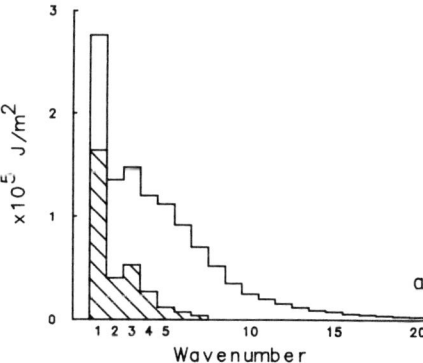

FIG. 7. The wavenumber spectra of the eddy kinetic energy, K_m, for (a) Mode 1 and (b) Mode 2 and the eddy available potential energy, A_m, for (c) Mode 1 and (d) Mode 2. In this and subsequent figures, the shading denotes the contribution to each term by the stationary component of the flow.

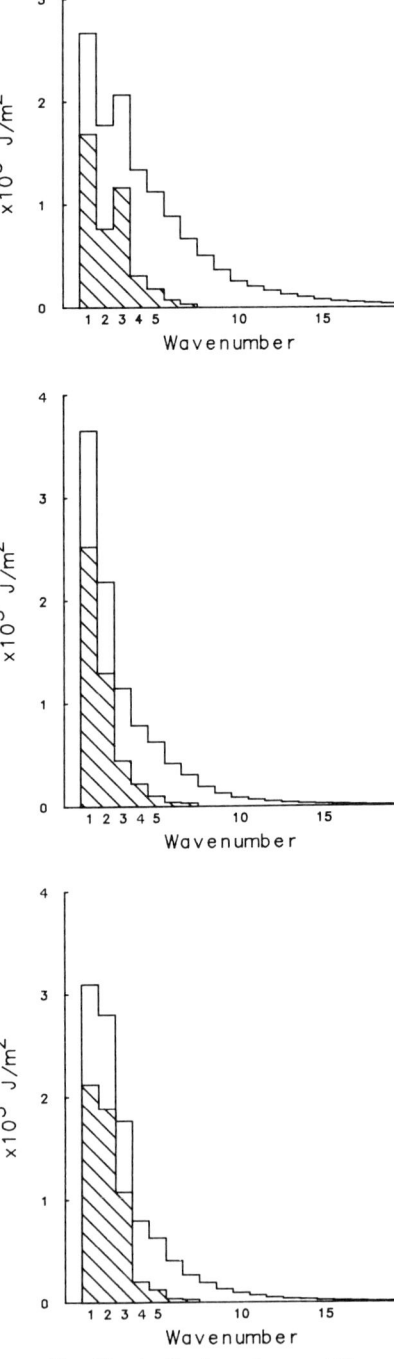

FIG. 7b–d. See legend on p. 117.

The spectra of the largest diagnosed energy conversions are illustrated in Fig. 8a–f. The $C(A_z, A_m)$ is markedly greater at wavenumbers 2 and 3 for Mode 2 compared to Mode 1. For our wave ensemble, $C(A_z, A_{2-4})$ increases 51%, with 100% of this increase due to the stationary component. This conversion, which depends primarily upon the correlation of northward eddy heat transport with the latitudinal gradient of zonal-mean temperature (see Saltzman, 1970), can be viewed as a baroclinic conversion resulting in creation of eddy available potential energy. The $C(A_z, A_m)$ for intermediate-scale wavenumbers 5–8 actually declines 20% for the Mode 2 average compared to their Mode 1 values. Likewise, the baroclinic kinetic energy conversion, $C(A_{2-4}, K_{2-4})$, is greater for Mode 2 (by 11%), with 79% of this increase due to the stationary component. The term $C(A_m, K_m)$ averages 15% smaller

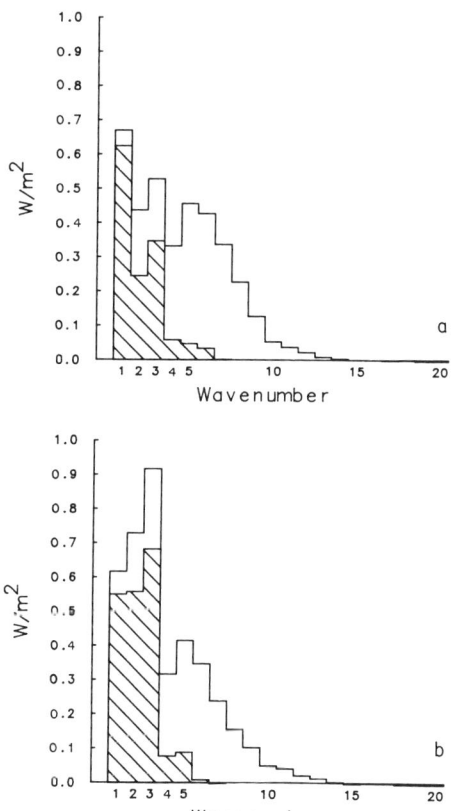

FIG. 8. Major energy budget terms for the two modes: (a) $C(A_z, A_m)$ for Mode 1, (b) $C(A_z, A_m)$ for Mode 2, (c) $C(A_m, K_m)$ for Mode 1, (d) $C(A_m, K_m)$ for Mode 2, (e) $C_K(m|n, l)$ for Mode 1, and (f) $C_K(m|n, l)$ for Mode 2.

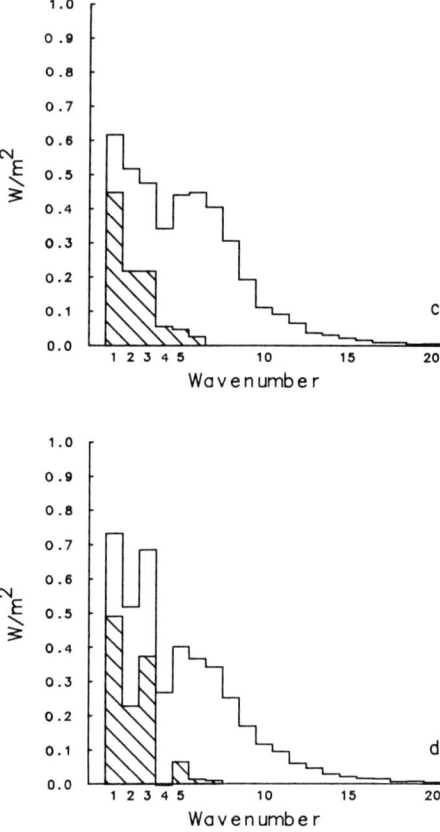

FIG. 8c and d. See legend on p. 119.

at intermediate wavenumbers for Mode 2 compared to Mode 1. Wave–wave interactions make a negligible contribution to the mean kinetic energy budget of wavenumbers 2–4 for both modes. Wavenumber 1 is the main recipient of energy from this process although there is some redistribution of energy among the stationary planetary waves (Fig. 7e and f).

The overall energy budget for the wavenumber 2–4 ensemble is given in Table II. The increase in K_{2-4} for Mode 2 must represent a balance between the various budget terms if the computed statistics are stationary. Note that the major terms in the kinetic energy budget are the baroclinic conversion, $C(A_{2-4}, K_{2-4})$, and the residual. This residual represents the combined effects of frictional dissipation and systematic errors in the data. If we assume that the residual represents primarily the frictional dissipation, $D(K_{2-4})$, and

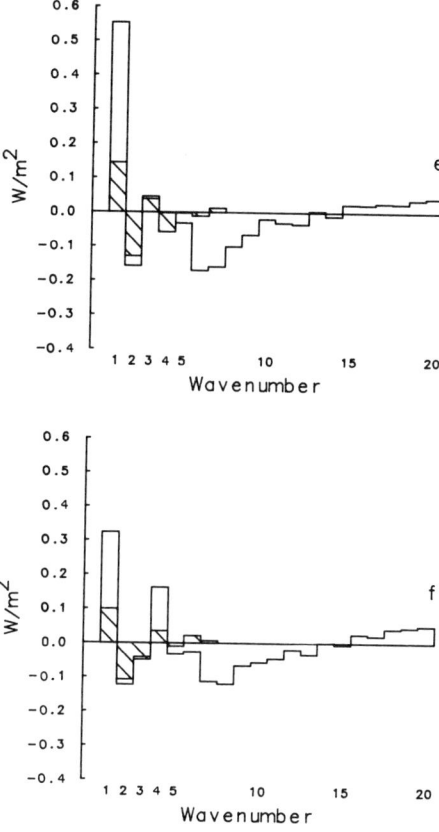

FIG. 8e and f. See legend on p. 119.

further assume that $D(K_{2-4})$ is linearly dependent upon the kinetic energy value, we can estimate a damping coefficient appropriate to our wave ensemble from

$$D(K_{2-4}) = -\kappa K_{2-4}$$

If our statistics are stationary, the kinetic energy budget must balance to within measurement error. So, assuming that the systematic errors are small, and setting the dissipation equal to the budget residual, we get

	K_{2-4} (J m^{-2})	Residual (W m^{-2})	κ (sec^{-1})
Mode 1	4.0×10^5	1.11	2.8×10^{-6}
Mode 2	5.2×10^5	1.47	2.8×10^{-6}

This value of $\kappa = 2.8 \times 10^{-6}$ sec^{-1} corresponds to a momentum dissipation time scale $\tau = 2/\kappa$ of roughly 8 days.

The major source term in the A_{2-4} budget is $C(A_z, A_{2-4})$ with the sink provided by $C(A_{2-4}, K_{2-4})$ leaving a small residual which represents data inaccuracies as well as the generation term.

Next, consider the enstrophy budget for each mode. The enstrophy spectra (Fig. 9) for the two modes are qualitatively similar to the kinetic energy spectra with increases in E_m for waves 2–4 for Mode 2 compared to Mode 1. The stationary components of E_m also increase for these wavenumbers but they make up a smaller fraction of the totals than do their counterparts in the

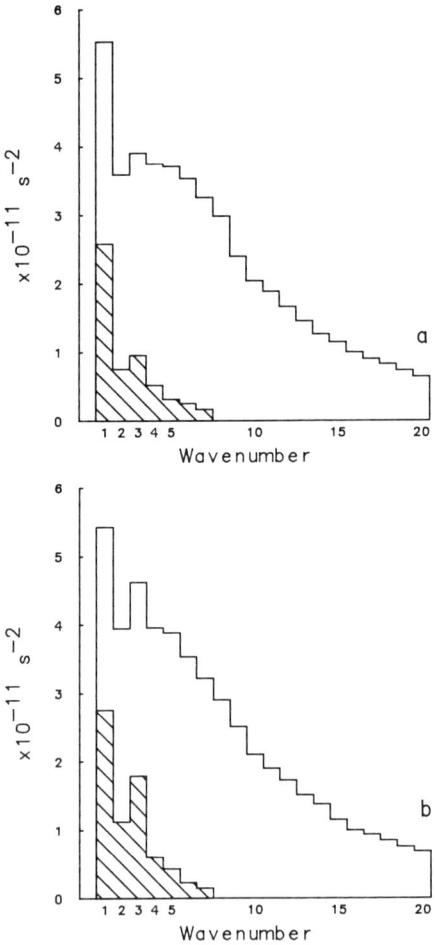

FIG. 9. The enstrophy spectrum, E_m, for (a) Mode 1 and (b) Mode 2.

energy spectra. Note that the E_m spectrum is relatively steep for wavenumbers 2–8 for Mode 2 compared to the flat spectrum of Mode 1 in this region. The major computed terms in the enstrophy budget are the generation, $G(E_m)$, and the wave–wave interaction (Fig. 10a–d). For Mode 1, $G(E_m)$ exhibits a pronounced maximum in the synoptic scale at wavenumber 7, whereas this value is reduced in Mode 2 and the value at wavenumber 3 increases to a magnitude comparable to that at wavenumber 7. The wave–wave interaction, $C_E(m|n, l)$, for Mode 1 is characterized by downscale cascades from intermediate wavenumbers to high wavenumbers with little net effect at the low-wavenumber end of the spectrum, while enstrophy is lost in this spectral region for Mode 2. Intermediate-scale wavenumbers lose less E_m during Mode 2 events. Tabulation of the enstrophy budget for wavenumbers 2–4 is given in Table III. Note that increased generation for Mode 2 compared to Mode 1 is compensated by an increased enstrophy sink due to wave–wave interactions. The residual is large for both modes.

4.3. Discussion

Given the 4-yr mean statistics, we might legitimately ask how significant the noted differences are in the energy and enstrophy budgets of the two

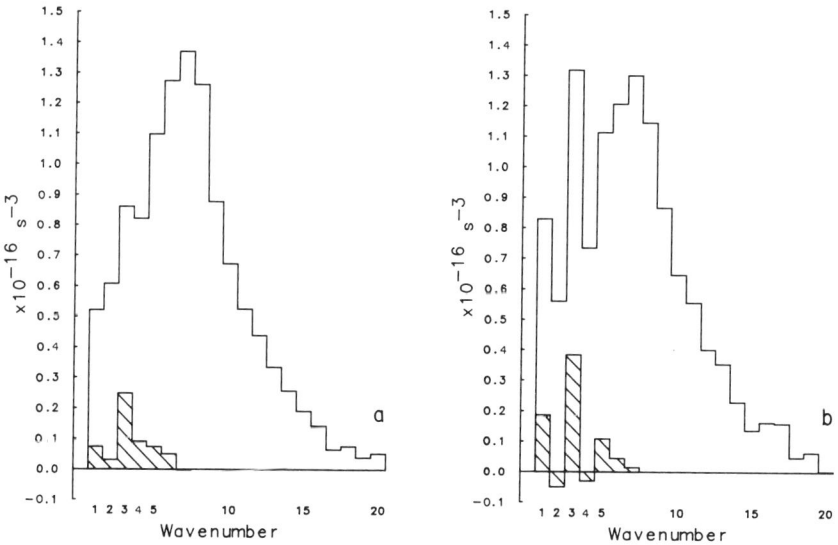

FIG. 10. Major enstrophy budget terms for the two modes: (a) $G(E_m)$ for Mode 1, (b) $G(E_m)$ for Mode 2, (c) $C_E(m|n, l)$ for Mode 1, and (d) $C_E(m|n, l)$ for Mode 2.

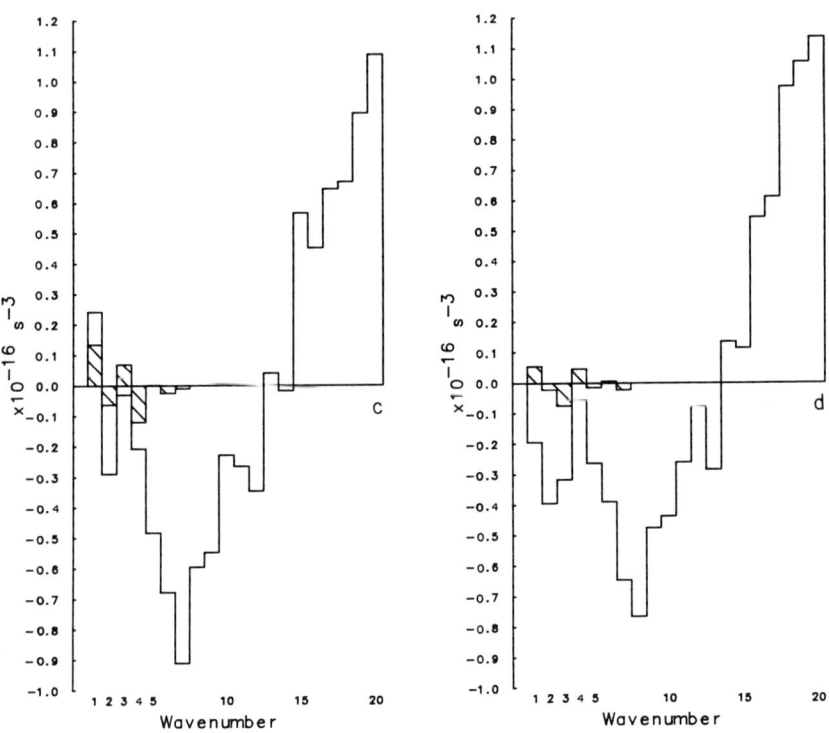

FIG. 10c and d. See legend on p. 123.

modes. To address this question, we examined similar budgets for the groups of Mode 1 and Mode 2 days in each of the 4 years individually. Qualitatively, similar differences appear consistently between the individual year's Mode 1 and Mode 2 composites, although interannual variability is present in the wavenumbers representing these differences.

As a representative example, consider $C(A_z, A_m)$ for both modes at wavenumbers 2–4 for each of the individual years (Table IV). The low-wavenumber increase in $C(A_z, A_m)$ in 1980/1981 occurs at both wavenumbers 2 and 3 with comparable changes in these waves' stationary contributions. In 1981/1982, the increases at wavenumbers 2 and 3 are quite spectacular (doubling of the total and quadrupling of the stationary contributions) while in 1982/1983, a very marked increase occurs at wavenumber 3 (more than doubling the total and tripling the stationary contribution), but little change occurs at wavenumber 2. In 1983/1984, the wavenumber 2 total contribution increases by over 50% and its stationary part doubles while little difference is evident at wavenumber 3. Wavenumber 4 does not make a

TABLE III. ENSTROPHY BUDGET OF TWO MODES FOR WAVENUMBERS 2–4

	E_{2-4} ($\times 10^{-10}$ sec^{-2})	$C(E_z, E_{2-4})$ ($\times 10^{-16}$ sec^{-3})	$G(E_{2-4})$ ($\times 10^{-16}$ sec^{-3})	$C_E(2-4)$[a] ($\times 10^{-16}$ sec^{-3})	β_{2-4} ($\times 10^{-16}$ sec^{-3})	$F(E_{2-4})$ ($\times 10^{-16}$ sec^{-3})	Residual ($\times 10^{-16}$ sec^{-3})
Mode 1[b]	1.12	−0.02	2.29	−0.53	−0.19	−0.02	−1.53
	(0.22)	(0.01)	(0.37)	(−0.11)	(−0.09)	(0.03)	(−0.21)
Mode 2[b]	1.25	0.07	2.61	−0.77	−0.21	0.04	−0.30
	(0.35)	(0.09)	(0.31)	(−0.05)	(−0.07)	(0.02)	(−0.30)

[a] $C_E(2-4)$ denotes the gain in E_{2-4} due to wave–wave interactions.
[b] The stationary contribution to each term is in parentheses.

TABLE IV. VALUES OF $C(A_z, A_m)$ FOR WAVENUMBERS 2–4

		$C(A_z, A_m)$			
		1980/1981	1981/1982	1982/1983	1983/1984
$m = 2$	Mode 1	0.40 (0.27)[a]	0.41 (0.20)	0.40 (0.18)	0.34 (0.32)
	Mode 2	0.70 (0.47)	0.96 (0.82)	0.46 (0.26)	0.82 (0.70)
	Increase[b]	75% (74%)	134% (310%)	15% (44%)	52% (119%)
$m = 3$	Mode 1	0.59 (0.43)	0.42 (0.18)	0.45 (0.24)	0.64 (0.52)
	Mode 2	0.99 (0.80)	0.85 (0.70)	1.07 (0.72)	0.73 (0.49)
	Increase	68% (86%)	102% (289%)	138% (200%)	14% (−6%)
$m = 4$	Mode 1	0.44 (0.04)	0.29 (0.06)	0.36 (0.08)	0.25 (0.05)
	Mode 2	0.34 (0.17)	0.24 (0.00)	0.40 (0.12)	0.28 (0.01)
	Increase	−22% (325%)	−17% (−100%)	11% (50%)	12% (−80%)

[a] The stationary contributions are in parentheses.
[b] The percentage increase for Mode 2 over Mode 1.

noteworthy contribution to the differences between the modes in any of the years.

In all 4 years, wavenumber 1 exhibits as strong stationary component for both modes. Years 1980/1981 and 1983/1984 have relatively strong stationary components at wavenumbers 2 and 3 for Mode 1, while in 1981/1982 and 1982/1983, the Mode 1 total and stationary components in the wavenumber 2–4 region are relatively weak. The diminution of cyclone-scale baroclinic conversions during Mode 2 events is experienced during all 4 years.

The total zonal to eddy available potential energy conversion and the eddy available to eddy kinetic energy conversion are only slightly greater in Mode 2 events compared to Mode 1, but there is a shift in the scale at which the conversions take place toward the low-wavenumber end of the spectrum. This results in corresponding increases in the kinetic energy and available potential energy content of the low wavenumbers, particularly in the stationary component of the flow.

Taken as a whole, examination of the results for the 4 individual years in our dataset reinforces the finding that systematic differences occur in the energy and enstrophy budgets between the low-wave-amplitude state and the large-amplitude state. That is, the large-amplitude state is *consistently* accompanied by marked increases in the energy and enstrophy associated with planetary waves and in baroclinic processes at length scales represented by wavenumbers 2 and 3. The majority of the increases in the energy budget terms is represented by increases in the stationary contributions to the large-amplitude mode. Interactions between these transient cyclones, and the wavenumber 2–4 ensemble, do not appear to be of importance to the time-mean energy budgets of the large-amplitude state, although wave–wave interaction represents a sink of E_{2-4} for Mode 2. However, this does not preclude the importance of nonlinear wave–wave interactions during transition periods. This aspect of the problem will be discussed further in a subsequent report.

5. Role of Cyclone Waves in Blocking

A number of studies have indicated the importance of interactions between synoptic-scale transient waves and planetary-scale waves during the onset or for the maintenance of winter blocking (Bengtsson, 1981; Hansen and Chen, 1982; Hoskins *et al.*, 1983; Shutts, 1983, Hansen and Sutera, 1984; Colucci, 1985; Brown *et al.*, 1986). As noted earlier, however, examination of synoptic charts for our dataset reveals Rex-type blocking episodes that are not included in the large-amplitude mode of our wave amplitude index. That is, blocking can occur on days otherwise designated as Mode 1.

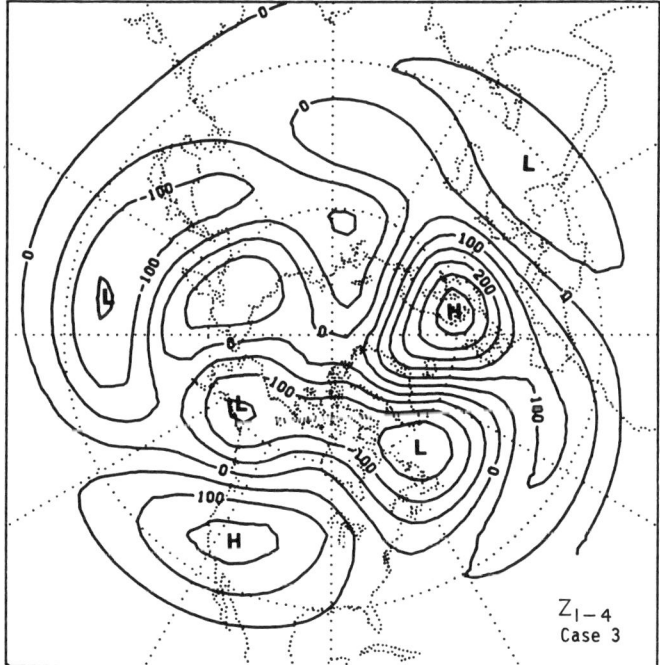

Fig. 11. Mean 500-mbar height synthesized from wavenumbers 1–4 for Case 3, 14 to 23 February 1982. Contour interval is 50 m.

Let us consider one such case, which, for the sake of discussion, we will designate as Case 3.

From 14 through 23 February 1982, inspection of daily synoptic charts reveals that a pattern satisfying Rex's traditional definition of blocking existed over Europe. Based on our wave amplitude index, these 10 days were classified as Mode 1 days. The 10-day mean 500-mbar height field synthe-

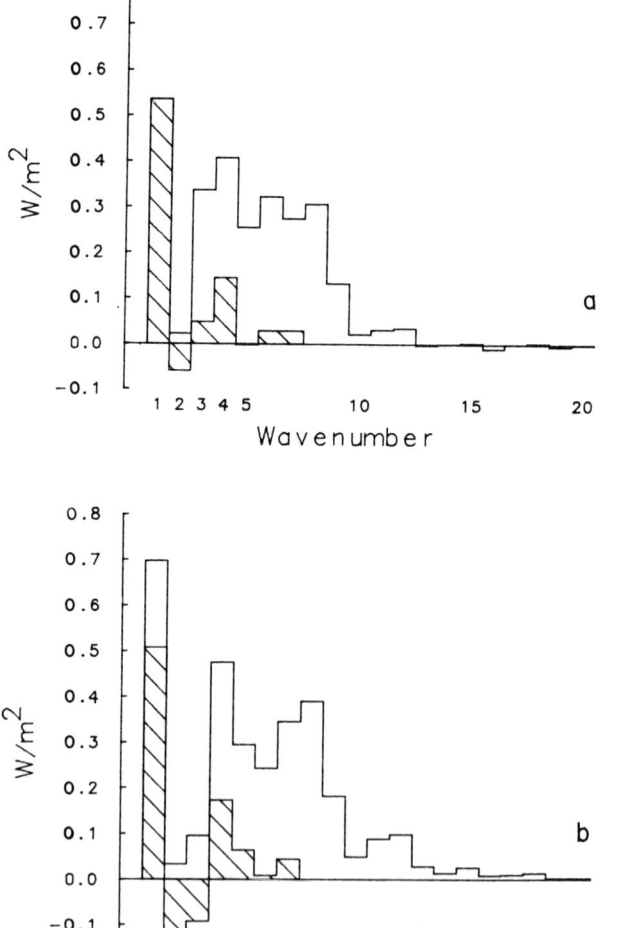

FIG. 12. Case 3 energy budget terms: (a) $C(A_z, A_m)$, (b) $C(A_m, K_m)$, and (c) $C_K(m|n, l)$.

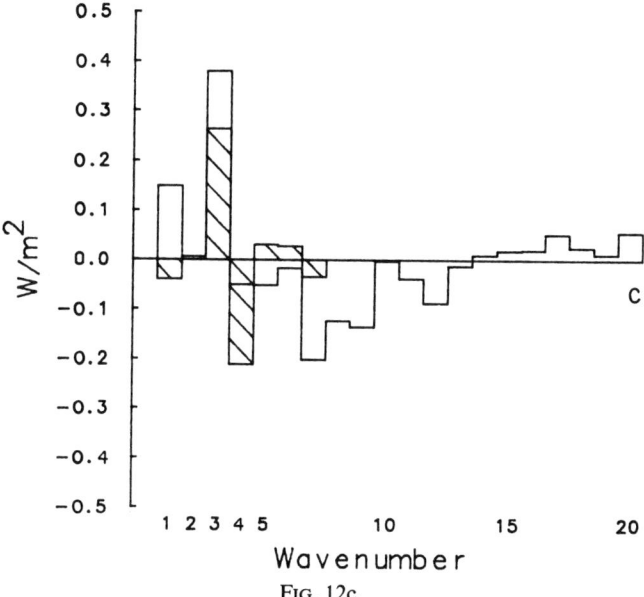

FIG. 12c.

sized from the planetary-scale waves (Fig. 11) illustrates a pattern quite different from that corresponding to Mode 2 (Fig. 5b). In particular, notice the very strong ridge over Europe. The pattern is superficially different and the ridge appears somewhat more regionally isolated than the Mode 2 wave pattern. An examination of the energetics of this case reveals sharply different results from those for Mode 2.

The A_m and K_m spectra for Case 3 (not shown) are very similar to the Mode 1 average. However, when we consider the leading terms in the A_m and K_m budgets, we see several interesting characteristics. At low wavenumbers, the baroclinic conversions $C(A_z, A_m)$ and $C(A_m, K_m)$ are quite small compared to the mean statistics for either Mode 1 or 2 (Fig. 12a and b). In particular, note that the stationary contribution determined from the 10-day mean fields to $C(A_z, A_m)$ at wavenumber 3 and the stationary contribution to $C(A_m, K_m)$ at waves 2 and 3 is *negative*. The chief source of kinetic energy at wavenumber 3 is the wave–wave interaction (Fig. 12c). Likewise, the enstrophy budget for Case 3 shows analogous features. The generation term shows a negative total generation at wavenumber 2 with negative stationary contribution for wavenumbers 2 and 3 (Fig. 13a). This is in direct contrast to Mode 2 where $G(E_m)$ has a prominent maximum at wavenumber 3. However, the nonlinear exchange of enstrophy provides the major source of enstrophy for wavenumbers 2–4, with a particularly large value at wave 4

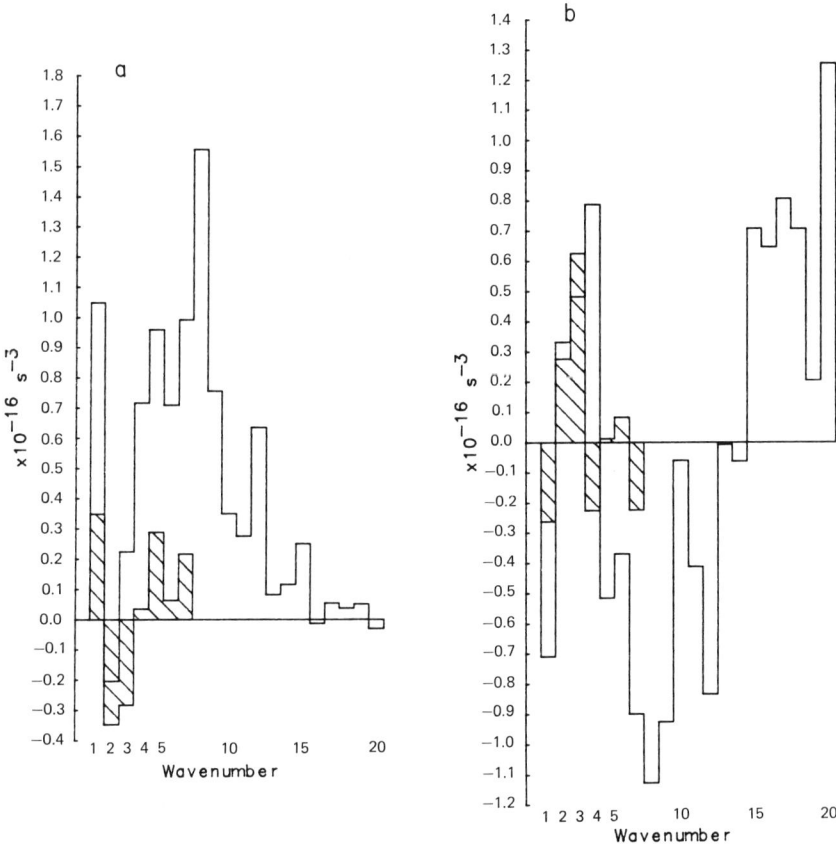

FIG. 13. Case 3 enstrophy budget terms: (a) $G(E_m)$ and (b) $C_E(m|n, l)$.

(Fig. 13b). Although this type of behavior is less common than Mode 2 events in the present dataset (roughly 20% of the days in our dataset can be subjectively identified as Rex blocking days while 46% belong to Mode 2), the general characteristic of the transient cyclones supplying energy and enstrophy to the wavenumber band representing the blocking is identical to that found by Hansen and Sutera (1984) in a different dataset. The role of the transient cyclones within the framework of the zonal wavenumber analysis used here may be analogous to the role of transients in the local potential vorticity budget analysis of Shutts (1983). Similar characteristics have been noted in general circulation model experiments (Fischer, 1984).

The marked differences in the budget statistics between this type of event and our Mode 2 events suggest that fundamental differences exist between the two. The Mode 2 wave pattern extends over a considerable part of the Northern Hemisphere midlatitudes and large-scale steady baroclinic processes appear to maintain the time-mean waves, while our Case 3 event and others like it are at least superficially more regional in character and appear more crucially associated with wave–wave interactions with intermediate-scale waves for their maintenance.

6. Conclusion

Analysis of data from ECMWF for the winters of 1980/1981, 1981/1982, 1982/1983, and 1983/1984 indicates that the probability density distribution of the amplitude of the planetary-wave ensemble represented by zonal harmonic wave numbers 2–4 is bimodal (Sutera, 1986). The mean 500-mbar height field for days in the large-amplitude mode exhibits markedly larger amplitude features over the Pacific Ocean and North America when compared to the low-amplitude mode. In most of the individual cases an enhanced ridge also appeared over the eastern Atlantic Ocean or Europe.

Spectral energetics analysis reveals that the large-amplitude mode is characterized by large, steady baroclinic energy conversions at wavenumbers 2 and 3, compared to the low-amplitude case. Baroclinic conversions at intermediate, cyclone-scale wavelengths are smaller for the large-amplitude wave ensemble. The time-mean energy cycle in wavenumbers 2–4 is characterized by a balance between baroclinic conversion and dissipation. Enhancement of low-wavenumber enstrophy generation also occurs for the large-amplitude mode. Neither of the two modes we have described represents "anomalous" behavior but rather their existence appears to be a fundamental aspect of the circulation.

In contrast to the stationary large-scale baroclinic waves of Mode 2, blocking events characterized by more regional synoptic patterns can occur independently of the more hemispheric-scale wave amplitude index. Energetically, events of this type gain kinetic energy and enstrophy through nonlinear interactions with cyclone waves, and large-scale, stationary baroclinic processes appear less important in their hemispherically averaged energy budget. Mode 2 events are often associated with flow patterns over the Pacific Ocean and also over the Atlantic Ocean and Europe that can be called blocking. However, Mode 1 and Mode 2 periods are not generally synonymous with nonblocking and blocking periods. Thus, blocking patterns can be associated with two quite different physical scenarios, and therefore it appears that there are at least two different classes of persistent, large-scale phenomena.

Acknowledgments

It is a pleasure to acknowledge Alfonso Sutera of the Center for the Environment and Man and Stefano Tibaldi of ECMWF for numerous stimulating conversations concerning this work and for invaluable assistance in the early stage of the computation. The helpful comments of Dr. H.-D. Schilling of the University of Munich on an earlier version of this paper are greatly appreciated. This study was supported by the Climate Dynamics Section of the National Science Foundation under Grant ATM-8403372.

References

Bengtsson, L. (1981). Numerical prediction of atmospheric blocking—A case study. *Tellus* **33**, 19–42.

Benzi, R., Hansen, A. R., and Sutera, A. (1984). On stochastic perturbation of simple blocking models. *Q. J. R. Meteorol. Soc.* **110**, 393–409.

Benzi, R., Malguzzi, P., Speranza, A., and Sutera, A. (1986). The statistical properties of general atmospheric circulation: Observational evidence and a minimal theory of bimodality. *Q. J. R. Meteorol. Soc.* **112**, in press.

Brown, J. A. (1964). A diagnostic study of tropospheric diabatic heating and the generation of available potential energy. *Tellus* **16**, 371–388.

Brown, P. S., Jr., Hansen, A. R., and Pandolfo, J. P. (1986). Circulation regime-dependent nonlinear interactions during Northern Hemisphere winter. *J. Atmos. Sci.* **43**, 388–397.

Charney, J. G., and DeVore, J. G. (1979). Multiple flow equilibria in the atmosphere and blocking. *J. Atmos. Sci.* **36**, 1205–1216.

Charney, J. G., and Eliassen, A. (1949). A numerical method for predicting the perturbations of the middle latitude westerlies. *Tellus*, **1**, 38–54.

Charney, J. G., and Strauss, D. M. (1980). Form-drag instability, multiple equilibria and propagating planetary waves in baroclinic, orographically forced planetary wave systems. *J. Atmos. Sci.* **37**, 1157–1176.

Charney, J. G., Shukla, J., and Mo, K. C. (1981). Comparison of a barotropic blocking theory with observation. *J. Atmos. Sci.* **38**, 762–769.

Chen, T. C. (1985). On the maintenance of enstrophy in the tropics during FGGE Northern summer. *Mon. Weather Rev.* **113**, 624–640.

Colucci, S. J. (1985). Explosive cyclogenesis and large-scale circulation changes: Implications for the onset of blocking. *J. Atmos. Sci.* **42**, 2701–2717.

Dole, R. M., and Gordon, N. D. (1983). Persistent anomalies of the extratropical Northern Hemisphere wintertime circulation: Geographical distribution and regional persistence characteristics. *Mon. Weather Rev.* **111**, 1567–1586.

Fischer, G. (1984). Spectral energetics analysis of blocking events in a general circulation model. *Beitr. Phys. Atmos.* **57**, 183–200.

Good, I. J., and Gaskins, R. A. (1980). Density estimation and bump-hunting by the penalized likelihood method exemplified by scattering and meteorite data. *J. Am. Stat. Assoc.* **75**, 42–56.

Hansen, A. R., and Chen, T. C. (1982). A spectral energetics analysis of atmospheric blocking. *Mon. Weather Rev.* **110**, 1146–1165.

Hansen, A. R., and Sutera, A. (1984). A comparison of the spectral energy and enstrophy budgets of blocking versus nonblocking periods. *Tellus* **36A**, 52–63.

Hart, J. (1979). Barotropic quasi-geostrophic flow over anisotropic mountains. *J. Atmos. Sci.* **36,** 1736–1746.

Hoskins, B. J., James, I. N., and White, G. H. (1983). The shape, propagation and mean flow interaction of large-scale weather systems. *J. Atmos. Sci.* **40,** 1595–1612.

Lorenc, A. (1981). A global three-dimensional multivariate statistical interpolation scheme. *Mon. Weather Rev.* **109,** 701–721.

Pedlosky, J. (1981). Resonant topographic waves in barotropic and baroclinic flows. *J. Atmos. Sci.* **38,** 2626–2641.

Rex, D. F. (1950). Blocking action in the middle troposphere and its effect upon regional climate. I. An aerological study of blocking action. *Tellus* **2,** 196–211.

Saltzman, B. (1957). Equations governing the energetics of the larger scales of atmospheric turbulence in the domain of wavenumber. *J. Meteorol.* **14,** 513–523.

Saltzman, B. (1970). Large-scale atmospheric energetics in the wavenumber domain. *Rev. Geophys. Space Phys.* **8,** 289–302.

Shutts, G. (1983). The propagation of eddies in diffluent jetstreams: Eddy vorticity forcing of "blocking" flow fields. *Q. J. R. Meteorol. Soc.* **109,** 737–762.

Smagorinsky, J. (1953). The dynamical influence of large-scale heat sources and sinks on the quasi-stationary mean motions of the atmosphere. *Q. J. R. Meteorol. Soc.* **79,** 343–366.

Sutera, A. (1986). Probability density distribution of large-scale atmospheric flow. *Adv. Geophys.* **29,** 227–249.

Temperton, C., and Williamson, D. L. (1981). Normal mode initialization for a multilevel gridpoint model. I. Linear aspects. *Mon. Weather Rev.* **109,** 729–743.

Tomatsu, K. (1979). Spectral energetics of the troposphere and lower stratosphere. *Adv. Geophys.* **21,** 289–405.

Williamson, D. L., and Temperton, C. (1981). Normal mode initialization for a multilevel gridpoint model. II. Nonlinear aspects. *Mon. Weather Rev.* **109,** 744–757.

A CASE STUDY OF EDDY FORCING DURING AN ATLANTIC BLOCKING EPISODE

G. J. SHUTTS

Meteorological Office
Bracknell, Berkshire RG12 2SZ
England

1. INTRODUCTION

Observed blocking flow patterns are not steady in time and fluctuate with the passage of baroclinic wave systems causing a weakening, then reintensification, of blocking anticyclones (Palmén and Newton, 1969). This replacement/displacement phenomenon has also been noted in Southern Hemisphere blocking (Wright, 1974).

As well as amplitude changes, blocking patterns frequently exhibit an overall zonal translation during their lifetime — typically toward the west though not exclusively. This complicates the problem of splitting fields into mean and eddy components and is only partially alleviated by time filtering, which is used to isolate the effect of cyclone waves.

Green (1977) put forward the hypothesis that the shape and orientation of cyclone waves in the drought summer of 1976 (over western Europe) led to mean anticyclone vorticity forcing which could, with subsidence, maintain high pressure at the surface against friction. Austin (1980) described a simple quasi-geostrophic model with prescribed vorticity forcing which could explain the warmth and equivalent barotropic structure of blocking anticyclones in the troposphere if the forcing increased sharply with height.

Illari (1982, 1984) calculated the vorticity and quasi-geostrophic potential vorticity budget over Western Europe and the eastern Atlantic during July 1976 using National Meteorological Center (NMC) analyses and found eddy forcing terms large (in the sense that the implied spin-up time is much less than the duration of the block) and spatially organized so as to support the block. Savijärvi (1977) found no coherent pattern of eddy vorticity flux divergence in his study of a blocking episode. He used height field data to calculate geostrophic vorticity and velocity so that the eddy vorticity flux divergence involves many numerical differentiations and is likely to have large errors.

Hansen and Chen (1982) carried out a case study of the energy conversions and transfer between cyclone-scale and ultralong waves in a blocking situation. They found strong transfer of energy from baroclinic, cyclone-scale waves to barotropic, ultralong wave scales of motion, particularly in cases of Atlantic blocking. The initiation of a blocking episode by intense cyclogenesis was thought to represent such an energy transfer process.

Hansen and Sutera (1984) compared the spectral transfer of energy and enstrophy during blocking and nonblocking periods and found a striking difference in the enstrophy transfer. Ultralong waves (wavenumbers 1–3) were normally found to lose enstrophy to smaller scales. In blocking episodes, however, enstrophy was transferred upscale to the ultralong waves in conjunction with a greater rate of expulsion of enstrophy from intermediate-scale waves. Their picture of anomalous energy and enstrophy cascade is consistent with the eddy straining hypothesis outlined in the following section.

Mahlman (1980) studied the evolution of blocks in a general circulation model and noted the important part played by eddies in their maintenance. Air of low potential vorticity was observed to flow northward from the tropics and to be deposited on the western flank of the blocking anticyclone. The barotropic experiments of Shutts (1983) strongly support this Lagrangian picture and show it to be associated with the east–west scale compression of the eddy field immediately upstream of the block. It was hypothesized that this scale collapse or enhanced enstrophy cascade provides a mean anticyclonic/cyclonic dipole forcing pattern which strengthens the existing dipole circulation. Essential to the process is an irreversible folding and diffusion of potential vorticity contours paralleling the wave breaking mechanism discussed by McIntyre and Palmer (1983) in connection with sudden stratospheric warmings.

The aim of this article is to verify the preceding conceptual model of the interaction of eddies with blocking flow patterns.

2. The Eddy Straining Mechanism

A brief review of the energy and potential vorticity arguments which form the basis of the eddy straining hypothesis will be presented here.

It was suggested that traveling cyclone-scale eddies approaching a split-jetstream flow pattern suffer a greater rate of deformation than when in a more typical (less diffluent) flow field and that this provides an enhanced energy cascade (Kraichnan, 1967; Rhines, 1979) to larger scales (i.e., the block flow field).

Synoptically, depressions become narrow, slow-moving cold frontal troughs with meridional orientation. This ensures deep excursions of air between low and high latitudes and results in strong heat and vorticity transport. Subsequently the trough may split into two smaller scale systems which travel separately around the block in the northern and southern jetstream branches.

From a potential vorticity viewpoint, the east–west scale compression corresponds to an irreversible cascade of potential enstrophy to smaller scales and ultimately dissipation. Following Holland and Rhines (1980), the equation for eddy potential enstrophy is obtained by multiplying the equation for conservation of potential vorticity (q) by the deviation (q') from a time mean ($\bar{q}$) and averaging in time giving:

$$(\partial/\partial t)\tfrac{1}{2}\overline{q'^2} + \overline{q'\mathbf{V}'} \cdot \nabla\bar{q} + \overline{\mathbf{V}} \cdot \nabla\tfrac{1}{2}\overline{q'^2} + \overline{\mathbf{V}' \cdot \nabla\tfrac{1}{2}q'^2} = \overline{F'q'}$$

$$(\overline{}) = \text{time mean} \quad (1)$$

where q may represent the quasi-geostrophic potential vorticity on an isobaric surface or the Ertel potential vorticity on an isentropic surface, $\mathbf{V}$ is the nondivergent wind on the appropriate surface, and F' is a forcing term due to irreversible physical processes.

Following Marshall and Shutts (1981), if the mean potential vorticity $\bar{q}$ is approximately conserved along the mean streamfunction contours or $\bar{q} = \bar{q}(\psi)$, where ψ is the streamfunction and $\mathbf{V} = \mathbf{k} \wedge \nabla\psi$ where $\mathbf{k}$ is the unit vector pointing vertically upward, then it can be shown that Eq. (1) simplifies to

$$\tfrac{1}{2}(\partial/\partial t)\overline{q'^2} + (\overline{q'\mathbf{V}'})_* \cdot \nabla\bar{q} + \tfrac{1}{2}\overline{\mathbf{V}' \cdot \nabla q'^2} = \overline{F'q'} \quad (2)$$

where

$$(\overline{q'\mathbf{V}'})_* = \overline{q'\mathbf{V}'} - \mathbf{k} \wedge \nabla\left(\frac{1}{2}\overline{q'^2}\frac{d\bar{\psi}}{d\bar{q}}\right)$$

Note that $\text{Div}(\overline{q'\mathbf{V}'})_* = \text{Div}(\overline{q'\mathbf{V}'})$.

The tendency term in Eq. (2) may be neglected for a sufficiently long averaging period and the triple correlation term will be assumed small. These assumptions are all formally valid in the linear experiments described in Shutts (1983). Equation (2) is then simply

$$(\overline{q'\mathbf{V}'})_* \cdot \nabla\bar{q} = \overline{F'q'} \quad (3)$$

and the so-called residual potential vorticity flux $(\overline{q'\mathbf{V}'})_*$ is downgradient if F' is dissipative. Illari and Marshall (1983) have used this technique of subtracting out a rotational component to the vector q flux in an analysis of the 1976 European block. The enhanced enstrophy cascade upstream of the block promotes strong downgradient $(\overline{q'\mathbf{V}'})_*$, which in turn implies a north–south dipole of eddy q flux divergence in a sense such as to reinforce the anomalous potential vorticity field.

From the Lagrangian viewpoint this is manifest as narrow tongues of high or low q moving southward or northward, respectively, in association with meridionally oriented eddies.

When the meridional wavelength is long, the time spent by a parcel of air in any phase of the eddy is inversely proportional to its zonal wavelength according to the classical Rossby formula. Consequently, large meridional excursions with respect to the mean flow are possible just upstream of the block where the deformation rate of the velocity field is greatest.

Enhanced enstrophy cascade at cyclone scales and the support of the mean vorticity field by eddy vorticity transport are consistent with the observed spectral transfer of enstrophy during blocking episodes found by Hansen and Sutera (1984). This process was investigated in Shutts (1983) by integrating linear and nonlinear barotropic models in β-plane channel geometry with a wavemaker to provide a source of eddies. In the linear experiments eddies were forced in a basic state flow field which contained a split jetstream. The resulting long-term eddy vorticity fluxes, their divergence, and the second-order correction to the flow induced by the eddy vorticity flux divergence were all calculated. Anticyclonic forcing was found on the western side of the blocking high and extending around the northern side of the block in the jet flow. The second-order induced flow clearly showed that the eddies tend to reinforce the basic state block pattern.

The nonlinear barotropic vorticity equation was integrated with a similar wavemaker from an initial state of uniform westerly flow. The evolution of the flow tended toward a statistically steady state due to the inclusion of an Ekman friction term. For sufficiently weak initial uniform flow, perpetual blocking ensued downstream of the wavemaker (Fig. 1). Blocking was due entirely to the eddy vorticity flux divergence pattern associated with the wavemaker-forced eddies since no other time-mean forcing terms exist in the nonlinear barotropic vorticity equations integrated (e.g., orographic terms). Residual vorticity fluxes $\overline{(q'\mathbf{V}')}_*$ are large and down the absolute

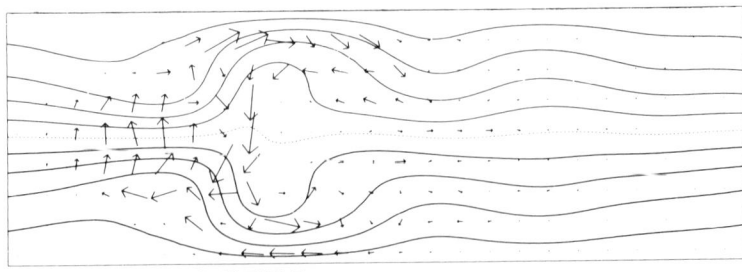

BASIC STATE HEIGHT FIELD

FIG. 1. Mean streamfunction field with residual eddy vorticity flux vectors; [Figure 6c of Shutts (1983)].

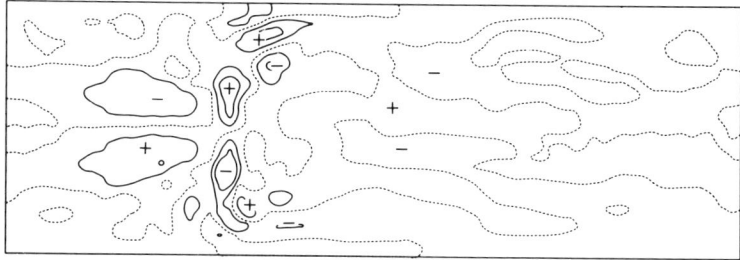

FIG. 2. Time-mean eddy vorticity flux divergence. Contour interval: 5×10^{-11} sec^{-2}. [Figure 6d of Shutts (1983).]

vorticity gradient at the point of jet splitting, as hypothesized. The accompanying eddy vorticity flux divergence field has a strong meridional dipole on the western side of the block in the sense required to strengthen the blocking pattern (Fig. 2).

A trajectory analysis revealed that low absolute vorticity air was being intermittently swept northward ahead of approaching troughs and into the center of the mean anticyclone. The circulation comprising the southern part of the blocking pattern was maintained in a similar way by the injection of high vorticity from the north. Confirmation that these eddy vorticity transfer processes occur in reality was sought from similar diagnostic analyses of real data.

3. Data Manipulation and the Synoptic Situation

Diagnostic quantities were calculated from global initialized analyses from the European Centre for Medium Range Weather Forecasts—for the period 5–22 February 1983 inclusive available at 6-hr intervals—the period of the forecast–analysis cycle. [For details of the data assimilation scheme see Lorenc (1981) and Lönnberg and Shaw (1983).] Analyzed data are stored as packed spherical harmonic coefficients with triangular truncation T80 (i.e., all harmonics of order 80 and less). For the purpose of this case study the fields are extracted on a 1.875° latitude-longitude grid and available at 1000, 850, 700, 500, 400, 300, 250, 200, 150, 100, 70, 50, and 30 mbar.

Vorticity is obtainable directly from the archive so that no loss of information need result through the use of finite difference expressions for the curl of the vector wind field. However, the isentropic vorticity required for the calculation of Ertel potential vorticity is derived by the latter method. Unlike other studies which use twice-daily datasets (e.g., Savijärvi, 1977; Illari, 1984), here we sample four times daily. Although most upper air data is reported at 00Z and 12Z, much of the small-scale information in an analysis comes from the 6-hr forecast (first guess field) and it seemed worthwhile to include the 06Z and 18Z analyses, particularly in view of the high-frequency contributions made to the vorticity flux by frontal systems.

Time filtering is used to emphasize the contribution made by traveling weather systems with time periods of 6 days or less. To this end, a high-pass filtered flux $(\overline{S'V'})_{HP}$ is defined such that

$$(\overline{S'V'})_{HP} = \frac{1}{N-12} \sum_{i=7}^{N-6} S'_i V'_i$$

$$\begin{pmatrix} S'_i \\ V'_i \end{pmatrix} = \begin{pmatrix} S_i \\ V_i \end{pmatrix} - \frac{1}{13} \sum_{j=i-6}^{j=i+6} \begin{pmatrix} S_j \\ V_j \end{pmatrix}$$

where S_i denotes the ith 6-hr values of some physical quantity S out of a sequence of N analysis values. Although this filter exhibits some undesirable features such as oscillation of the root-mean-square frequency response about unity, it does attenuate low frequencies effectively (e.g., the root-mean-square response at periods of 6 days is ≈ 0.4). This is particularly desirable for a short time series of data known to be dominated by low-frequency fluctuations (Blackmon et al., 1977).

Spatial smoothing is sometimes desirable since many of the fields of interest have pronounced small-scale structure. The eddy vorticity flux divergence is such a field for which smoothing helps to bring out its contribution to forcing large-scale circulation anomalies. A smoothing scheme similar to those suggested by Sardeshmukh and Hoskins (1984) is used. Global field values are decomposed into a sum of orthogonal spherical harmonics with triangular truncation T80. Coefficients in the expansion with $n > n_1$, are multiplied by $\exp\{-[(n-n_1)/(n_2-n_1)]^2\}$ where n is the degree of the associated Legendre function and n_1 and n_2 are constants taken here to be 20 and 25, respectively. Sardeshmukh and Hoskins show that such a class of filters, which depend only on the total wavenumber of each mode through n, are equivalent to an isotropic spatial filter in the spherical domain. It is tacitly assumed that isotropic smoothing of a two-dimensional field is the most appropriate one for the problem at hand.

The blocking episode began with the building of a mid-Atlantic ridge (30°W) on 5 February and a sympathetic cyclonic development between Norway and Scotland. As the ridge extended northward to Iceland on 6 February, the cold trough likewise extended southward into Holland, reaching Italy by 7 February accompanied by an upper cold cut-off vortex. Further cyclonic activity between Greenland and Norway on 9 and 10 February led to the establishment of a strong northerly airstream over the British Isles with a strong (1042 mbar) surface anticyclone at 55°N 25°W. A vigorous depression, which began forming off the eastern seaboard of the United States on 12 February, had developed to 987 mbar off Newfoundland by 12Z, 13 February. As the storm reached its maximum intensity (975 mbar) south of Greenland by 12Z, 14 February, a meridional elongation of the associated upper trough began. Simultaneously, the sea level pressure of the blocking anticyclone between Scotland and Southern Norway began to rise (1023 mbar at 12Z, 13 February to 1030 mbar 24 hr later). By 12Z, 15 February, the depression was centered near southern Greenland with a central pressure of 992 mbar and with a trough extending to a separate low center near 40°N 30°W (see Fig. 7). At that time the blocking anticyclone was centered north of Scotland with a central pressure of 1037 mbar but

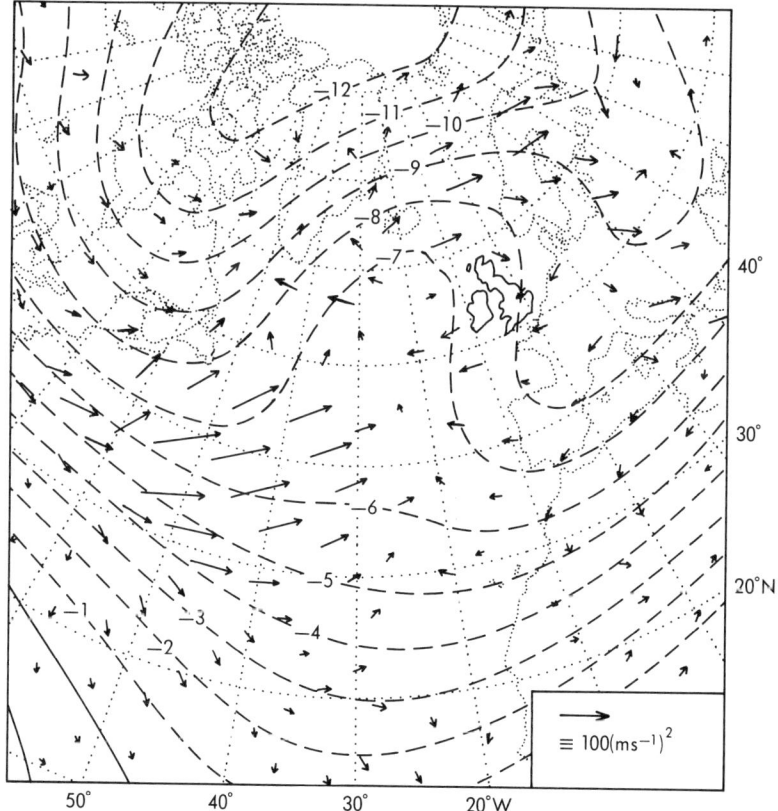

FIG. 3. High-pass filtered E vectors superimposed on the mean streamfunction field (centered at 30°W).

continued to rise to 1043 mbar by 17 February. A further depression developed off the eastern seaboard of the United States on 16 February and became slow moving in the central Atlantic. The situation remained rather static until 22 February, by which time the surface anticyclone had moved into central Europe and cyclonic activity dominated the North Atlantic.

The time-mean block appears as strong mid-Atlantic ridge tilting northeastward with a trough extending southward into Europe and thence westward to Iberia. A weak broad trough also exists to the south of the main ridge at 30° W, forming the weaker cyclonic component of the block dipole (Fig. 3).

4. E Vectors and the Sense of Momentum Forcing

The usual concept of eddy momentum forcing as the tensor divergence of Reynolds stresses arises from splitting the velocity field in the Eulerian form of the momentum equations into mean and eddy parts and then averaging. For large-scale atmospheric flow the covariance statistics involving u and v, the zonal and meridional components of velocity, respectively, are the dominant Reynolds stresses.

Traditionally, the horizontal transport of momentum has been represented in the zonal-mean sense by $\overline{u'v'}$.

Only recently has the problem of the local forcing of large-scale circulation by eddies been considered. Eddies are then conventionally defined as the deviation from a time mean so that momentum forcing becomes a vector quantity defined at each point in space. Finding a vector quantity to generalize the notion of $\overline{u'v'}$ as a momentum flux has turned out to be intimately related to concepts of group velocity, wave action flux, and wave/mean flow interaction. Hoskins *et al.* (1983) introduced the quasi-vector $\mathbf{E} = (\overline{v'^2} - \overline{u'^2}, -\overline{u'v'})$ appropriate to Cartesian geometry whose divergence is a measure of the time-mean forcing of westerly momentum exerted by the eddy field and whose direction is related to the direction of the group velocity vector for Rossby waves. If $\mathbf{E}$ makes a small angle with the x-axis (in Cartesian β-plane geometry) then this angle is one-half of that made between the group velocity vector (relative to the mean flow) and the x-axis. Since observationally, the high-pass filtered velocity covariance statistics typically satisfy the inequality $\overline{v'^2} > \overline{u'^2} > |\overline{u'v'}|$, the corresponding $\mathbf{E}$ vectors tend to point eastward along the storm tracks. $\mathbf{E}$ can be regarded as an effective flux of *westward* momentum since convergence regions of $\mathbf{E}$ experience a net force from east to west.

It is helpful at this point to recall the theoretical motivation for E. In spherical geometry, the equations governing the horizontal motion field in pressure coordinates are

$$\frac{\partial u}{\partial t} + \frac{u}{a\cos\theta}\frac{\partial u}{\partial \lambda} + \frac{v}{a}\frac{\partial u}{\partial \theta} + \omega\frac{\partial u}{\partial p} - \frac{uv\tan\theta}{a}$$
$$- 2\Omega\sin\theta\cdot v + \frac{1}{a\cos\theta}\frac{\partial \Phi}{\partial \lambda} = 0 \quad (4)$$

$$\frac{\partial v}{\partial t} + \frac{u}{a\cos\theta}\frac{\partial v}{\partial \lambda} + \frac{v}{a}\frac{\partial v}{\partial \theta} + \omega\frac{\partial v}{\partial p} + \frac{u^2\tan\theta}{a}$$
$$+ 2\Omega\sin\theta\cdot u + \frac{1}{a}\frac{\partial \Phi}{\partial \theta} = 0 \quad (5)$$

$$\frac{1}{a\cos\theta}\left(\frac{\partial u}{\partial \lambda} + \frac{\partial(v\cos\theta)}{\partial \theta}\right) + \frac{\partial \omega}{\partial p} = 0 \quad (6)$$

where Φ is the geopotential, θ is latitude, λ is longitude, ω is the pseudovertical velocity of pressure coordinates, and a and Ω are the radius and angular speed of rotation of the earth. It can be shown that the time average of Eqs. (4) and (5) give

$$\overline{\mathbf{V}}_H \cdot \nabla \overline{u} + \overline{\omega}\frac{\partial \overline{u}}{\partial p} - \frac{\overline{uv}\tan\theta}{a} - 2\Omega\sin\theta\cdot \overline{v} + \frac{1}{a\cos\theta}\frac{\partial \overline{\Phi}}{\partial \lambda}$$
$$+ \frac{1}{a\cos^2\theta}\left[\frac{\partial \overline{u'^2}\cos\theta}{\partial \lambda} + \frac{\partial \overline{u'v'}\cos^2\theta}{\partial \theta}\right] + \frac{\partial \overline{u'\omega'}}{\partial p} = 0 \quad (7)$$

and

$$\overline{\mathbf{V}}_H \cdot \nabla \overline{v} + \overline{\omega}\frac{\partial \overline{v}}{\partial p} + \frac{\overline{u}^2\tan\theta}{a} + 2\Omega\sin\theta\cdot \overline{u} + \mathbf{\frac{1}{a}\frac{\partial \overline{\Phi}}{\partial \theta}} + \frac{1}{a\cos\theta}\frac{\partial \overline{u'v'}}{\partial \lambda}$$
$$+ \mathbf{\frac{1}{a}\frac{\partial \overline{v'^2}}{\partial \theta}} + \frac{\partial \overline{v'\omega'}}{\partial p} + (\overline{u'^2} - \overline{v'^2})\frac{\tan\theta}{a} = 0 \quad (8)$$

where $\mathbf{V}_H = \overline{u}\mathbf{i} + \overline{v}\mathbf{j}$.

Hoskins *et al.* (1983) suggest that the boldface terms in Eq. (8) are in approximate balance (based on observational analysis), the eddy forcing term $a^{-1}(\partial \overline{v'^2}/\partial \theta)$ causing a slight departure from geostrophic balance. Although it is a small term compared with the pressure gradient term, its influence cannot be neglected since unlike the latter it is an important source of mean vorticity (though not a *net* source, of course).

We can replace this force in the **j** direction by a conservative force plus a force in the **i** direction as follows:

$$\frac{1}{a}\frac{\partial \overline{v'^2}}{\partial \theta}\mathbf{j} = \nabla_H \overline{v'^2} - \frac{1}{a\cos\theta}\frac{\partial \overline{v'^2}}{\partial \lambda}\mathbf{i}$$

where

$$\nabla_H = \frac{\mathbf{i}}{a\cos\theta}\cdot\frac{\partial}{\partial\lambda} + \frac{\mathbf{j}}{a}\frac{\partial}{\partial\theta}$$

The conservative force may be combined with the pressure gradient term so that Eqs. (7) and (8) may be reexpressed as

$$\overline{\mathbf{V}}_H \cdot \nabla \overline{u} + \overline{\omega}\frac{\partial \overline{u}}{\partial p} - \frac{\overline{uv}\tan\theta}{a} - 2\Omega\sin\theta\cdot\overline{v} - \frac{1}{a\cos^2\theta}\left[\frac{\partial}{\partial\lambda}(\overline{v'^2} - \overline{u'^2})\right.$$

$$\left.\times\cos\theta - \frac{\partial}{\partial\theta}(\overline{u'v'}\cos^2\theta)\right] + \frac{\partial}{\partial p}\overline{u'\omega'} + \frac{1}{a\cos\theta}\cdot\frac{\partial}{\partial\lambda}(\overline{\Phi} + \frac{1}{2}\overline{v'^2}) = 0$$
(9)

and

$$2\Omega\sin\theta\cdot\overline{u} + \frac{1}{a}\frac{\partial}{\partial\theta}(\overline{\Phi} + \frac{1}{2}\overline{v'^2}) \doteq 0 \qquad (10)$$
(approximated)

In effect, the important eddy forcing terms capable of generating mean vorticity are "condensed" into a term representing a force in the east–west direction. Conservative eddy forces, like the pressure gradient force, serve only to balance the time-mean mass and wind fields.

The eddy forcing of zonal momentum as given by Eq. (9) can be written as

$$-(\text{Div }\mathbf{E} + \partial\overline{u'\omega'}/\partial p)$$

with $\mathbf{E} = [(\overline{v'^2} - \overline{u'^2})\cos\theta, -\overline{u'v'}\cos\theta]$ in spherical polar geometry.

Hoskins *et al.* (1983) found that, typically, the high-pass filtered eddy field is elongated in the meridional direction corresponding to $\overline{v'^2} > \overline{u'^2}$ so that **E** points eastward. "Bowing" of trough lines about the jetstream implies a "fanning out" of the **E** vectors in the storm tracks with an equatorward bias reflecting the dominance of poleward momentum transport in the zonal mean.

Figure 3 shows the distribution of high-pass filtered **E** (using the same filter as Hoskins *et al.*, 1983) at 300 mbar for the period 5–22 February. Two regions of strong eddy momentum forcing of the block are evident.

Strong convergence of **E** into the jetstream split region implies deceleration of westerlies there of the order of 10 msec^{-1}/day while in the northern jetstream branch, divergence of **E** implies acceleration of the flow. The former region on the western side of the block agrees with the vorticity forcing picture described earlier in which eddies propagating into a split jetstream become extended meridionally, thereby tending to force an anticyclonic/cyclonic dipole flow field. In contrast, the divergent **E** vectors at 65°N (near the Greenwich Meridian) are associated with baroclinic development and have no counterpart in the aforementioned numerical experiments.

The formation of depressions near Iceland and their subsequent intensification over Scandinavia are quite commonly associated with Atlantic blocking ridges leading to strengthening of northerlies or northeasterlies over western Europe. Also evident in Fig. 3 are westward-pointing **E** vectors similar to those found by Hoskins *et al.* (1983) in a case study of Atlantic blocking though of weaker intensity.

5. Eddy Vorticity Flux Divergence Patterns

Perhaps the most direct approach to quantifying the forcing effect of eddies due to momentum transfer is to calculate the eddy vorticity flux divergence and infer its effect from the mean vorticity equation. Large conservative forces disappear on taking the curl of the momentum equation, though this is only at the expense of creating a more highly differentiated (and therefore spatially detailed) diagnostic quantity. The point-by-point variation of eddy vorticity flux divergence is of minor importance compared to the overall pattern which determines the response in the large-scale pressure field. Indeed, point values are likely to be inaccurate in view of the numerical differentiation involved.

Figure 4a shows the eddy vorticity flux divergence at 300 mbar, for the period with no spatial smoothing and high-pass filtered in time using the method defined in Section 3. In spite of the rather complex pattern, individual features are quite well resolved since the horizontal spacing of gridpoints is only 1.875°. Anticyclonic forcing maxima appear at 50°N 40°W just to the north of the mean jetstream split, in the mean anticyclone near 20°W and over Scandinavia. Cyclonic forcing exists over much of Europe and to some extent south of blocking anticyclone ridge in the Atlantic. Upstream of the block the eddy vorticity flux divergence takes on a banded appearance with some evidence of antisymmetry in the pattern about the jetstream axis.

Spatial smoothing makes a tangible improvement to the ease of interpretation of the eddy vorticity flux divergence fields. The smoothed version of Fig. 4a (Fig. 4b) shows the dramatic change in eddy vorticity forcing characteristics from the storm track pattern off the eastern seaboard of the United States with a band of cyclonic forcing sandwiched between two bands of anticyclonic forcing, to that in the blocking region with anticyclonic forcing

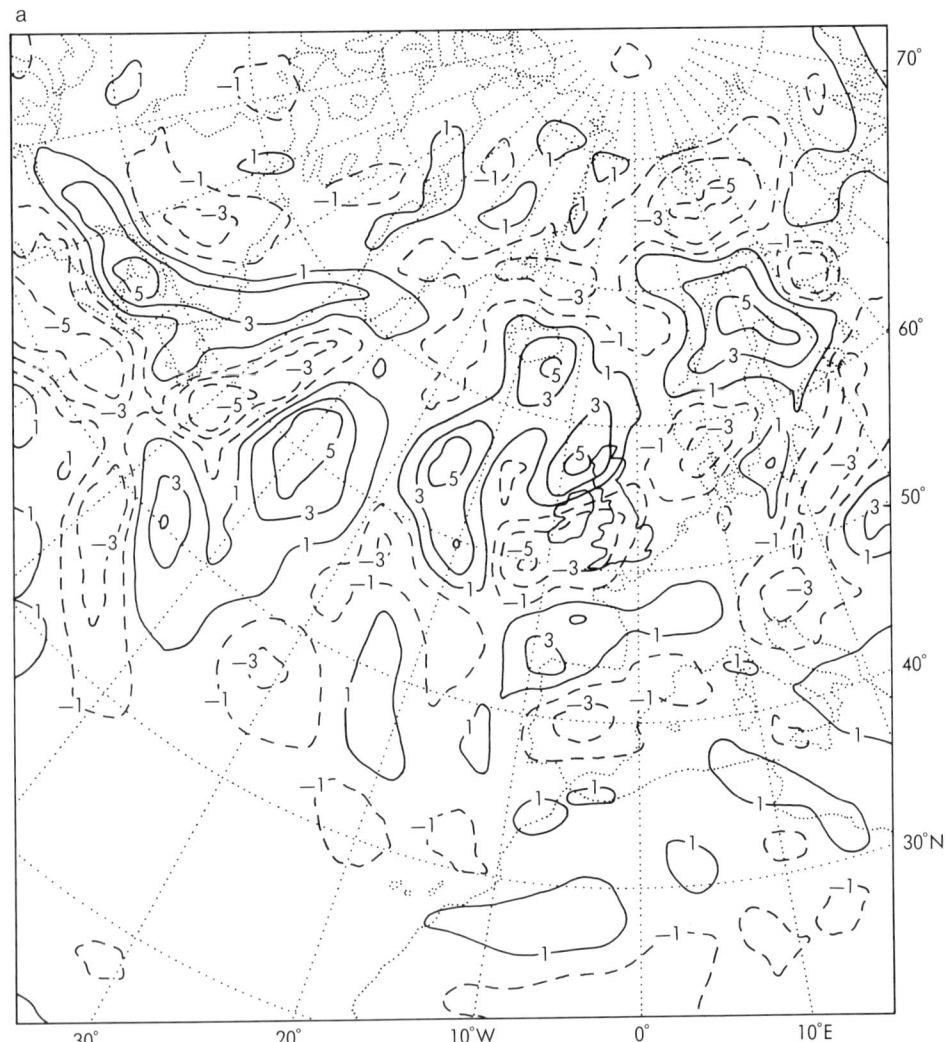

FIG. 4. (a) Unsmoothed eddy vorticity flux divergence at 300 mbar (high-pass filtered in time) for the North Atlantic and Europe. Contour interval: 2×10^{-10} sec^{-2}. (b) Smoothed version of (a) except for a larger area (100°W 100°E). Contour interval: 1×10^{-10} sec^{-2}.

extending north–east to south–west following the axis of the block ridge. With magnitudes of vorticity forcing of -2×10^{-10} sec^{-2}, eddies would be capable of spinning up the observed mean vorticity anomaly in about 2 days, in the absence of other physical processes represented in the vorticity equation.

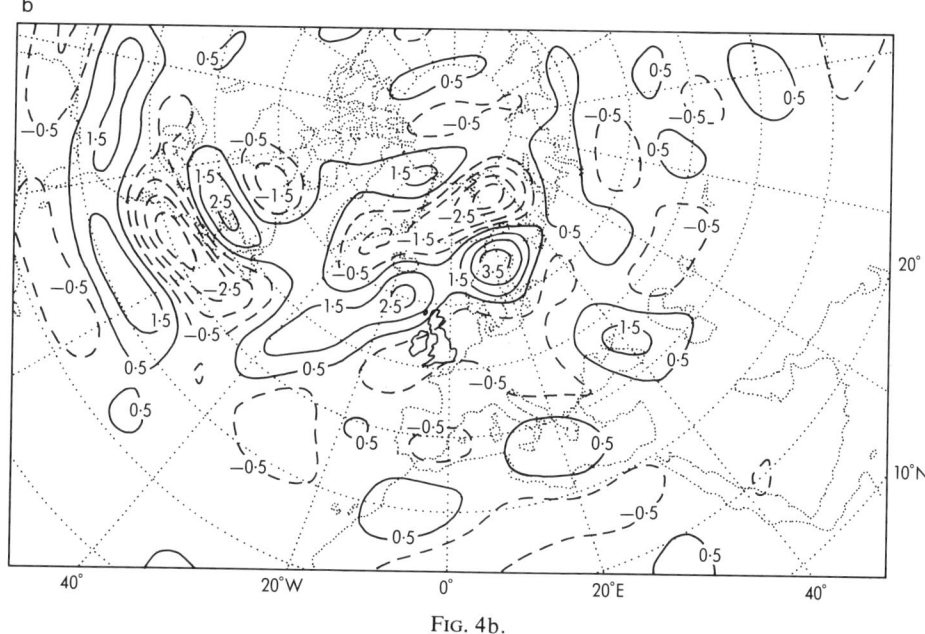

FIG. 4b.

Although it would be rather audacious to claim any great likeness between these patterns and those simulated in the barotropic model discussed earlier, there are some definite points of agreement. For instance, the normal sense of dipole eddy vorticity forcing [eddy vorticity flux convergence (divergence) to the north (south)] is clearly visible centered near 38°N 65°W (Fig. 4b) on the axis of the 300 mbar jetstream, but ends abruptly at the jetstream split to be replaced by a reversed dipole pattern. Another strong "normal" dipole of eddy vorticity forcing lies in the northern jetstream branch of the block between Iceland and Scandinavia, consistent with the **E** vector pattern there (Fig. 3). As a consequence, anticyclonic eddy vorticity forcing exists throughout most of the mean anticyclonic ridge at 300 mbar.

6. Ertel Potential Vorticity Analysis

In order to obtain a Lagrangian perspective on the maintenance of blocking anticyclones by eddies, a time sequence of Ertel potential vorticity Q maps were produced. Ertel potential vorticity is a conserved quantity in adiabatic, inviscid flow so that by studying maps of Q in isentropic surfaces it is possible to understand much about air movement. This approach has in the past provided considerable insight into the three-dimensional nature of cyclone-wave development, frontogenesis, and the associated exchange of chemical tracers between troposphere and stratosphere (Reed, 1955; Danielsen, 1968). Recently McIntyre and Palmer (1983) have advocated isentropic analysis of Q to reveal Rossby wave breaking processes in the upper subtropical troposphere, similar to those found in sudden warming events in the stratosphere. The eddy enstrophy cascade mechanism discussed in Section 2 can also be regarded as a wave breaking process in which the east-west scale of eddies collapses and Q is irreversibly mixed.

In an exhaustive review of the concept of Ertel potential vorticity, Hoskins *et al.* (1985) show that this single quantity and its behavior in isentropic surfaces provide a physical basis, at least conceptually, for unifying our understanding of most large-scale dynamical phenomena in meteorology (e.g., baroclinic instability, Rossby wave propagation, and frontogenesis) when a suitably defined notion of balance exists. They argue that isentropic Q maps are the key to sharpening our dynamical insight into the full-three dimensional structure of real atmospheric motion systems and present examples of the formation and maintenance of cut-off lows and blocking anticyclones using such an analysis technique. On a more cautionary note, if we accept that systems of equations incorporating a level of balance higher than quasi-geostrophy permit discontinuous behavior, then the potential vorticity approach may be undermined. In these cases the Lagrangian conservation of more fundamental physical quantities such as gas entropy and momentum are required (Cullen, 1983; Cullen and Purser, 1984). These objections are unlikely, however, to detract from the usefulness of isentropic Q maps for observational purposes.

The isentropic coordinate expression for Ertel potential vorticity adopted here is

$$Q = -(\zeta_\theta + f) \cdot (1/\theta)(\partial\theta/\partial p)$$

where ζ_θ is the relative vorticity of the horizontal wind vector and $(1/\theta)\partial\theta/\partial p)$ is the static stability both measured on a chosen isentropic surface. Temperature fields at 850, 700, 500, 400, 300, 250, 200, and 150 mbar are converted to potential temperature and interpolated using a cubic spline fitting procedure to find p_*, the pressure corresponding to the level of the isentropic surface θ_*. The vertical profile of potential temperature for each gridpoint is then represented by a fifth-order polynomial in pressure and differentiated for the static stability at p_*. The wind components are interpolated to the isentropic surface and ζ_θ is found by the standard centered-difference expression for curl.

Vertical interpolation for p_* is sensitive to the nature of the θ profile, and the cubic splines routine often fails if an isentropic layer is detected. In these cases, linear interpolation is used to find p_*. Unfortunately, isentropic layers are quite common beneath the tropopause and near the surface over subtropical deserts. Superadiabatic layers are sometimes found in the European Centre for Medium Range Weather Forecasts archived temperature data and arise from the interpolation of model fields from sigma to pressure levels (A. Hollingsworth, 1983; personal communication). Isentropic and superadiabatic layers were adjusted to make $(-\partial\theta/\partial p)$ small and positive so as to provide a monotonically increasing sequence of θ values for the interpolation routine.

The 320 K isentropic surface was selected since it is one of the lowest surfaces which remain above the 1000 mbar level (except perhaps in subtropical desert zones). Since the 320 K surface usually forms part of the polar front, it is an active region of meridional air mass exchange and is ideally located for the study of eddy processes at the level of maximum amplitude of

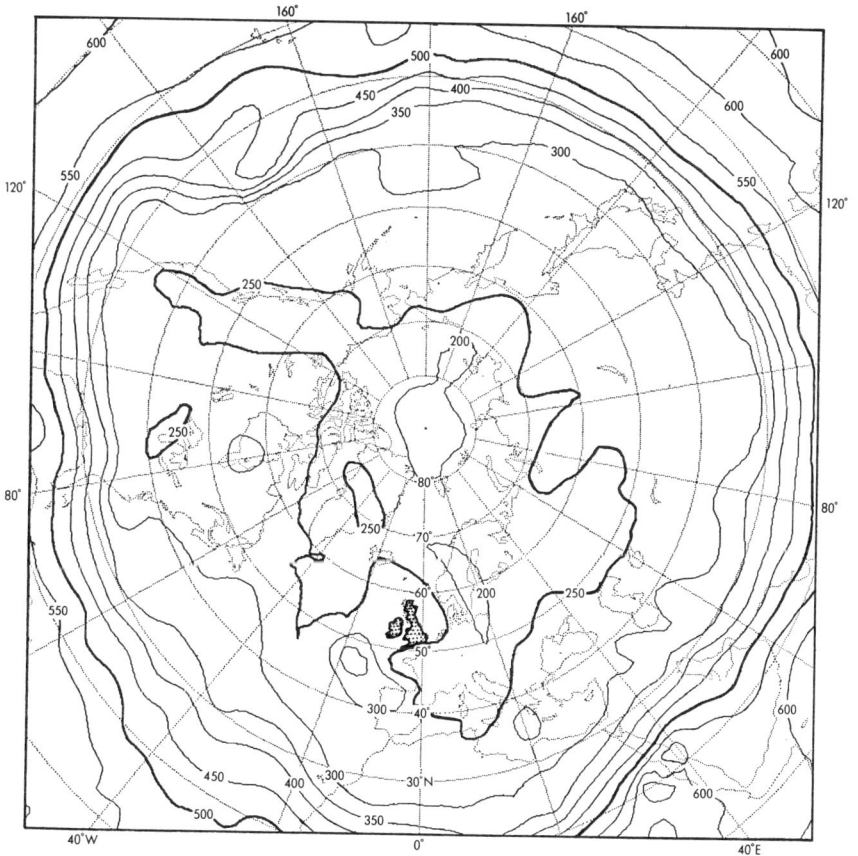

FIG. 5. Isobars (millibars) on the 320 K isentropic surface for 5 February, 12Z.

the block (about 300 mbar). On average it slopes from 200 mbar over the pole to 600 mbar in the tropics. Figure 5 shows isobars of the 320 K surface on the first day of the period. In middle and high latitudes the 320 K surface is sufficiently high for nonconservative effects associated with latent heat release in condensation of water vapor to be unimportant.

Gradients of Q become very large where the surface intersects the tropopause and enters the lower stratosphere. For the purposes of display, the fourth root of Q is plotted so as to give more emphasis to gradients in the subtropics and middle latitudes. It goes without saying that any function of Q is conserved following the motion of a parcel if Q is conserved. The rather arbitrary choice of this function here highlights the possibility that there

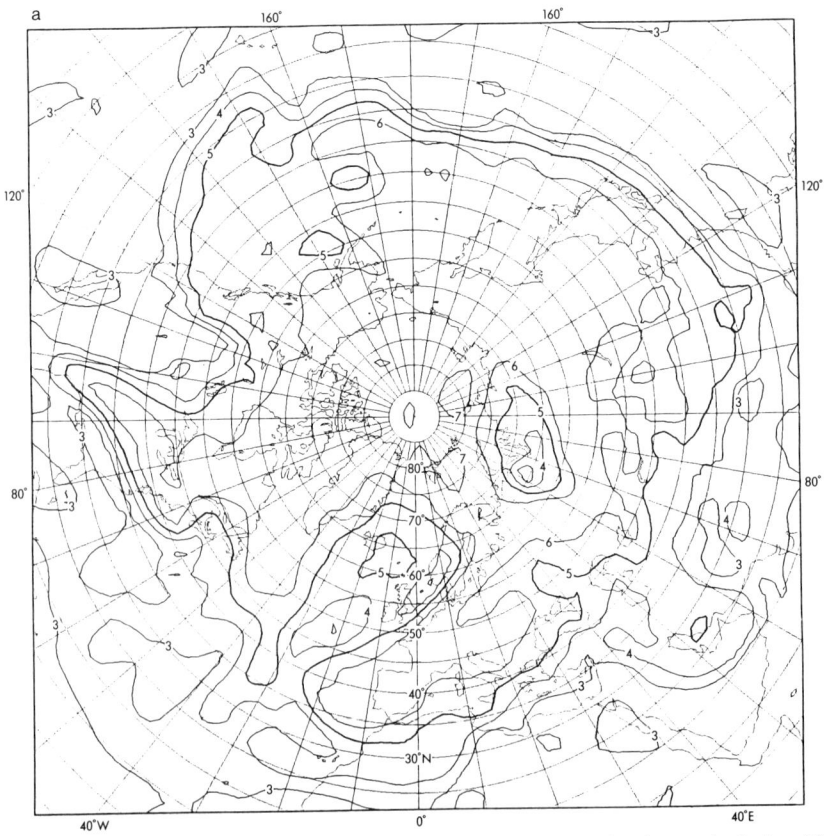

FIG. 6. (a)–(e) Daily sequence of Q maps for the period 12–16 February inclusive. The plotted contours are actually the fourth root of the Ertel potential vorticity as defined in the text. Using MKS units, the contour interval is 1×10^{-3} and the bold contour represents 5×10^{-3}.

exists some optimal functional form. B. J. Hoskins (1983; personal communication) has suggested that Q^{-1} might be a useful choice since the diabatic source term in the isentropic equation for Q^{-1} can be written in a flux form, $(\partial/\partial p)(H/Q)$, where H is the diabatic heating. For the period 5–22 February Q maps have been produced four times daily, though only five are reproduced here. A twice-daily selection is to be collated as a Meteorological Office Technical Note available on request.

Figures 6a–6e are a sequence of maps of $Q^{1/4}$ (hereafter we shall refer to $Q^{1/4}$ as Q) at 12Z for 12–16 February inclusive during which time a major injection of low Q into the block occurs.

The first Q map shows the blocking anticyclone as a region of low Q to the west of the British Isles and extending northward to Iceland. Lowest Q values occur in a band along the eastern flank of the region from southern Norway

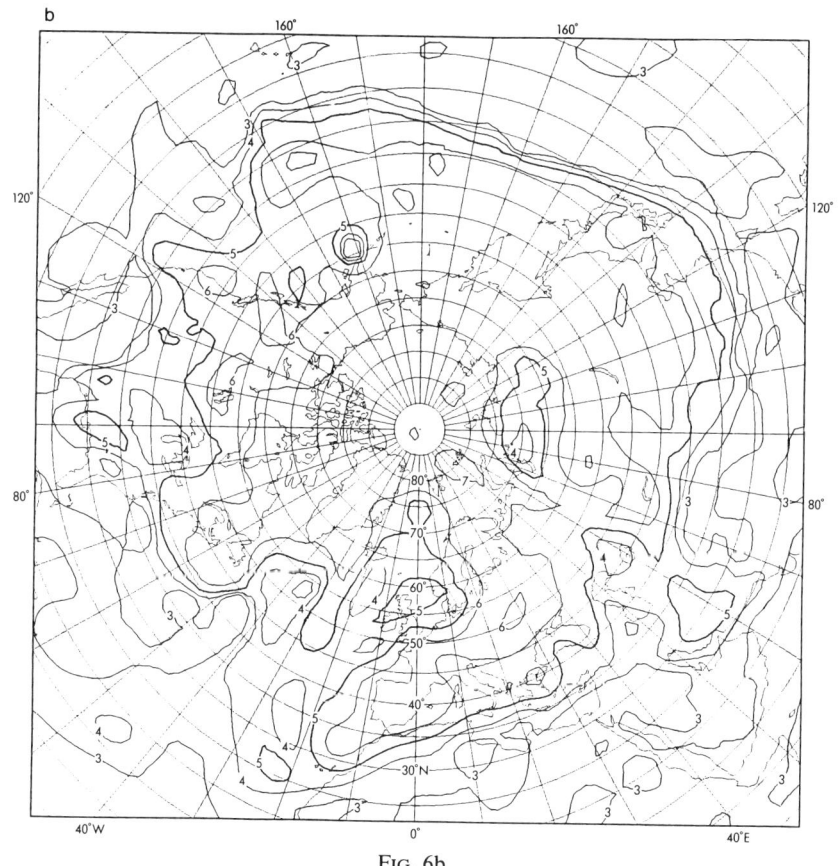

FIG. 6b.

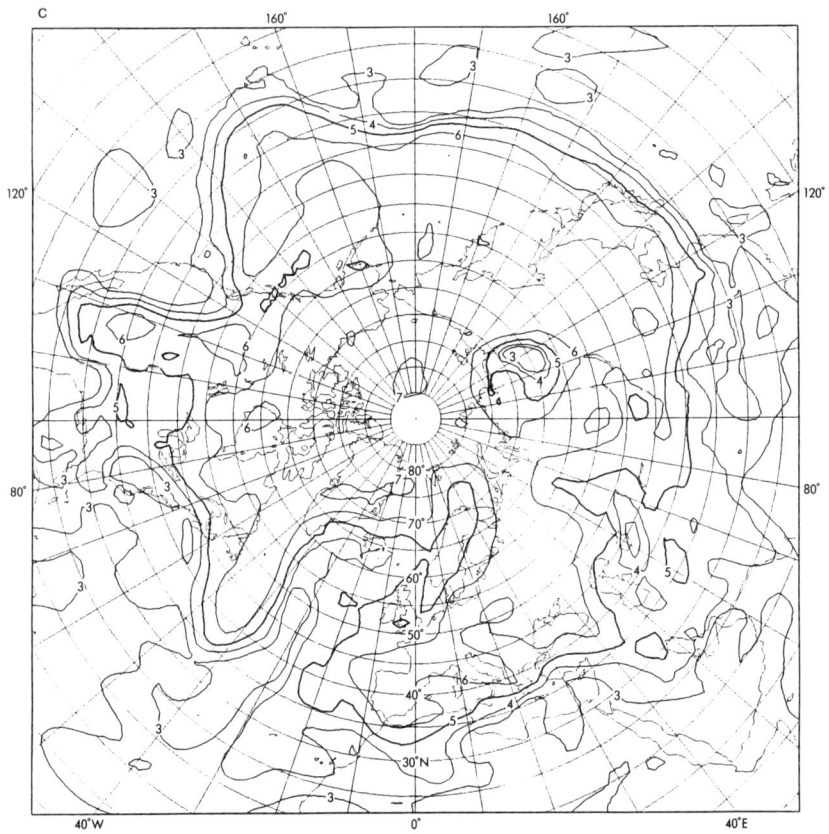

Fig. 6c. See legend on p. 150.

to northern England and near 48°N 22°W. The former air mass can be traced back to the subtropics several days earlier and was transported northeastward ahead of a developing depression. High Q air forming the southern portion of the block dipole extends across Europe to the west of Portugal. A cut-off anticyclone (low Q) at 65°N 60°E originally broke off the block near Norway on 9 February and drifted eastward with little change in intensity, rather like a Gulf Stream ring. The tongue of high Q lying northwest–southeast ahead of the block is swept northeastward and squeezed into a narrow feature at 25°W on 13 February (Fig. 6b). At the same time, tightening Q gradients near 45°N 50°W are associated with the northward advance of low Q air ahead of a newly developing weather system. By 14 February the high Q tongue at 25°W has disappeared and the developing system has

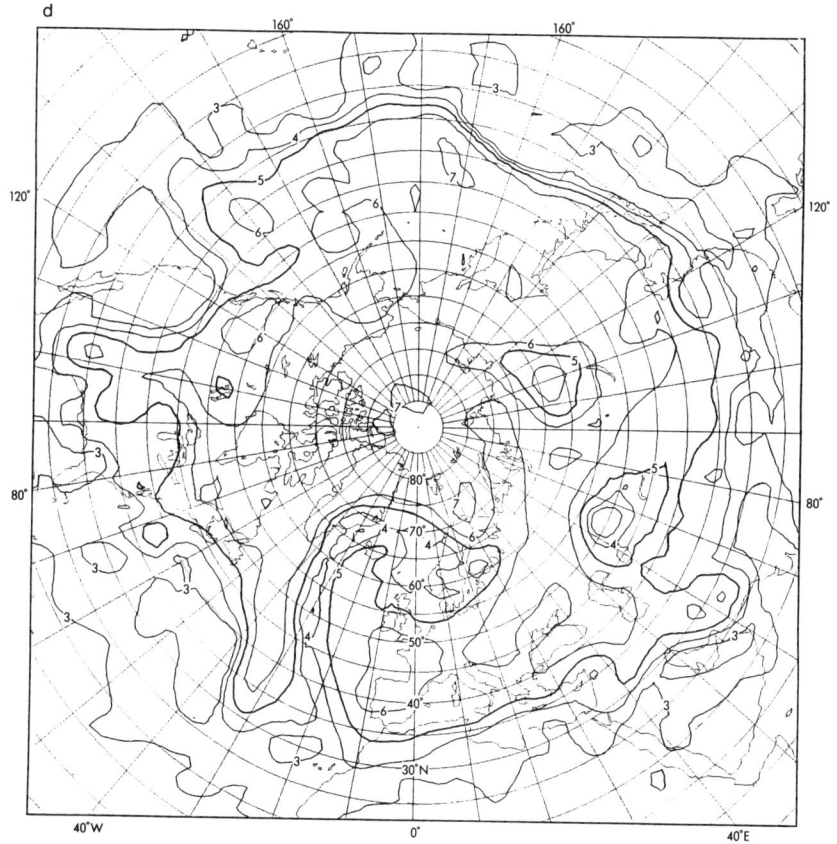

Fig. 6d. See legend on p. 150.

successfully engaged the subtropical air with a tongue of low Q extending from the central North Atlantic to Iceland.

From the point of view of the sea-level pressure and 500-mbar height fields, 15 February was the climax of the blocking event with a vigorous north–south dipole pattern (Fig. 7). A rather beautiful Ertel potential vorticity field (Fig. 6d) accompanies this pressure pattern, with the narrow strip of low Q now extending northward from the Canary Islands to Greenland, then curling eastward into the anticyclone. An adjacent tongue of high Q to the west slides southward to 30°N and forms a cut-off on 16 February. By this time, the low Q strip has been swept around into the blocking anticyclone to the north of the British Isles and the intensified anticyclonic circulation associated with this advects high Q air over France and Spain north-

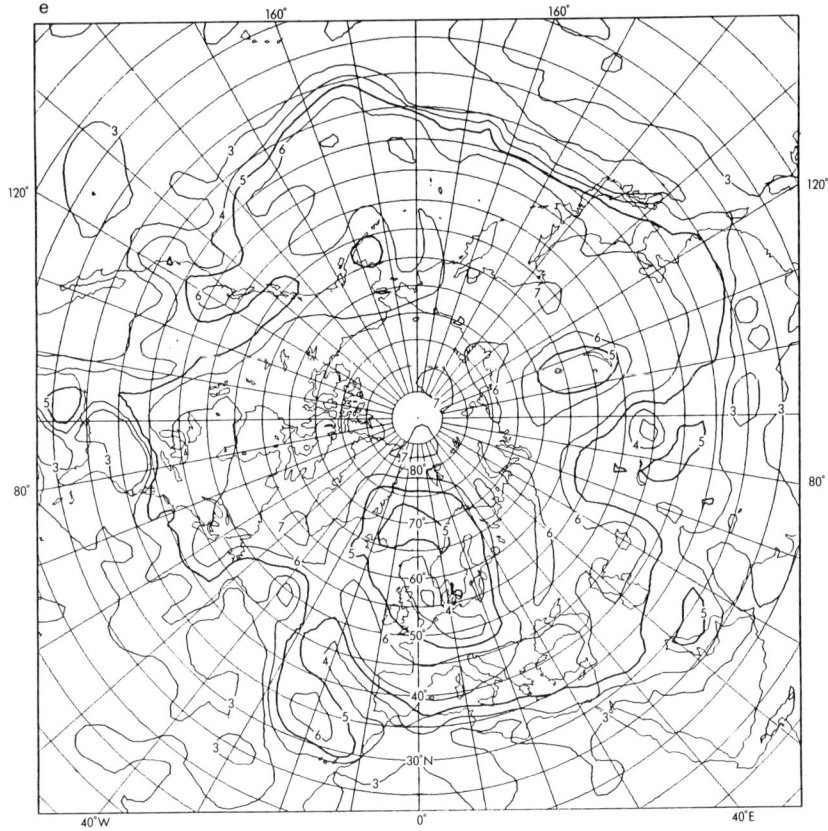

FIG. 6e. See legend on p. 150.

westward. (This is evident from plots of the isentropic wind vectors that are not shown here.)

The Q pattern on 17, 18, and 19 February becomes more complicated, with the low Q air (forming the blocking anticyclone) drifting eastward across northern Europe by 19 February and high Q over the eastern seaboard of the United States, pushing east across the Atlantic and linking up with that over southern Europe. In this way the intense Q gradient normally found at 30° to 40°N reestablishes across the Atlantic and a more cyclonic pattern ensues.

The Q maps confirm the hypothesized role of eddies (Mahlman, 1980; Shutts, 1983) whereby approaching weather systems carry subtropical air of low potential vorticity northward and inject it into the circulation of blocking anticyclones. To a lesser extent, high Q is drawn southward and cut off to

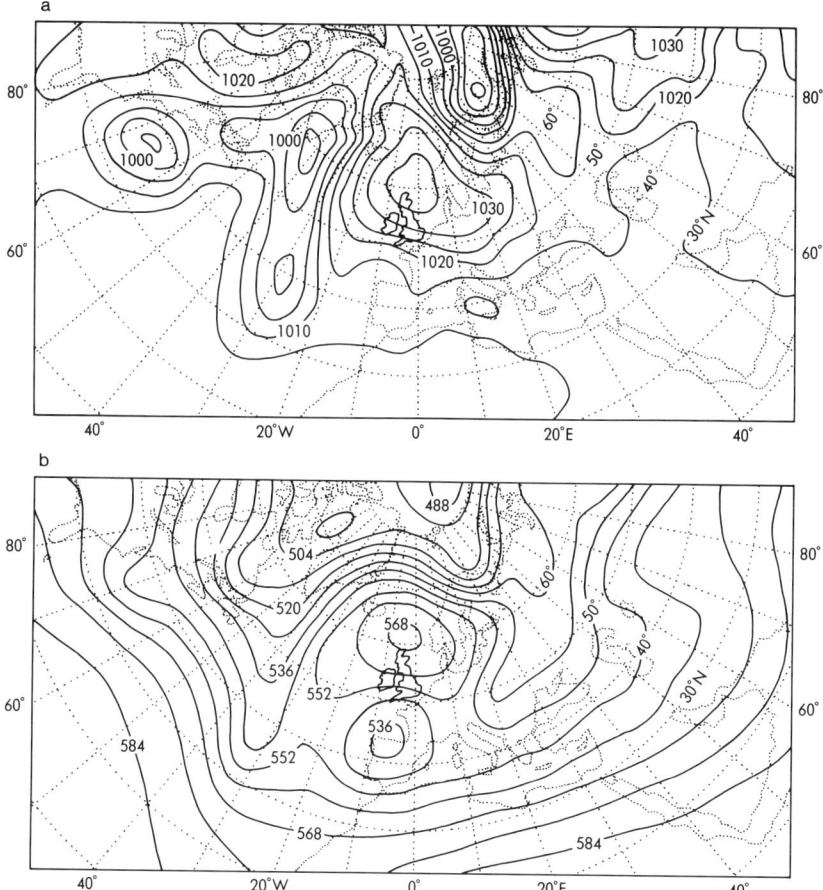

FIG. 7. (a) Mean sea level pressure field for 15 Feburary, 12Z. Contour interval: 5 mbar. (b) Height contours of the 500-mbar surface for 15 February, 12Z. Contour interval: 8 dam.

the south of the blocking anticyclone, where it dissipates quite rapidly. As a general rule, cut-off blobs of low Q are more persistent than blobs of high Q (Hoskins *et al.*, 1985). For instance, the low Q vortex formed on 9 February at 68° N 20° E persists for 8 days without much change in intensity, whereas a high Q cutoff formed on 8 February near 38° N 45° W lasted for about 2 days only.

Figure 8 shows the time-mean Ertel potential vorticity field during the period, with mean isentropic winds superimposed. In spite of the existence of diabatic and eddy sources of mean potential vorticity, the mean flow tends to follow mean Q contours. To this extent steady, nonlinear flow models of

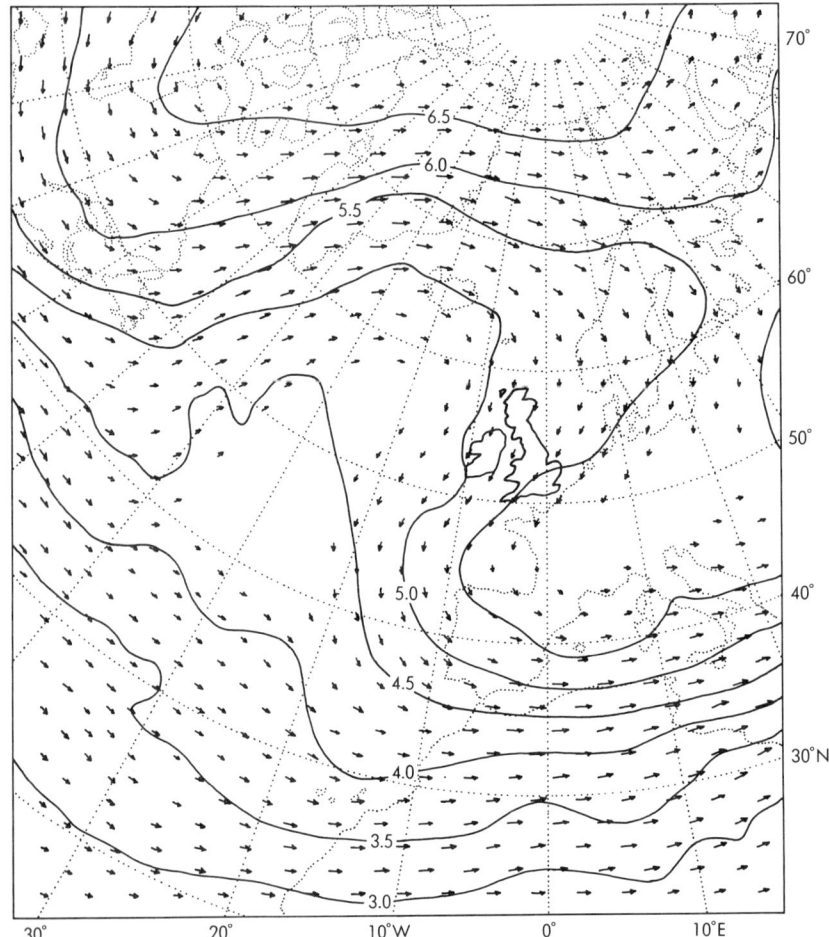

FIG. 8. Time-mean Q map for the period 5–22 February inclusive plus mean wind vectors on the 320 K surface. Contour interval: 1×10^{-3} MKS units.

blocking (e.g., McWilliams, 1980; Mitchell and Derome, 1983) are highly revelant.

In order to obtain some quantitative measure of the high-pass filtered eddy forcing of the mean potential vorticity field, the eddy potential vorticity flux divergence was calculated (Fig. 9). Upstream of the block, near the east coast of the United States, a band of eddy Q flux divergence lies to the north of a band of convergence and both extend toward the jetstream split region. The region of the mean block ridge is characterized by divergence of the eddy Q flux consistent with the synoptic picture of periodic intrusions of low Q.

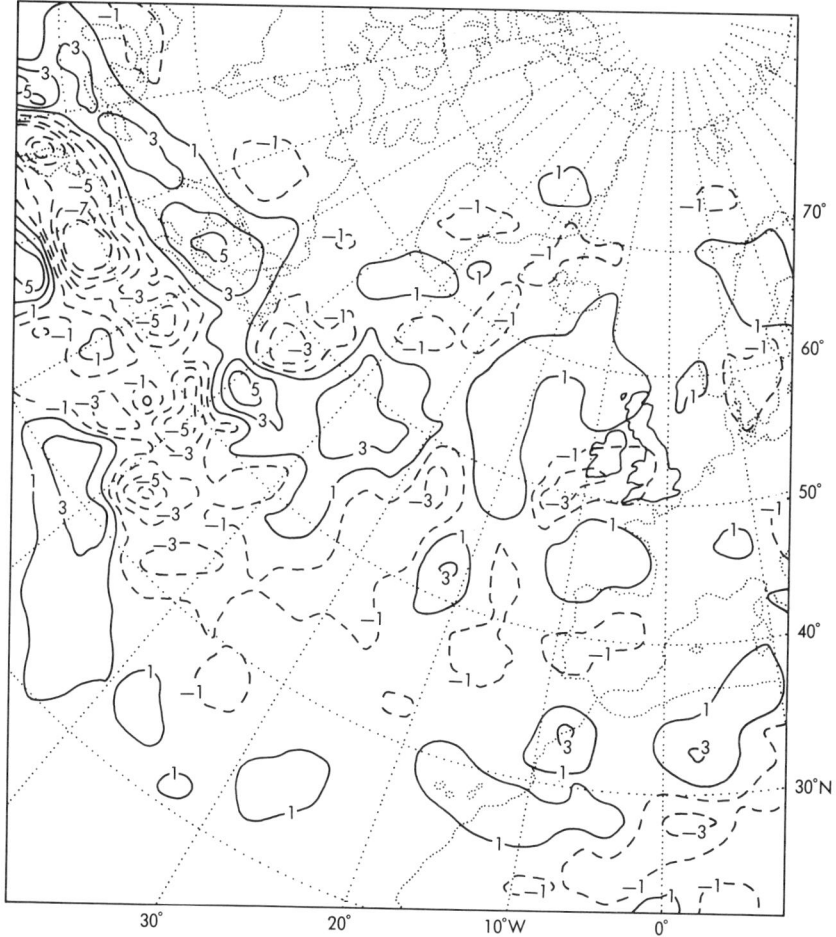

FIG. 9. Unsmoothed eddy Q flux divergence (high-pass filtered in time). Contour interval: 1×10^{-9} MKS units.

Convergence of eddy Q flux exists to the southwest of the British Isles as in the relative vorticity forcing picture (Fig. 4b).

Downgradient transfer of Q by cyclone-scale eddies is responsible for the dipole structure of the eddy Q flux divergence pattern in the storm track region. There is also some evidence that this dipole is enhanced in the jetstream split region as required by the eddy straining mechanism (Section 2) [see also Illari and Marshall (1983)].

Figure 10 shows the mean Q contours with eddy Q flux vectors superimposed. A small rotational flux of the form $C\mathbf{k} \wedge \nabla \overline{Q'^2}$ has been subtracted out using an empirically determined value of C. The resulting eddy Q flux

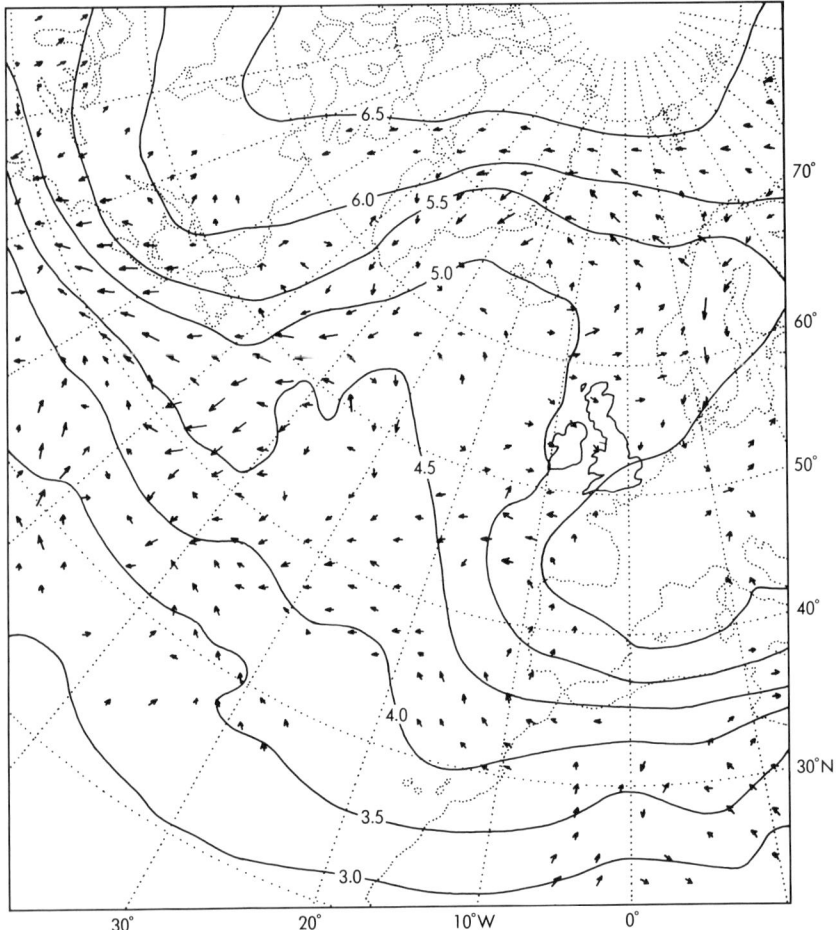

FIG. 10. Time-mean Q contours with eddy Q flux vectors. (A small rotational component has been subtracted out—see text.)

vectors should, in accordance with the analysis of the eddy enstrophy equation given in Section 2, be downgradient (upgradient) in regions of eddy enstrophy dissipation (generation). The eddy potential vorticity flux is strongly downgradient in the storm track and jetstream split regions.

7. Summary and Discussion

An attempt has been made to expose the role played by synoptic time-scale eddies in forcing a blocking anticyclone, using initialized data archived at European Centre for Medium Range Weather Forecasts. A variety of diagnostics have been presented to quantify this eddy forcing as a term in the momentum, vorticity, and Ertel potential vorticity equations and to compare these patterns of eddy forcing with those suggested by a theoretical model.

Time-mean eddy covariance diagnostics are combined with a daily sequence of Ertel potential vorticity maps to provide both Eulerian and Lagrangian viewpoints. All aspects of the "mechanical" forcing (i.e., eddy momentum and vorticity forcing) confirm that eddies, particularly when high-pass filtered, spin up the upper-level anticyclone. **E** vectors are strongly convergent into the jetstream split region, signifying the deceleration of westerlies there; they are divergent out of the northern jetstream branch, implying acceleration of the jet.

Convergent, eastward-pointing **E** vectors are the hallmark of the proposed barotropic eddy straining mechanism and represent the collapse of the east – west scale of incident eddies through deformation by the ambient flow. The high-pass filtered eddy vorticity flux divergence reveals strong anticyclonic forcing in the region of the block sufficient to spin up the observed anticyclonic anomaly in about 2 days. Spatial smoothing helps to emphasize those components of the eddy vorticity forcing pattern which force synoptic-scale pressure anomalies. The smoothed eddy vorticity forcing picture helps one to visualize the type of circulation pattern being forced by eddies whereas the **E** vectors provide extra information concerning the magnitude and orientation of eddies together with their source and sink regions.

The sequence of Ertel potential vorticity maps provides a clear picture of meridional exchange of air masses on an isentropic surface and reveals the block as a region of reversed meridional gradient of Q. Synoptic-scale disturbances approaching the block inject low Q air into the anticyclone and high Q into the cyclonic region to the south, thereby reinforcing the reversed gradient and leading to a dipole pressure pattern.

Cut-off areas of high and low Q form frequently, with the latter being much more persistent typically. Calculation of the high-pass filtered eddy Q flux divergence shows that the low Q region of the blocking anticyclone coincides with a region of eddy Q flux divergence, consistent with the intermittent intrusion of low Q air carried northward from the subtropics. A region of strong Q flux divergence just north of the jetstream split offers further support for the eddy straining hypothesis.

The fundamental question which cannot be answered by observational analysis alone is "To what extent does the blocking flow field depend on the existence of eddy forcing?" We have shown that eddy forcing is an important term insofar as it could create the necessary momentum or vorticity. It would, however, be difficult to *disprove* the importance of eddies, since even a small forcing effect in a near-resonant system can give rise to a large flow response if dissipation is small. One could, given the time-mean flow, try to calculate the linearized response to the observed eddy forcing pattern at, say, 300 mbar using a barotropic model. As a measure of the effect of eddies this perturbation response would prove superior to the streamfunction forcing (Hoskins et al., 1983) or the smoothed eddy vorticity flux divergence. It would of course rely on the stability of the time-mean flow pattern. Even if the calculation is possible it is still artificial, since the time-mean flow is fictitious and actually contains contributions from the real eddy field. A better approach has been taken by J. Egger, W. Metz, and G. Müller (this volume) in which observed, time-dependent eddy forcing was introduced into a linearized, barotropic model and the statistical distribution of modeled blocking events found.

Perhaps the most important unaddressed question in this paper is "How does the eddy forcing mechanism fit in with the observed geographical and seasonal variability of blocking?" The numerical experiments discussed in Section 2 suggested that sufficiently weak westerlies were a necessary condition for blocking. They need only be weak in a certain longitude sector, which, restated, implies that the quasi-stationary planetary wave field adjusts to create a region of greater than normal diffluence locally. Eddies then enhance this diffluence by overturning the meridional potential vorticity gradient.

In this sense, the cause of blocking is directly linked to the cause of the anomalous ultralong wave field. Multiple equilibrium models based on orographic forcing (Charney and DeVore, 1979) naturally provide a framework for such planetary wave anomalies, though it is preferable to keep an open mind to the possibility that other physical mechanisms for changing the large-scale circulation are important. Seasonal variability of blocking frequency must result from the changing planetary wave field associated with the seasonal change of thermal forcing. For instance, the rapid weakening of the meridional temperature gradient near the surface over Europe in spring favors increased upper-level diffluence, which might explain the spring maximum of blocking there. The development of models which can explain the seasonal pattern of variation in blocking is clearly a most desirable goal.

ACKNOWLEDGMENTS

This work was carried out as part of a joint project between the Synoptic Climatology Branch of the Meteorological Office and the Department of Meteorology, Reading University, using archived data and computing facilities at European Centre for Medium Range Weather Forecasts. I received useful help and advice from members of all three institutions. Dr. Michael McIntyre provided considerable encouragement to produce the fine-resolution maps of Ertel potential vorticity.

References

Austin, J. F. (1980). The blocking of middle latitude westerly winds by planetary waves. *Q. J. R. Meteorol. Soc.* **106**, 327–350.

Blackman, M. L., Wallace, J. M., Lau, N.-C., and Mullen, S. L. (1977). An observational study of the Northern Hemisphere wintertime circulation. *J. Atmos. Sci.* **34**, 1040–1053.

Charney, J. G., and DeVore, J. G. (1979). Multiple flow equilibria in the atmosphere and blocking. *J. Atmos. Sci.* **36**, 1205–1216.

Cullen, M. J. P. (1983). Solutions to a model of a front forced by deformation. *Q. J. R. Meteorol. Soc.* **109**, 565–573.

Cullen, M. J. P., and Purser, R. J. (1984). An extended Lagrangian theory of semigeostrophic frontogenesis. *J. Atmos. Sci.* **41**, 1477–1497.

Danielson, E. F. (1968). Stratospheric-tropospheric exchange based on radioactivity, ozone and potential vorticity. *J. Atmos. Sci.* **25**, 502–518.

Green, J. S. A. (1977). The weather during July 1976: Some dynamical considerations of the drought, *Weather* **32**, 120–126.

Hansen, A. R., and Chen, T.-C. (1982). Spectral energetics analysis of atmospheric blocking. *Mon. Weather Rev.* **110**, 1146–1165.

Hansen, A. R., and Sutera, A. (1984). A comparison of the spectral energy and enstrophy budgets of blocking versus non-blocking periods. *Tellus* **36**, 52–63.

Holland, W. R., and Rhines, P. B. (1980). An example of eddy-induced ocean circulation. *J. Phys. Oceanogr.* **10**, 1010–1031.

Hoskins, B. J., James, I., and White, G. (1983). The shape, propagation and mean flow interaction of large-scale weather systems. *J. Atmos. Sci.* **40**, 1595–1612.

Hoskins, B. J., McIntyre, M. E., and Robertson, A. W. (1985). On the significance of isentropic potential-vorticity maps. *Q. J. R. Meteorol. Soc.* **111**, 877–946.

Illari, L. (1982). A diagnostic study of a warm blocking anticyclone. Ph.D. thesis, University of London.

Illari, L. (1984). Diagnostic study of the potential vorticity in a warm blocking anticyclone. *J. Atmos. Sci.* **41**, 3518–3526.

Illari, L., and Marshall, J. C. (1983). On the interpretation of eddy fluxes during a blocking episode. *J. Atmos. Sci.* **40**, 2232–2242.

Kraichnan, R. H. (1967). Inertial ranges in two-dimensional turbulence. *Phys. Fluids* **10**, 1417–1423.

Lönnberg, P., and Shaw, D., eds. (1983). ECMWF Data Assimilation. Scientific Documentation, Reading, Pennsylvania.

Lorenc, A. (1981). A global three dimensional multivariate statistical interpolation scheme. *Mon. Weather Rev.* **109**, 701–721.

McIntyre, M. E., and Palmer, T. N. (1983). Breaking planetary waves in the stratosphere. *Nature (London)* **305**, 593–600.

McWilliams, J. C. (1980). An application of equivalent modons to atmospheric blocking. *Dyn. Atmos. Oceans* **5**, 43–66.

Mahhman, J. D. (1980). Structure and interpretation of blocking anticyclones as simulated in a GFDL general circulation model. *Proc. 13th Stanstead Seminar Bishops Univ., Lennoxville, Quebec, Canada July 9–13, 1979*, pp. 70–76. Dept. of Meteorology, McGill University, Montreal.

Marshall, J. C., and Shutts, G. J. (1981). A note on rotational and divergent fluxes. *J. Phys. Oceanogr.* **11**, 1677–1680.

Mitchell, H. L., and Derome, J. (1983). Blocking-like solutions of the potential vorticity equation: Their stability at equilibrium and growth at resonance. *J. Atmos. Sci.* **40**, 2522–2536.

Palmén, E., and Newton, C. W. (1969). "Atmospheric Circulation Systems." Academic Press, New York and London.
Reed, R. J. (1955). A study of a characteristic type of upper-level frontogenesis. *J. Meteorol.* **12**, 226–237.
Rhines, P. B. (1979). Geostrophic turbulence. *Ann. Rev. Fluid Mech.* **11**, 401–441.
Sardeshmukh, P. D., and Hoskins, B. J. (1984). Spatial smoothing on the sphere. *Mon. Weather Rev.* **112**, 2524–2529.
Savijärvi, H. (1977). The interaction of the monthly-mean flow and large-scale transient eddies in two different circulation types. Part II. *Geophysica* **14**, 207–229.
Shutts, G. J. (1983). The propagation of eddies in diffluent jetstreams: Eddy vorticity forcing of blocking flow fields. *Q. J. R. Meteorol. Soc.* **109**, 737–761.
Wright, A. D. F. (1974). Blocking action in the Australian region. Technical Report 10, Bur. Meteorol., Australia.

Part III

THEORY

THE EFFECT OF LOCAL BAROCLINIC INSTABILITY ON ZONAL INHOMOGENEITIES OF VORTICITY AND TEMPERATURE

R. T. Pierrehumbert

Geophysical Fluid Dynamics Laboratory
National Oceanic and Atmospheric Administration
Princeton University
Princeton, New Jersey 08542

1. Introduction

The central problem in the theory of persistent anomalies in the atmosphere is to account for the magnitude of high-amplitude anomalies and for the persistence of such amplitudes. Any attack on the problem is immediately faced with the existence of a vigorous spectrum of transient synoptic-scale baroclinic eddies; there are no observations which suggest that such eddies disappear during blocking episodes. It is therefore of great importance to understand the effects of synoptic eddies on the large-scale environment through which they propagate. This is equally true whether one thinks of the anomalies as resulting from strongly nonlinear processes — as in the various multiple-equilibria theories — or from essentially linear free and forced Rossby waves. The nature of the eddy effects is also relevant to the theory of the structure of the climatological stationary waves and of the associated storm tracks.

Generally speaking, we shall be concerned with the feedback of eddies on a large-scale diffluent jet pattern consisting of a region of weak winds and weak baroclinicity downstream of a zone of intense winds and baroclinicity, such as pictured in Fig. 1. Since the synoptic eddies draw their energy from the meridional temperature gradient, it is expected that mixing of this gradient will occur in response. The fundamental question is whether the mixing occurs predominantly in the highly baroclinic zone, where it would tend to destroy the large-scale pattern, or downstream of the baroclinic zone, where it would tend to accentuate the zonal inhomogeneity. Vorticity fluxes generated as a byproduct of the eddy evolution may also act to maintain or destroy the basic state.

The two Northern Hemisphere blocking regions are at the end of the climatological storm tracks, hinting at a direct connection between blocking

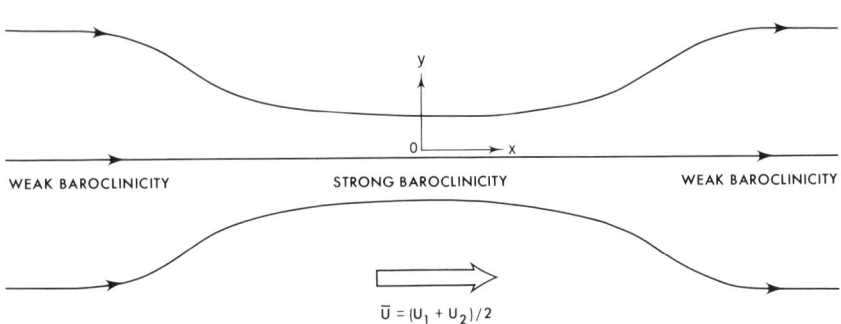

FIG. 1. Plan view illustrating general character of large-scale flow patterns under consideration. The contours represent either upper level streamlines or midlevel potential temperature. The lower layer wind is considered to be weak. The origin of the coordinate system is fixed at the center of the highly baroclinic zone.

and eddies. There is especially good observational evidence that Atlantic blocking is partly maintained by eddy fluxes (Illari and Marshall, 1983; Hoskins *et al.*, 1983). On the theoretical side, Kalnay-Rivas and Merkine (1981) have shown that transient eddies can enhance the amplitude of an orographically forced Rossby wave, and Shutts (1983) has demonstrated a mechanism whereby transient eddies forced by a wavemaker in a barotropic model can act to create and maintain a diffluent jet. We shall see that flux patterns similar to those appearing in Shutts' calculation can be obtained within a self-consistent baroclinic theory, in which the eddies are generated by baroclinic instability. In addition, the baroclinic theory admits several important eddy-induced circulations that are not represented in the barotropic model. The mechanism to be described below is partly complementary to that of Shutts, and it is likely that both can coexist in the atmosphere.

We will look at the eddy-mean flow interaction problem in terms of the approximate theory of local baroclinic instability of zonally varying flow developed in Pierrehumbert (1984). Attention will be restricted to the two-layer quasi-geostrophic model, which is adequate for illustrating the basic phenomena involved; extensions to continuous models are possible, though technically difficult. In Section 2 we derive the zonally inhomogeneous eddy-mean flow interaction equations for the two-layer model. Section 3.1 contains a review of the properties of local baroclinic instability of zonally varying flow, as developed in Pierrehumbert (1984). The structure of the eigenmodes and the relation to the observed structure of storm tracks are discussed in Section 3.2. In Section 3.3 we compute the eddy fluxes associated with local instability. Some comments concerning the relation between eddy forcing and the circulation induced by that forcing are offered in Section 3.4. Our conclusions are summarized in Section 4.

2. Eddy Fluxes in the Two-Layer Model

Consider quasi-geostrophic flow of two layers of fluid of depth D, denoting upper layer quantities with the subscript 1 and lower layer quantities with the subscript 2. In nondimensional units with velocity scale U and length scale $L_D = [gD\delta\rho/(\rho_0 f^2)]^{1/2}$, the equations of motion are

$$\partial_t q_j + \nabla \cdot \mathbf{v}_j q_j = R_j \tag{1}$$

where the potential vorticity is

$$q_j = \nabla^2 \psi_j + (-1)^j(\psi_1 - \psi_2) + \beta y \tag{2}$$

In Eq. (2) ψ_j is the streamfunction and $\beta = \beta_{\text{dim}} L_D^2/U$. The term R_j in Eq. (1) represents the net forcing and dissipation. Next consider an ensemble of solutions to Eq. (1), denoting the average of a quantity A over this ensemble by $\langle A \rangle$ and the deviation from the average by A'. Then, the ensemble average tendency is

$$\partial_t \langle q_j \rangle = -\nabla \cdot \langle \mathbf{v}'_j q'_j \rangle + \{\langle R_j \rangle - \nabla \cdot \langle \mathbf{v}_j \rangle \langle q_j \rangle\} \tag{3}$$

Suppose now that $\langle R_j \rangle$ exactly balances the mean advection, so that the term in curly brackets vanishes. This amounts to the assumption that the ensemble average state is a steady state in the absence of the eddies. Under this condition, the first term on the right-hand side gives the *initial* vorticity tendency that would result from perturbing the steady state with an ensemble of eddies. Given the potential vorticity tendencies, one can easily solve for the geopotential height and temperature tendencies. Initial tendency calculations have proven useful in diagnosing transient eddy effects in the real atmosphere (Lau and Holopainen, 1984; Hoskins *et al.*, 1983) and therefore serve as a convenient vehicle for comparison between theory and observation. However, one must be aware that the initial tendencies do not represent the ultimate effect of the eddies on the mean flow. We shall return to this point in Section 3.4.

The interpretation of the tendency is simplified by splitting it into a barotropic and a baroclinic part. To obtain the former, we add the upper and lower layer equations, resulting in

$$\partial_t q_B = -\nabla \cdot \mathbf{F} \tag{4a}$$

where

$$q_B = \langle \nabla^2(\psi_1 + \psi_2) \rangle \tag{4b}$$

$$\mathbf{F} = \langle (u'_1 \nabla^2 \psi'_1 + u'_2 \nabla^2 \psi'_2) \rangle e_x$$
$$+ \langle (v'_1 \nabla^2 \psi'_1 + v'_2 \nabla^2 \psi'_2) \rangle e_y \tag{4c}$$

and e_x and e_y are the unit vectors in the zonal and meridional direction. The vector $\mathbf{F}$ is simply the vertically integrated horizontal flux of relative vorticity, and Eq. (4a) states that the initial rate of change of vertically integrated relative vorticity is opposite to the divergence of the integrated relative vorticity flux.

The baroclinic structure of the tendency is obtained by subtracting the lower layer tendency from the upper layer tendency. We obtain

$$\partial_t q_T = -\{\nabla \cdot \mathbf{G} - \nabla \cdot \mathbf{H}\} \tag{5a}$$

where

$$q_T = \nabla^2 \langle(\psi_1 - \psi_2)\rangle - \langle(\psi_1 - \psi_2)\rangle \tag{5b}$$

$$\mathbf{G} = \langle u_1' \nabla^2 \psi_1' - u_2' \nabla^2 \psi_2' \rangle e_x$$
$$+ \langle v_1' \nabla^2 \psi_1' - v_2' \nabla^2 \psi_2' \rangle e_y \tag{5c}$$

$$\mathbf{H} = \langle (u_1' + u_2')(\psi_1' - \psi_2') \rangle e_x$$
$$+ \langle (v_1' + v_2')(\psi_1' - \psi_2') \rangle e_y \tag{5d}$$

According to Eq. (5b), the streamfunction of the thermal wind can be recovered from q_T by solving a linear elliptic equation; because of the minus sign in the second term on the right-hand side of Eq. (5b), the Greens' function of the problem is exponentially decaying, with characteristic length equal to the radius of deformation. The tendency of q_T has contributions from the divergence of $\mathbf{G}$ and $\mathbf{H}$; $\mathbf{G}$ is the difference in the relative vorticity fluxes in the two layers while $\mathbf{H}$ is twice the horizontal heat flux. The contribution of the vorticity flux to the thermal wind arises because the flux sets up mean circulations which create dynamic heating via mean vertical motions. The necessity of solving an elliptic equation to find the thermal wind (even in the absence of $\mathbf{G}$) arises because heat and vorticity fluxes create mean circulations which affect the winds via the Coriolis force. When the spatial scale of the ensemble-averaged fluxes is large compared to the radius of deformation, Eq. (5a) reduces to

$$\partial_t \langle \psi_1 - \psi_2 \rangle = \nabla \cdot \mathbf{G} - \nabla \cdot \mathbf{H} \tag{6}$$

which gives the temperature tendency (and hence thermal wind tendency) in terms of the divergence of heat and vorticity fluxes. The $\mathbf{G}$ is not *a priori* negligible compared to $\mathbf{H}$, as the eddies themselves could have scales comparable to the radius of deformation even when the associated rectified fluxes vary slowly in space. However, we shall see that in theory as in observation, the synoptic eddies have a vertical structure such that the first term in the right-hand side of Eq. (6) is negligible, whence the temperature tendency is approximately proportional to the heat flux convergence.

3. Fluxes and Tendencies Associated with Local Baroclinic Instability

3.1. Summary of Properties of Local Baroclinic Instability

We will attack the problem of eddy-anomaly feedback by computing the pattern of eddy fluxes associated with the most unstable baroclinic eigenmode occuring in the system linearized about the given zonally inhomogeneous flow. The results will be used in an attempt to determine whether the eddies act to maintain or destroy the deviations of the basic state from its zonal mean. This approach suffers from two deficiencies: (1) the fluxes during the transient stage before the eigenmode emerges are ignored, and (2) changes in the character of the eddies arising from the nonlinear process of maturation and decay are neglected. Of course, linear theory gives no information on the overall magnitude of the fluxes. Nevertheless, it is of interest to see how far we can get in explaining the qualitative aspects of the observations with linear theory alone, in the hope that discrepancies may bring into relief the kinds of nonlinear effects that are most important.

Instead of dealing with the exact zonally inhomogeneous stability problem, we will make use of the approximate theory developed in Pierrehumbert (1984), the rudiments of which we will outline here. This theory relies on a separation in scale between the size of the eddies and the scale of variation of the basic state; although it was developed within the framework of the two-layer quasi-geostrophic model without meridional shear, generalizations are possible. Let $U_1(x)$ be the upper layer basic-state wind, $U_2(x)$ be the lower level wind, $U_m = (U_1 + U_2)/2$, and $DU = (U_1 - U_2)$. Further, assume that the maximum shear occurs at $x = 0$ and is normalized such that $DU(0) = 1$; let the minimum shear occurring downstream of the highly baroclinic zone be denoted by DU_{min}. This situation is depicted in Fig. 1. The key to understanding the instability of such flows is the concept of "absolute growth rate." For zonally independent flows in a domain of infinite zonal extent, the absolute growth rate is the growth rate observed at a fixed point in space (as opposed to moving along with the unstable wave packet). The absolute growth rate can be obtained from the conventional dispersion relation; it is always less than or equal to the familiar maximum normal mode growth rate and generally decreases monotonically when U_m is increased while holding DU fixed. For a fuller exposition, see Pierrehumbert (1984), wherein the following results concerning local instability were obtained. Local modes, which are defined as eigenmodes that decay to zero at large $|x|$, exist provided two conditions are met: (1) the flow must be absolutely unstable at the point of maximum baroclinicity, and (2) DU_{min} must be less than a certain critical value which depends on β and the pattern of U_m. When local modes exist, their growth rates are equal to the absolute growth rate evaluated at the point of maximum baroclinicity. Since absolute growth rate falls to zero for sufficiently large U_m, it is not difficult to stabilize the flow against absolute baroclinic instability in physically plausible circumstances; in such a flow, all unstable disturbances eventually propagate away leaving nothing behind. Thus, when a flow is absolutely stable, eddies may propagate away before they have time to affect the anomaly. The peak amplitude of a local eigenmode occurs downstream of the site of maximum baroclinicity, and the downstream shift increases with increasing U_m.

For $x > 0$, a local eigenmode has the WKB form

$$\psi'_j = A_j(x)[A(x)e^{i\alpha}] \cos(ly) \tag{7a}$$

where

$$A(x) = \exp\left[-\left(\int_0^x k_i dx\right)\right] \exp(\omega_i t) \tag{7b}$$

and

$$\alpha = \int_0^x k_r dx - \omega_r t \tag{7c}$$

The complex zonal wavenumber $k(U_1(x), U_2(x), \omega)$ is determined by solving the familiar two-layer dispersion relation at each x for k in terms of the (fixed) complex frequency of the eigenmode. Similarly, the vertical structure coefficients A_j are identical to those of the conventional two-layer eigenmodes corresponding to wavenumber k and winds U_1 and U_2. The A_j are indeterminate to the extent of an overall multiplicative constant; in the following we shall adopt the convention $|A_1|^2 + |A_2|^2 = 1$. The solution in Eq. (7) will form the basis of our discussion of the pattern of eddy transports of heat and vorticity.

3.2. Structure of the Eigenmodes

Figure 2 summarizes the three-dimensional structure of the most unstable eigenmode on the profile $DU(x)$ shown by the solid curve. In this calculation, we set U_2 identically to zero, approximating the situation in the real atmosphere. The remaining parameters are $\beta = 0.25$ and $l = 0$. Results are shown only for $x > 0$, as the mode has negligible amplitude at negative x (recall that $x = 0$ represents the center of the highly baroclinic zone of the basic-state jet). It was shown in Pierrehumbert (1984) that the wind field upstream of the point of maximum baroclinicity is essentially immaterial to the structure of the modes. First, we note that the maximum amplitude A occurs well downstream of the site of maximum baroclinicity. This is consistent with the observed pattern of synoptic eddy variance reported by Blackmon *et al.*

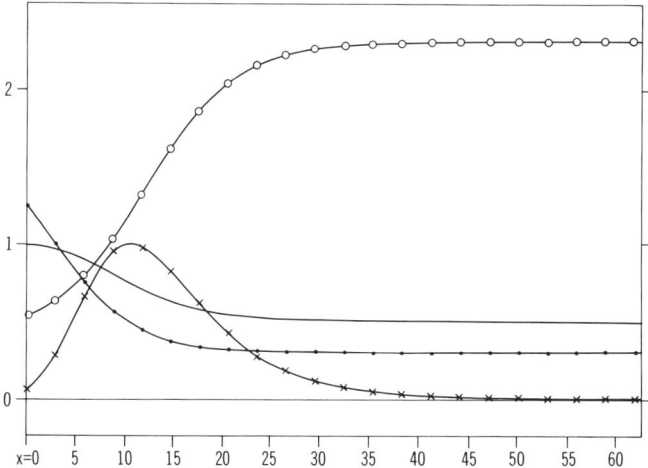

FIG. 2. Structure of the eigenmodes for $x > 0$. (—), DU; (×), A; (●), phase shift; and (○), A_1/A_2.

(1977). As the maximum eddy activity occurs in the diffluence region, it cannot effectively disrupt the strongly baroclinic zone. Next, we note that the vertical phase tilt is everywhere westward; it decreases monotonically from a value of nearly $\pi/2$ at $x = 0$ to a small value in the region of weak baroclinicity. Thus, the eddies become increasingly barotropic with distance along the storm track. This is precisely the pattern observed by Lau (1979, Fig. 5a). It is also consistent with the three-dimensional **E** vector diagnosis of observed synoptic eddies reported in Hoskins et al. (1983), in which the zone of strong phase tilt at the beginning of each storm track shows up as a region of "vertical propagation of eddy activity," in accordance with the definition of the vertical component of the **E** vector [see Eq. (41) and Fig. 13 of their paper]. Finally, we find that $|A_1/A_2|$ is small at the start of the storm track, but becomes large toward the end. Thus, the eddy activity is surface trapped near the start of the storm track, but moves into the upper troposphere as we progress downstream. This sort of behavior has been most often associated with the mature nonlinear stage of the life cycle of a baroclinic disturbance, as in Simmons and Hoskins (1978); we see here that similar effects can be obtained through zonal inhomogeneity, without recourse to nonlinearity.

Farrell (1983) has also sought to explain the storm track structure in terms of linear theory. In Farrell's theory, the longitude–height structure of a storm track is identified with the longitude–height structure of the long-time asymptotic form of an unstable wave packet propagating through a *zonally homogeneous* flow. This wave packet is not an eigenmode; in contrast to the modes we have considered, different parts of the packet grow at different rates and the whole pattern propagates downstream with time. An observer at a fixed point in space would see first one part of the pattern and then another. It thus seems that Farrell's theory cannot account for the fixed spatial structure of the storm tracks.

3.3. Eddy Fluxes and Tendencies

In order to compute the fluxes associated with local instability, one need only substitute the real part of the WKB eigenmode [Eqs. (7a)–(7c)] into the flux expressions [Eqs. (4c), (5c), and (5d)] and carry out the indicated ensemble averages. The proper definition of the ensemble average is a matter of some uncertainty; here, we shall make use of the random-phase ensemble average introduced by Frederiksen (1983), in which the ensemble is taken to consist of all phases of the eigenmode with equal probability. In the context of Eqs. (7a)–(7c), this amounts to an unweighted average of quadratic quantities over the phase α. Thus, if $p = P \exp(i\alpha)$ and $q = Q \exp(i\alpha)$, in which P and Q are independent of α, the ensemble average of the product of their real parts is

$$\langle \mathrm{Re}(p)\mathrm{Re}(q) \rangle = (1/2)\mathrm{Re}(PQ^*) \tag{8}$$

where the asterisk denotes complex conjugation. The construction of the local modes was carried out explicitly only for the case $l = 0$ in Pierrehumbert (1984). However, when l is small the lowest order form of the modes can be obtained by simply multiplying the $l = 0$ solutions by the appropriate sinusoidal modulation in y. As a matter of expedience, this approximation was used in producing the quantitative results presented below.

Upon carrying out the indicated procedure, the components of the barotropic vorticity flux are found to be

$$F_x = -(lA^2/2)[k_r^2 + l^2 - k_i^2] \sin(ly) \cos(ly) \tag{9a}$$

$$F_y = -(k_i A^2/2)[k_r^2 + k_i^2 - l^2] \cos^2(ly) \tag{9b}$$

The zonal flux is antisymmetric in y. North of the centerline of the storm track, F_x is negative, provided the spatial amplification rate k_i is not so large as to dominate the real part of the wavenumber. At each y, the maximum magnitude of F_x occurs near the site of maximum amplitude of the eigenmode, since k_i vanishes and A is maximized there. The meridional flux is symmetric about $y = 0$. Since $k_i < 0$ upstream of the site of peak eigenmode amplitude and $k_i > 0$ downstream, the meridional flux points northward upstream of the peak and southward downstream of the peak provided the meridional wavenumber is not too large. It is noteworthy that the meridional flux is nonzero even though the eigenmode [Eqs. (7a)–(7c)] has purely sinusoidal *y*-structure. The meridional flux is made possible by the spatial amplification in the zonal direction, which gives each eddy a trapezoidal shape. This feature demonstrates that the presence of meridional vorticity flux should not be taken as evidence for either barotropic instability or meridional propagation of eddies; the model under consideration supports neither, and yet has nonzero meridional flux.

The barotropic flux pattern for $l = 0.1$ is shown for $x > 0$ in Fig. 3a; it

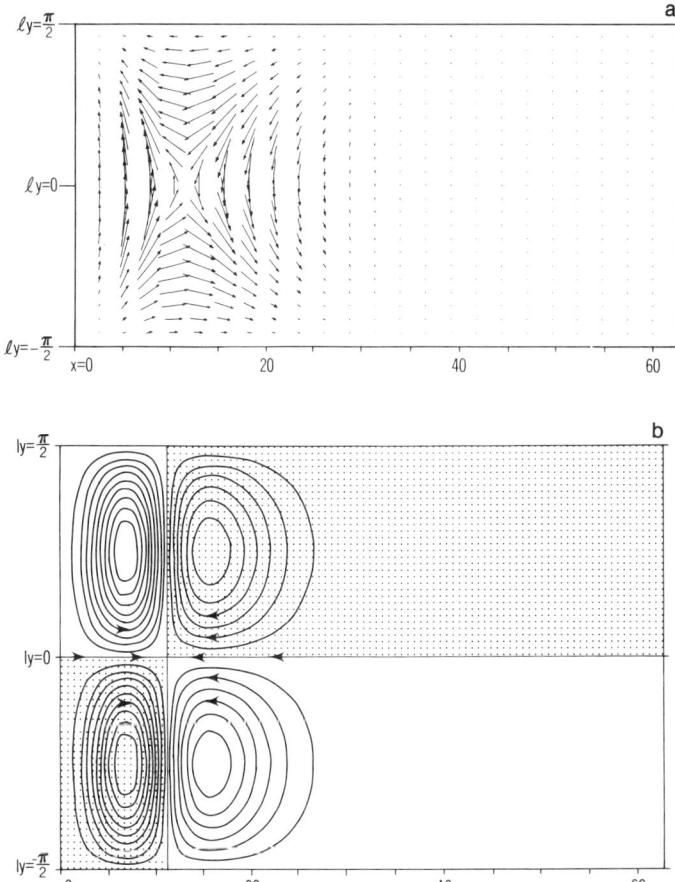

FIG. 3. (a) Barotropic vorticity flux vector **F** in the domain $x \geq 0$, $\pi/2 \geq ly \geq -\pi/2$. Note that the site of maximum basic-state baroclinicity is located at $x = 0$, at the left-hand edge of the figure. (b) Contours of barotropic vorticity tendency $-\nabla \cdot \mathbf{F}_j$ in same domain as Fig. 3a. Maximum value is 1.0 in arbitrary units and contour interval is 0.1. Negative values are shaded, and arrows indicate the sense of the induced circulation.

bears a superficial resemblance to an irrotational deformation field with axes of dilation and contraction passing through the site of maximum modal amplitude at 45° angles to the x-axis. In reality, it is made up of a quadrupole of sources and sinks disposed about the col at the center. This is seen most clearly in the divergence of the field, which takes the form

$$\partial_x F_x + \partial_y F_y = k_i k_r^2 l A^2 \sin(ly) \cos(ly) \tag{10}$$

upon dropping terms that are negligible within the WKB approximation. The barotropic vorticity tendency [i.e., $-(\partial_x F_x + \partial_y F_y)$] is shown in Fig. 3b; the arrows in this figure give the sense of the induced circulation. It is evident that the eddies initially act to accelerate the barotropic component of the jet at the upstream end of the storm track and decelerate it at the downstream end, which would tend to extend the basic-state jet downstream and sharpen the split.

The acceleration at the upstream end of the storm track is in accord with the observed effects of synoptic eddies, as analyzed by Hoskins et al. (1983, Fig. 13). In order to compare the theoretical results with the diagnosed circulations described in Lau and Holopainen (1984), the vorticity-induced circulations shown in their Figs. 3 and 4 must be split into a vertically averaged part (corresponding to the circulation induced by **F**) and the deviation from the vertical average (corresponding to **G**). In fact, the observed circulations have little vertical structure below 400 mbar, so that they project almost entirely onto **F**. (There is also a moderately strong peak in the lower stratosphere, however, which is not reproduced by the linear theory.) Again, marked accelerations of the barotropic component of the mean flow are observed at the upstream ends of both storm tracks. The downstream deceleration is more problematic both with regard to theory and observation. Hoskins et al. include it in their summary picture of synoptic eddy effects (their Fig. 13), but in the climatological analysis the deceleration is very weak at the end of the Atlantic storm track and essentially absent at the end of the Pacific storm track. Lau and Holopainen (1984) also find no deceleration in the Pacific and a slight deceleration in the Atlantic (see especially their Fig. 4b). In any event, the theory certainly overpredicts the magnitude of the downstream deceleration as compared to the upstream acceleration. We shall have more to say about this matter shortly.

Shutts (1983) obtained a similar quadrupole pattern by means of a completely *barotropic* process (see his Fig. 6b). The reason for the similarity in results between such disparate models may be understood in terms of the E vector formalism developed by Hoskins et al. (1983). The E vector and its influence on the zonal flow are given by

$$\mathbf{E} = \langle v'^2 - u'^2 \rangle e_x + \langle -u'v' \rangle e_y \tag{11a}$$

$$\partial_t \langle u \rangle = \partial_x E_x + \partial_y E_y \tag{11b}$$

In our model **E** is predominantly zonal and the eddies are meridionally extended. Thus, E_x is positive with peak amplitude at the site of peak eigenmode amplitude, implying upstream acceleration and downstream deceleration. We have argued in Pierrehumbert (1984) that nonlinearity would tend to reduce the spatial decay rate downstream of the peak; this effect would mitigate the excessive deceleration alluded to in the preceding paragraph. Some of the differences between the Pacific and Atlantic cases may also be due to the aspects of storm track coupling discussed in Pierrehumbert (1984); in essence, the decay of eddies downstream of the Pacific track is inhibited because the entrance to the Atlantic baroclinic zone is quite close the the exit of the Pacific storm track. In contrast, the entrance to the Pacific baroclinic zone is nearly a hemisphere away from the exit of the Atlantic track, leaving more room for the eddies to decay.

In Shutts' model, the eddies are forced by a localized wavemaker, and grow in amplitude across the region of forcing. This leads to growth of E_x and consequent acceleration much as in the baroclinic model. The downstream deceleration, on the other hand, is created by distortion of the shape of the eddies by a large-scale deformation field—a mechanism without counterpart in our baroclinic model. Essentially, the deformation field expels eddy energy from the centerline of the strom track, leading to a decrease in E_x, and shears out the eddies in such a way as to create E_y patterns that accentuate this deceleration. This exercise at least shows that totally unrelated mechanisms can lead to similar flux patterns.

The baroclinic model includes a number of eddy–mean flow interactions which are entirely absent from barotropic models. These are represented in the vectors **G** and **H**. The components of the heat flux vector are

$$H_x = (l/2)A^2(|A_1|^2 - |A_2|^2) \sin(ly) \cos(ly) \tag{12a}$$

$$H_y = (1/2)A^2 \cos^2(ly)(2k_r|A_1\|A_2| \sin(\theta) - k_i(|A_1|^2 - |A_2|^2)) \tag{12b}$$

where θ is the phase shift between the upper and lower level wave (positive equals westward shift of the trough with height). Because the upper and lower level amplitudes are nearly the same near the maximum of A, **H** is predominantly meridional and points to the north. Its maximum magnitude occurs somewhat upstream of the maximum of A, because the vertical phase shift has its maximum at $x = 0$. The pattern is shown in Fig. 4a and is similar to the observations (see Fig. 3 of Lau and Wallace, 1979). From the standpoint of feedback on the anomalies, it is noteworthy that the maximum meridional mixing of temperature occurs well downstream of the maximum basic-state baroclinicity, so that the eddies do not particularly act to destroy the baroclinic zone.

The divergence of the heat flux is given by

$$\partial_x H_x + \partial_y H_y = -2k_r lA^2|A_1\|A_2| \sin(\theta) \sin(ly) \cos(ly) \tag{12c}$$

Note that the heat flux divergence is proportional to the sine of the vertical phase shift even though zonal fluxes have been taken into account. The other contribution to the tendency of q_T is the divergence of $\mathbf{G}$; this was found to be an order of magnitude smaller than the heat flux term, and will not be discussed here. This is consistent with the diagnostic results of Lau and Holopainen (1984), who found that for synoptic eddies the heat flux contribution to the thermal tendency overwhelms the vorticity flux contribution in the midtroposphere. The temperature tendency ($-\nabla \cdot \mathbf{H}$) is shown in Fig. 4b; downstream of the highly baroclinic zone it warms the northern air and

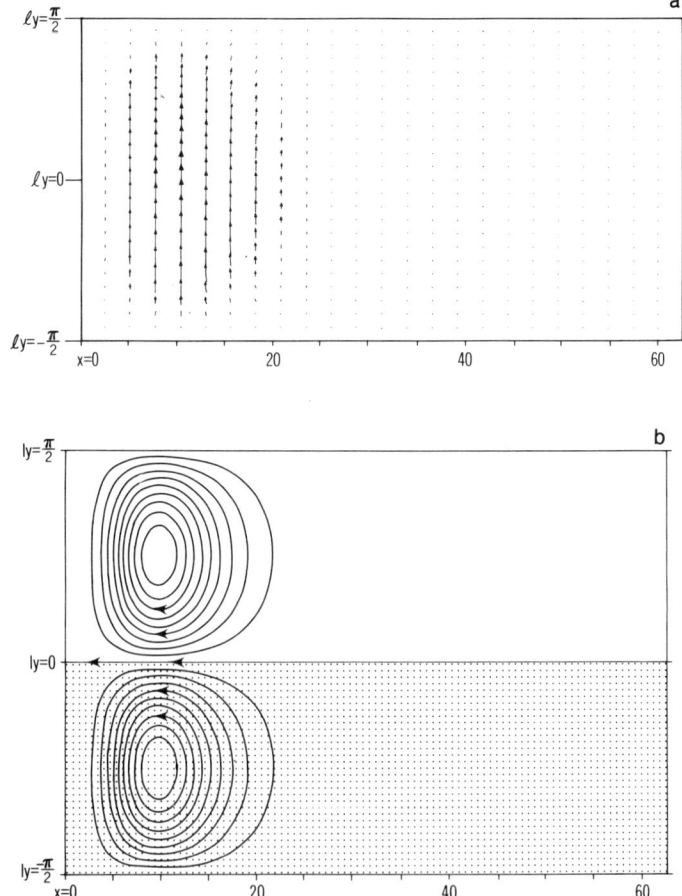

FIG. 4. (a) As in Fig. 3a, but for heat flux vectors. (b) As in Fig. 3b, but for temperature tendency $-\nabla \cdot \mathbf{H}$. Maximum value is 4.25 and contour interval is 0.425, in same arbitrary units as Fig. 3b. (c) As in Fig. 3b, but for total tendency of q_T.

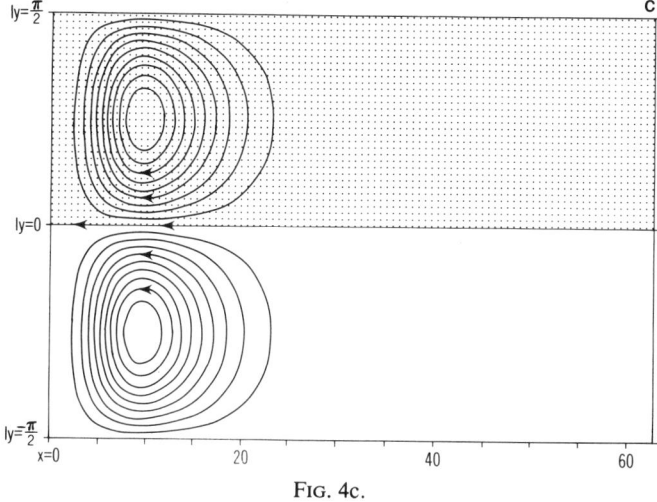

FIG. 4c.

cools the southern air, acting to reduce the thermal wind in the region in which it is already relatively weak. The full tendency of q_T is shown in Fig. 4c, and is virtually indistinguishable from $\nabla \cdot \mathbf{H}$; it too is indicative of a reduction of the thermal wind. Lau and Holopainen (1984) report that synoptic eddies in the real atmosphere act to reduce the thermal wind within the storm tracks, in agreement with the theory (see their Fig. 3a and b).

3.4. Tendency versus Response

At this point we must sound a cautionary note with regard to the interpretation of the preceding tendency results. One cannot answer the question of whether the eddies act to maintain the anomaly, without first determining what they must maintain it *against*. By way of illustration, let us consider the steady-state response of the barotropic vorticity equation to the barotropic vorticity forcing shown in Fig. 3b. If the dominant balance is between advection and forcing (and the β-effect is negligible) the response is in quadrature with the forcing and takes the form of a constriction in the westerlies blowing through the forcing, as shown in Fig. 5a. If, on the other hand, advection is weak and the dominant balance is between forcing and Ekman friction, the shape of the response is identical to that of the forcing, as illustrated in Fig. 5b. A third, more subtle, possibility exists as well: the downstream dipole can serve to maintain a high-amplitude coherent structure with closed streamlines against weak dissipation, as discussed in Pierrehumbert and Malguzzi (1984). The hypothetical response is shown in Fig. 5c. In this case advection

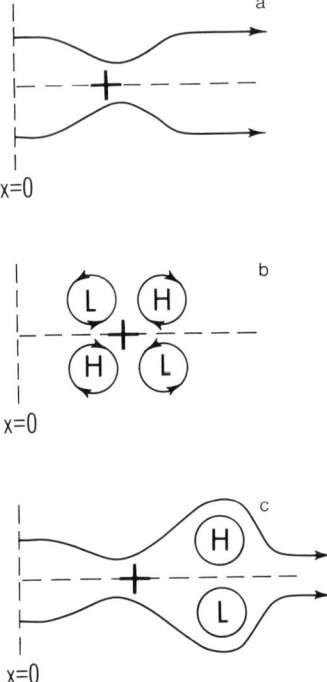

FIG. 5. Three possible barotropic responses to barotropic forcing shown in Fig. 3b. Dominant balance is (a) advection with forcing, (b) Ekman friction with forcing, and (c) weak dissipation of an inviscid nonlinear coherent structure with forcing.

disappears from the balance at the downstream side because the potential vorticity of the coherent structure is nearly constant on streamlines. It is likely that the responses in Fig. 5a and c could exist as multiple equilibria corresponding to the same forcing.

Next, let us consider the response of the baroclinic model to forcing by the heat flux divergence, neglecting advection effects and beta but retaining Ekman friction in the lower layer. Properly speaking, the frictional effects should also be taken into account in the determination of the eddy structure. We will ignore this effect, positing that the friction is too small to create drastic changes in the eddy pattern; alternatively, one could think of the theoretical heat flux pattern as a surrogate for the observations, since the two are quite similar.

Neglecting **G**, the long-wave approximation [Eq. (6)] implies that

$$\langle \psi_1 \rangle - \langle \psi_2 \rangle = -(\nabla \cdot \mathbf{H})t \tag{13}$$

while the equation for q_B in the presence of Ekman friction becomes

$$\partial_t \langle \nabla^2(\psi_1 + \psi_2) \rangle = -r\nabla^2 \psi_2 \qquad (14)$$

Substituting Eq. (13) into Eq. (14) yields

$$\partial_t \langle \nabla^2 \psi_2 \rangle + (r/2)\langle \nabla^2 \psi_2 \rangle = \nabla^2 \nabla \cdot \mathbf{H} \qquad (15)$$

The solution to Eq. (15) relaxes to

$$\langle \psi_2 \rangle = (2/r)\nabla \cdot H \qquad (16)$$

with time constant $2/r$. Without friction, the heat flux would tend to accelerate the low-level wind and decelerate the upper level wind by equal amounts, leading to no net vertically averaged effect. With friction, the low-level wind equilibrates at a westerly value while the upper layer easterlies increase without bound. Note that according to Eq. (13) the upper layer circulation at any given time is made stronger by increasing r. This effect is mediated by frictionally induced meridional circulations, which decelerate the upper level flow via the Coriolis force. In the limit of large friction, the circulation induced by the heat flux appears entirely in the upper layer, and takes the form of a dipole pattern. As it is located in the midst of the diffluent region, this dipole is precisely what is needed to maintain the upper level split against upper level advection; strong friction allows this to be achieved without disrupting the weak low-level flow. Again, the dipole could also be used to maintain a high-amplitude coherent structure, as discussed in the preceding paragraph.

Addition of an accelerating barotropic force to Eq. (14) does not substantially change the argument, as the lower level westerlies then equilibrate at a somewhat larger value while the upper level easterlies continue to grow linearly without bound. Hence the fact that the net acceleration in Fig. 4c of Lau and Holopainen is everywhere positive at the beginning of the storm tracks does not compromise the effect, since the westerly acceleration decreases with altitude throughout most of the troposphere. The question of whether the barotropic acceleration overcomes the baroclinic deceleration at upper levels can only be answered with reference to the particular balance that eventually comes into play to prevent the upper level easterlies from becoming indefinitely stronger.

Of course, we are solving just small pieces of a complicated problem here. A particularly stark illustration of the subtleties involved emerges in Held *et al.* (1986), in which it is shown that some forms of dissipation can act to increase the amplitude of planetary waves by catalyzing release of zonal-mean available potential energy. An energetic analysis of this situation would yield the correct but misleading result that the eddies draw energy out of the planetary waves. In order to determine the actual response to the eddy forcing, one should simultaneously take into account advection, friction, and the basic-state potential vorticity gradient. The preceding results nonetheless suggest that the combination of synoptic eddy heat flux and surface friction are conducive to the formation of diffluent jets. The barotropic vorticity fluxes primarily act to extend the basic-state jet somewhat further downstream; to the extent that they have zero zonal average as in the theory, they do not strongly interfere with the maintenance of the split, and under some circumstances may even contribute to it.

4. Conclusions

The summary picture of a storm track as deduced from the linear local baroclinic instability theory is as follows. The peak amplitude of the eddies is situated well downstream of the region of maximum baroclinicity of the basic-state flow. Although the maximum westward vertical phase shift occurs at the point of maximum baroclinicity and decreases monotonically with downstream distance, the greatest meridional heat flux occurs in the diffluent region of the jet, somewhat upstream of the site of maximum eddy amplitude. The heat fluxes tend to weaken the large-scale thermal wind in the diffluent region. The vorticity fluxes act to accelerate the barotropic component of the jet at the beginning of the storm track and decelerate it at the downstream end.

The principal successes of the theory with respect to the observations are the location of the site of maximum eddy amplitude, the distribution of heat flux along the storm track, and the acceleration of the barotropic component of the jet at the beginning of the strom track. The latter effect is simply due to the growth in eddy amplitude along the storm track, and is quite robust. The main shortcoming of the theory is its overprediction of the strength of the downstream barotropic deceleration relative to the upstream acceleration. We have argued that nonlinearity may ameliorate this problem. It is also possible that the difference between the Pacific and Atlantic storm tracks with regard to downstream deceleration is partly due to the coupling between the Pacific and Atlantic regions; the fact that the Atlantic storm track is in some sense more "isolated" may account for some of the differences between the roles of synoptic eddies in Atlantic and Pacific blocking. Finally, we note that meridional shear of the basic-state wind has been neglected in our model; the effects of this shear could appreciably alter the vorticity fluxes.

Concerning blocking, the main implications are as follows. First, although the atmosphere is highly unstable in the usual normal mode sense, the relevant growth rate characterizing the zonally inhomogeneous problem is the absolute growth rate, which is typically much lower; thus, there may be some utility in thinking of the atmosphere as a weakly unstable system. Second, because the maximum eddy amplitudes and heat fluxes occur in the diffluent region rather than in the jet maximum, they cannot effectively eliminate the zonal inhomogeneity. In fact, given sufficient Ekman friction, the heat flux can induce a circulation which tends to decelerate the upper level westerlies and thus maintain the diffluence, provided the effect is not overwhelmed by the acceleration due to the barotropic fluxes. Indulging in some speculation, one could view blocking as a manifestation of the fluctuation of the large-scale flow occurring in association with the life cycles of baroclinic eddies developing on that zonally inhomogeneous flow: one might begin with a weakly diffluent jet and weak eddies. As the eddies grew, the split would intensify in response to the heat-flux forcing and the jet streak would extend further eastward in response to the vorticity forcing. The eddies would then equilibrate and decay, whereafter the cycle would commence anew.

The eddy flux patterns just described rely on the penetration of strong eddy activity into the region of weak baroclinicity. The extent of the penetration, in turn, depends on the configuration of the large-scale basic-state wind in the manner described in Pierrehumbert (1984). If the vertically averaged wind is too strong compared to the vertical shear, the flow will not be absolutely unstable, local modes will not develop, and eddies will propagate away from the region of interest before they have an opportunity to appreciably affect it. On the other hand, weak or easterly low-level winds favor absolute instability. Thus, the weakened westerlies or low-level easterlies typically found in blocking regions serve to enhance the localization of eddy activity in the block, with attendant enhancement of the diffluent pattern. In consequence, certain anomalies consisting of a region of weak winds and baroclinicity downstream of a region of strong winds and baroclinicity have a natural tendency to amplify under the influence of synoptic eddies.

Absolute instability at some point in the flow is a requirement for the existence of true local modes. At present it is uncertain whether appreciable absolute instability can occur in the real atmosphere in the absence of a region of easterly winds. Results for the Charney model (Farrell, 1983) indicate that the Charney modes become absolutely unstable only when the surface winds are easterly, in contrast with the two-layer model, which permits short-wave, absolute instability for westerly surface winds. The effects of meridional shear have not yet been studied, and may eliminate the necessity of easterlies. It is clear, however, that the weak surface winds characterizing the real atmosphere place it at least near the border of absolute instability.

Under these circumstances, we expect slowly moving disturbances to give rise to patterns which resemble local modes for a long period of time, even if they eventually become delocalized. The effects of such "meta-local" modes are apt to be very similar to those of true local modes.

Acknowledgment

The author is indebted to Eero Holopainen for a number of useful suggestions concerning the relation of the theoretical results to the observations.

References

Blackmon, M. L., Wallace, J. M., Lau, N., and Mullen, S. (1977). An observational study of the Northern Hemisphere wintertime circulation. *J Atmos. Sci.* **34**, 1040–1053.
Farrell, B. F. (1983). Pulse asymptotics of three-dimensional baroclinic waves. *J. Atmos. Sci.* **40**, 2202–2210.
Frederiksen, J. S. (1983). Disturbances and eddy fluxes in Northern Hemisphere flows: Instability of three-dimensional January and July flows. *J. Atmos. Sci.* **40**, 836–855.
Held, I., Pierrehumbert, R. T., and Panetta, R. L. (1986). Dissipative destabilization of external Rossby waves. *J. Atmos. Sci.,* submitted.
Hoskins, B. J., James, I., and White, G. (1983). The shape, propagation and mean-flow interaction of large-scale weather systems. *J. Atmos. Sci.* **40**, 1595–1612.
Illari, L., and Marshall, J. C. (1983). On the interpretation of eddy fluxes during a blocking episode. *J. Atmos. Sci.* **40**, 2232–2242.
Kalnay-Rivas, E., and Merkine, L. O. (1981). A simple mechanism for blocking. *J. Atmos. Sci.* **38**, 2077–2091.
Lau, N. C. (1979). The structure and energetics of transient disturbances in the Northern Hemisphere wintertime circulation. *J. Atmos. Sci.* **36**, 982–995.
Lau, N. C. (1979). The observed structure of tropospheric stationary waves and the local balances of vorticity and heat. *J. Atmos. Sci.* **36**, 996–1016.
Lau, N. C., and Holopainen, E. O. (1984). Transient eddy forcing of the time-mean flow as identified by geopotential tendencies. *J. Atmos. Sci.* **41**, 313–328.
Lau, N. C., and Wallace, J. M. (1979). On the distribution of horizontal transports by transient eddies in the Northern Hemisphere wintertime circulation. *J. Atmos. Sci.* **36**, 1844–1861.
Pierrehumbert, R. T. (1984). Local and global baroclinic instability of a zonally varying flow. *J. Atmos. Sci.* **41**, 2141–2162.
Pierrehumbert, R. T., and Malguzzi, P. (1984). Forced coherent structures and local multiple equilibria in a barotropic atmosphere. *J. Atmos. Sci.* **41**, 246–257.
Shutts, G. J. (1983). The propagation of eddies in diffluent jetstreams: Eddy vorticity forcing of blocking flow fields. *Q. J. R. Meteorol. Soc.* **109**, 737–761.
Simmons, A. J., and Hoskins, B. (1978). The life cycles of some nonlinear baroclinic waves. *J. Atmos. Sci.* **35**, 414–432.

FORCING OF PLANETARY-SCALE BLOCKING ANTICYCLONES BY SYNOPTIC-SCALE EDDIES

J. Egger, W. Metz, and G. Müller

Meteorology Institute
University of Munich
D-8000 Munich 2
Federal Republic of Germany

1. Introduction

The flow patterns typical of blocking are of large scale. Correspondingly, most theories of blocking concentrate on the large-scale aspect of blocking. The impact of synoptic-scale waves is excluded. For example, Egger (1978) used a truncated semispectral model of barotropic channel flow where only the largest modes where retained to demonstrate that blocking-type flow patterns can be created through a nonlinearly modified resonance mechanism. Tung and Lindzen (1979) advanced the view that blocking can be explained through linear resonance of forced planetary-scale waves. Schilling (1982) provided evidence that baroclinic energy conversion between large-scale flow modes can lead to blocking. These theoretical studies found some support from data analyses (e.g., Colucci *et al.,* 1981) and from the analysis of global circulation model (GCM) data (Chen and Shukla, 1983).

However, there is ample evidence that this basic view of blocking as a phenomenon which is almost completely dominated by the dynamics of the largest atmospheric modes is not adequate. It has been realized for some time that traveling weather systems interact with the block. Green (1977), for example, pointed out that the transfer of momentum by the synoptic-scale eddies played an important role in maintaining the block linked to the British drought of 1976 (see also Illari and Marshall, 1983). Hansen and Chen (1982), when studying a case of blocking in the Atlantic, found that this block was forced by the nonlinear interaction of intense baroclinic synoptic-scale waves with barotropic ultralong waves. Fischer (1984) arrived at similar conclusions through an analysis of data obtained in numerical experiments with a GCM. Hansen and Sutera (1984) provided evidence that the transfer of enstrophy between these two wave regimes was different for blocking and nonblocking situations. These findings have been taken into account by modelers. Egger (1981), using a simplified version of the model proposed by

Charney and DeVore (1979), prescribed the interaction of large-scale modes with shorter waves as a white-noise forcing and studied the impact of this forcing on the flow climatology (see also Benzi et al., 1984). Shutts (1983) was able to show that dipole blocking patterns can be generated in barotropic flow, where smaller eddies are excited by a wave generator (see also Kalnay-Rivas and Merkine, 1981). However, it is a shortcoming of these theoretical efforts that the choice of the forcing is largely intuitive and not based on observations. The synoptic-scale forcing has been determined by Egger and Schilling (1983) on a day-by-day basis from data. Egger and Schilling used the resulting time series of the forcing as input in a model. They were not concerned with blocking but with the variability of the atmosphere and found that a large part of the observed long-term variability of the atmosphere can be explained as a response of the planetary-scale modes to this forcing. Since blocking is a phenomenon with relatively large time scales one might expect that the origin and maintenance of blocking patterns can be explained at least partly by the same mechanism. Here we want to find out if this view is correct. We prescribe the observed forcing in a barotropic model of planetary flow. Numerical experiments are carried out in order to see if blocking can be generated in this model.

2. Stochastically Forced Planetary Modes

Our modeling strategy is fairly similar to that described in Egger and Schilling (1983, 1984) so that only a brief outline is given here. We assume that the barotropic component of atmospheric large-scale flow can be described by the vorticity equation

$$(\partial/\partial t)(\nabla^2 - \lambda^2)\psi + J(\psi, \nabla^2\psi + 2\Omega a^{-1} \sin \varphi) = k\nabla^4\psi \quad (1)$$

where ψ is the streamfunction, φ is latitude, λ^{-1} is a radius of deformation, a the earth's radius, and Ω the earth's rotation rate. A damping term is included on the right-hand side of Eq. (1). We partition the streamfunction in a planetary part ψ_r to be resolved and treated explicitly and an unresolved synoptic-scale part ψ_u:

$$\psi = \psi_r + \psi_u \quad (2)$$

The equation for the planetary part is

$$(\partial/\partial t)(\nabla^2 - \lambda^2)\psi_r + J_r(\psi_r, \nabla^2\psi_r + 2\Omega a^{-1} \sin \varphi) - k\nabla^4\psi_r$$
$$= -J_r(\psi_r, \nabla^2\psi_u) - J_r(\psi_u, \nabla^2\psi_r) - J_r(\psi_u, \nabla^2\psi_u) \quad (3)$$

The subscript r at the symbol J denotes the projection of this Jacobian on the resolved flow. The terms on the right-hand side of Eq. (3) describe the interaction of the planetary-scale modes and the synoptic-scale modes. Of course, there exists a similar prognostic equation for ψ_u. This latter equation will not be solved, however. Instead we lump all the three Jacobians together and prescribe them as forcing. We split the streamfunction into a mean flow and a deviation thereof:

$$\psi_r = \bar{\psi} + \psi' \tag{4}$$

where $\bar{\psi} = -U_0 a \sin \varphi$ is the superrotational mode with a zonal wind profile $\bar{u} = U_0 \cos \varphi$.

It is convenient to project ψ' onto spherical harmonics

$$\psi' = \frac{1}{2} \sum_{m=-M}^{M} \sum_{n=1}^{N} \psi_{mn} P_{m+2n-1}^{m}(\sin \varphi) \exp(im\tilde{\lambda}) \tag{5}$$

where P is the corresponding associated Legendre function and $\tilde{\lambda}$ is longitude. It is seen that we admit only those modes which are antisymmetric with respect to the equator. We select the modes with $m \leq 5$, $n \leq 5$ as planetary modes so that $M = N = 5$. This choice is somewhat arbitrary and the results depend on it.

We shall mainly present results obtained with a linearized version of Eq. (3), where the linearization is made with respect to the mean state described by $\bar{\psi}$. After performing the linearization, Eq. (3) can be transformed into a set of equations for the expansion coefficients

$$(d/dt)\psi_{mn} + \alpha \psi_{mn} = X_{mn} \tag{6}$$

where α contains the effects of damping and wave propagation and X_{mn} is the forcing by synoptic-scale eddies. Eleven years of geopotential height data have been used to construct forcing time series for each of the planetary modes. The statistical characteristics of these time series are discussed in Egger and Schilling (1983, 1984). We quote here the main result that the autocovariance of the real (and imaginary) part of the forcing can be approximated by

$$R_{XX}(\tau) = Z \exp(-s|\tau|) \cos(p\tau) \tag{7}$$

where Z, s, and p are the fitting parameters to be determined from data for each mode and τ is the lag. Typically, $s \sim 0.3 - 1.2\,\text{day}^{-1}$; $p \sim 0.5 - 1.4\,\text{day}^{-1}$. Given Eq. (7) we can solve Eq. (6) for the covariances $R_{X\psi}(\tau)$ of the response ψ and the forcing X (Egger and Schilling, 1984). For intercomparison $R_{X\psi}(\tau)$ must be determined from data as well. The result is displayed in Fig. 1 for two modes. The mode with $m = 3$, $n = 2$ is a low retrograde mode with an Rossby frequency of $\omega \simeq -0.7 \times 10^{-1}\,\text{sec}^{-1}$.

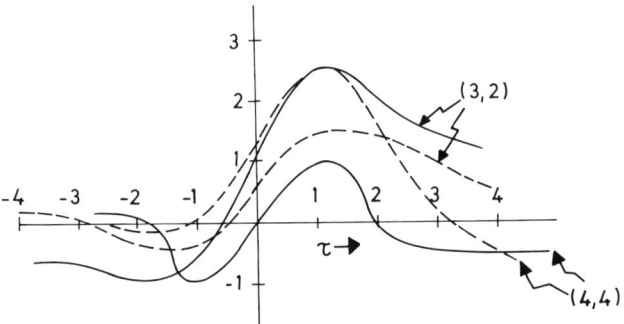

FIG. 1. Covariance of forcing and response $R_{X\psi}(\tau)$ as observed (solid lines) and as determined theoretically (dashed lines) for two modes of the planetary flow; τ in days, R in 10^6 m^4 sec^{-3}. Adapted from Egger and Schilling (1984).

The agreement of the theoretical curve and the observations is quite good. We have negative values of the covariance for $\tau < 0$ and a sharp increase of $R_{X\psi}$ when τ approaches the origin. There is a maximum of $R_{X\psi}$ after about 1 day and a slow decay afterwards. It is easy to understand the curve for $\tau > 0$. A positive forcing X at $\tau = 0$ creates a positive response, of course. Friction and movement with the Rossby phase speed limit the increase of the response with increasing lag. Furthermore, the forcing changes rapidly according to Eq. (7). It is not so easy to understand the negative values of R for $\tau < 0$ and the reader is referred to Egger and Schilling (1984) for a discussion of this phenomenon. The mode with $m = 4$, $n = 4$ (Fig. 1) is progressive and is close to the boundaries of the planetary wave group. There the agreement of observations and theory is less satisfactory. A more exhaustive comparison led Egger and Schilling (1984) to conclude that the agreement is particularly good if we look at the slow quasi-stationary modes. These modes experience the fluctuations of the vorticity transfer from synoptic-scale modes as an independent stochastic forcing. Since blocking is a relatively slow quasi-stationary process and since the planetary-scale response to stochastic forcing is particularly realistic for the slow modes, we feel confident that the forcing of blocks by synoptic-scale modes can be studied by aid of Eq. (6). Moreover, Schilling (1986) demonstrates that the kinetic energy of waves with $m \leq 5$ is increased before the onset of blocking by this forcing.

3. Results

In almost all runs the observed winter mean value $U_0 = 12$ m sec^{-1} is prescribed for the mean flow. The forcing X_{mn} for each mode of the resolved flow is inserted as observed at every day of the numerical integration ($m \leq 5$,

$n \leq 5$). Most of the results have been obtained for winter cases. These runs are started with the undisturbed superrotation $\bar{\psi} = -U_0 a \sin \varphi$ as initial state and the flow evolves as a response to the forcing. The forcing X_{mn} is prescribed for 11 winters according to the observations. The experiments are started on 20 November. The evaluation of the flow patterns begins on 1 December and the runs are terminated after 3 more months. The integrations have been carried out first with the linear version of Eq. (3). They have been repeated with all nonlinear terms included. In addition there has been a nonlinear experiment where the integration has been extended over 2 years and where an annual cycle was prescribed for U_0 (Müller, 1984).

The evaluation of the numerical experiments has been done mainly in terms of blocking events. Blocking in the model is defined as follows: one considers the streamfunction of the resolved flow in the latitude band 40°N to 75°N. Maps of the resolved flow are prepared every day and the extrema of the deviation of the streamfunction from climatology are searched for. Such an extremum is called a blocking anticyclone when the difference $\Delta = \psi(x, y, t) - \bar{\psi}^t(x, y)$ ($-t$, time mean) satisfies the amplitude criterion $\Delta > 1.5 \times 10^7$ m^2 sec^{-1} and if the "anomaly" can be found for more than 4 consecutive days. In addition, slow anomalies are selected by requiring that the shift of the location of the anomalies is less than 5° longitude per day. If the blocking definition is applied to data we use a vertically averaged streamfunction. More details of the selection criteria can be found in Metz (1986).

There are two problems with such blocking definitions. First, it is not clear if an observed block still looks like a block when only resolved modes are admitted to describe the flow. In Fig. 2 we show an example of an observed block (Fig. 2a) and its counterpart in the space of resolved modes. It is seen that much detail is lost in the truncated version but the overall flow pattern is fairly similar in both representations. In particular, the blocking patterns in the Pacific and Atlantic are resistent with respect to this truncation. Second, one cannot be sure if our definition of blocking yields about the same results as have been obtained by other authors (e.g., Treidl et al., 1981). To check this possibility we took the flow observations for the 11 winters and searched for blocks using the criterion. The result is displayed in Fig. 3. There is a pronounced maximum of blocking activity in the Atlantic and Pacific. This result is in satisfactory agreement with the findings of others (e.g., Treidl et al., 1981). The latitudinal distribution peaks at 55°, but there are blocks well within the whole latitudinal band under search. These features also come out clearly in Fig. 4, where the mean locations of blocking highs are shown. The majority of blocks is situated in the Atlantic and Pacific. Note the almost complete absence of blocks over North America.

Let us now have a look at the modeling results. Although it is straightforward to apply the blocking definition to the model results this need not be a meaningful procedure. The only source of wave energy in our model is the

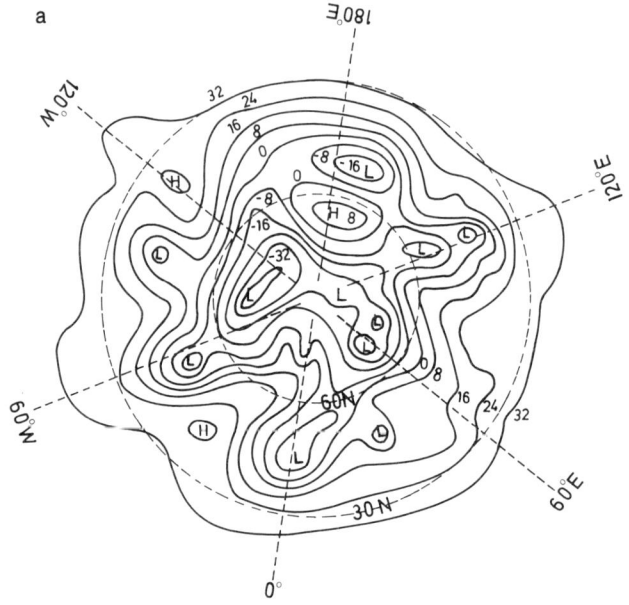
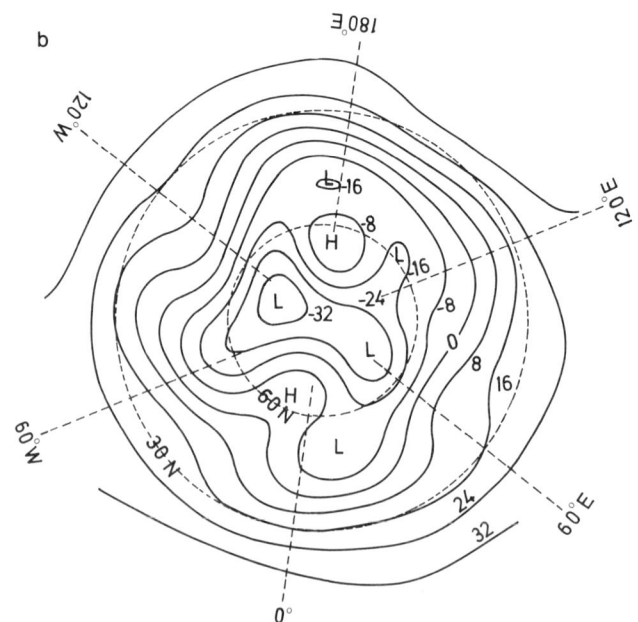

FIG. 2. 500-mbar analysis as provided by the German Weather Service on 14 May, 1972; a, original analysis redrawn; b, resolved modes only. Contour interval, 8 gdam. Adapted from Müller (1984).

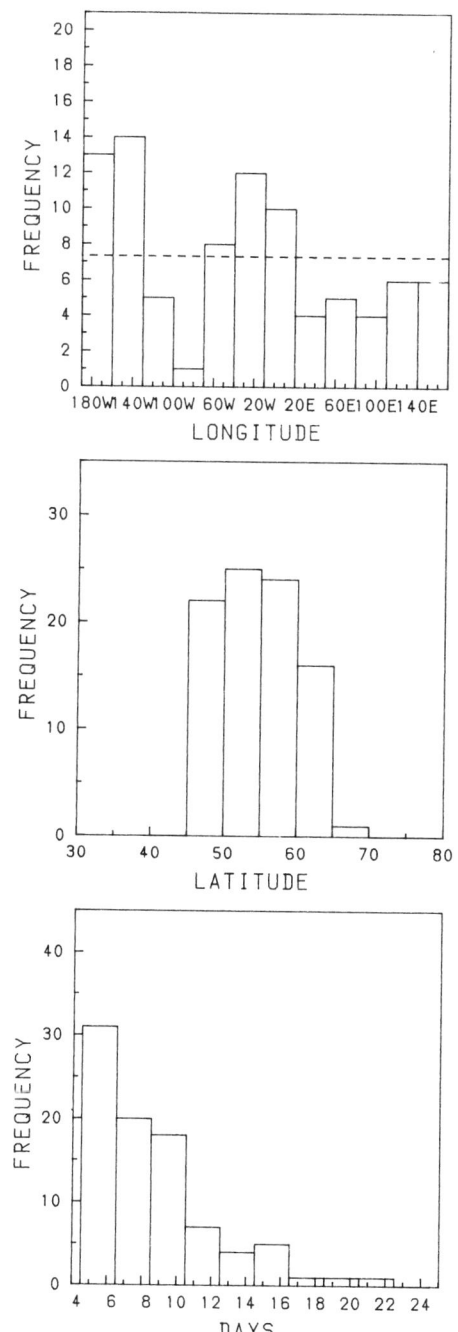

FIG. 3. Frequency distribution of blocking highs with longitude, latitude, and duration. Adapted from Metz (1986).

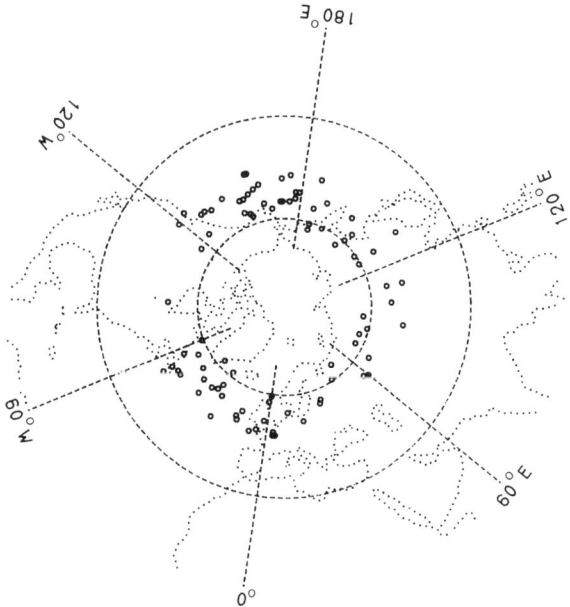

FIG. 4. Mean positions of blocking highs. Adapted from Metz (1986).

synoptic-scale forcing. In the atmosphere topographic effects will generate planetary-scale waves. Moreover, the planetary modes can extract energy from the mean flow. So the wave amplitudes in the model should be smaller than those observed. In these circumstances blocks that can satisfy the amplitude criterion will be rare. We circumvent this problem by tuning the damping in the model so as to have wave amplitudes that are close to those observed. The value adopted is $k = 6.0 \times 10^5$ m^2 sec^{-1}. This value is certainly within the range of what one would consider acceptable for barotropic large-scale flow.

In Fig. 5 we show the distribution of the linear model blocks over the Northern Hemisphere for comparison with Fig. 3. The model is correct in predicting the maximum position density over the Atlantic but tends to predict blocking action in areas where none is observed. There is also a good agreement of the duration of the blocks, although the model blocks tend to be longer lived.

It is not only in the gross statistical characteristics that the model comes reasonably close to reality. If we look at the spatial structure of the simulated blocks we also see quite a good agreement. We show first the evolution of a blocking ridge at the Greenwich Meridian that recedes westward at the end of the blocking period (Fig. 6), as obtained by Müller (1984) in the run with

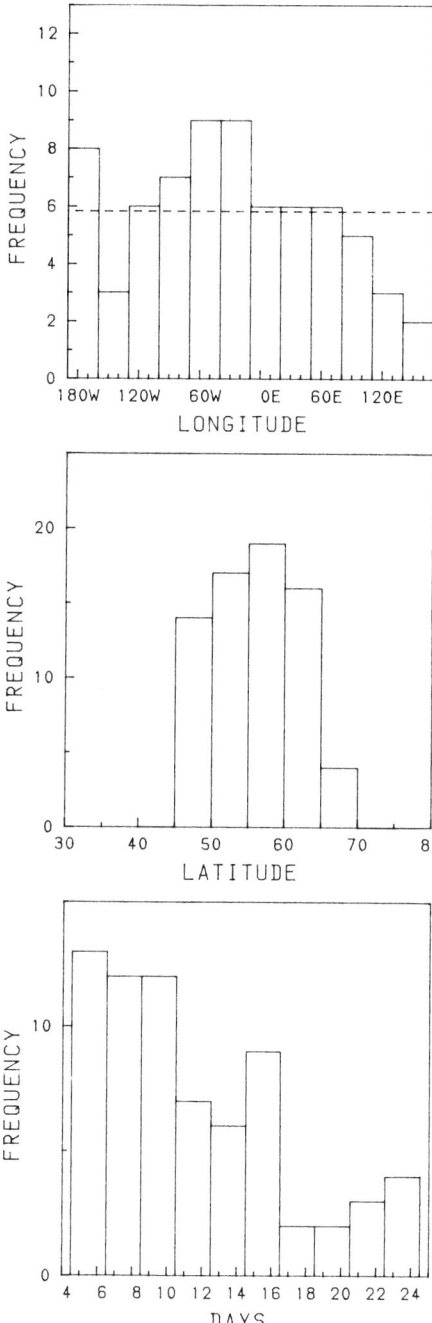

FIG. 5. Frequency distribution of blocking highs with longitude, latitude, and duration as obtained in the stochastically forced linear model. Adapted from Metz (1986).

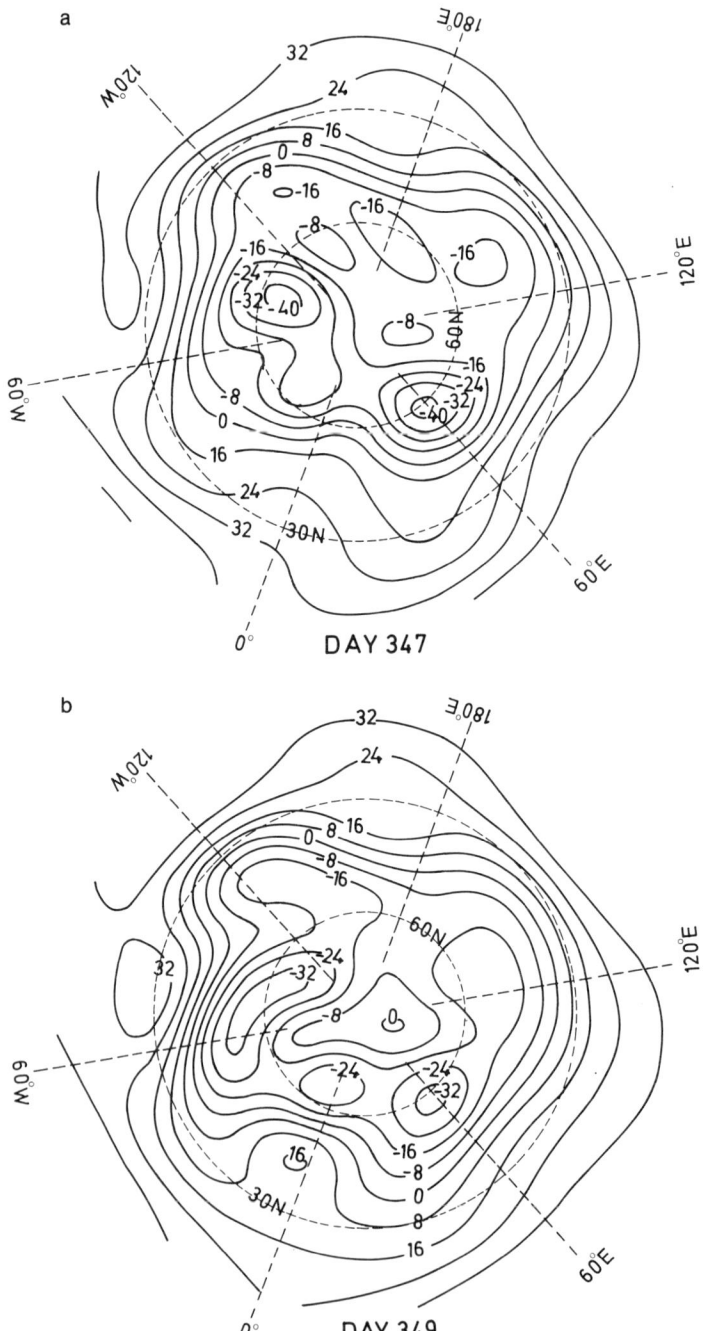

FIG. 6. Blocking event as obtained in a nonlinear integration of the stochastically forced model. Adapted from Müller (1984).

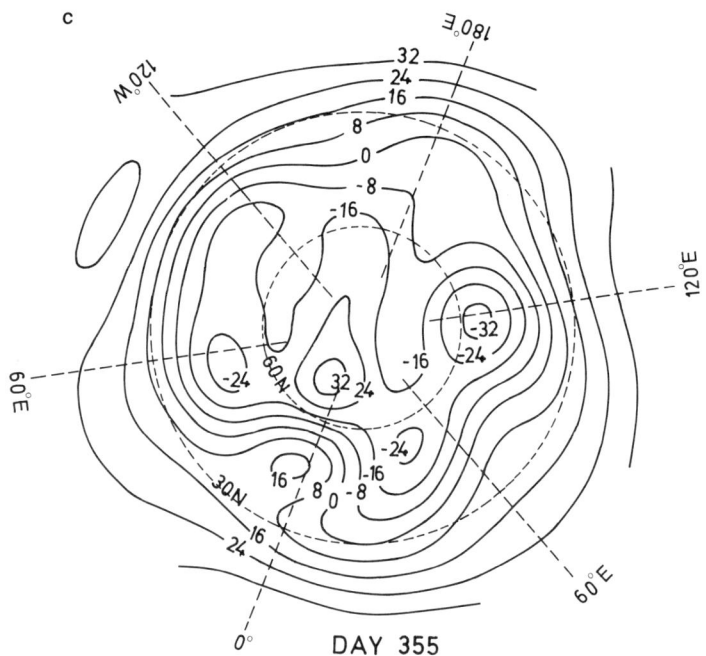

DAY 355

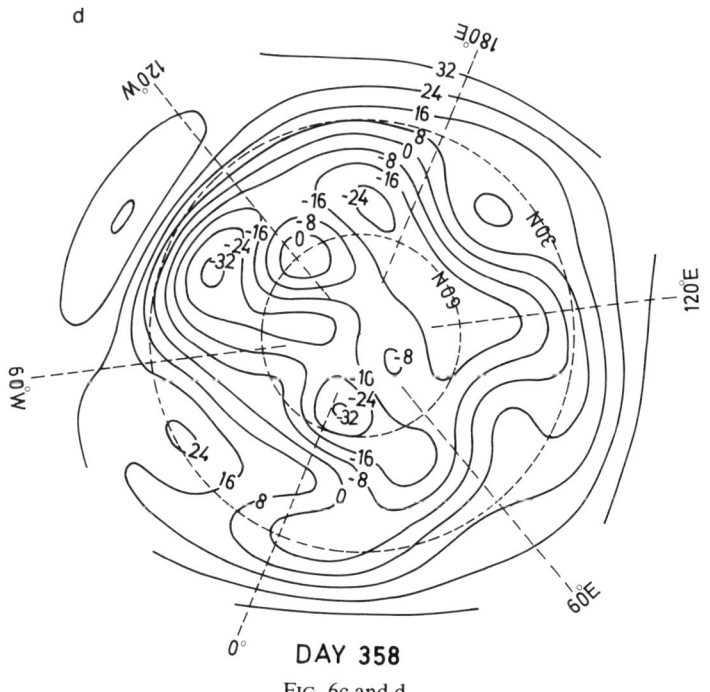

DAY 358

FIG. 6c and d.

the annual cycle. To establish the typical blocking flow pattern in the model Metz (1986) has computed the mean maps of the streamfunction anomalies averaged over all Atlantic and all Pacific blocking periods separately. The result for the Atlantic case and for the linear run is shown in Fig. 7 and is to be compared to the corresponding map derived from the 11 winters. The agreement is quite good, although the amplitudes are somewhat too low in the model-derived map. The corresponding map for the nonlinear run is similar. The agreement for the Pacific cases is not so good but is still satisfactory (Metz, 1986). All this demonstrates that a majority of the most prominent characteristics of atmospheric blocking are captured by the model. Strictly speaking, this does not prove that atmospheric blocks can be caused by the synoptic-scale forcing. After all, the flow evolution in the model is not the same as that in the atmosphere, and the forcing acts in flow fields which are not those observed. However, if we can demonstrate that the model flows are

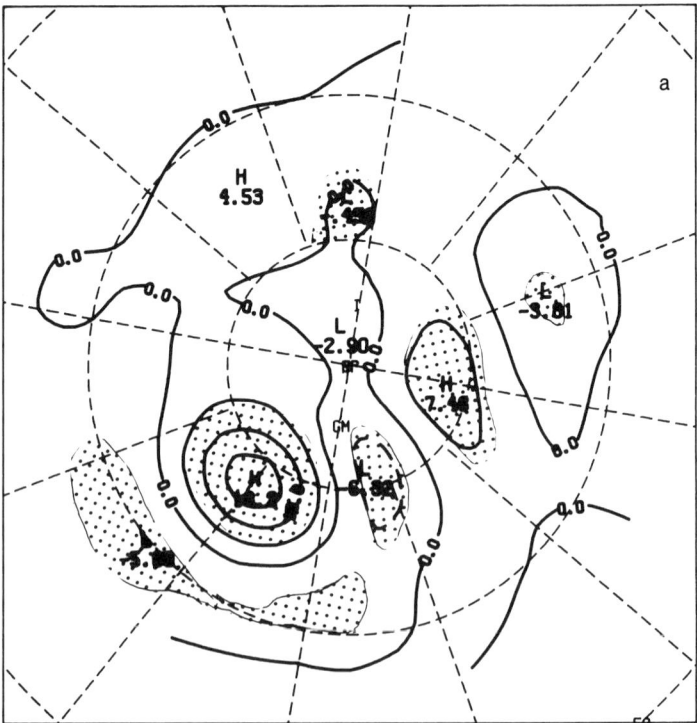

FIG. 7. Mean streamfunction (10^6 m^2 sec^{-1}) anomaly for wintertime blocking in the Atlantic (40°W to 10°W); a, observations; b, linear model result. Stippling: 95% confidence area. Adapted from Metz (1986).

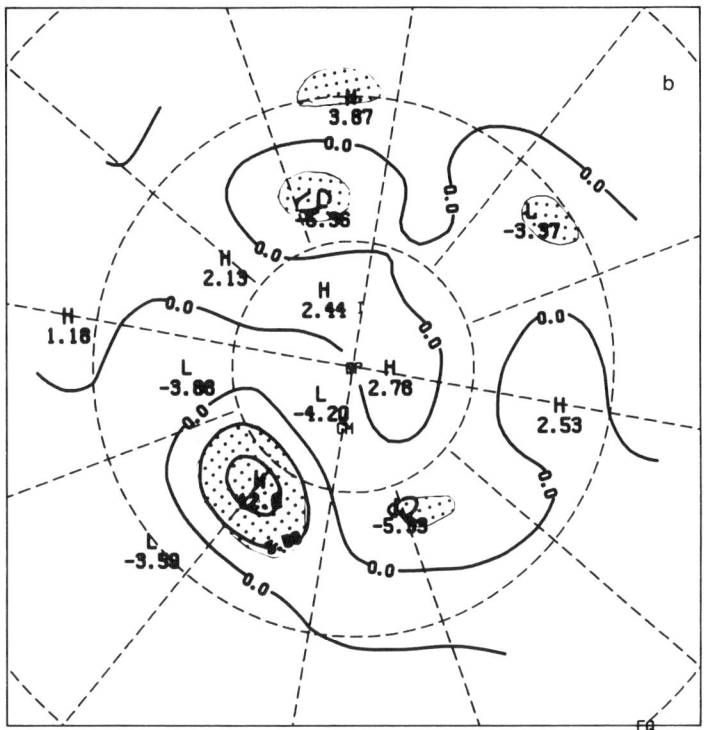

Fig. 7b.

similar to the observed flow over time, this will be a strong argument in favor of the forcing hypothesis. Metz (1986) has computed mean model flow maps averaged over all those days wherein Atlantic (Pacific) blocks have been observed in the atmosphere. For the Pacific blocks the result was negative. The mean flow obtained hardly resembles the observed mean flow for Pacific blocking. However, the agreement for the Atlantic cases is surprisingly good both for the linear and the nonlinear run (Fig. 8). We obtain the familiar high over the Atlantic almost at the correct position. This means that many of the Atlantic model blocks occur on the same days as in the atmosphere. As for the Pacific, possibly a data problem exists since the forcing fields are subject to large errors in data-sparse areas. It is also conceivable that synoptic-scale forcing is not so important for the Pacific blocks.

The results presented lead to tentative conclusions:

1. The barotropic vorticity equation can be used even in a linearized form to study the impact of synoptic-scale forcing on the planetary flow.
2. The stochastic forcing creates blocking patterns in linear and nonlinear

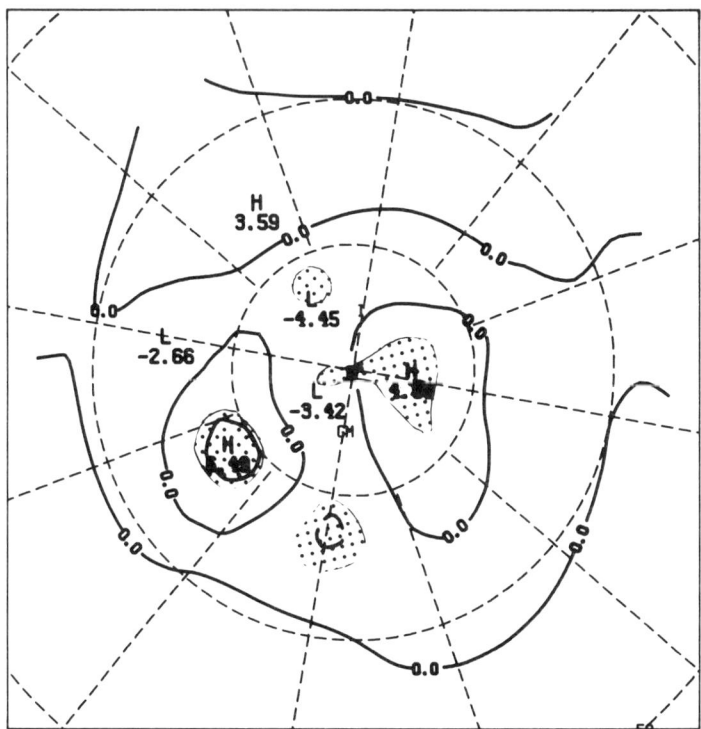

Fig. 8. Mean streamfunction anomaly (10^6 m^2 sec^{-1}) in the model when the model flow is averaged over episodes of observed blocking. Stippling: 95% confidence areas. Adapted from Metz (1986).

planetary-scale flow. Location and duration of these enforced blocking ridges show reasonably good agreement with the observations, although local preferences are not as pronounced in the model results.

This suggests that the forcing of planetary-scale modes by the nonlinear interaction with synoptic-scale weather systems contributes considerably to the creation and maintenance of blocking in the Northern Hemisphere. This is not to say that other mechanisms are unimportant. After all, it has been shown in case studies that this nonlinear transfer is not the only mechanism. Moreover, we have to keep in mind that the response of the model to the forcing may be too large since we chose relatively small damping rates. With stronger damping only a few of the blocking situations in the model would satisfy the amplitude criterion. Additional mechanisms would be required in order to create strong blocking anticyclones.

Let us close with a comment on the forcing problem. The stochastic forcing in the numerical experiment is prescribed and, therefore, not influenced at all by the planetary flow in the model. In particular, the blocks in the model are enforced by the synoptic-scale eddies but these eddies do not "feel" the block in the model. Nevertheless, the blocking climatology of the model has been found to be reasonably realistic. This suggests that the influence of the planetary flow on the synoptic-scale forcing may be minor. However, it would be premature to draw such a conclusion. A significant interdependency of forcing and large-scale flow may be found in reality despite the fact that the model produces reasonably realistic results with independent forcing. This is suggested by the observed strong local preference of the eddy forcing and by instability calculations like those of Frederiksen (1983).

References

Benzi, R., Hansen, A., and Sutera, A. (1984). On stochastic perturbation of simple blocking models. *Q. J. R. Meteorol. Soc.* **110**, 393–409.
Charney, J., and DeVore, J. (1979). Multiple flow equilibria in the atmosphere and blocking. *J. Atmos. Sci.* **37**, 1157–1176
Chen, T.-C., and Shukla, J. (1983). Diagnostic analysis and spectral energetics of a blocking event in the GLAS climate model simulation. *Mon. Weather Rev.* **111**, 3–22.
Colucci, S., Loesch, A., and Bosart, L. (1981). Spectral evolution of a blocking episode and comparison with wave interaction theory. *J. Atmos. Sci.* **38**, 2092–2111.
Egger, J. (1978). Dynamics of blocking highs. *J. Atmos. Sci.* **35**, 1788–1801.
Egger, J. (1981). Stochastically driven large-scale circulations with multiple equilibria. *J. Atmos. Sci.* **38**, 2606–2618.
Egger, J., and Schilling, H.-D. (1983). On the theory of the long-term variability of the atmosphere. *J. Atmos. Sci.* **40**, 1073–1085.
Egger, J., and Schilling, H.-D. (1984). Stochastic forcing of planetary scale flow. *J. Atmos. Sci.* **41**, 779–788.
Fischer, G. (1984). Spectral energetics analyses of blocking events in a general circulation model. *Cont. Atmos. Phys.* **57**, 183–200.
Frederiksen, J. (1983). Disturbances and eddy fluxes in Northern Hemisphere flows in stability of three-dimensional January and July flows. *J. Atmos. Sci.* **40**, 836–855.
Green, J. (1977). The weather during July 1976: Some dynamical considerations of the drought. *Weather* **32**, 120–128.
Hansen, A., and Chen, T.-C. (1982). A spectral energetics analysis of atmospheric blocking. *Mon. Weather Rev.* **110**, 1146–1159.
Hansen, A., and Sutera, A. (1984). A comparison of the spectral energy and enstrophy budgets of blocking compared to non-blocking periods. *Tellus* **36A**, 52–63.
Illari, L., and Marshall, L. (1983). On the interpretation of maps of eddy fluxes. *J. Atmos. Sci.* **40**, 2232–2242.
Kalnay-Rivas, E., and Merkine, L. O. (1981). A simple mechanism for blocking. *J. Atmos. Sci.* **38**, 2077–2091.
Metz, W. (1986). Transient cyclone-scale vorticity forcing of blocking highs. *J. Atmos. Sci.*, submitted.

Müller, G. (1984). Blockierungen in einem Modell mit stochastischer Forcierung. Diplomarbeit, Meteorol Inst. Univ. München.

Schilling, H.-D. (1982). A numerical investigation of the dynamics of blocking waves in a simple two-level model. *J. Atmos. Sci.* **39,** 998–1017.

Schilling, H.-D. (1986). On atmospheric blocking types and blocking numbers. *Adv. Geophys.* **29,** 71–99.

Shutts, G. (1983). The propagation of eddies in diffluent jet-streams: eddy vorticity forcing of "blocking" flow fields. *Q. J. R. Meteorol. Soc.* **109,** 737–761.

Treidl, R. A., Birch, E., and Sajecki, P. (1981). Blocking action in the northern hemisphere: A climatological study. *Atmos. Ocean* **19,** 1–23.

Tung, K.-K, and Lindzen, R. (1979). A theory of stationary long waves. Part I: A simple theory of blocking. *Mon. Weather Rev.* **107,** 714–734.

DETERMINISTIC AND STATISTICAL PROPERTIES OF NORTHERN HEMISPHERE, MIDDLE LATITUDE CIRCULATION: MINIMAL THEORETICAL MODELS

A. Speranza*

*Consiglio Nazionale Ricerche–
Istituto Fisica Bassa e Alta Atmosfera
and Institute of Physics
University of Bologna
I-40126 Bologna, Italy*

1. Introduction

Although the notion of "weather types" is quite old in meteorology, the physical discussion concerning the definition, interpretation, and modeling of "anomalous" atmospheric circulations is relatively recent. The papers which addressed the fundamental questions concerning the structure of phase space in the atmospheric system all appeared in 1979: in a sequence of works, Lindzen and Tung (Tung and Lindzen, 1979a,b; Tung, 1979) give a full account of the linear theory of three-dimensional, neutral planetary waves in the middle latitude circulation. They discussed the possible interpretation of blocking anomalies in terms of neutral Rossby waves of particularly large amplitude modulated by zonal nonhomogeneities in the boundary conditions (topography) or in the external forcing (heat sources and sinks). Low-frequency variability was interpreted to be caused by the linear "adjustment" of the circulation to changes in the basic zonal wind, the evolution of which was conceived as determined by independent dynamics. In statistical terms, the picture that emerged was that of an atmospheric circulation performing "small" (linear) oscillations around an equilibrium point essentially determined by the dynamics of the zonal flow (not discussed in the above-mentioned papers).

Charney and DeVore (1979) and Wiin-Nielsen (1979) studied essentially the same linear dynamics of planetary-wave excitation by zonal nonhomogeneities, only with the addition of an explicit law of zonal wind time evolution in terms of the amplitudes of the waves themselves. Minimal truncations of the motion equations were shown to produce a more complex structure of the phase space, characterized by the presence of different equi-

---

* Most of the work presented here has been performed in cooperation with Dr. R. Benzi of the IBM Scientific Center in Rome, Italy.

librium basins. The meteorological implication of the presence of multiple equilibria was suggested to be the existence of different persistent types of atmospheric circulations.

Successive papers aimed at extending the theory to more realistic model atmospheres (Charney and Straus, 1980; Charney et al., 1981) showed how the complexity of the problem increases considerably with the number of degrees of freedom. This impression was confirmed by later papers. Reinhold and Pierrehumbert (1982), for example, extending the model of Charney and Straus to include all additional baroclinic waves in the truncation, found an aperiodic vacillation between two distinct weather regime states, not located near any of the stationary equilibria of the Charney–Straus type of truncated model.

In view of the difficulty of the problem, it appears advisable to reexamine minimal models of general circulation in the light of known observational phenomenology, to try to organize a hierarchy of models in which the increase in complexity is maintained at each step at the minimum necessary to represent the essential physical symmetries needed to satisfy observational requirements.

To begin with, we concentrate on the statistical properties of the general circulation, leaving other properties (detailed space–time structure of anomalies, etc.) as a subject of future work. In Section 2, we reexamine some properties of the Charney and DeVore (CDV) model which play a key role in determining its statistical properties. In Section 3, we propose, still within the context of barotropic theory, some modifications of CDV theory, aimed at clarifying several of the obscure points concerning the statistical properties of Northern Hemisphere circulation. We discuss an extension to incorporate baroclinity in Section 4 and in Section 5 we draw some tentative conclusions. Observational studies regarding the statistical properties discussed here are presented in the companion articles by A. Sutera and A. Hansen in this volume.

2. A Reexamination of CDV

Although a number of interesting questions could be raised concerning the procedure of projection and truncation of the barotropic equation leading to the CDV model, we will assume this procedure known and start from the CDV system written in the unidimensional formulation outlined by Hart (1979) and extensively used by Buzzi et al. (1984).

Our preference for unidimensional formalism is not only due to its intrinsic simplicity, but also to the fact that it provides us with a field theory characterized by some useful mathematical properties. In particular, the stationary solutions of the unidimensional problem we intend to examine are solutions of the barotropic equation. The same is not true for the stationary solutions of the original CDV system. As a matter of fact, this difficulty is encountered every time, following a classical procedure, the equations of motion are projected onto the eigenfunctions of the Laplace operator. Such eigenfunctions, in fact, form a complete set, but their linear combinations are not necessarily solutions of the stationary nonlinear problem: the stationary solution of any truncated version of the projected equation is not, in general, a solution of the original stationary problem.

In the case of unidimensional formalism, on the other hand, we shall deal with stationary solutions that are solutions of the full barotropic equation. This property may prove to be crucial to an appropriate representation of localized (soliton type) solutions.

Given these premises, we proceed with the analysis of the unidimensional CDV model. Topography is assumed to be of the form

$$h(x) = h_0 \cos K_m x \tag{1}$$

The streamfunction is composed of a uniform zonal flow plus a wave of the same shape as the topography:

$$\psi(x, t) = -U(t)y + A(t) \cos K_m x + B(t) \sin K_m x \tag{2}$$

The wave is maintained by zonal flow over the topographic slope and dissipated by friction:

$$\dot{A} + (v/K_m)A + (U - \beta/K_m^2)B = 0$$
$$\dot{B} - (U - \beta/K_m^2)A + (v/K_m)B + (f_0 h_0/K_m^2 H)U = 0 \tag{3}$$

and the zonal flow is forced by some momentum convergence mechanism U^*, dissipated by linear friction, and accelerated or decelerated by mountain drag:

$$\dot{U} = (f_0 h_0 K_m/4H)B - v(U - U^*) \tag{4}$$

Equilibrium solutions of Eqs. (3) and (4) can be found easily by solving the linear system [Eq. (3)] for B (the amplitude of the wave component out of phase with respect to topography) as a function of U and by searching for the intersection of the resulting curve $B(U)$ with the linear relationship corresponding to the stationary version of Eq. (4), as shown in Fig. 1.

Once the appropriate points in the plane B, U are selected, A is also determined along the curve $A(U)$. By means of stability analysis, equilibrium E_1 was identified by CDV as zonal, E_2 as blocked, and the third, E_3, as an intermediate equilibrium, unstable with respect to a new kind of instability,

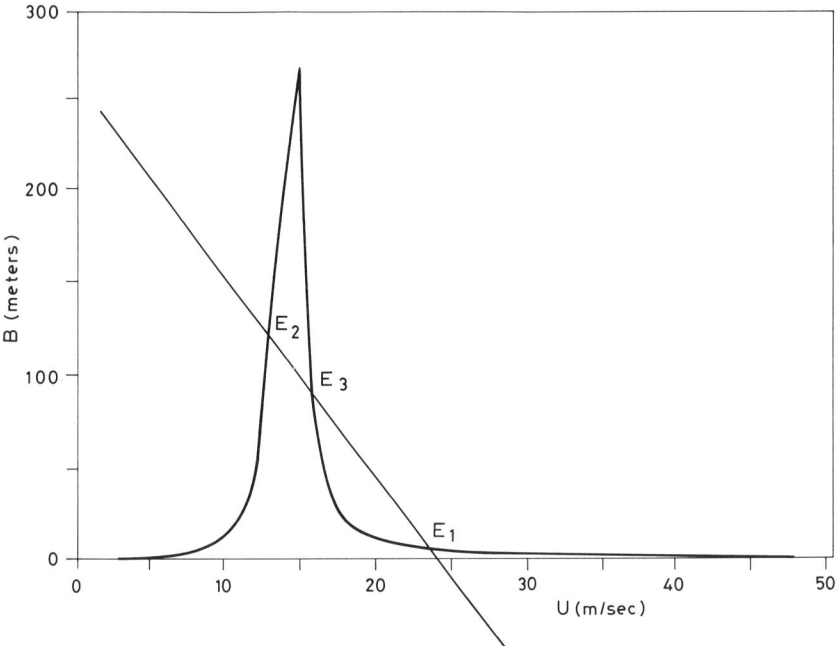

FIG. 1. Out-of-phase amplitude of the resonant wave for the one-dimensional model, Eq. (3). The amplitude is expressed in meters and the resonant wind is about 12 m sec^{-1}. The straight lines in the figure represent the stationary version of the form–drag equation, Eq. (4), as in CDV for different value of the momentum forcing. The E_1, E_2, and E_3 are stationary (respectively "zonal," "blocked," and "intermediate") solutions.

named "orographic" because of the role played in its growth by the orographic drag action appearing in Eq. (4).

It is clear that the presence of multiple equilibria is due to a simple folding of the state surface in the direction U^* in parameter space. Although the physics is simple and interesting, one already notices some of the theory's inadequacies, even at this early stage. First, not all the locations of the equilibria along the U axis are realistic[1]: this zonal flow corresponds to speeds which are never observed in the real atmosphere (we shall return to this point later). Moreover, the phase of the blocked wave is negative (troughs over ridge), again, against observational evidence. Such distortions of the observed dynamics can perhaps be tolerated in such a simple model. However, even more serious inadequacies are found in the statistical properties of the CDV model atmosphere.

---

[1] Notice that in Fig. 1 the slope in the B,U plane of "form–drag" relationship Eq. (4) exceeds by a factor of 4 that of CDV because of the different choice of latitudinal zonal flow shape and is, therefore, already improved with respect to the original CDV formulation.

Given the equilibrium structure of phase space illustrated in Fig. 1, it is obvious that the asymetric time behavior of the time-dependent system [Eqs. (3) and (4)] is characterized by only two possibilities: orbits in phase space can, in a certain span of time, die either in E_1 or in E_2. In order to produce transitions between zonal and blocked states, we have to superimpose on the stationary solution some sort of perturbation. This can be done by increasing the dimensionality of the model or, more simply, by taking into account the effects of small random noise on the deterministic CDV dynamics. This was done by Egger (1981), who integrated numerically the Fokker–Plank equation for the evolution of probability density in phase space of the stochastically perturbed CDV system; Egger obtained the intuitive result that the asymptotic statistics of occupation of states in the CDV system are characterized by maxima centered at the stable equilibria. The same problem was also studied by Benzi *et al.* (1984a), with an analytic technique for estimating the expectation values of exit times. From the potential wells of the CDV theory, use of this analytical technique, instead of numerical integration, makes the overall picture of the model's statistical properties more controllable.

The estimate is based on the assumption

$$1/v \ll 1/\epsilon \tag{5}$$

where $1/v$ is the time typical of dissipative processes and $1/\epsilon$ is the time scale of random noise.

In the limit Eq. (5) it can be proved that the behavior of the entire system is approximated by the single stochastic differential equation for the zonal wind:

$$dU = -(\partial V/\partial U)\, dt + \epsilon^{1/2}\, dW \tag{6}$$

where $V(U)$ is the equivalent of a potential for the motion of perturbed CDV system in the phase space, and W is a Wiener process with amplitude $\epsilon^{1/2}$. The potentials are shown in Fig. 2. The corresponding stationary probability density is

$$P(U) = N \exp[-2V(U)/\epsilon] \tag{7}$$

where N is a normalization constant.

Figure 3 shows the probability distributions corresponding to the potentials of Fig. 2. In the limit of $\epsilon \to 0$, the expectation values $E[\ ]$ of the exit times $\tau(\ ,\ )$ can be estimated as

$$E[\tau(E_2, E_2)] \lesssim \pi |V''(U_2) \cdot V''(U_3)|^{-1/2} \exp(2\Delta_1 V/\epsilon)$$
$$E[\tau(E_2, E_2)] \lesssim \pi |V''(U_1) \cdot V''(U_3)|^{-1/2} \exp(2\Delta_2 V/\epsilon) \tag{8}$$

where $E_{1,2,3}$ represent the equilibria and $\Delta_1 V = V_3 - V_2$, $\Delta_2 V = V_3 - V_1$,

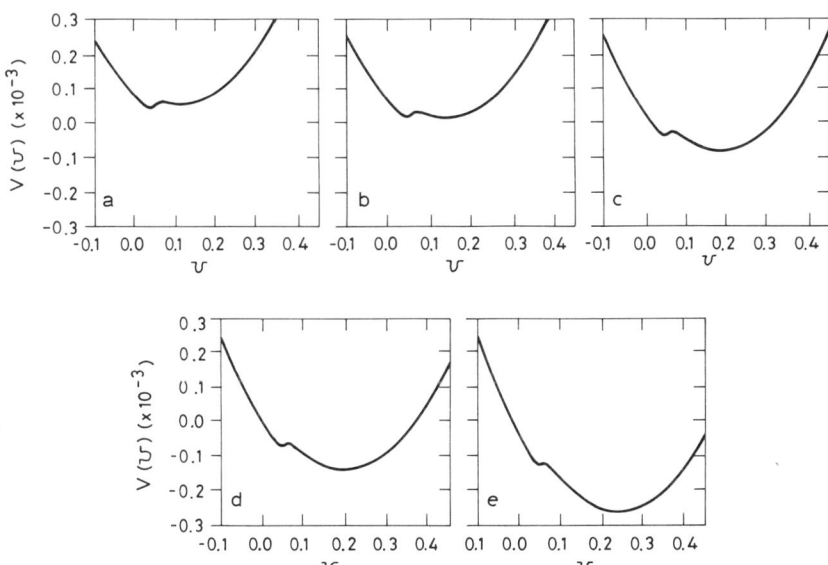

FIG. 2. The CDV model potential $V(U)$, for $h_0/H = 0.05$ as a function of the zonal forcing U^*. (a) $U^* = 0.12$; (b) $U^* = 0.14$; (c) $U^* = 0.18$; (d) $U^* = 0.20$; (e) $U^* = 0.24$. (From Benzi et al., 1984a.)

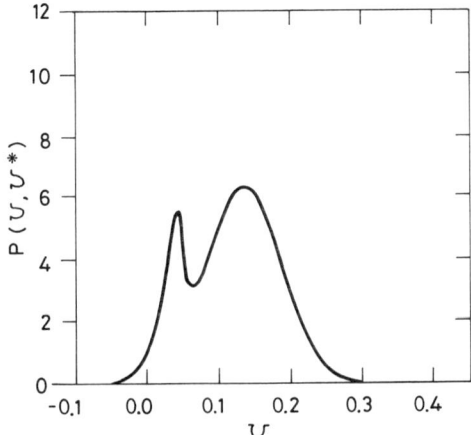

FIG. 3. Normalized probability distribution for $U^* = 0.14$, $h_0/H = 0.05$, and $\nu = 0.01$. $P(U)$ from Eq. (9) with $\epsilon = 5 \times 10^{-5}$. (From Benzi et al., 1984b.)

the potential barriers to be overcome in order to escape from the blocked and zonal potential wells.

Barrier heights and exit times are shown in Table I. It is clear that dramatic changes take place when the external forcing U^* changes. The structural dependence on U^* can be stabilized if we include red noise of amplitude E_1 and decorrelation time $1/\Delta\omega$ in the zonal momentum equation. The effect of the inclusion of red noise is in general to stabilize the unstable state. The corresponding probability distribution can be proved to be

$$P(U) = \frac{1}{R^2 + g^2(U)} \exp\left(\frac{4\Delta\omega}{\epsilon_1} \int \frac{\overline{\overline{F}}}{R^2 + g^2(U)} dU\right) \quad (9)$$

where $R = 2\Delta\omega\epsilon/E_1$ and $\overline{\overline{F}}$ and g are functions of U defined in Benzi et al. (1984).

In this case, the probability distributions take the form shown in Fig. 4. It can be seen that the inclusion of red noise reduces the occupation of the zonal state, while leaving the probability of the blocked state relatively unchanged. This behavior is due to the presence of a minimum of the function $g(U)$ at the resonant value of the zonal wind and to the fact that the resonance value is located between the zonal and the blocked ones.

The general impression at this point is that we are dealing with a mechanism of zonal-flow evolution which is basically unrealistic. The wave equation, on the other hand, seems to be, within the limits of a minimal theory,

TABLE I. BARRIER HEIGHTS AND EXIT TIMES FOR THE CDV MODEL AS A FUNCTION OF THE ZONAL FORCING U^{*a}

U^* (N.D.)	$V(1) - V(2)$ ($\times 10^{-5}$)	$V(2) - V(3)$ ($\times 10^{-5}$)	τ(block) (days)	τ(zonal) (days)
0.10	2.17	0.004	76.6	180.1
0.12	1.76	0.61	18.1	43.6
0.14	1.48	1.79	10.8	49.8
0.16	1.24	3.43	7.6	79.2
0.18	1.05	5.51	5.8	159.5
0.20	0.88	8.02	4.6	395.6
0.22	0.73	11.00	3.9	1.2×10^3
0.24	0.61	14.30	3.4	4.3×10^3
0.26	0.49	18.10	3.0	1.8×10^4
0.28	0.40	22.20	2.8	9.6×10^4
0.30	0.31	26.80	2.7	5.8×10^5

[a] The nondimensional (N.D.) height of topography is $h_0/H = 0.05$. The indices 1, 2, and 3 refer to blocked, zonal, and intermediate states; τ(block) and τ(zonal) are times for the blocking-to-zonal and zonal-to-blocking transitions, respectively. (From Benzi et al., 1984a.)

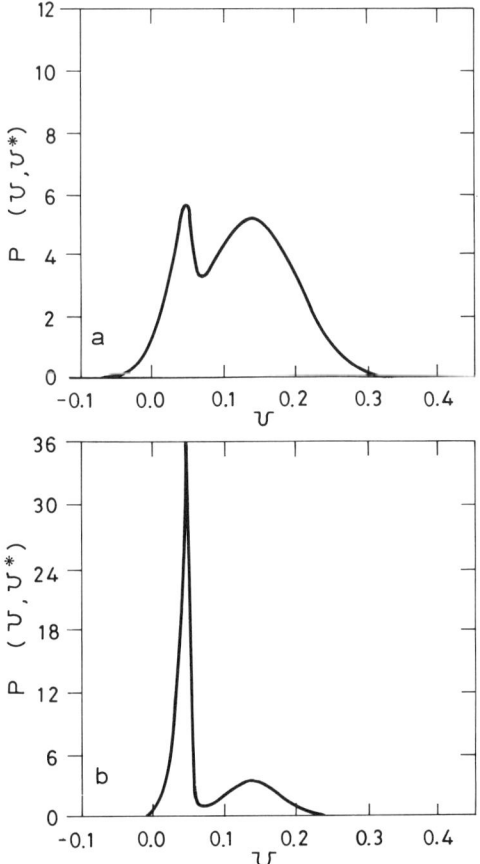

FIG. 4. Same as Fig. 3, but with the addition of red noise. (a) $P(U)$ with $\epsilon = 5 \times 10^{-5}$, $\epsilon_1 = 2 \times 10^{-4}$, and $\Delta\omega = 0.02$; (b) $P(U)$ with $\epsilon_1 = 2 \times 10^{-4}$, $\Delta\omega = 0.02$, but with $\epsilon = 0$. (From Benzi et al., 1984a.)

acceptable. A possible explanation of such difference between the wave and the zonal-flow equations is that the presence of a resonance in the wave equation makes it easily approximable with a one-wave representation, while the zonal-momentum equation has no such resonance and is, therefore, not suitable for a one-wave representation. The fact that the one-mode representation is a poor approximation of the surface drag integral can be easily verified with the use of observational data.

To give a more physical idea of the nature of the problem, we have represented in Fig. 5 the distributions of instantaneous states in the CDV parameter space B,U computed from observations of 500-mbar height during two winters (1977/1978 and 1978/1979). In order to follow the analogy

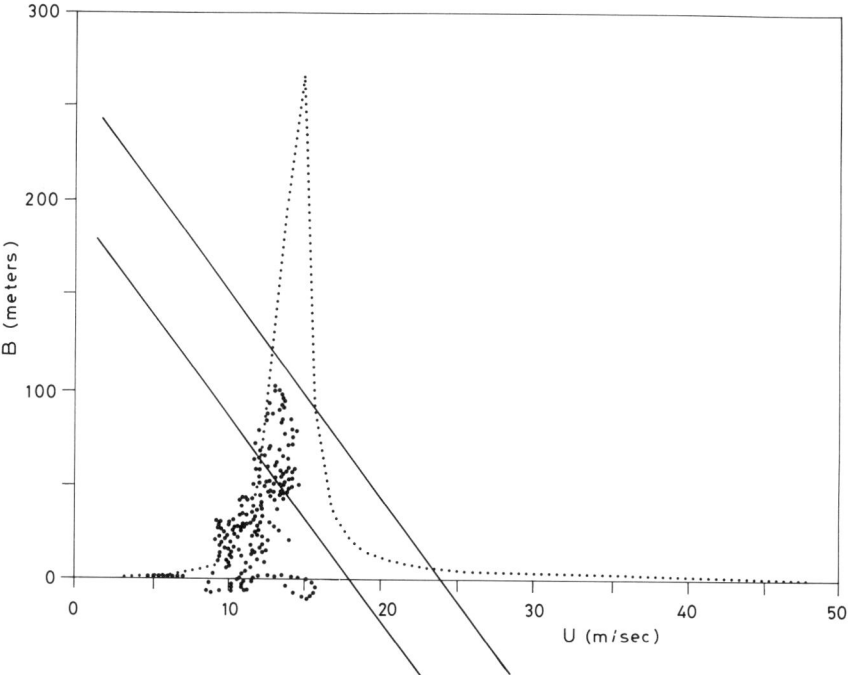

FIG. 5. Plot of the linear resonance of the CDV model and the form–drag law for different values of U^*. The circles represent the amplitude of wavenumber 3, computed from observed data of winters 1977/1978 and 1978/1979.

with the quasi-unidimensional model, the variables are integrated between 30°N and 70°N. The distribution of observed values of zonal wind appears much less dispersed than in CDV. No similar disagreement with the range of observed variability can be noticed in the wave amplitude.

It is clear that, if we accept the observed zonal-momentum distribution as a phenomenological law and still insist on the idea of multiple equilibria, we need a mechanism capable of equilibrating finite amplitude of the wave in a limited range of zonal wind values.

This is not possible in CDV because of the linear structure of the wave equation, which does not allow the existence of different wave amplitudes for the same zonal wind U. However, insertion of nonlinearity in the wave equation can easily "bend" the resonance. The easiest[2] way of producing resonance bending consists of introducing weak nonlinearity into the solu-

---

[2] More complex forms of nonlinearity can produce quite realistic "localized" stationary flows, as shown by Malanotte Rizzoli and Malguzzi (1984) and Pierrehumbert and Malguzzi (1984). The theory we discuss here is, however, essentially "global."

tions of the stationary, inviscid equation (the MS model: Malguzzi and Speranza, 1981).

It is not difficult to prove that in the quasi-unidimensional limit, assuming a cubic functional $q(\psi)$, we obtain an equation of the form

$$d_{xx}\psi + \beta y + (f_0/H)h = -(\beta/U)\psi - b\psi^3 \tag{10}$$

Assuming, however, topography and nonlinearity to be both small and of the same order, we can stabilize at finite amplitude the zero-order (linear, free wave) solution of Eq. (10):

$$\psi^{(0)} = A \exp(i\sqrt{\beta/U}x) + (*) - Uy \tag{11}$$

obtaining for the slowly varying ($X = \epsilon x$), slightly detuned (with respect to resonant wavenumber $\Delta K = K_m - \sqrt{\beta/U}$) amplitude

$$A(X, y) = \Psi_0(y) \exp(i \Delta KX) + (*) \tag{12}$$

a cubic equation

$$\Psi_0^3 + \left[U^2 y^2 - \frac{2\sqrt{\beta/U}\Delta K}{3b} \right]\Psi_0 + \frac{f_0 h_0}{3Hb} = 0 \tag{13}$$

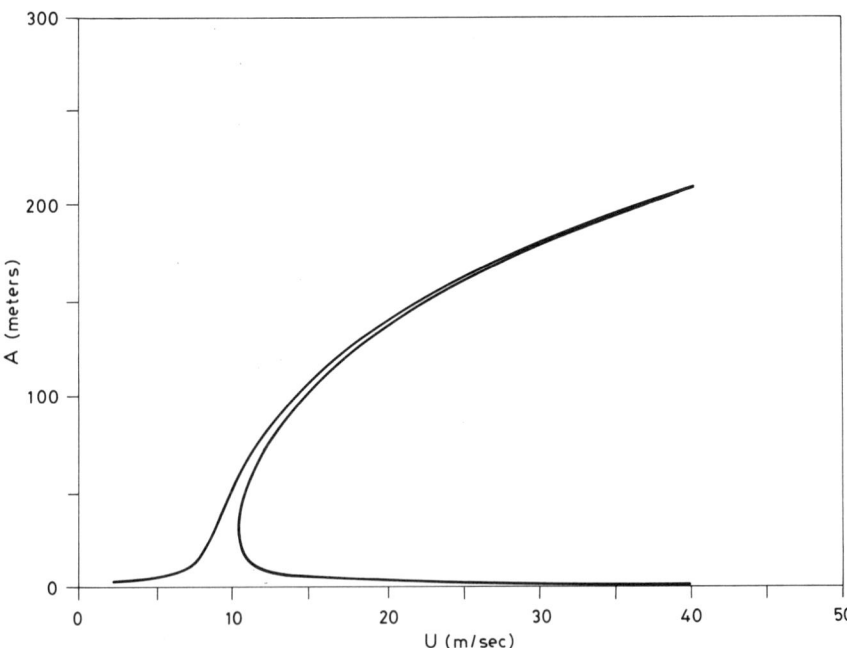

FIG. 6. Amplitude of the resonant wave for Eq. (10) as a function of the zonal wind. (After Malguzzi and Speranza, 1981.)

which can easily be solved by giving the type of bending shown in Fig. 6.

The MS model, however, is inadequate for direct incorporation into the CDV scheme since it is essentially inviscid. In this case, in fact, the wave component out of phase with respect to topography is identically zero. The form-drag relationship Eq. (4) for zonal wind has consequently no projection on the parameter space shown in Fig. 6.

In the next section, we shall see how we can produce a theory which incorporates the realistic features of both CDV and MS and produces the right statistics in the theoretical framework of Benzi et al. (1984a).

3. Modification of the CDV Wave Equation

Before proceeding with the necessary modification of CDV, it is useful to consider briefly the nature of the different nonlinearities appearing in the quasi-geostrophic equations.

As we have seen, the CDV nonlinearity, characterizing the feedback on zonal momentum of mountain waves, produces a folding of the surface of equilibrium wave amplitude in the U^* direction in parameter space, but not, as required by observed data, along the U axis. The nonlinearity of the functional in the inviscid barotropic equation, discussed in the previous section, produces folding in U, but only of the wave component in phase with respect to topography.

Nonlinearity of the Jacobian operator has two distinct parts: wave-wave interaction and wave-zonal flow interaction. The latter has very frequently been considered in theoretical studies (see, for example, Pedlosky, 1981), while the former has usually been neglected. This is because, in the usual one-wave representation of the wave field near neutral stability as an eigenfunction of the Laplace equation, direct wave self-interaction vanishes. In the following theoretical treatment of equilibrium solutions of the barotropic equation (Benzi et al., 1984b), however, we prefer to consider as basic nonlinearity (that bends the resonance in the B,U plane) the self-interaction of waves. Here again, of crucial importance is the choice of the eigenfunction in terms of which the streamfunction field is written: our choice is that of separated modes of the form $A(x, t)g(y)$ with a latitudinal structure such as to guarantee wave self-interaction. In mathematical terms, we expand the barotropic wave equation with dissipation in terms of functions of the form

$$\psi(x, y, t) = \sum_n A_n(x, t)g_n(y)$$
$$h(x, y) = \sum_n h_n(x)g_n(y)$$
(14)

and, multiplying by $g_n(y)$, integrating in y and keeping only one mode, obtain

$$\{\partial_t + U\partial_x\}(A_{1xx} + \alpha A_1) + (f_0/H)Uh_x + \beta A_{1x} + \tfrac{3}{2}\delta A_{1x}A_1$$
$$- \nu(A_{1xx} + \alpha A_1) = \gamma\{A_1[A_{1xx} + (f_0/H)h] - A_1[A_{1xx} + (f_0/H)h_x]\} \quad (15)$$

where the nonlinear self-interaction coefficients have been defined as ($\langle\ \rangle \equiv \int dy$):

$$\alpha \equiv \langle g_1 g_{1yy}\rangle$$
$$\gamma \equiv \langle g_1^2 g_{1yy}\rangle \quad (16)$$
$$\delta \equiv \langle g_1 g_{1y} g_{1yy}\rangle$$

The nature of the coefficients defined in Eq. (16) clarifies the meaning of the preceding remarks concerning nonlinearity in our model. This nonlinearity depends on the latitudinal structure of the modes: it would vanish for an eigenfunction of Laplace operator of the kind considered by Pedlosky (1981) and many others. It can be proved (Speranza et al., 1985) that inclusion of local topography, for example, is able to produce, in the quasi-neutral modes, the latitudinal structure needed to produce self-interaction. The ordering of parameters that turns out to be necessary for bending the resonance in CDV parameter space is (ϵ is a small parameter)

$$\delta = 0(\epsilon); \quad \delta \to \epsilon\delta$$
$$h_0 = 0(\epsilon^2); \quad h_0 \to h_0\epsilon^2 \quad (17)$$
$$\nu = 0(\epsilon^2); \quad \nu \to \nu\epsilon^2$$

which, in physical terms, means that Jacobian interaction is stronger than topographic action and frictional dissipation.

Introducing long time and space scales

$$T = \epsilon^2 t; \quad X = \epsilon^2 t \quad (18)$$

and assuming the amplitude of the wave mode is of the form

$$A_1(x, X, t, T) = A^{(0)}(x, X, T) + \epsilon A^{(1)} \quad (19)$$

we obtain[3], from Eq. (14), at zero order in ϵ

$$(U_\alpha + \beta)A_x^{(0)} + U A_{xxx}^{(0)} = 0 \quad (20)$$

The solution of Eq. (19) is

$$A^{(0)} = B(X, T)\exp(iK_s x) + (*) \quad (21)$$

---

[3] Notice that we substitute here $\delta \to \delta\epsilon$, $h_0 \to h_0\epsilon^2$, and $\nu \to \nu\epsilon^2$ according to Eq. (17).

where

$$K_s^2 = (U\alpha + \beta)/U \qquad (22)$$

that is a modified wavenumber of the stationary Rossby wave [Eq. (20)]. At first order in ϵ, we obtain

$$(\beta + U\alpha)A_x^{(1)} + UA_{xx}^{(1)} + \delta A^{(0)} A_x^{(0)} = 0 \qquad (23)$$

A particular solution of Eq. (22) is

$$A^{(1)} = -\frac{\delta|B|^2}{\beta + \alpha U} - \frac{\delta|B|^2 \exp(i2K_s x)}{2[(\beta + U\alpha) - K_s^2 U]} \qquad (24)$$

At the second order in ϵ, secular terms appear and the consequent solution condition reads:

$$\partial_T[(\alpha^2 - K_s^2)B] + iK_s U \frac{f_0}{H} h - 2(\beta + \alpha U)B_X - \frac{7}{6}\frac{i\delta^2 K_s B|B|^2}{(\beta + \alpha U)}$$
$$= -\nu(\alpha^2 - K_s^2)B \quad (25)$$

The introduction of slight detuning $\Delta K = K_m - K_s$ allows us to write

$$h(X) = h_0 \exp(i \Delta KX) + (*) \qquad (26)$$

which, introduced into Eq. (24), gives the stabilization equation

$$\partial_T[(\alpha^2 - K_s^2)B] + iK_s U \frac{f_0}{H} h - 2i\Delta K(\beta + \alpha U)B - \frac{7}{6} i \frac{\delta^2 K_s B|B|^2}{\beta + \alpha U}$$
$$= -\nu(\alpha^2 - K_s^2)B \quad (27)$$

The stationary version of this equation is a cubic condition, similar to that of Eq. (13) of the preceding section, which produces resonance bending in the CDV parameter space (as shown in Fig. 7). Although the expansion outlined above is based on *ad hoc* assumptions, the phenomenology of resonance bending it describes is very robust and can easily be observed in many physical systems. Moreover, with particular reference to the barotropic equation, such phenomenology is confirmed by numerical experiments with both finite-difference (Rambaldi and Mo, 1984) and spectral (Legras and Ghil, 1983) models of very high spatial resolution. Figure 8 shows, for example, the shape of a nonlinear resonance appearing in the stationary solutions obtained by Legras. Since Legras resonance is in the wave energy–zonal energy space, it must be compared to our resonance in an analogous parameter space (Fig. 8b). The similarity is obvious. Also, Rambaldi's results for barotropic flow in a channel confirm the fact that the solutions identified by means of perturbation techniques are good approximations or real solutions to the full field equations. A thorough analysis

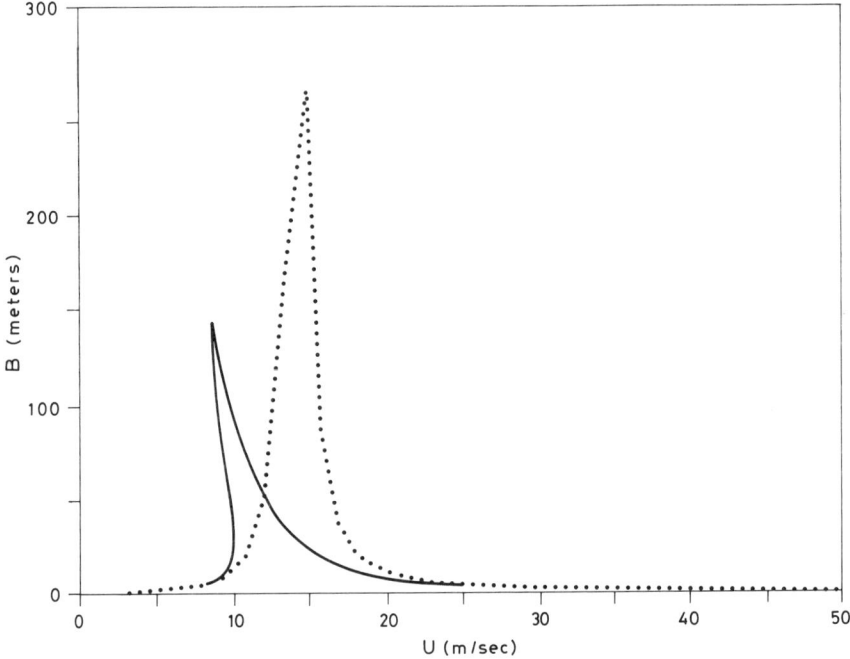

FIG. 7. Amplitude of the resonant wave of Eq. (25) (component out of phase with respect to topography), computed from Eq. (27). The resonant wind is the same as in Fig. 5. The dashed curve is the linear resonance.

shows, however, that both the Rambaldi–Mo and the Legras–Ghil results essentially depend on the wave–zonal flow interaction.

The structure of equilibrium solutions that we have drawn above is obviously apt to produce a time-dependent behavior with statistical properties in essential agreement with observations. However, in order to write a consistent dynamical system, we must close Eq. (3) with a prognostic equation for zonal wind. Here, we face the essential difficulty of all barotropic theories: in order to produce orographic waves of the observed amplitudes, we must postulate an orographic drag much larger than the observed one. The problem is most easily expressed in terms of the energetics of the barotropic models. For a nonlinear wave equation of the type Eq. (15), with the addition of some forcing and/or dissipation F_A

$$\partial_t(A_{xx} + \alpha A) + UA_{xxx} + \alpha UA_x + f_0/Hh_x + \beta A_x + \tfrac{3}{2}\delta AA_x = F_A \quad (28)$$

closed with a form–drag type of equation for zonal flow, again with the addition of forcing and/or dissipation F_U:

$$\partial_t U + (f_0/H)\overline{h_x A}^x = F_U \quad (29)$$

we obtain the energy equations

$$\partial_t\left(-\frac{\overline{A_x^2}^x}{2} + \alpha\frac{A^2}{2}\right) + \frac{f_0}{H}\overline{Uh_xA}^x = \overline{F_AA}^x$$

$$\partial_t\left(\frac{U^2}{2}\right) + \frac{f_0}{H}\overline{h_xA}^xU = \overline{F_UU}^x \tag{30}$$

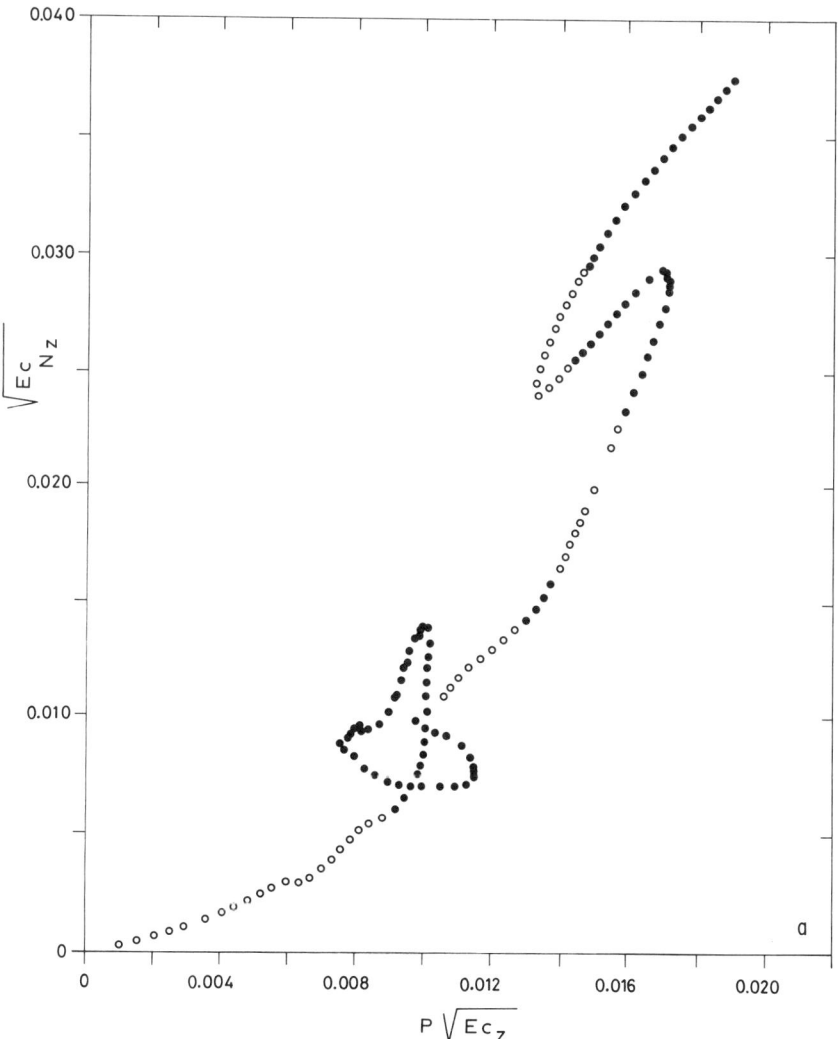

FIG. 8. (a) Equilibrium solutions of the barotropic model on the sphere of Legras; the energy of the states is plotted against a nondimensional measure of zonal forcing (from Legras and Ghil, 1983). (b) Total amplitude of the resonant wave of Eq. (25) as a function of zonal wind.

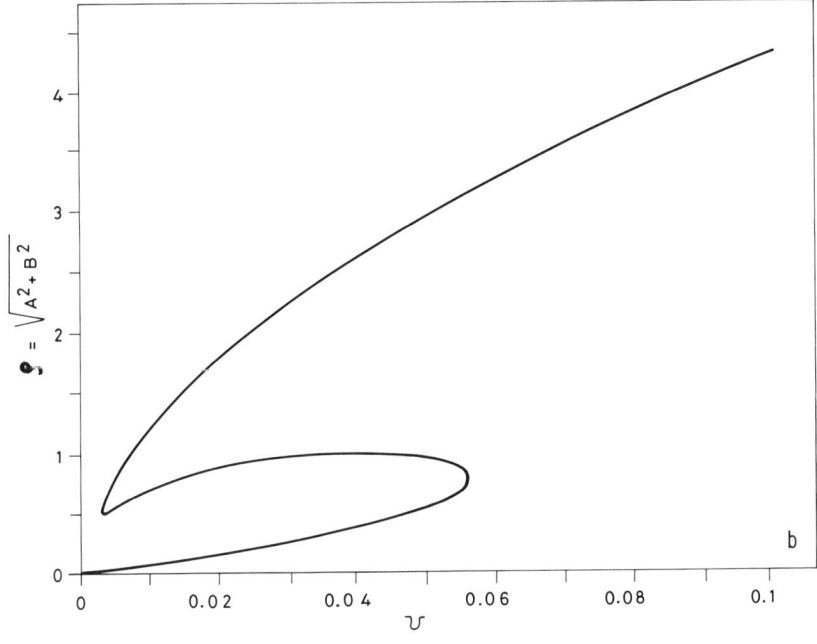

FIG. 8b. See legend on p. 213.

Remembering that the total (kinetic zonal and kinetic nonsymmetric), global (volume average) energy is

$$\mathscr{E}_T = \frac{\overline{A_x^2}^x}{2} - \alpha \frac{\overline{A^2}^x}{2} + \frac{U^2}{2} \tag{31}$$

we obviously find

$$\dot{\mathscr{E}}_T = \frac{\overline{A_x^2}^x}{2} - \alpha \frac{\overline{A^2}^x}{2} + \frac{U^2}{2} \tag{32}$$

where the two mountain drag terms of Eq. (30) cancel each other. It is clear at this point that the form–drag solution is exactly what we energetically require in order to transfer energy from the zonal flow to the wave: in the absence of potential energy sources or direct wave forcing (i.e., in the CDV type of energetics) the only way of amplifying planetary waves is through the action of mountain drag. However, observational analysis of this term, both for the seasonal average waves (Holopainen, 1970) and their time fluctuations (H.-D Shilling, personal communication), shows that its role is consistently negligible ($\sim 10^{-2}$ W m^{-2}) in maintaining waves against dissipation (~ 1 W m^{-2} required).

More recent analysis of winter data in general by Chen (1982) and Chen and Marshall (1984), and, more specifically, of blocking versus nonblocking cases by A. Hansen and H.-D. Shilling (both of which see in this volume) confirm the above results, together with the general observation that the source of energy for low-frequency variations (10 to 40 days) of the ultralong waves is available potential.

However, the conclusion must not be drawn here that the mountain drag term does not play an important role in the real atmosphere. As we shall see in the next section, form–drag can operate as an essential catalyst for baroclinic conversion, although it directly forces only irrelevant amounts of energy in the ultralong waves.

4. Baroclinic Energetics

The barotropic model has served well the purpose of isolating the mechanism of nonlinear stabilization required to produce theoretical multiple-equilibrium states in the observed variability range of Northern Hemisphere circulation. However, as we have seen, the energetics of the observed variability is dominated by baroclinic conversion. We shall now attempt to outline here how to extend the concepts concerning nonlinear stabilization to a baroclinic model atmosphere. For mathematical convenience we make use of the quasi-geostrophic, two-level model in the standard formulation (Pedlosky, 1964):

$$\partial_t q_1 + J(\psi_1, q_1) = F_1 \qquad q_1 = \nabla^2 \psi_1 + F(\psi_2 - \psi_1) + \beta y$$
$$\partial_t q_2 + J(\psi_2, q_2) = F_2 \qquad q_2 = \nabla^2 \psi_2 + F(\psi_1 - \psi_2) + \beta y + h/R_0 H_2 \tag{33}$$

where F_1 and F_2 represent any kind of forcing and/or dissipation and we have assumed $F_1 = F_2 = F$. It is useful to write Eq. (33) in terms of the barotropic [$\phi = (\psi_1 + \psi_2)/2$] and the baroclinic [$\theta = (\psi_1 - \psi_2)/2$] flow components

$$\partial_t q_\phi + J(\phi, q_\phi) + J(\vartheta, q_\vartheta) = F_\phi$$
$$\partial_t q_\vartheta + J(\theta, q_\vartheta) + J(\phi, q_\vartheta) = F_\vartheta$$
$$q_\phi = \nabla^2 \phi + \beta y + h/2R_0 H_2 \tag{34}$$
$$q_\theta = \nabla^2 \vartheta - 2F\vartheta - h/2R_0 H_2$$

where $F_\phi = (F_1 + F_2)/2$ and $F_\theta = (F_1 - F_2)/2$ obviously represent forcing and/or dissipation in the barotropic and the baroclinic flow component.

Here, again, we consider a one-wave solution of the form
$$\phi = -Uy + A(x,t)g(y)$$
$$\vartheta = -my + B(x,t)g(y) \qquad (35)$$
for flow over a mountain $h = h(x)g(y)$.

After substitution into Eq. (34) and projection on $g(y)$, we obtain for the wave equation

$$\left\{\frac{\partial}{\partial t} + U\frac{\partial}{\partial x}\right\}(A_{xx} + \alpha A) + \beta A_x + m\frac{\partial}{\partial x}(B_{xx} + \alpha B)$$
$$= \frac{3}{2}\delta\frac{\partial}{\partial x}(A^2 + B^2) - \frac{1}{2R_0 H_2}(U - m)\frac{\partial h}{\partial x} + \langle F_\phi g \rangle \qquad (36)$$

$$\left\{\frac{\partial}{\partial t} + U\frac{\partial}{\partial x}\right\}[B_{xx} + (\alpha + 2F)B] + \beta B_x + m\frac{\partial}{\partial x}(A_{xx} + \alpha A) + 2mFA_x$$
$$= 3\delta\frac{\partial}{\partial x}(AB) + \frac{1}{2R_0 H_2}(U - m)\frac{\partial h}{\partial x} + \langle F_\vartheta g \rangle$$

It is clear that, besides the usual linear terms of dispersion and topographic drag, we have now nonlinear self- and mutual interaction of the barotropic and baroclinic flow components. The interesting point is that such terms in turn permit us to fit the observed baroclinic energetics. To demonstrate this, we need to close energetically the system Eq. (36) by adding equations for the evolution of U and m. These are, respectively, the two-level version of the momentum equation, Eq. (4), and the equation for the time evolution of the average baroclinicity

$$dU/dt = -\overline{(A-B)h_x}^x + \nu U^* - \nu U \qquad (37)$$
$$(1 + 2F\langle y^2\rangle)(dm/dt) = \overline{(A-B)h_x}^x + 2F\overline{A_x B}^x + \nu m^* - \nu m \qquad (38)$$

where we have assumed that linear friction and forcing are operating on the zonal flow components. It is now interesting to compute specifically the total energy of the system Eqs. (37) and (38). The total energy density can be written in terms of the barotropic and baroclinic flow components,

$$e(x,y) = \tfrac{1}{2}(\phi_x^2 + \phi_y^2 + \vartheta_x^2 + \vartheta_y^2) + 2F\vartheta^2 \qquad (39)$$

which, once expressed in terms of the one-wave solution Eq. (35) and integrated over latitude, becomes

$$\mathscr{E}(x) = \int e(x,y)\,dy = \frac{1}{2}(A_x^2 + B_x^2) + \frac{U^2}{2} + \frac{1 + 2F\langle y^2\rangle m^2}{2}$$
$$- \frac{\alpha}{2}(A^2 + B^2) + FB^2 \qquad (40)$$

The global energy is the integral in longitude of (x):

$$E = \int \mathcal{E}(x)\, dx = \overline{\frac{A_x^2 + B_x^2}{2}}^x + \overline{\frac{U^2}{2}}^x + \overline{\frac{1 + 2F\langle y^2 \rangle}{2}}^x m^2$$

$$-\frac{\alpha}{2}\overline{(A^2 + B^2)}^x + F\overline{B^2}^x \quad (41)$$

The time derivative of this quantity can be recovered from Eqs. (32) and (33) by multiplying respectively by A, B, U, and M and integrating over longitude:

$$-\frac{1}{2}\frac{\partial}{\partial t}(\overline{A_x^2}^x - \alpha^2\overline{A^2}^x) + \frac{1}{2R_0H_2}(U - m)\overline{h_x A}^x = \overline{\langle F_\phi g \rangle A}^x$$

$$-\frac{1}{2}\frac{\partial}{\partial t}[\overline{B_x^2}^x - (\alpha^2 - 2F)\overline{B^2}^x] - \frac{1}{2R_0H_2}(U - m)\overline{h_x B}^x$$

$$+ 2mF\overline{A_x B}^x = \overline{\langle F_\theta g \rangle B}^x \quad (42)$$

$$\frac{1}{2}\frac{d\overline{U^2}^x}{dt} = -U\overline{(A - B)h_x}^x + \nu U^*U - \nu U^2$$

$$\frac{1 + 2F\langle y^2 \rangle}{2}\frac{dm^2}{dt} = m\overline{(A - B)h_x}^x - 2Fm\overline{A_x B}^x + \nu m^*m - \nu m^2$$

where contributions from nonlinear terms have obviously disappeared. By adding Eq. (42), the conservation of Eq. (41) for unforced inviscid flow can be checked.

Since the contributions from nonlinear terms do not appear in the global energetics, we can discuss our problem in terms of linear orographic waves. Assuming that dissipation is of Laplacian form (ν_S and ν_D are respectively the dissipation coefficients of the barotropic and baroclinic part), for the energetic balance of the stationary-wave solutions of Eq. (36), we obtain

$$\frac{1}{2R_0H_2}(U - m)\overline{\frac{h_x}{2}(A - B)}^x + 2Fm\overline{BA_x}^x - \nu_S(\overline{A_{xx}A}^x - \alpha\overline{A^2}^x)$$

$$- \nu_D(\overline{B_{xx}B}^x - \alpha\overline{B^2}^x) = 0 \quad (43)$$

where the first term represents orographic conversion from kinetic energy of the zonal flow, the second term is the divergence of the latitudinal heat flux and is associated with the conversion of available energy of the zonal flow, and the last two terms represent dissipation of the wave flow component.

The linearized, stationary Eq. (36), with the simplifying assumption $\alpha = 0$

reads

$$UA_{xxx} + \beta A_x + \frac{1}{2R_0H_2}(U-m)\frac{h_0}{2} + mB_{xxx} = -v_S A_{xx}$$

$$UB_{xxx} + 2FUB_x + \beta B_x - \frac{1}{2R_0H_2}(U-m)\frac{R_x}{2} + mA_{xxx} \quad (44)$$

$$- 2FmA_x = -v_D B_{xx}$$

If topography is sinusoidal $h(x) = h_0 e^{iK_m x} + (*)$, wave solutions of the form

$$A(x) = A_0 e^{iK_m x} + (*)$$
$$B(x) = B_0 e^{iK_m x} + (*) \quad (45)$$

have complex amplitudes:

$$A_0 = \{h_0(U-m)[(\beta - UK_m^2) + 2FU + iK_m v_D] + K_m^2 m\}/4R_0 H_2 \Delta$$
$$B_0 = \{h_0(U-m)(K_m^2 + 2F)m - [(\beta - K_m^2 U) + iK_m v_S]\}/4R_0 H_2 \Delta \quad (46)$$

where Δ is the determinant of the homogeneous linear system. The corresponding heat flux divergence,

$$\phi = \overline{A_x B}^x = iK_m(A_0 B_0^* - A_0^* B_0)$$
$$= -4k_m|A_0||B_0|\sin\left[\text{atg}\frac{\text{Im}(A_0/B_0)}{\text{Re}(A_0/B_0)}\right] \quad (47)$$

is essentially determined by

$$\frac{A_0}{B_0} = \frac{\beta - UK_m^2 + 2FU + K_m^2 m + iK_m v_D}{(K_m^2 + 2F)m - \beta + K_m^2 U + iK_m v_S} \quad (48)$$

and goes to zero as

$$\text{Im}\left(\frac{A_0}{B_0}\right)$$
$$= \frac{-[\beta - (U-m)K_m^2 + 2FU]K_m v_S + K_m v_D[-\beta + (U+m)K_m^2 + 2Fm]}{[-\beta + (U+m)K_m^2 + 2Fm]^2 - (K_m v_S)^2} \quad (49)$$

which is clearly independent of h. In fact, the conversion term Eq. (47) only depends on the dissipation $v_{S,D}$. It is therefore clear that regimes are possible in the baroclinic atmosphere in which topography only fixes the structure and the phase of the wave, while the maintenance against dissipation is guaranteed by baroclinic conversion. In fact, such regimes are found in models that are linear in the wave field (see Charney and Straus, 1980, p. 1170, for an explicit comment on this property).

5. Resonance Bending in a Baroclinic Model Atmosphere

At this stage, one may be tempted to conclude that the nonlinear bending due to self-interaction of the wave field is not needed in the baroclinic dynamics since the energetic problems of CDV can be solved quite naturally in the linear wave theory. Here, however, other properties come into play.

If we insist in postulating a linear wave dynamics, we deduce from stability analysis of the solutions of the stationary problem Eq. (44), that the solutions located on the subresonant side in the plane $U - m$ are stable, while the ones on the superresonant side are unstable (see Buzzi *et al.*, 1984, for a discussion of the general stability properties of linear stationary solutions). Multiple solutions come in sets which are distributed on different sides of the resonance, with the large wave amplitude ("blocked") solution always on the subresonant side. The attracting character resulting from the stability of subresonant states causes all trajectories in phase space to end eventually in the blocked state (Yoden, 1983). Again, as in the barotropic case, we are in the situation of needing resonance bending in order to be able to locate different equilibria on the same side of the resonance.

The extension of the ideas concerning wave self-interaction and resonance bending to the baroclinic case is relatively straightforward, although requiring a considerable amount of algebra. The full problem is discussed in detail in a paper by Benzi *et al.* (1985). Here we want only to prove with a specific example that wave nonlinearity can produce two stable states, differing very little in zonal flow and average baroclinicity, both stable with respect to orographic baroclinic instability.

For this purpose it is convenient to consider the baroclinic Eqs. (36)–(38) in the layer version,

$$\partial_t[A_{1xx} + F(A_2 - A_1)] + U_1 A_{1xxx} + \beta A_{1x} + F(U_1 A_{2x} - U_2 A_{1x})$$
$$- \delta A_1 A_{2x} = -\nu A_{1xx}$$

$$\partial_t[A_{2xx} + F(A_1 - A_2)] + U_2 A_{2xxx} + \beta A_{2x} + F(U_2 A_{1x} - U_1 A_{2x})$$
$$- \delta A_2 A_{2x} + U_2 h_x = -\nu A_{2xx} \quad (50)$$

$$\partial_t U = -\overline{A_2 h_x}^x - \nu(U - U^*)$$

$$C\partial_t m = 2F\overline{A_2 A_{1x}}^x + \overline{A_2 h_x}^x - \nu(m - m^*)$$

where we have introduced a Laplacian friction with equal relaxation time $1/\nu$ in all the processes and defined the constant $C = 1 + 2F\langle Y^2 \rangle$.

Since we want to produce solutions that are characterized by small interaction with topography, we shall assume that the topographic term $U_2 h_x$ is small. In analogy with the barotropic case we will also assume the nonlinear and frictional terms to be small. Since we are dealing with small fluctuations around equilibrium we assume also time derivatives to be small. We will also introduce a slow space modulation in order to balance secularities in the expansion.

The exact ordering of the perturbative terms that produces the balance we are interested in comes out to be as follows:

First-order small quantities: nonlinearity

$$\delta = O(\epsilon) \qquad \delta \to \delta\epsilon$$

Second-order small quantities: dissipation, flow in the lower layer, slow space modulation, slow time modulation

$$\nu = O(\epsilon^2) \qquad \nu \to \nu\epsilon^2$$
$$U_2 = O(\epsilon^2) \qquad U_2 \to U_2\epsilon^2$$
$$X^2 = \epsilon^2 x$$
$$\tau = \epsilon^2 t$$

A zero order in the small quantities we obtain:

$$U_1 A^{(0)}_{1xxx} + \beta A^{(0)}_{1x} = 0 \qquad (51)$$
$$A^{(0)}_2 = 0$$

This is the equation for a stationary, free Rossby wave confined in the upper layer that we will assume of the form

$$A^{(0)}_1 = a(X, t)e^{iKx} + (*) \qquad (52)$$

where X is the slow space scale and $K_s = \sqrt{\beta/U_1}$ is the "resonant" wavenumber.

At first-order, nonlinearity affects the structure of the Rossby wave:

$$U_1 A^{(1)}_{1xxx} + \beta A^{(1)}_1 - \frac{\delta}{2}(A^{(0)}_1)^2_x = 0 \qquad (53)$$

exactly as in the barotropic case [see Eqs. (23)–(24)].

At the second order, the other terms (time and slow space modulation, dissipation, and flows in the lower layer) enter the picture:

$$\partial_\tau(A^{(0)}_{1xx} - FA^{(0)}_1) + U_1 A^{(0)}_{1xxx} + \beta A^{(2)}_{1x} + 3U_1 A^{(0)}_{1xxX} + \beta A^{(0)}_{1X}$$
$$+ F(U_1 A^{(2)}_{2x} - U_2 A^{(0)}_{1x}) - \delta A^{(1)}_1 A^{(0)}_{1x} - \delta A^{(0)}_1 A^{(1)}_{1x} = -\nu A_{1xx} \qquad (54)$$
$$F\partial_\tau A^{(0)}_1 + \beta A^{(2)}_{2x} + F(U_2 A^{(0)}_{1x} - U_1 A^{(2)}_{2x}) + U_2 h_x = 0$$

Remembering that we are dealing with sinusoidal topography $h = h_0 e^{iK_m x} + (*)$ and defining a deturning parameter $\Delta K = K_s - K_m$, we obtain from the conditions of solvability of Eq. (54) the following form of the equation for the modulation $a(X, t)$ of the quasi-stationary wave Eq. (52) [the wave amplitude in the lower layer can be computed explicitly from that in the upper layer by means of the second equation of Eq. (54)]:

$$\left[F + K_s^2 + \frac{F^2 U_1}{\beta - FU_1}\right]\frac{\partial a}{\partial \tau} = -\left[2iK_s^2 \Delta K U_1 + \frac{iK_s F\beta U_2^{(0)}}{\beta - FU_1} + \nu K^2\right] a$$

$$- \frac{iK_s F h_0 U_1 U_2}{\beta - FU_1} - i\frac{7}{6}\frac{K_s^2}{6\beta}\delta^2 a|a|^2 \quad (55)$$

The stationary version of Eq. (55) is a cubic that gives three solutions in certain parameter ranges.

After solving Eq. (55) for a, the corresponding values of flows in the lower layer and average baroclinicity can be deduced respectively from the second part of Eq. (54), the form–drag and the average baroclinicity equations. One specific example, corresponding to realistic values of the variables, is listed in Table II together with the barotropic limiting Case $m = 0$. It can be noticed that in the solution with m variable amplitude of the wave field changes appreciably (the order of magnitude is 100 geopotential meters), while the variation of the zonal flow U and the average baroclinicity m remain confined (the order of magnitude in 1 m sec^{-1}) as in the linear case of the preceding section.

In the barotropic limit $m = 0$, a similar change of wave amplitude requires a much larger change in the zonal flow (order of magnitude 10 m sec^{-1}) as discussed in Section 3. It is interesting now to analyze the stability of the equilibrium states computed with the aid of the nonlinear Eq. (55).

In order to substantiate our qualitative remarks at the beginning of this section, it will suffice here to consider the inviscid limit. The matrix of linear

TABLE II. PARAMETERS AND VARIABLES OF MULTIPLE STABLE STATIONARY SOLUTIONS OF THE NONLINEAR WAVE EQ. (55)

U^* (m sec^{-1})	m^* (m sec^{-1})	U (m sec^{-1})	m (m sec^{-1})	A_1 ($m \times 100$)	A_2 ($m \times 100$)	Type of equilibrium
\multicolumn{7}{c}{Baroclinic state ($m^* = 15.0$ m sec^{-1})}
23.8	15.0	21.0	14.0	$1.14 + i0.40$	$0.34 + i0.11$	Blocked
23.8	15.0	23.8	14.9	$-0.17 + i0.006$	$-0.02 + i0.002$	Zonal
\multicolumn{7}{c}{Barotropic state ($m = 0$)}
40.6	15.0	21	0	$2.14 + i0.41$	$4.28 + i0.78$	Blocked
40.6	15.0	40.5	0	$-0.32 + i0.004$	$-0.29 + i0.06$	Zonal

perturbation of our dynamical system is

$$\begin{bmatrix} -(K_s Q/F)\sigma & (QU_2 + 2\beta \, \Delta K) + \delta K_s a_r^2 & 0 \\ -(QU_2 + 2\beta \, \Delta k) - 3K_s \delta a_r^2 & -\dfrac{QK}{F}\sigma & Qa_r + QH \\ -2HQK_s\beta & -2U_2QHK_s^2/\beta & -\sigma \end{bmatrix} \quad (56)$$

where we have defined $Q = K_s^3 F/(K_s^2 - F)$, $H = h_0/K_s^2$, and σ is the eigenvalue of the linear perturbations (assumed proportional to $e^{\sigma t}$). Notice that only the wave and form–drag equations have been perturbed since $\partial m/\partial t$ is constant in the orographic baroclinic instability (see Buzzi et al., 1984). Note also that only the real part of the stationary wave solution (a_r) appears since in the inviscid limit the imaginary part vanishes ($a_i = 0$).

The dispersion relation derived from the vanishing of the determinant of Eq. (56) can be written in the form

$$\delta K_s a_r^3 + (QU_2 + 2\beta \, \Delta K)a_r + QU_2 H = 0 \quad (57)$$

where use has been made of the fact that a_r must be a stationary solution of the nonlinear problem, Eq. (55).

We can analyze the stability of multiple-equilibrium solutions by taking the limit of Eq. (57) for $a_r \to \infty$ (large-amplitude "blocked" and "intermediate" states) and $a_r \to 0$ (the small-amplitude "zonal" state). In the limit of large wave amplitude we obtain

$$\sigma^2 = \frac{2U_2 H(\delta - Q^2 K_s^2/\beta F)}{QK_s} a_r \quad (58)$$

which, for sufficiently large nonlinearity, $\delta > Q^2 K_s^2/\beta F$, implies stability for $a_r > 0$ ("blocked" branch) and instability for $a_r < 0$ ("intermediate" branch). In the opposite limit, σ^2 becomes large and positive, implying stability. In conclusion, the stability properties are as envisaged in the beginning of this section.

At this point, any further quantitative statement concerning the physical processes we are dealing with must be based on a mathematical analysis of the baroclinic equilibrium states and their stability properties in the presence of resonance folding. However, whatever the physical process in question, we expect mountain drag, even if small, to be correlated with the fluctuations of the ultralong waves. This correlation, in fact, is observed at periods of about 20 days as shown in Fig. 9. As mentioned at the end of the preceding section, the form–drag is too small to convert energy directly from the zonal wind but, acting as a catalyst, can produce baroclinic conversion on the scale of topography which, in our case, is much larger than the natural scale of "slantwise convection," i.e., the Rossby deformation radius.

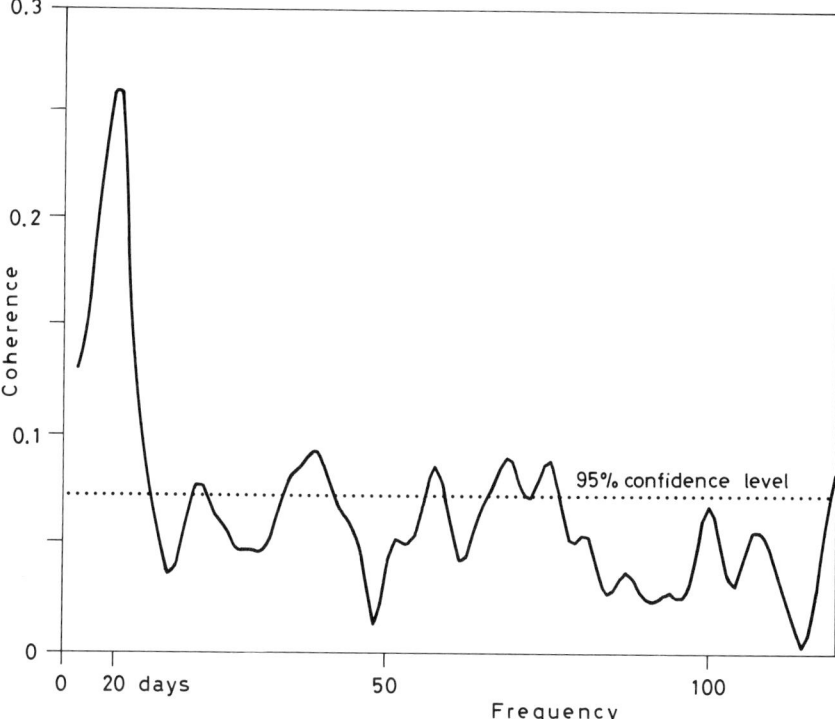

FIG. 9. Spectral cross-correlation between the amplitudes of the wavenumber 2 and the zonal wind computed from a 500-mbar geopotential height of 14 years.

6. Summary and Conclusions

Without going into detail regarding the problem of determining, from available observations, the statistical properties of the general circulation (for this purpose, see, in this volume, the previously mentioned papers by Hansen and Sutera), we analyzed some basic inadequacies of the CDV theory in explaining the atmospheric circulation.

We have shown that some of these inadequacies, in particular those connected with the excessive separation of equilibrium states in the zonal wind, can be removed in the context of a barotropic theory introducing the wave self-interaction as stabilizing nonlinearity.

However, the energetic requirements emerging from observational studies cannot be satisfied by the dynamics of a barotropic model atmosphere. It is therefore necessary to consider a baroclinic model atmosphere.

Stationary baroclinic solutions in the presence of orographic boundary modulation display the correct energetics (large baroclinic conversion and small mountain drag) even in the linear dynamics. However, the stability properties of multiple baroclinic equilibrium states obtained with a linear resonance theory (as in Charney and Straus, 1980) are not of the type required: the only state stable with respect to orographic baroclinic instability is the subresonant ("blocked") state where the minimal dynamical is ultimately bound to reside independently of the initial condition.

The same concepts (resonance folding due to wave self-interaction, etc.) that appear of key importance in allowing the fitting of the observed statistics in the barotropic case can be applied to baroclinic models by means of straightforward, albeit somewhat lengthy, calculations. In such a model the mountain forcing does not play a direct role but acts only as a symmetry-breaking boundary modulation, catalyzing baroclinic conversion at ultralong wavelengths.

Introduction of wave nonlinearity in the minimal baroclinic system allows not only production of multiple-equilibrium states (which, although differing appreciably in wave amplitude, are confined within a limited range of variation of mean momentum and baroclinicity), but also adjustment of (through resonance bending) the stability properties to the required characteristics of stability: two stable extreme states and one intermediate, unstable state with respect to orographic baroclinic instability.

Such considerations are not intended to signify that the type of minimal systems just sketched can exhaust the problematics of low-frequency variability. Many more problems have to be faced, in particular those connected with the interaction of ultralong waves with synoptic-scale disturbances.

However, we think we have enough evidence to conclude that the full range of nonlinear effects has yet to be thoroughly analyzed and that these effects should be further studied before they are used to draw final conclusions about what goes under the name of "multiple-equilibria theory."

References

Benzi, R., Hansen, A. R., and Sutera, A. (1984a). On stochastic perturbation of simple blocking models. *Q. J. R. Meteorol. Soc.* **110**, 393–409.

Benzi, R., Malguzzi, P., Speranza, A., and Sutera, A. (1984b). On the theory of stationary waves and blocking. *Q. J. R. Meteorol. Soc.*, July.

Benzi, R., Speranza, A., and Sutera, A. (1985). A minimal baroclinic theory of low-frequency variability in the Northern-Hemisphere. *J. Atmos. Sci.*, submitted.

Buzzi, A., Speranza, A., and Trevisan, A. (1984). Instabilities of a baroclinic flow related to topographic forcing. *J. Atmos. Sci.* **41**, 637–650.

Charney, J. G., and DeVore, J. G. (1979). Multiple flow equilibria in the atmosphere and blocking. *J. Atmos. Sci.* **36**, 1205–1216.

Charney, J. G., and Straus, D. M. (1980). Form-drag instability, multiple equilibria and propagating planetary waves in baroclinic, orographically forced, planetary wave systems. *J. Atmos. Sci.* **37**, 1157–1176.

Charney, J. G., Shukla, J., and Mo, K. C. (1981). Comparison of a barotropic blocking theory with observation. *J. Atmos. Sci.* **38**, 762–779.

Chen, T. C. (1982). A further study of spectral energetics in the winter atmosphere. *Mon. Weather Rev.* **110**, 947–961.

Chen, T. C., and Marshall, H. G. (1984). Time variation of atmospheric energetics in the FGGE winter. *Tellus* **36A**, 251–268.

Egger, J. (1981). Stochastically driven large-scale circulation with multiple equilibria. *J. Atmos. Sci.* **38**, 2606–2618.

Hart, J. E. (1979). Barotropic, quasi-geostrophic flow over anisotropic mountains. *J. Atmos. Sci.* **36**, 1736–1746.

Holopainen, E. O. (1970). An observational study of the energy balance of the stationary disturbances in the atmosphere. *Q. J. R. Meteorol. Soc.* **96**, 626–644.

Legras, B., and Ghil, M. (1983). Blocking and variations in atmospheric predictability. *In* "Predictability of Fluid Motion" (G. Holloway and B. J. West, eds.). American Inst. of Physics, N. 106.

Malanotte Rizzoli, P., and Malguzzi, P. (1984). Nonlinear stationary Rossby waves over nonuniform zonal winds and atmospheric blocking. Part I: The analytical theory. *J. Atmos. Sci.* **41**, 2620–2628.

Malguzzi, P., and Speranza, A. (1981). Local multiple equilibria and regional atmospheric blocking. *J. Atmos. Sci.* **38**, 1939–1948.

Pedlosky, J. (1964). The stability of currents in the atmosphere and the ocean. Part I. *J. Atmos. Sci.* **21**, 201–219.

Pedlosky, J. (1981). Resonant topographic waves in barotropic and baroclinic flows. *J. Atmos. Sci.* **38**, 2626–2641.

Pierrehumbert, R. T., and Malguzzi, P. (1984). Forced coherent structures and local multiple equilibria in a barotropic atmosphere. *J. Atmos. Sci.* **41**, 246–257.

Rambaldi, S., and Mo, K. C. (1984). Forced stationary solutions in a barotropic channel: Multiple equilibria and theory of non-linear resonance. *J. Atmos. Sci.* **41**, 3135–3146.

Reinhold, B. B., and Pierrehumbert, R. T. (1982). Dynamics of weather regimes: Quasi-stationary waves and blocking. *Mon. Weather Rev.* **110**, 1105–1145.

Speranza, A., Buzzi, A., Trevisan, A., and Malguzzi, P. (1985). A theory of deep cyclogenesis in the lee of the Alps: Modifications of baroclinic instability by localized topography. *J. Atmos. Sci.* **42**, 1521–1535.

Tung, K. K., and Lindzen, R. S. (1979a). A theory of stationary long waves. Part I: A simple theory of blocking. *Mon. Weather Rev.* **107**, 714–734.

Tung, K. K., and Lindzen, R. S. (1979b). A theory of stationary long waves. Part II: Resonant Rossby waves in the presence of realistic vertical shears. *Mon. Weather Rev.* **107**, 735–750.

Yoden, S. (1983). Nonlinear interactions in a two-layer, quasi-geostrophic, low-order model with topography. *J. Meteorol. Soc. Jpn* **61**, 1–18.

Wiin-Nielsen, A. (1979). Steady states and stability properties of a low order barotropic system with forcing and dissipation. *Tellus* **31**, 375–386.

PROBABILITY DENSITY DISTRIBUTION OF LARGE-SCALE ATMOSPHERIC FLOW

ALFONSO SUTERA

The Center for Environment and Man
Hartford, Connecticut 06117, and
Department of Geology and Geophysics
Yale University
New Haven, Connecticut 06511

1. INTRODUCTION

Whether the large-scale atmospheric motion undergoes temporal variations about more than one dynamical regime is a question with both theoretical and practical relevance. In fact, if we suppose that more than one regime occurs, then the mean state of the atmosphere would not be adequately represented by its long-term average, but rather by the set of average states obtained by considering, separately, each individual regime. As another consequence of this hypothesis, it is apparent that transition from one regime to the others would be conditioned by the initial state, so that long-term forecasts would be more strongly dependent on initial conditions than in the case in which one single mode occurs (Moritz and Sutera, 1981; Leith, 1978). This question has received more renewed attention recently since simple models (for instance Charney and DeVore, 1979) of atmospheric motions have shown the possibility that more than one equilibrium regime is consistent with the equations considered. Efforts to connect these theoretical studies with observations have been described by a few authors (White, 1980; Hartman and Gahn, 1980; Charney *et al.,* 1981; Dole and Gordon, 1983). All of these agreed that the observational evidence supporting the theoretical models in their essential prediction—i.e., the existence of multiple regimes—was sparse. In this article, we address the same question. We find, contrary to these recent studies, that in the context of the data as here considered, there is evidence that the large-scale dynamics of the atmosphere are consistent with the presence of more than one regime (in some sense that will be explained later). In one regime the planetary-scale waves have large amplitude; in another the same waves do not show appreciable amplitude. *A posteriori,* we find that events falling into the first group can be associated with what from a synoptic and subjective point of view would resemble major blocking episodes. This result does not contradict, when examined in detail, the previously mentioned studies. When we consider the specific state variables of those studies, we confirm the results of their authors. Hence, the different conclusions we have reached have their source in our having considered different variables. In Section 2, we describe the process that has led us to consider this particular set of variables rather than any other. The data source and its processing are described in Section 3. In Section 4, we illustrate the technique employed to calculate the probability density associated with our random sample set. In Sections 5 through 8, we present our principal results. Conclusions and some speculations drawn from them are offered in Section 9.

2. Theoretical Background

To illustrate the ideas which guided our observational study, it is worthwhile to recall a few basic concepts about the theory of orographically induced stationary waves in the atmosphere, within the framework of a simple model that we borrow from Benzi et al. (1986) (see also A. Speranza, this volume), and that we slightly modify for our present purposes. Let us consider a barotropic atmosphere confined to a β-channel and flowing over a topography that will be here considered the only relevant inhomogeneity of the earth's surface. The governing equation expresses the conservation of potential vorticity. In this case, the potential vorticity, q, is defined as

$$q = \Delta\psi + \beta y + (f_0/H)h \qquad (1)$$

where ψ is the streamfunction, β the latitudinal derivative of the Coriolis force, f_0 the Coriolis parameter at 45°N latitude, H the equivalent atmospheric depth, and h the surface elevation. Hence the equation of motion is

$$\partial_t q + J(\psi, q) = \text{dissipation} \qquad (2)$$

If we split the streamfunction into its zonal average $\bar{\psi}$ and the departure ϕ, we have

$$\psi = \bar{\psi} + \phi \qquad (3)$$

Let us assume that $\bar{\psi} = -Uy$. In this case the governing equation is

$$\underbrace{\partial_t \Delta\phi}_{\text{I}} + \underbrace{J(\phi, \Delta\phi)}_{\text{II}} + \underbrace{\beta\partial_x\phi}_{\text{III}} + \underbrace{U\partial_{xxx}\phi}_{\text{IV}} + \underbrace{U\partial_{xyy}\phi}_{\text{V}}$$

$$+ \underbrace{J[\phi, (f_0 h/H)]}_{\text{VI}} + \underbrace{U(f_0/H)\,\partial_x h}_{\text{VII}} = \text{dissipation} \qquad (4)$$

If we neglect in Eq. (4) all but the linear terms I, III, IV, and VII, we get

$$\partial_t \Delta\phi + \beta\,\partial_x\phi + U\,\partial_{xxx}\phi + U\,\partial_{xyy}\phi + (Uf_0/H)\,\partial_x h = \text{dissipation} \qquad (5)$$

Equation (5) reduces to the Charney–Eliassen model (1949) when ϕ_{yy} is set to zero. At the stationary state, we get resonant response as shown in Fig. 1 (dotted curve), where ϕ has been expressed through one component of its spectral expansion, namely

$$\phi = Ae^{i\mathbf{K}\mathbf{x}} \tag{6}$$

where $\mathbf{K}$ is the total wavenumber. If we integrate Eq. (4) over latitude, $\mathbf{K}$ will become the zonal wavenumber. The Charney–DeVore (CDV) theory would close Eq. (4) by adding the form–drag equation for U. Typically, at the stationary state, this equation will express a relationship between A and U.

In Fig. 1, as in CDV, this relationship is represented by a straight line. As is evident from the figure, three regimes are possible: (1) one in which the wave amplitude is large, (2) one intermediate regime which is dynamically unstable, and (3) a regime in which the wave has a small amplitude. For each of the individual regimes, a distinct value of the zonal wind is predicted. If the model were realistic in its predictions, one would expect the observations to show multimodal probability density distributions for both variables; viz., A and U. Hence, a verification of the theory previously described would consist of finding multimodal distribution of both these variables.

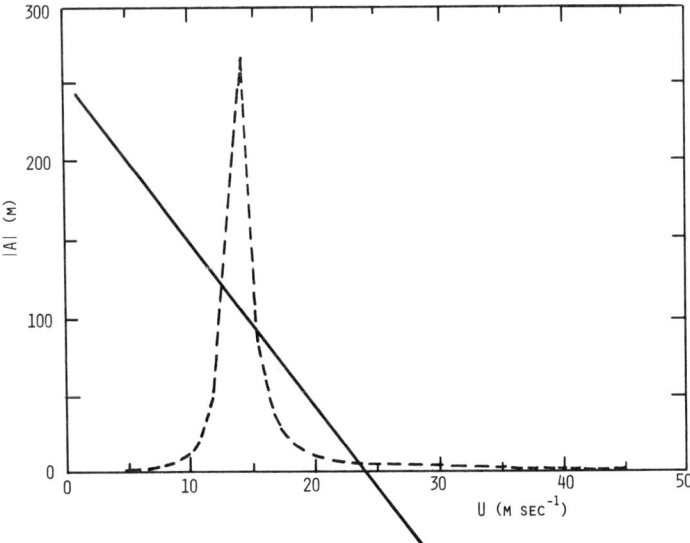

FIG. 1. Phase diagram for CDV model. The straight line represents the form–drag relationship.

3. THE DATA

The Charney–Eliassen equation, Eq. (4), has long been thought to be a good representation of the gross features of the Northern Hemisphere's midlatitude winter circulation; hence, we considered data including four winters where winter is artificially defined as the period extending from 1 December through 28 February. Equation (4) has been often applied to the 500-mbar surface (Charney and DeVore, 1979). So we consider data at the 500-mbar level. Finally, since Eq. (4) assumes geostrophic motion, the streamfunction is properly represented by the geopotential gZ, where g is gravity. Consequently, our analyses will be concerned with samples deduced from the 500-mbar geopotential height. The data, originally archived in the form of spherical harmonic coefficients, were retrieved and then interpolated onto a regularly spaced latitude–longitude grid with a 5.625° resolution in each direction. After the data were so prepared, the following preprocessing was performed:

1. For each day an average over a latitude band (in our case a straight algebraic mean) was calculated, obtaining a 64-dimensional array. It was then projected onto Fourier space by employing a standard fast Fourier transform. Although the results described hereafter were deduced by latitude averaging between 15°N and 75°N, we studied the effects of the width of the latitude band on our results by considering other (narrower) latitudinal bands. Besides obvious effects on the total amplitude of the waves, no other significant differences were noticed.

2. For the same day, the mean zonal wind was evaluated geostrophically from the geopotential height gradient in the latitudinal band considered above. We noted that employing the classic Rossby (two-dimensional) formula, the resonant stationary wavenumber oscillated between zonal wavenumbers 2, 3, and 4, so we decided to consider the variance associated with the combined amplitude of these three wavenumbers for each day: $A = (\Sigma_i |c_i|^2)^{1/2}$, c_i = complex amplitude of wavenumber i, $i = 2, 3,$ or 4.

3. At the completion of these operations, we were left with two samples of size $N = 1552$: the mean zonal wind U_n and the variance of the combined amplitudes of wavenumbers 2, 3, and 4, here designated A_n ($n = 1, 2, \ldots, N$).

For our purposes, we needed to further process the samples so that each sample point could be considered to be statistically independent of all the others. This is obviously not true for the raw dataset, since it is known that atmospheric variables decorrelate over a finite time period.

In addition, our sample sets might contain other longer trends that could not be directly connected with the natural variability of the system but owe their behavior to variations of the external forcing itself. Such trends might be easily isolated on the lower end of the frequency spectrum and they would correspond to well-known frequencies such as the annual cycle and its subharmonics. To remove these trends we proceeded as follows.

1. For each variable set, we determined the autocorrelation function and its Fourier transform.
2. From the observed decay of the autocorrelation function, we estimated the decorrelation time. This turned out to be about 5 days for both variables. This is in agreement with previous analyses (see, for example, Leith, 1978).
3. From the power spectrum, we noted a strong annual cycle and three detectable subharmonics.

To eliminate the short-frequency variability, we designed a low-pass filter with a cutoff frequency of 5 days. Then the data were resynthesized. To eliminate the low-frequency variability, we detrended the data by using the following method. For each day of the winter periods, we considered

$$\{\cdot\}_i^l = \frac{1}{2r+1} \sum_{-r}^{+r} {}_J\{\cdot\}_{i+j} \qquad (7)$$

where $r = 45$ (see Priestley, 1981). By employing this method we detrended any variability longer than 90 days, and, hence, the year and its detectable subharmonics. So done, we considered the five distinct datasets—the first four, one for each year, contained $N = 90$ observations, while the last one contained $N = 360$ observations, which were obtained by grouping together the winters after the individual winter means were subtracted. The last step was performed to avoid any residual long-term variability.

The two datasets so prepared allowed us to study (1) the probability density distribution induced by the data for each winter and (2) the probability density distribution induced by all the data available.

4. Nonparametric Probability Density Estimation

Since the general form of the probability density for our sample is unknown, we must use the nonparametric estimation theory. Among nonparametric estimators of unknown probability density associated with a random sample, the oldest and most frequently used is the histogram. A histogram is a maximum likelihood (ML) estimator in the sense that it maximizes the likelihood function with good consistency properties (see, for example, Scott et al., 1977). A disadvantage of such an estimation is that it becomes very noisy when the mesh size tends to zero (number of bins tends to infinity), and moreover does not tend in this case to the Dirac-delta at the sample point.

Since, once we have deduced our estimate of the density, we would like to perform some statistical evaluations of "the goodness of the fit," this undesirable property of the histogram turns out to be a major handicap.

A different approach to density estimation, that has the same advantages as the histogram but avoids its shortcomings, is the method of maximum penalized likelihood (MPL) (see Scott et al., 1977). This technique allows us to combine maximization of both the likelihood function and a penalizing smoothness function by taking explicitly into account the smoothness of the density through a free parameter that we shall call α.

The MPL method consists in finding the maximum of the following score functions:

$$L(W) = \prod_i x_i \exp\left(-\alpha \int \frac{d^{i2}W}{dt^2} dt\right) \qquad (8)$$

with the constraint

$$\int W \, dt = 1$$

where x_i are the observations and W is any density. The discrete approximation to Eq. (1) can be solved by employing an algorithm implemented and incorporated into the International Mathematical Statistical Library (IMSL), and in this article we used their routine NDMPLE on our data.

The external parameters to be provided to the algorithm (besides the data) are (1) the mesh points at which the density must be evaluated and (2) the hyperparameter α. We decided to choose 41 mesh points when we evaluated the density corresponding to the variable A, while 15 mesh points were employed for the variable U. This choice was based by assuming an experimental error in determining A and U of 1 m and 40 cm sec^{-1}, respectively.

Once a probability density has been estimated, as already mentioned, it will depend on the smoothness parameter α. The choice of α is arbitrary, and it is not deduced from the data (in other words, it is specified *a priori*). However, for any α, the corresponding theoretical probability density can be compared to the sample probability density through classical tests such as the Kolmogorov–Smirnov (K–S) test. For the sake of clarity, we recall that the K–S test consists of evaluating the quantity:

$$K = \mathop{\text{Sup}}_{x_i}[S(x_i) - p(x_i, \alpha)]$$

$$\mathop{\text{Sup}}_{x_i} = \text{Supremum over all } x_i \qquad (9)$$

where S and p are the sample probability density and the theoretical probability density, respectively, at the sample point x_i. It is known that the statistic K defined by Eq. (9) follows the K–S distribution and we can evaluate p_K, i.e., the probability of an occurrence of K that is less than that of a random variable with that distribution. It follows by this definition that as $p(x_i, \alpha)$ takes on a greater resemblance to S, we would have better confidence in p. Taken to the extreme, if $p(x_i, \alpha)$ were found by the combination of as many Dirac-deltas as there were data points, the fit, as defined by examining K, would be perfect. Of course such a p would not be representative of the true density because such a probability would predict that if x_i is an outcome, then the next outcome will be x_i with probability 1. Now it is easy to show that the MPL estimation tends to a Dirac-delta at the sample points as $\alpha \to 0$ (the estimation optimizes the maximum likelihood), hence a K–S test alone would help to distinguish between an oversmooth p and one closer to the data. But, on the contrary, the estimation would be insufficient to bound from below the optimal α. For this reason, we decided to calculate different densities by decreasing the value of the hyperparameter α up to the point on which the confidence level given by the K–S test would reach the value of 95%.

5. Results

In this section we will show our main results on the nature of the density distribution of the data described in Section 3.

We first start by considering each individual winter period in the set of the values of the amplitude A. We recall that, for this variable, we used 41 mesh points in evaluating the density in accordance with the previously discussed experimental error. In Figs. 2a–2d, we show the densities that satisfy the

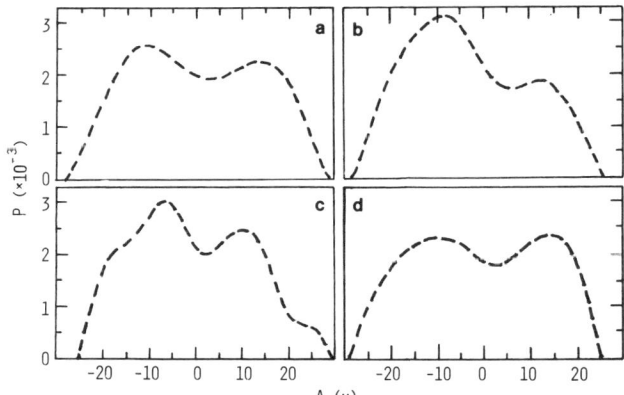

FIG. 2. Density distributions of A for (a) 1980/1981, (b) 1981/1982, (c) 1982/1983, and (d) 1983/1984.

TABLE I. CONFIDENCE LEVELS OF THE
KOLMOGOROV–SMIRNOV TEST (p_K) FOR A[a]

α	p_K			
	1981	1982	1983	1984
0.5×10^{13}	0.9	0.96	0.99	0.6
0.2×10^{13}	0.8	0.91	0.98	0.4[b]
0.8×10^{12}	0.5[b]	0.76	0.97	0.13
0.3×10^{12}	0.2	0.39[b]	0.90	0.05
0.1×10^{12}	0.1	0.06	0.6[b]	0.005
0.5×10^{11}	0.03	0.0055	0.5	0.0013
0.2×10^{11}	0.01	0.0053	0.1	0.0012

[a] The behavior of p_K is a function of the hyperparameter α for the variable A.
[b] For these values of α, or lower, the density is always bimodal.

criterion of a 95% confidence level for the K–S test. Apparently, each density is bimodal with a clear minimum separating the two maxima that are about 20 m apart.

We believe that no additional comment is necessary other than to note that each of the datasets on its own does not represent a satisfactorily large sample. It is interesting to consider Table I in which we report the confidence level of the K–S test for decreasing values of α. Note the values of α at which the density is always bimodal. It appears that if we would assume that the density is unimodal, we must be content with a mere 70% confidence level at the most. Considering that usually we would accept a confidence level for the K–S test of 95% or higher, we think that, at least statistically (unfortunately, the only way possible), we have proved that the variable A distributes itself

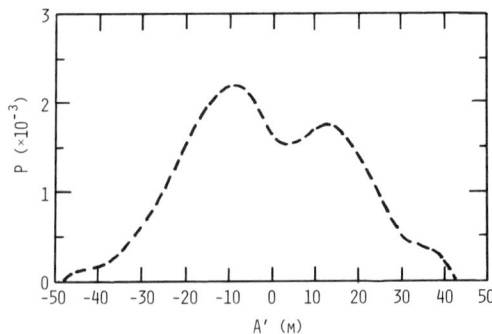

FIG. 3. Density distribution of A, collecting the 4 years together.

DENSITY DISTRIBUTION OF ATMOSPHERIC FLOW 235

TABLE II. CONFIDENCE LEVELS OF THE
KOLMOGIROV–SMIRNOV TEST (p_K):
4-YR SUMMARY[a]

α	p_K
0.1×10^{14}	0.3^b
0.5×10^{13}	0.22
0.25×10^{13}	0.19
0.12×10^{13}	0.15
0.6×10^{12}	0.11
0.3×10^{12}	0.08
0.15×10^{12}	0.06
0.7×10^{11}	0.05
0.3×10^{11}	0.04

[a] The behavior of p_K as a function of the hyperparameter α for the variable A.
[b] For this value of α, *or lower,* the density is always bimodal.

following a bimodal density. The same results are obtained when we consider A as defined by dataset {2}. For this case, since we have more observations, we have even removed the assumption that each individual sample point be independent, and we have calculated the one-sided K–S test by assuming we have only 72 independent observations, as a nonfiltered dataset would suggest. The result is presented in Fig. 3, and Table II contains the behavior of the test for different values of the hyperparameter; as in Table I, note the value of α after which we have a bimodal density. In summary, it appears that the measure of planetary-scale waves that we have studied show a marked bimodality which is independent of the known externally forced periodicities of the system.

6. CONNECTION WITH PATTERNS OF THE 500-mbar GEOPOTENTIAL HEIGHT

Up to now we have demonstrated that a global indicator (the latitudinally averaged geopotential field has been Fourier decomposed) of large-scale atmospheric behavior has a bimodal distribution. This result, while interesting in itself, might be of marginal physical significance if we could not connect each mode to anything peculiar in the character of the "weather" map from which it has been deduced.

To connect the result to actual geopotential patterns, we can select those days corresponding to each mode and consider, for example, the mean maps representative of each of the two modes. In Figs. 4a–4d (Mode 1) and 5a–5d

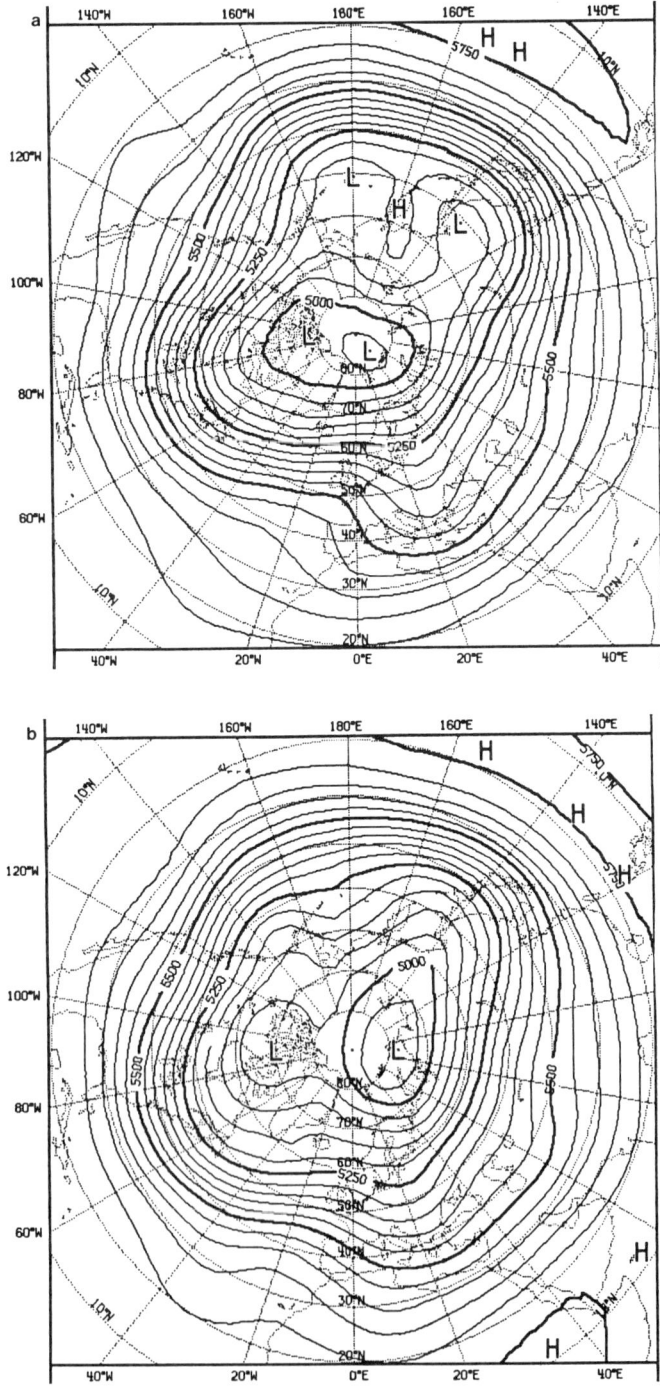

FIG. 4. Mean 500-mbar maps for Mode 1 events in each individual winter. Notice that the mean maps are obtained using the full fields and not only the wavenumber group 2–4. (a) 1981, (b) 1982, (c) 1983, and (d) 1984.

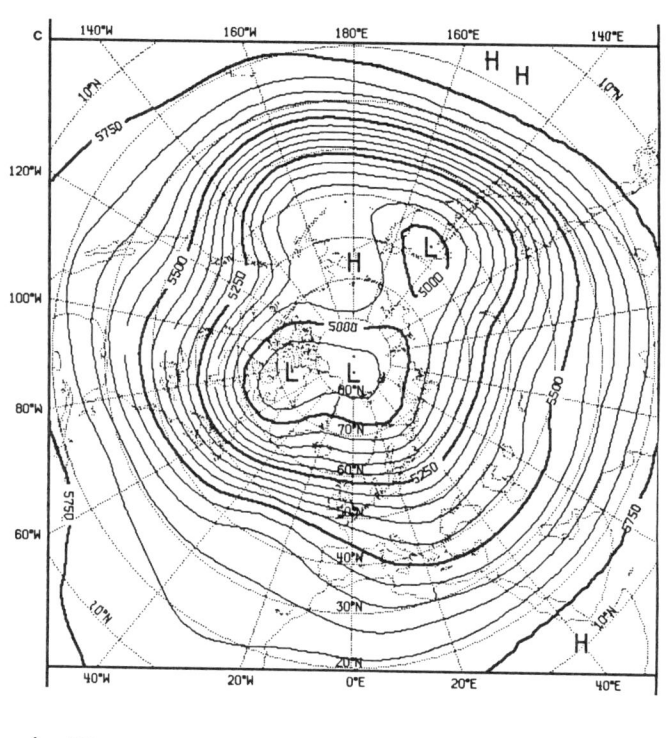

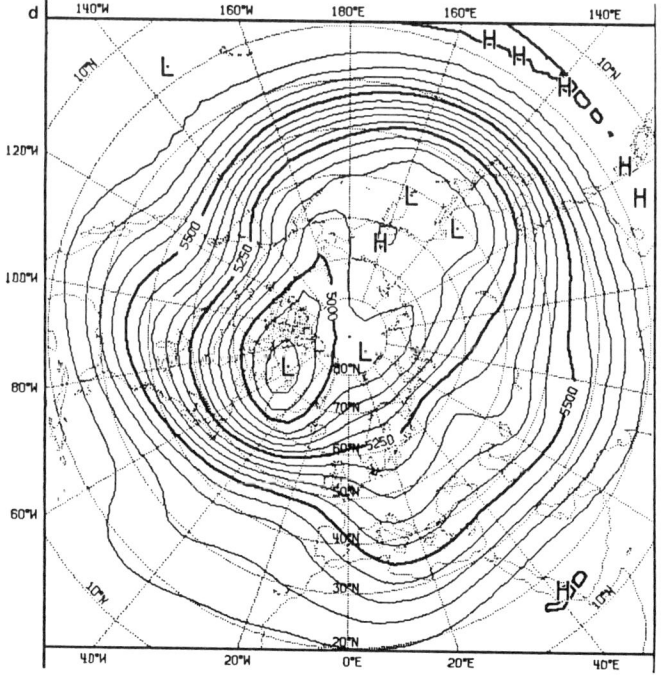

Fig. 4c and d.

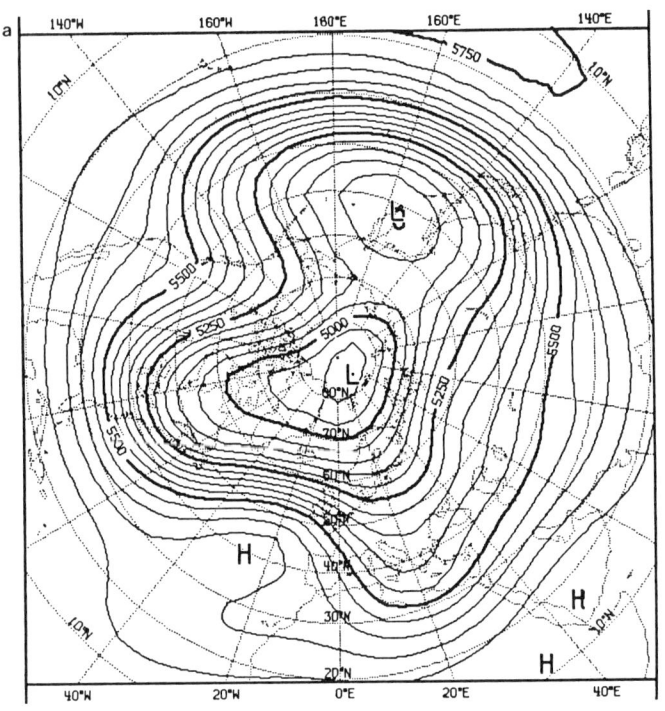

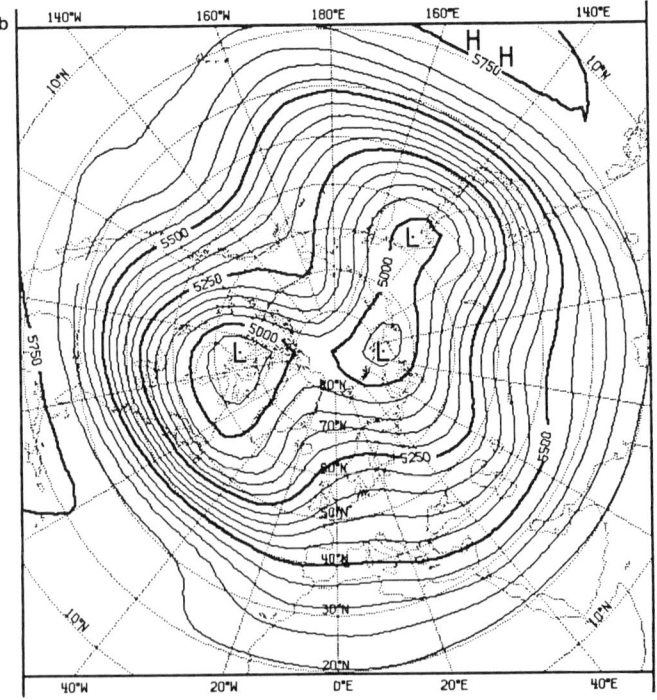

FIG. 5. (a–d) As in Fig. 4a–d, for Mode 2.

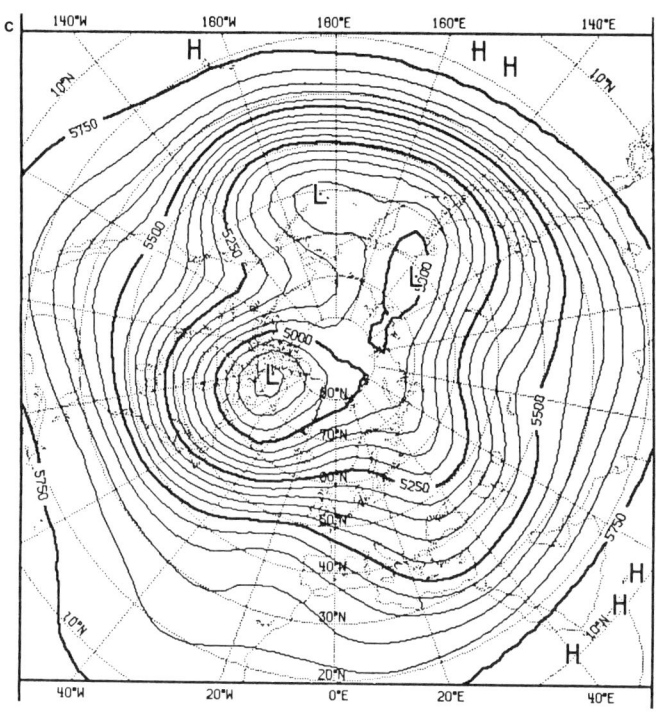
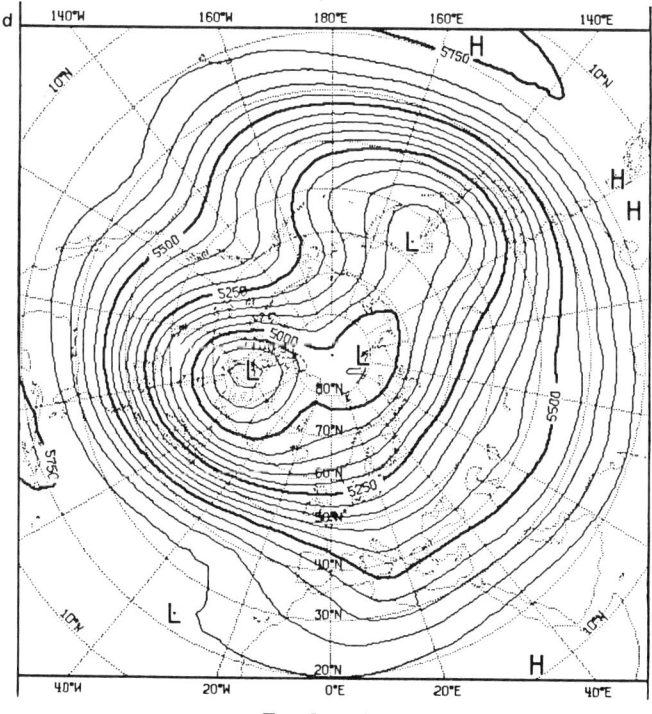

FIG. 5c and d.

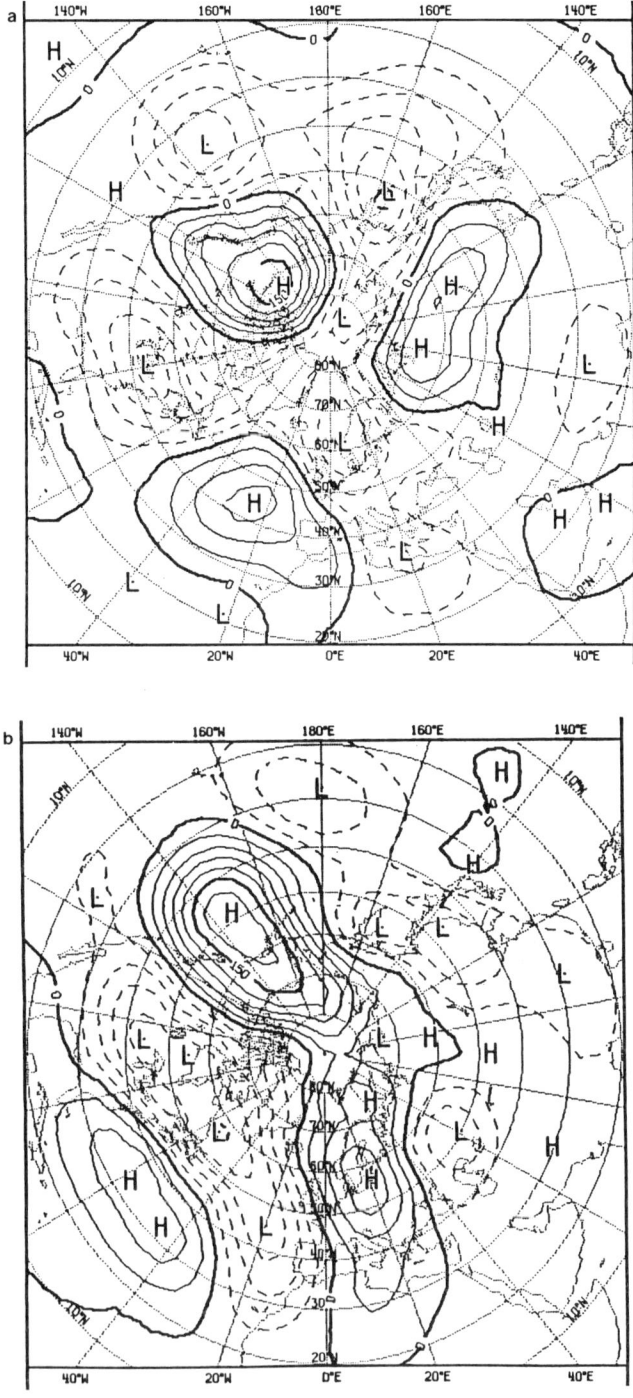

FIG. 6. (a–d) Mean differences (Mode 2, Mode 1) in each year (1981–1984). (e) Mean differences ("blocking"–"nonblocking") for 15 January days using the Charney–Shukla–Mo (1981) definition.

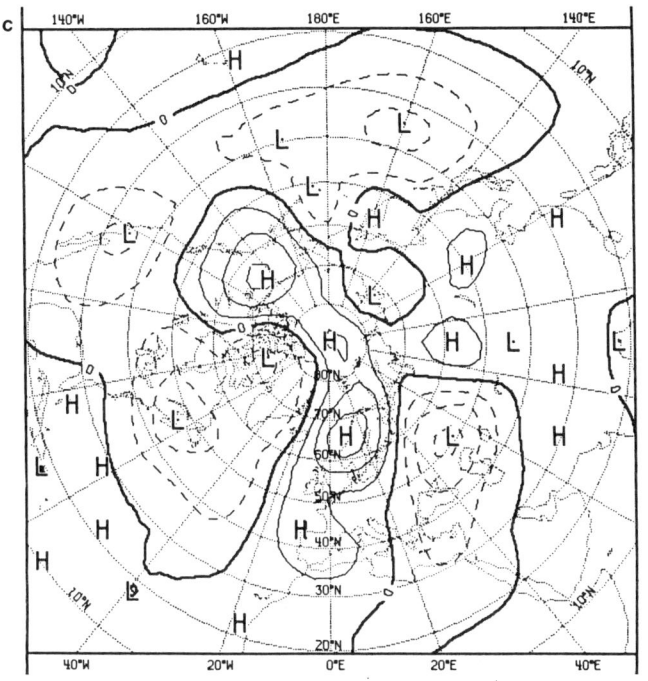

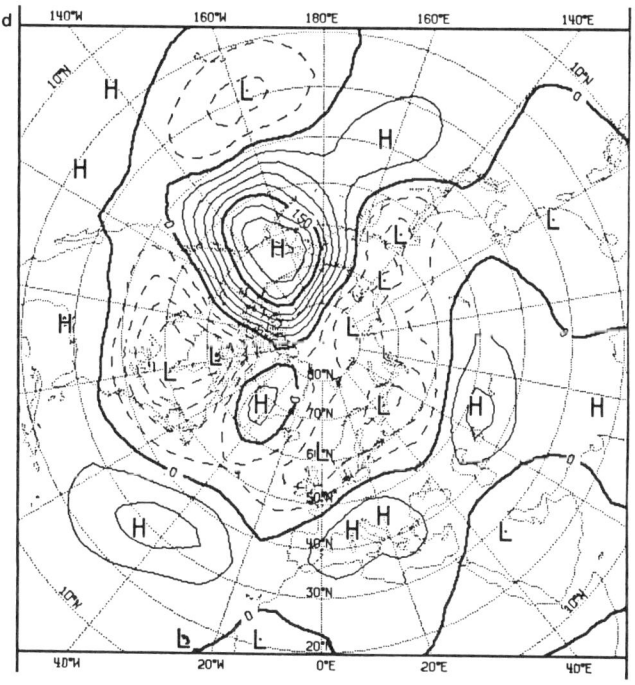

Fig. 6c and d.

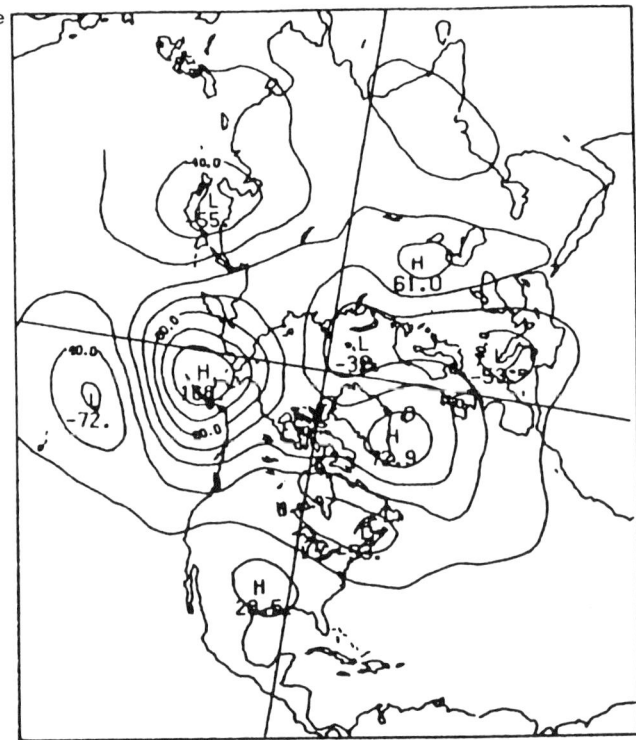

FIG. 6e. See legend on p. 240.

(Mode 2), we present the results for each year, while in Figs. 6a–6d the differences between Mode 1 and Mode 2 mean maps are shown.

Figure 6 shows that the large-scale difference pattern between Mode 1 and Mode 2 has features resembling major blocking events (or more exactly, large-amplitude wave events). Moreover, inspection of events included in each mode revealed, in a subjective analysis, that indeed most of the "blocking" events contribute to Mode 2, while all apparently "high-index" days fall into Mode 1. As another argument supporting these findings, we present an additional difference map (Fig. 6e). This map consists of the difference between the mean 500-mbar heights, 15 January days falling in the positive-anomaly group as defined by Charney *et al.* (1981), minus the mean field. The similarity is quite convincing. Since Charney *et al.* (1981) divide the days by following a different objective definition of blocking, the observed similarities between our difference map and theirs adds confidence to our claim that Mode 2 includes only blocking events. Moreover, since the Charney *et al.* maps are based upon a larger dataset, we feel we can speculate that

TABLE III. Days in which the Variable A Is in Mode 2

Winter	From	To
1980/1981	18 Dec	24 Dec
	2 Jan	24 Jan
	30 Jan	13 Feb
1981/1982	17 Dec	29 Dec
	8 Jan	23 Jan
	2 Feb	11 Feb
1982/1983	2 Dec	10 Dec
	27 Dec	3 Jan
	15 Jan	26 Jan
	5 Feb	11 Feb
	22 Feb	28 Feb
1982/1983	1 Dec	6 Dec
	14 Dec	29 Dec
	5 Jan	21 Jan
	24 Jan	2 Feb

our smaller dataset is fairly representative. Moreover, we remark that analogous calculations for a period running from 1965 to 1979 seem to support the general conclusions so far discussed (Benzi and Speranza, 1986).

In Table III we summarize for each winter the days selected as Mode 2 for the four winters here considered.

7. Discussion

So far, we have shown that, even with a limited sample size (only 4 years), the measure of active large-scale wave amplitude that we have considered exhibits a probability density distribution that is apparently bimodal. To some extent, this result is consistent with the theoretical properties of the model treated by Charney and DeVore. We will now discuss the extent of this consistency.

First of all, our density estimation is connected with a particular projection of the true probability density in a particular direction. We firmly believe that the true probability density is a multidimensional one. Hence, in order not to draw misleading conclusions, some care should be taken. For instance, the fact that we have two maxima does not necessarily imply that large-scale flow patterns possess two *stable* stationary states, as the CDV theory would predict. Within each mode the state of the variable might be highly fluctuating around the corresponding maximum of the density and still the net result of the projection would be that we have a maximum in the probability density. Moreover, by projecting the phase space of a complex dynamical system (such as the atmosphere) onto a particular direction (as in our representation of the large-scale features), any possible fine structure (such as fractal dimensionality or more exotic features) of the true density has been distorted and discarded in favor of gross representation into two global maxima. On the other hand, the unequivocal information that can be deduced by our density estimation is that some instability is present in large-scale atmospheric flow, which has induced the *minimum* of the density distribution. Hence, the correct interpretation of the results, as presented thus far, is that there exists an instability in the complete phase space, leading to the fact that its projection on our coordinate axes induces a minimum in the probability density. (In other words, the property shared by the multidimensional probability density and our projection estimation is the minimum rather than the two maxima.) In fact, the regression equation that would have our density distribution as an asymptotic property is

$$dx = (x - \gamma_1)(x - \gamma_2)(x - \gamma_3) + \epsilon^{1/2}\, dw$$
$$w = \text{Wiener's process}$$
(10)

where γ_1, γ_2, and γ_3 are the values of the density at the first maximum, minimum, and second maximum, respectively; the instability is self-evident. These further aspects of the interpretation of our results will be addressed in a forthcoming paper.

Second, at the present stage we cannot conclude that the instability detected is global (spatially) in its nature. In fact, the preparation of the dataset (i.e., averaging over latitude and especially using the spatial Fourier projection) might have smeared out any local property of the instability into a global feature. The only evidence in favor of the interpretation of the instability as a global property comes from the difference maps presented in the previous section. There an educated inspection could reveal a global wave train intensifying in the region of exit of the jet. On the other hand, some support on behalf of the local nature of the instability could come from the very same maps that show a weaker wave train in the Atlantic sector than that seen in the Pacific sector. Since this feature is less evident than the previous one, we would opt with no firm commitment to the global interpretation.

Finally, it is possible that the same density would be explained by a random interference between stationary waves and westward-propagating global free Rossby waves (R. S. Lindzen, this volume). Since we have not separated the standing part of the waves contributing to the indicator $\{A\}$ from the traveling contribution, we cannot disregard this alternative interpretation. However, an argument (weak, and not conclusive) against this interpretation can be produced on the following basis. If it is the interference between stationary and free westward- propagating Rossby waves that produces the bimodal density, then some memory of the time periodicity of these traveling waves should be detected in the data; not exact, of course, but the duration of the events should be, at least, Gaussian around the period of the Rossby wave. As already mentioned, we did not find any spectral peak for frequency shorter than 90 days. This excludes the exact periodicity requirement.

In disfavor of the Gaussian distribution, we present the following preliminary evidence. In Fig. 7 is shown the density distribution of the duration of the events in Mode 1 and Mode 2 as deduced from considering the same indicator here as discussed for the 13 winters spanning 1966 to 1979 (Benzi and Speranza, 1986). Apparently, this density fails to show any sign of Gaussian nature, while it would fit very well the Poisson distribution. It has been proven (see Benzi and Sutera, 1985) that the distribution of the persistence in a basin of attraction of a fixed point follows a Poisson distribution. Hence, we would conclude that the observations do not lend support to the previously mentioned interpretation. Of course, the density just mentioned has not been carefully estimated and a more thorough analysis could very well prove that other distributions might fit the data.

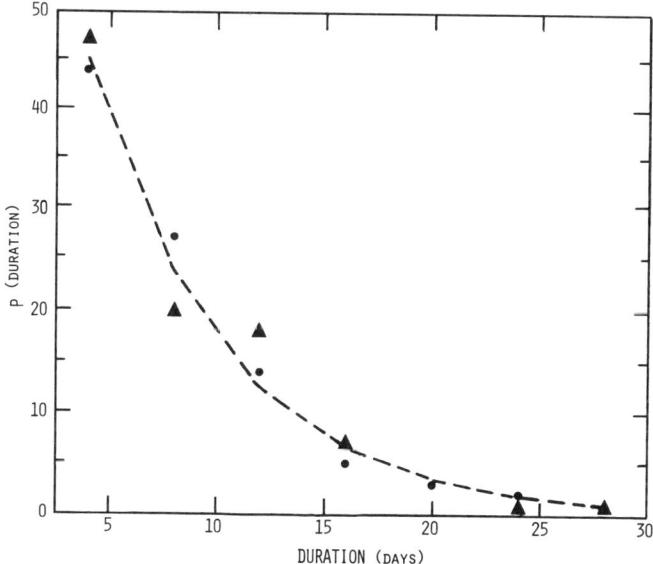

FIG. 7. Density distribution (histogram) of the duration of events in Mode 2 for statistics of 13 winters (1966–1979). (After Benzi et al., 1984.)

8. The Zonal Wind

As mentioned in Section 2, CDV theory predicts that the zonal wind must be bimodal also. In the next section, using the same methods developed so far, we will address this question.

We present in Table IV the results of the density estimation for the variable U. In Figs. 8a–8d we present the corresponding densities that satisfy the 95% confidence level. It is evident that no sign of bimodality is present in any of the estimations considered. The density distributions are always unimodal and strongly resemble Gaussian, slightly skewed density distributions. However, it appears that the skewness of the distribution varies from one year to another. It implies that if we were to consider the distribution of the four winters together, as we have done for the variable A, we might obtain a spurious bimodal distribution. In fact, in this case, the correct interpretation would be that the bimodal nature of the density is a consequence of the interannual variability — i.e., of time scales much longer than the one here considered (periods shorter than 90 days). Of course, this variation of the zonal wind is worthy of further investigation, and we hope to do it in the near future with a larger sample. However, on the basis of the yearly individual density, a regression equation, with such an asymptotic density deduced from the data, is

$$dU = -\hat{\beta}(U - U^*)\, dt + \epsilon^{1/2}\, dw \tag{11}$$

where $\hat{\beta}$ is the decorrelation time for U (about 6 days). At the stationary state (possible only if ϵ were 0; see Sutera, 1981 for example) we will have $U = U^*$. If we plot this relationship in the same phase plane of the CDV resonance

TABLE IV. Confidence Levels of the Kolmogorov–Smirnov Test (p_K) for U[a]

	p_K			
α	1981	1982	1983	1984
0.1×10^4	0.9	0.99	0.98	0.99
0.4×10^3	0.8	0.96	0.94	0.99
0.1×10^3	0.7	0.8	0.8	0.75
0.6×10^2	0.55	0.7	0.5	0.3
0.2×10^2	0.5	0.5	0.2	0.005
0.1×10^2	0.4	0.3	0.07	0.015
0.4×10^1	0.20	0.1	0.02	0.013
0.1×10^1	0.05	0.05	0.002	0.005
0.6×10^0	0.03	0.01	0.0001	0.0008

[a] The behavior of p_K as a function of the hyperparameter α for the variable U.

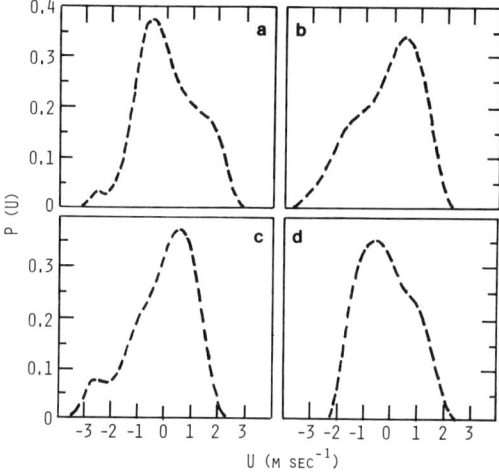

FIG. 8a–d. As in Figs. 2a–d for the variable U.

curve, we see that for $U^* \simeq 10$ m (i.e., the mean value deduced from observations), only one intersection between Eq. (11) and the resonance curve is possible. This result would not allow any interpretation of the observed wave instability discussed in the previous section as being consistent with the orographic instability of the model proposed by CDV. However, we can note that if the resonance curve were a multivalued function of U in the region of the U variability, the consistency could be restored. In Benzi et al. (1986) and Speranza (this volume), by including nonlinear effects in a CDV formulation of the barotropic orographic instability, it is demonstrated that such a multivalued resonance curve can be deduced by an appropriate folding of the resonance curve.

9. Conclusions

We have studied the essential behavior of a measure of the large-scale wave amplitude as deduced from the 500-mbar analysis for the last four winters (the only analyses available from the European Centre for Medium Range Weather Forecasts). Our findings are as follows:

1. The amplitude of the waves has a highly statistically significant bimodal distribution.
2. The presence of a large-scale instability mechanism is implied.
3. This instability is consistent with a theory which connects the instability with an orographic effect [in the sense discussed in Benzi et al. (1986)],

although an alternative explanation might be advocated (see Lindzen, this volume).

4. We are aware of the possibility that our results might be dependent on the sample size. Extension of the present study to more years is required and we hope to do it in the near future.

5. Other limitations of the present study have been discussed, particularly in assessing the nature of the multidimensional probability density induced by atmospheric motion. This has led us to be cautious in interpreting the modes of the probability density as a sign of the existence of stable steady states.

6. Apparently the modes can be identified with atmospheric patterns resembling high-index and amplified large-scale wave patterns.

7. The zonal wind failed to show any bimodality.

8. As a consequence of item 7, we emphasize the role of the nonlinear processes if the instability found in the data has its theoretical counterpart in the orographic instability mechanism.

Acknowledgments

It is my pleasure to thank Drs. L. Bengtsson and D. Burridge and the staff of ECMWF for their hospitality and frequent discussions on the topic addressed in this article. Professor B. Saltzman and a reviewer helped in improving the manuscript.

References

Benzi, R., and Sutera, A. (1985). The mechanism of stochastic resonance in climate theory. In "Turbulence and Predictability in GFD and Climate Theory" (M. Ghil, R. Benzi, and G. Parisi, eds.). North Holland Publ., Amsterdam.

Benzi, R., Malguzzi, P., Speranza, A., and Sutera, A. (1986). The statistical properties of general atmospheric circulation: Observational evidence and a minimal theory of bimodality. *Q. J. R. Meteorol. Soc.,* in press.

Benzi, R., and Speranza, A. (1986). The statistical properties of low frequency variability in the Northern Hemisphere. *Mon. Weather Rev.,* submitted.

Charney, J. G., and DeVore, J. G. (1979). Multiple flow equilibria in the atmosphere and blocking. *J. Atmos. Sci.* **36,** 1205–1216.

Charney, J. G., and Eliassen, A. (1949). A numerical method for pedicting the perturbation of the middle latitude westerlies. *Tellus* **1,** 38–54.

Charney, J. G., Shukla, J., and Mo, K. C. (1981). Comparison of a barotropic blocking theory and observation. *J. Atmos. Sci.* **38,** 762–779.

Dole, R. M., and Gordon, N. D. (1983). Persistent anomalies of the extratropical Northern Hemisphere wintertime circulation: Geographical distribution and regional persistence chacteristics. *Mon. Weather Rev.* **111,** 1567–1586.

Hartmann, D. L. and Ghan, S. J. (1980). A statistical study of the dynamics of blocking. *Mon. Weather Rev.* **108,** 1144–1159.
International Mathematical and Statistical Library (1982). *Routine NDDMPL NKS1.*
Leith, C. (1978). Objective methods for weather prediction. *Annu. Rev. Fluid Mech.* **10,** 107–128.
Moritz, R. E., and Sutera, A. (1981). The predictability problems: Effects of stochastic perturbations in multiequilibrium systems. *Adv. Geophys.* **23,** 345–383.
Priestley, M. B. (1981). "Spectral Analysis and Time Series," Vol. 1. Academic Press, New York.
Scott, D. W., Tapia, R. A., and Thompson, J. R. (1977). Kernel density estimation revisited. *Nonlinear Anal. Theory, Methods Appl.* **1,** 339–372.
Sutera, A. (1981). On stochastic pertubation and long term climate behaviour. *Q. J. R. Meteorol. Soc.* **107,** 137–151.
White, G. H. (1980). Skewness, kurtosis and extreme values of Northern Hemisphere geopotential heights. *Mon. Weather Rev.* **108,** 1446–1455.

STATIONARY PLANETARY WAVES, BLOCKING, AND INTERANNUAL VARIABILITY

R. S. Lindzen

Center for Meteorology and Physical Oceanography
Massachusetts Institute of Technology
Cambridge, Massachusetts 02139, and
Physical Research Laboratory
Ahmedabad, 380 009, India

1. Introduction

This paper is a somewhat didactic discussion of various current trends in the study of large-scale planetary waves and their anomalies. The intensity of current activity on this topic stems from the belief that the anomalies can be unusually long lived and may, therefore, represent particularly predictable features of weather. The word "blocking" is sometimes associated with these anomalies. Much of the present activity stems from the winter of 1976/1977, when the period December to mid-February was, on the average, unusually cold; this was associated with a stationary wave pattern that was significantly amplified over its climatological mean (Namias, 1978).

It seems to me that much of the current approach is unbalanced at best. Far more work is being devoted to accounting for anomalies than to accounting for the climatology itself. Moreover, the contention that the anomalies are persistent may itself be misleading. By persistence I suppose is meant slow time variation compared to the expected periods of oscillation of the atmosphere. Such persistence, I hope to show, accounts for a relatively small part of the planetary-wave anomalies.

I will begin, in Section 2, by examining some aspects of the observed time behavior of anomalies. It appears that there is some modest interannual variability of stationary wave patterns, but little evidence of unusual persistence in the larger anomalies. Indeed it appears that a significant part of interannual variability is simply a statistical residue of short-period anomalies.

Given the above conclusions, it is awkward to discuss the various explanations put forward to account for the allegedly persistent anomalies. However, the observational picture is not yet convincingly clear. There is some evidence, that on rare occasions, long-lived (> 1 month) significant anomalies do occur. The period 1962–1963 is sometimes cited in this regard (Tung and Lindzen, 1979). (Even in such events it seems possible that we are simply dealing with shorter events "running into each other.") Two approaches have been particularly popular: multiple equilibria, and teleconnections between the tropics and midlatitudes. The former, which is amply represented in this volume, considers the possibility that stationary wave patterns have several equilibrium configurations and that persistent anomalies arise when the atmosphere passes from one equilibrium state to another. If multiple-equilibria theories were correct, then the nonexistence of persistent anomalies would be a problem. As I will show in Section 3, the most common multiple-equilibria theories appear inappropriate to the atmosphere.

The role of teleconnections is more complicated. Horel and Wallace (1981) noted that patterns of geopotential heights involving both the tropics and midlatitudes accounted for significant (about 20%) portions of the winter variability of monthly means. Thus, it was conjectured that tropical occurrences like El Niño should have anticipatable midlatitude consequences. Given the modest role of this pattern it is difficult to believe that it can offer much improvement in long-term forecasts. Certainly, it does not appear to be a mechanism for the prominent short-period anomalies. In Section 4, I will simply note the poor correlation between El Niño years and North American winter climate. I will show some recent data analyses by Plumb (1985) which appear to suggest a very small role for the tropics in forcing the climatological stationary waves of northern midlatitudes. The clear conclusion that one is forced to is that the tropics are certainly not the major factor in determining midlatitude stationary waves, though they may be one contributor among others to interannual variability. This is consistent with the recent work of Lau and Phillips (1984) and Dole (1985), who found that composites of planetary-wave anomalies in midlatitudes showed no evidence of a tropical precursor. On the contrary, there was a suggestion of tropical response (i.e., time lag). Nevertheless, a large number of theoretical papers have purported to find large midlatitude responses to tropical forcing—duplicating the Horel–Wallace teleconnection pattern.

We will discuss the theoretical papers in Section 5 where, however, I concentrate on recent results by Jacqmin and Lindzen (1984) on the climatological stationary wave pattern and its sensitivity to changes in both the zonally averaged flow and in the thermal forcing.

Finally, in Section 6, some recent work on the atmosphere's free transient Rossby waves is described. It has been found that these oscillations are planetary in scale, of significant amplitude, and characterized by periods of 10–20 days—periods not unlike those identified with unusual persistence.

2. How Persistent Are Anomalies?

Although it is commonly taken for granted that anomalies of unusually long persistence are an important feature of weather, I would like to suggest that this may, in fact, be untrue. Consider Fig. 1, taken from the *New York Times* of 9 January 1983, which summarizes New York City's weather in 1982 (there was no particular reason for choosing 1982 other than the convenient availability of this figure). As always, the temperature vacillates above and below the "normal" range. Of particular note is December 1982, an unusually warm month. Note that the warmness of the month as a whole was due to a sequence of warm episodes, none of which lasted longer than a week. Note also that this warm month also had episodes of below-normal temperature. The point is perhaps obvious: namely, a warm month does not represent an anomaly which persisted for a month.

This is consistent with Fig. 2 taken from Dole and Gordon (1983). This figure shows number of anomalies (for different anomaly thresholds) versus duration. We see that the expected duration of any anomaly, once observed, is less than 6 days.

To be sure, there are times when 6 days seems anomalously long. However, for persistence to have a physical meaning, "long" must be long compared to the normal time scales for transiency. The most obvious comparison, here, is with the time scales of the free *planetary-scale* Rossby waves. As we shall describe in Section 6, such waves are prominent in the data and commonly have periods of from 10 to 20 days, consistent with the theory of such waves. In comparison with these periods, the persistence time scales are certainly not long.

While the bulk of planetary-wave anomalies are not "persistent" in a physically meaningful way, there is in fact some significant interannual variability in "stationary" waves. Figure 3 from van Loon *et al.* (1973) shows

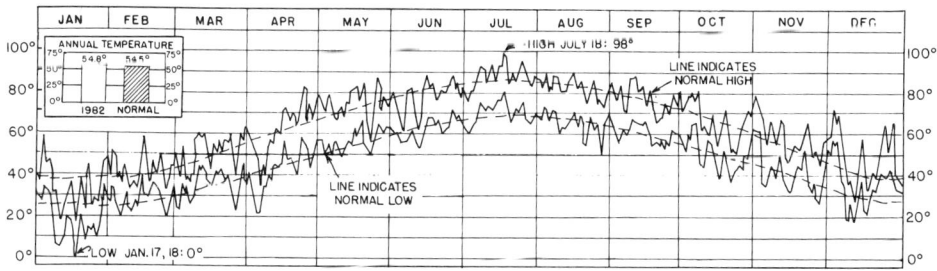

Fig. 1. Temperature in New York City as a function of time for 1982. (From *New York Times* of 9 January 1983.)

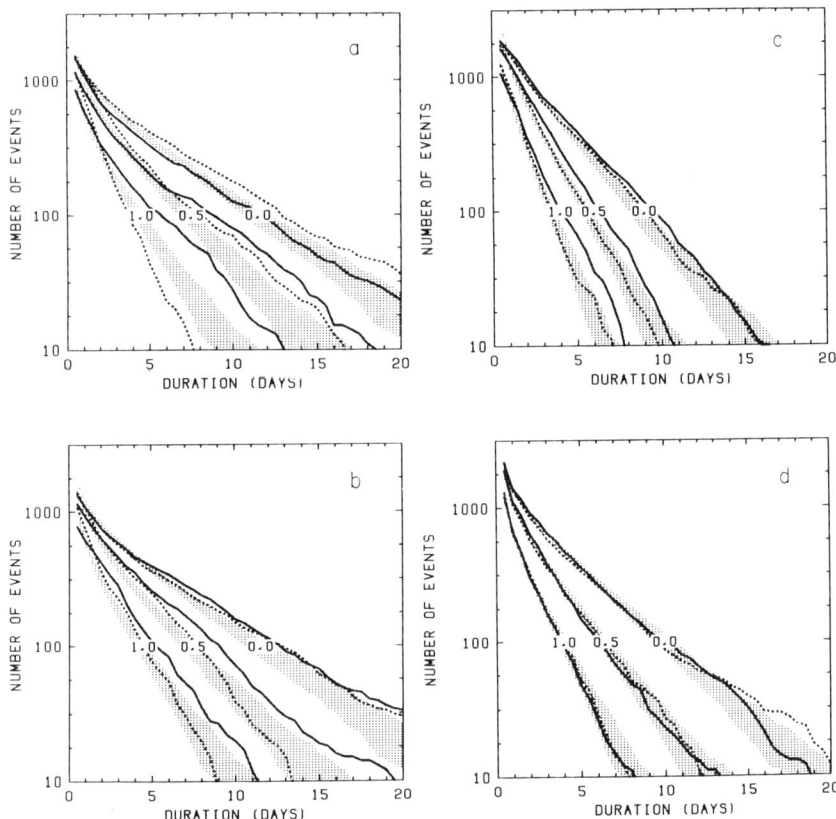

FIG. 2. Histograms of number of persistent anomalies versus duration, for different areas and for positive (solid lines) and negative (dotted lines) anomalies. (a) For area PAC; (b) for area AME; (c) for area ATL; (d) for area EAS. (After Dole and Gordon, 1983.)

January mean planetary-wave response for wavenumbers 1, 2, and 3 at fixed latitudes and for a range of years. There are variations but they are *not* $O(1)$ and judging from Fig. 1 they are significantly due to short-period episodes. This is supported by Table I, which shows the interannual variability of 3-month winter means of planetary-wave height fields at 50°N (where variability is usually largest). The variability is still smaller than found by van Loon for January means. Figure 4 shows the wavenumber 1–3 sums (based on Table I) for four of the most deviant years. There are indeed meteorologically significant differences but gross patterns and magnitudes remain similar.

Thus, we are left with a picture wherein strong planetary-wave anomalies are associated with short periods, while interannual variability, though real,

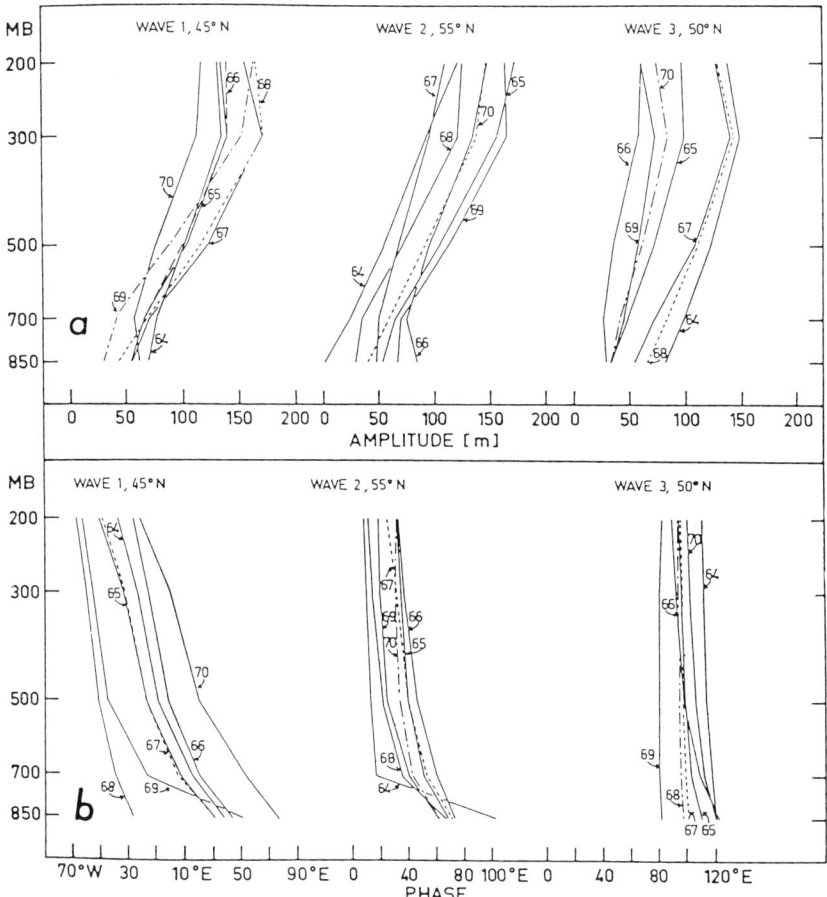

FIG. 3. Wavenumber decomposition of January mean height field for the years 1964–1970. (After van Loon et al., 1973.)

is modest. It should be added that modest changes in stationary waves involving phase shifts of a few hundred kilometers and amplitude changes of 20% are still of major importance to the weather of specific regions.

3. MULTIPLE EQUILIBRIA?

Under the impression that unusually persistent stationary anomalies existed (something we question in Section 2), models were developed suggesting that there existed a multiplicity of stationary wave states, each in equilib-

TABLE I. AMPLITUDE AND PHASE OF WAVENUMBERS 1, 2, AND 3[a]

Year	Wave 1 Amplitude (m)	Wave 1 Phase (ridge longitude)	Wave 2 Amplitude (m)	Wave 2 Phase (ridge longitude)	Wave 3 Amplitude (m)	Wave 3 Phase (ridge longitude)
1	105	2	85	91	115	−29
2	108	−6	81	73	96	−84
3	85	−5	131	97	40	−94
4	125	−9	87	91	95	−63
5	81	−18	64	78	100	−66
6	110	−26	81	65	49	−116
7	124	−5	90	110	84	−62
8	80	2	97	63	69	−56
9	72	6	116	57	71	−53
10	114	−6	92	73	75	−34
11	93	6	91	81	59	−49
12	117	1	92	79	88	−43
13	116	−9	66	70	116	−52
14	103	0	135	103	80	−41
Average	102 ± 17		93 ± 20		81 ± 22	
Phase		−4.8		80.8		−60.1
Range	28° long. ±14° long. ±14° phase		27° long. ±13.5° long. ±27° phase		28° long. ±14° long. ±42° phase	

[a] Contributions to winter average height field at 500 mbar for 1964–1977.

rium with the topography and the forcing of the zonally averaged basic state. The pioneering effort along these lines was due to Charney and DeVore (1979). Among the successor efforts, several are discussed in these proceedings. All such models of multiple equilibria depend on the existence of stationary free waves (normal modes) which are resonantly forced by stationary forcing. Clearly, the existence of resonance is crucially dependent on the mean zonal wind. In addition, in each model, stresses due to the resonant stationary wave ("form–drag") play a major role in determining the mean flow. The interplay of these two dependences (the dependence of resonance and hence form–drag on the mean flow and the dependence of mean flow on form–drag) is at the heart of the existence of multiple equilibria. In most approaches, different equilibria are associated with different mean flows and different stationary wave amplitudes. However, recent work by Speranza (this volume) has considered nonlinear resonance curves where two different stationary amplitudes can be associated with almost identical mean flows.

The purpose of this section is to briefly note some objections to the existing theories of multiple equilibria. [Other objections are described in Tung and

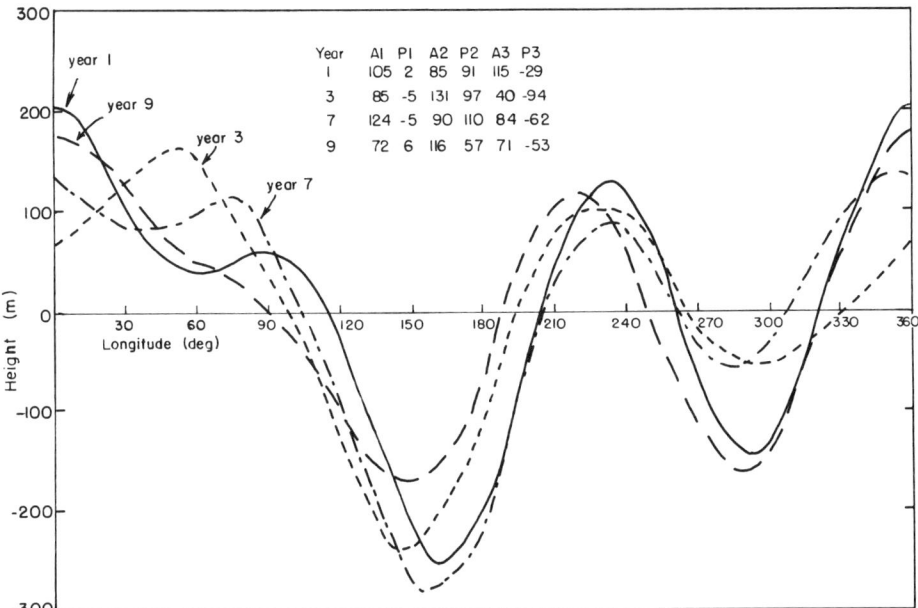

FIG. 4. Winter (Dec–Feb) average height fields versus longitude at 50°N for various years.

Rosenthal (1984).] The objections fall into two broad categories. The first deals with the existence of suitable resonances, the second with the effect of stationary waves on the mean flow.

As concerns the existence of resonance, several difficulties exist. The only well-documented planetary-scale free oscillations of the atmosphere are the external modes associated with global scales (zonal wavenumbers 1–3 with comparably small meridional wavenumbers) and very large westward phase speeds [$0(70$ m sec^{-1})]. The most recent description of these waves is in Lindzen et al. (1984). These waves are only mildly altered by realizable changes in the mean flow; there are *no* observed mean flows which reduce their phase speed to zero. Thus they cannot be resonantly forced by stationary forcing. The only free waves with suitably low phase speeds would be internal modes. Since the atmosphere does not have a lid, the only internal modes arise from internal reflections from surfaces in the atmosphere where its index of refraction changes sign (Tung and Lindzen, 1978; Held, 1984). Resonances arise from the repeated reflection of waves from such surfaces where incident and reflected waves are in phase so that amplification may take place. In idealized models the reflecting surface is perfectly horizontal and the necessary constructive interference is readily achieved. Unfortunately, in reality this surface (when it exists at all) is quite bent and distorted

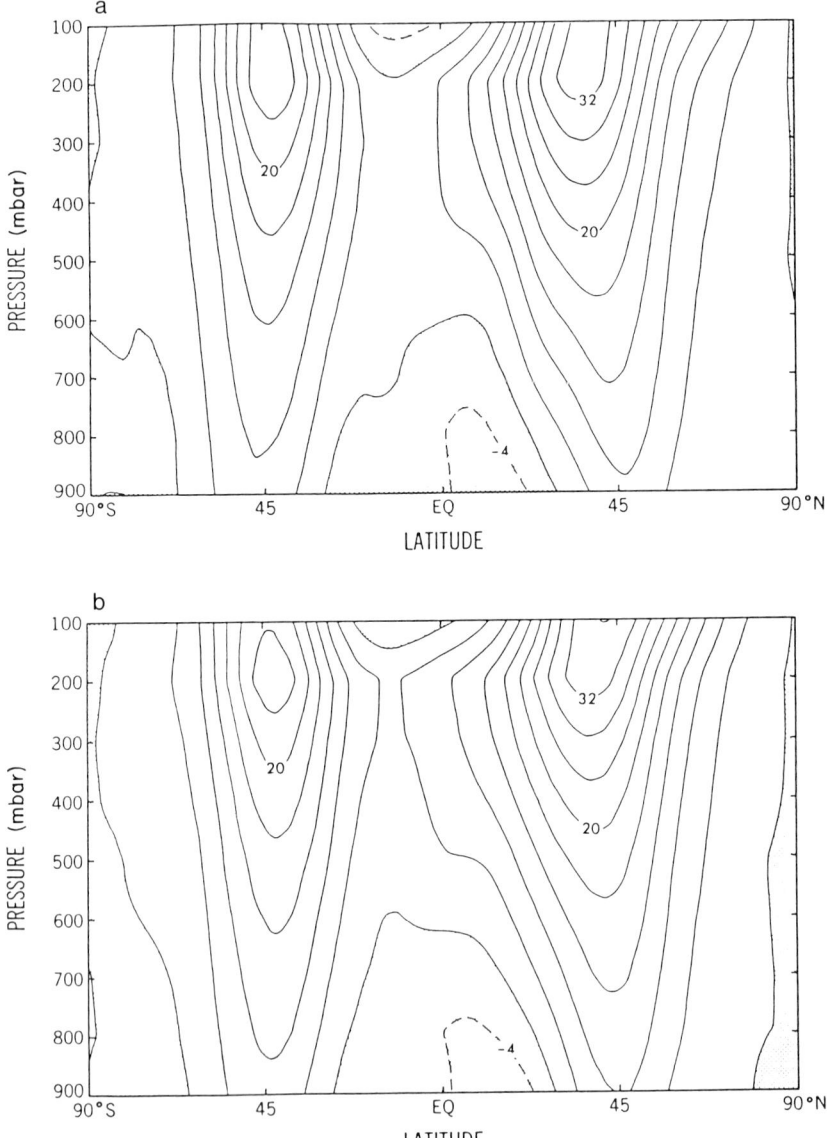

Fig. 5. Zonally averaged mean zonal wind in GCMs without (a) and with (b) mountains. (After Nigam, 1983.)

and reflected waves do not coincide with incident waves; resonant amplification, therefore, is profoundly limited. A clear example of this effect is given by Lindzen and Hong (1974). The situation is rendered even more negative by the fact that stationary resonant modes would also encounter zero wind surfaces. Although it has been argued that nonlinearity can lead to reflection at such surfaces (Tung, 1979), it is also true that in the neighborhood of such surfaces damping becomes very important and is inimical to resonance. Even normal damping will severely restrict resonance. It has recently been suggested by McIntyre and Palmer (1984) that wave steepening near such surfaces leads to irreversible mixing so that the damping limit may, in fact, be intrinsic. [However, P. Killworth and M. E. McIntyre (private communication) recently noted that this may not be uniquely associated with wave absorption.]

For all the above reasons one may reasonably conclude that the resonance called for in multiple-equilibria theories is unlikely under realistic circumstances. This view is supported by the recent work of Jacqmin and Lindzen (1984) wherein stratospheric wind configurations suggested by Tung and Lindzen (1979) as favorable for resonance were used, and no anomalous tropospheric response was obtained.

We turn, finally, to the influence of stationary waves on the mean flow — an influence essential to most multiple-equilibria theories. Nigam (1983) recently ran a general circulation model (GCM) at Geophysical Fluid Dynamics Laboratory (GFDL), where zonal wind distributions were calculated in models with and without mountains (the major forcers of stationary waves at midlatitudes). His results are shown in Fig. 5. Clearly the zonal winds (except at the surface) are almost identical.

Thus on both essential grounds, there seems to be little theoretical support for the possibility of proposed multiple equilibria.

4. Teleconnections — The Tropical Connection?

In the search for a means to anticipate years with anomalous stationary wave patterns, teleconnections have played a prominent role. The term was introduced by J. Namias to describe the apparent correlation of weather at remote points. In recent years the idea has been closely associated with the notion that tropical heating plays a major role in forcing high-latitude stationary waves (Wallace and Gutzler, 1981; Horel and Wallace, 1981). Figure 6 shows the Pacific–North American (PNA) pattern from Wallace and Gutzler (1981) which is found to account for about 18% of the planetary-wave variance. Clearly, this is not a dominant factor. Moreover, the omission of the tropical part of the PNA pattern would not greatly reduce the variance it accounts for. Despite this, great emphasis has been placed on this

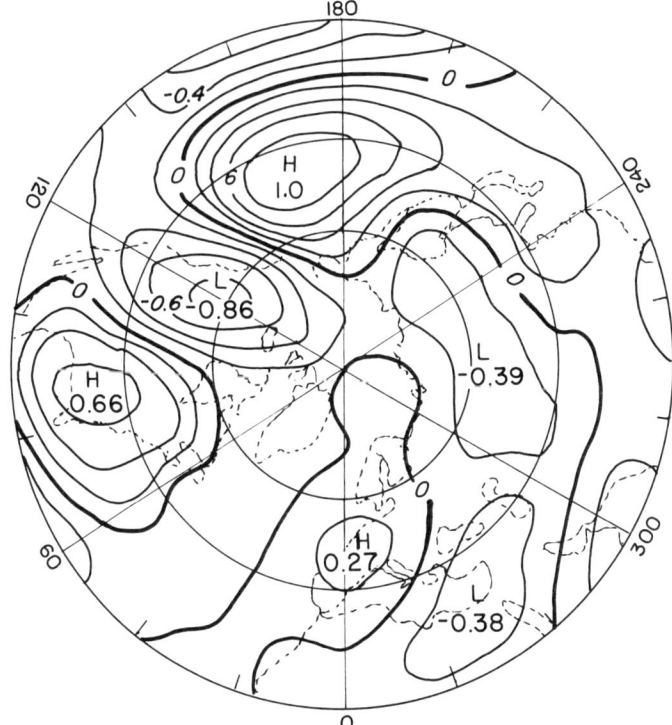

FIG. 6. Coherence map for PNA pattern. (After Wallace and Gutzler, 1981.)

pattern — with the promise that the sea surface temperature anomalies (and the resulting anomalies in cumulus convection distribution and latent heating) associated with El Niño might have predictable consequences for North American weather.

Figure 7 shows the distribution of winter temperature in the continental United States for 8 El Niño years. There is little more similarity in these patterns than may be found among arbitrarily chosen years. The figure shows regions of normal, above-normal, and below-normal temperature for the various winters. These designations are such that one-third of all winters fall into each of the categories. The category limits for selected cities are given in Table II.

It should be realized that anomalous heating associated with El Niño is primarily a redistribution of heating which occurs normally, rather than some source of additional heating. Thus, if El Niño were a profound influence on midlatitude stationary waves, one would also expect that the tropics would be a major source for climatological stationary waves. Plumb (1985)

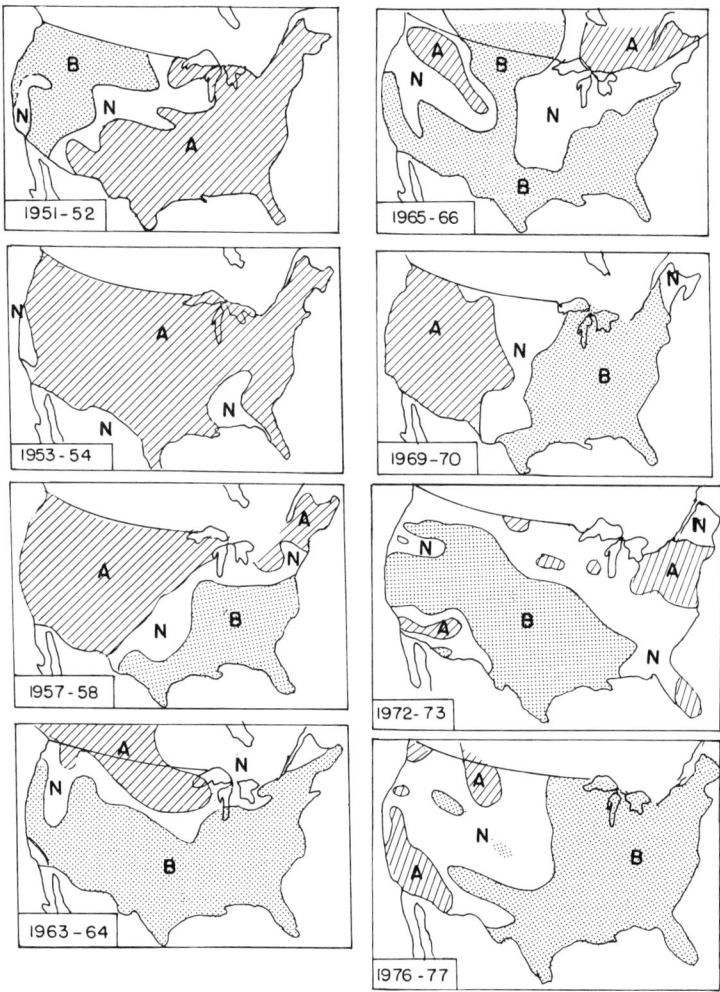

FIG. 7. Winter temperature pattern over North America for 8 El Niño years. (From NWS climate alert.)

TABLE II. WEATHER SERVICE CLASS LIMITS FOR 90-DAY WINTER MEANS

Boston	30.9° ± 1.3°F
New York City	33.6° ± 1.3°F
Miami	67.8° ± 1.2°F
New Orleans	54.4° ± 1.9°F
Des Moines	22.8° ± 1.6°F
Denver	31.7° ± 1.5°F
Los Angeles	55.3° ± 1.2°F
San Francisco–Oakland	49.7° ± 1.0°F
Seattle	40.2° ± 1.2°F
Great Falls	21.1° ± 2.9°F

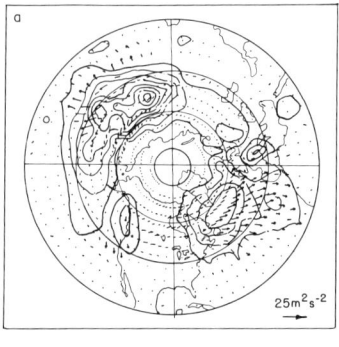

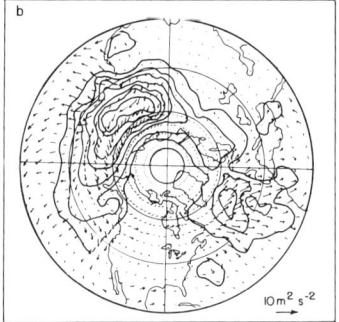

FIG. 8. Total stationary-wave flux at (a) 500 and at (b) 150 mbar. (After Plumb, 1985.)

has recently developed a three-dimensional variant of the E–P (Eliassen–Palm) flux in order to ascertain the geographic origin of climatological stationary waves. His approach is not without ambiguities. However, his results, shown in Fig. 8, are certainly suggestive. Not only do they fail to show the tropics as a source of stationary wave flux, they even suggest that the tropics are a modest sink of such fluxes.

Finally, Dole (1985) has recently performed a composite analysis of persistent anomalies in the North Pacific. This study suggests that the tropical part of the PNA pattern actually lags behind the more prominent northern part. This result was also obtained by Lau and Phillips (1984) in a study of satellite cloud data.

The above results are compatible with Jacqmin and Lindzen's (1985) recent stationary-wave calculations, which will be discussed next.

5. Linearized Response to Stationary Forcing

The most common approach to studying stationary waves has been to calculate them as forced linearizable perturbations on a zonally symmetric basic state. There are a number of worrisome questions about such an approach. Although the most obvious question concerns the validity of linear theory itself, there are still many uncertainties remaining even if linearity itself were acceptable. At the most basic level is the question of the forcing. Although orographic forcing is well known, thermal forcing is more ambiguous. The question of thermal forcing is discussed in some detail in Jacqmin and Lindzen (1985); the contribution of latent heating is to some extent documented in atlases of rainfall (viz., Schutz and Gates, 1972; Dorman and Burke, 1979); the contribution of sensible heat flux, while likely to be smaller (Charney, 1973), is not at all well known. There is also the possibility that transient disturbances traveling along geographically preferred storm paths might contribute to the modification of stationary waves (Youngblut and Sasamori, 1980; Niehaus, 1980; Frederiksen, 1979). An interesting test of this situation was made in the previously mentioned work of Nigam (1983). Nigam, working with his advisor, I. M. Held, developed a simplified, but physically complete, general circulation model. The model was run long enough for time averages to delineate the model's stationary waves. The same model was then linearized and its response to the stationary components of the forcing was calculated. The differences in the results are presumably due to the above factors (transients, nonlinearity, etc.).

Some results are shown in Fig. 9. This figure shows the January mean of the 500-mbar height taken from a 20-year run of the GCM (Fig. 9a), the same mean obtained from observations (Fig. 9b), the linearized response to topography and the GCM's January diabatic heating using the GCM's January zonally averaged zonal wind (Fig. 9c), the response to topography alone (Fig. 9d), the response to diabatic heating alone (Fig. 9e), and the stationary waves forced by the GCM's transient disturbances (Fig. 9f). The most important point is that the GCM and linearized results, while not identical, are both quantitatively and qualitatively close. It is also clear that a significant part of the small difference is due to the forcing of stationary waves by the GCM's transients. Finally, we see that the GCM and linearized results are equally close matches to the data.

The above leaves us with some confidence that linearized results are meaningful. This is of considerable importance since neither of Nigam's models is entirely adequate from the point of view of resolution. While the costs of suitable resolution in a GCM might prove prohibitive, they are easily acceptable in a linearized model. Such a high-resolution linearized model was developed by Jacqmin and Lindzen (1984). In this model [characterized by vertical resolution $0\ (1\ \text{km})$ and meridional resolution $0\ (1°)$], the response of an atmosphere with realistic distributions of zonal wind and zonally averaged temperature to realistic orography and to heating derived from rainfall atlases (latent heating is likely to be the dominant thermal forcing) is calculated. There is no point in giving a detailed discussion of the model here, but in general the response was in tolerably good agreement with the

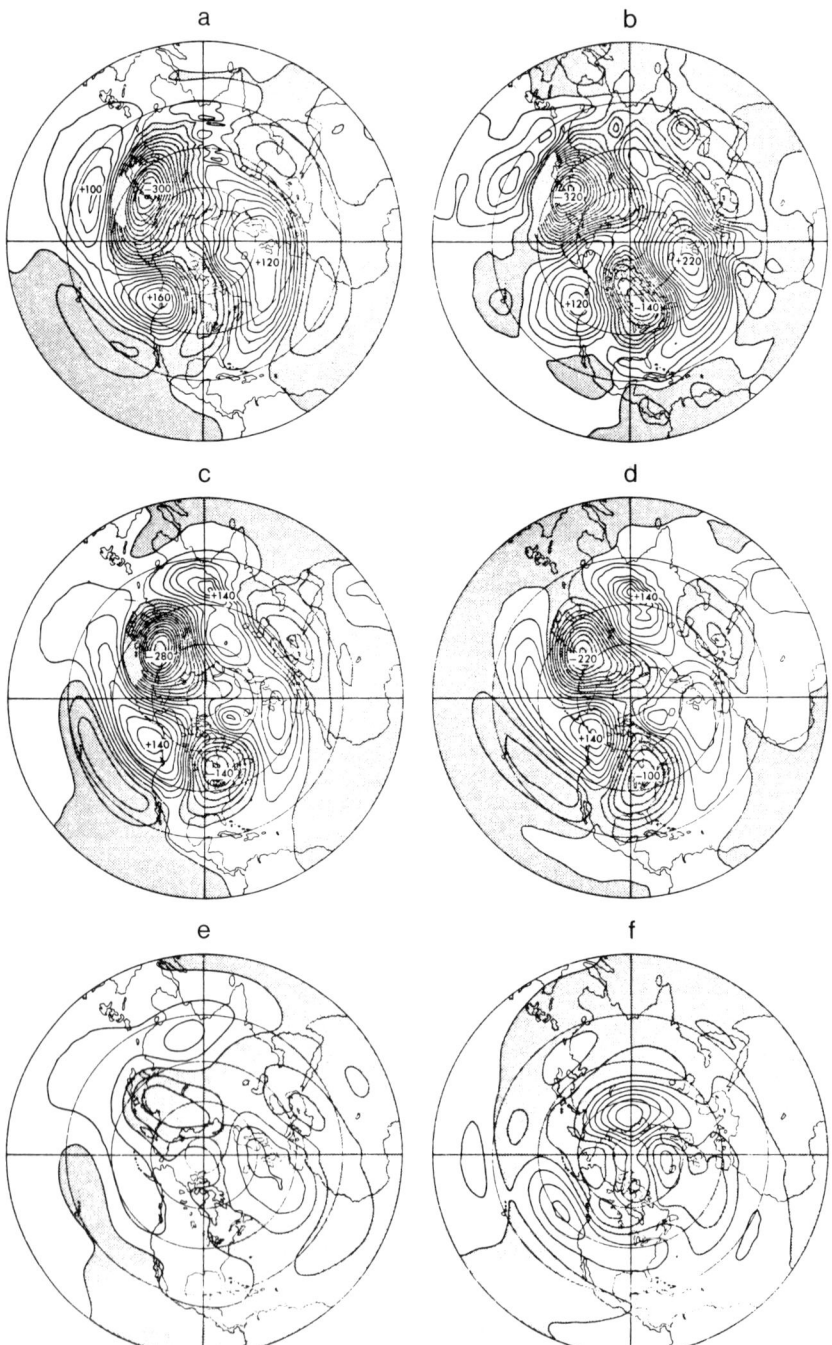

FIG. 9. Comparison of stationary wave patterns obtained by time averaging full GCM results and data and results from a linearized version of the GCM. (a) GCM; (b) atmosphere; (c) total; (d) topography; (e) heating; (f) transients. See text for details. (From Nigam, 1983.)

stationary-wave climatology of van Loon *et al.* (1973). One might, therefore, hope that the model might in fact tell us something about how stationary waves in the atmosphere work. Figure 10 shows the calculated stationary-wave height contours at various levels. Figure 11 shows the contours due to topographic forcing alone, while Fig. 12 shows the contours due to thermal forcing alone. Clearly Figs. 10 and 11 are almost identical; the contribution from thermal forcing is relatively small. It should be noted, moreover, that

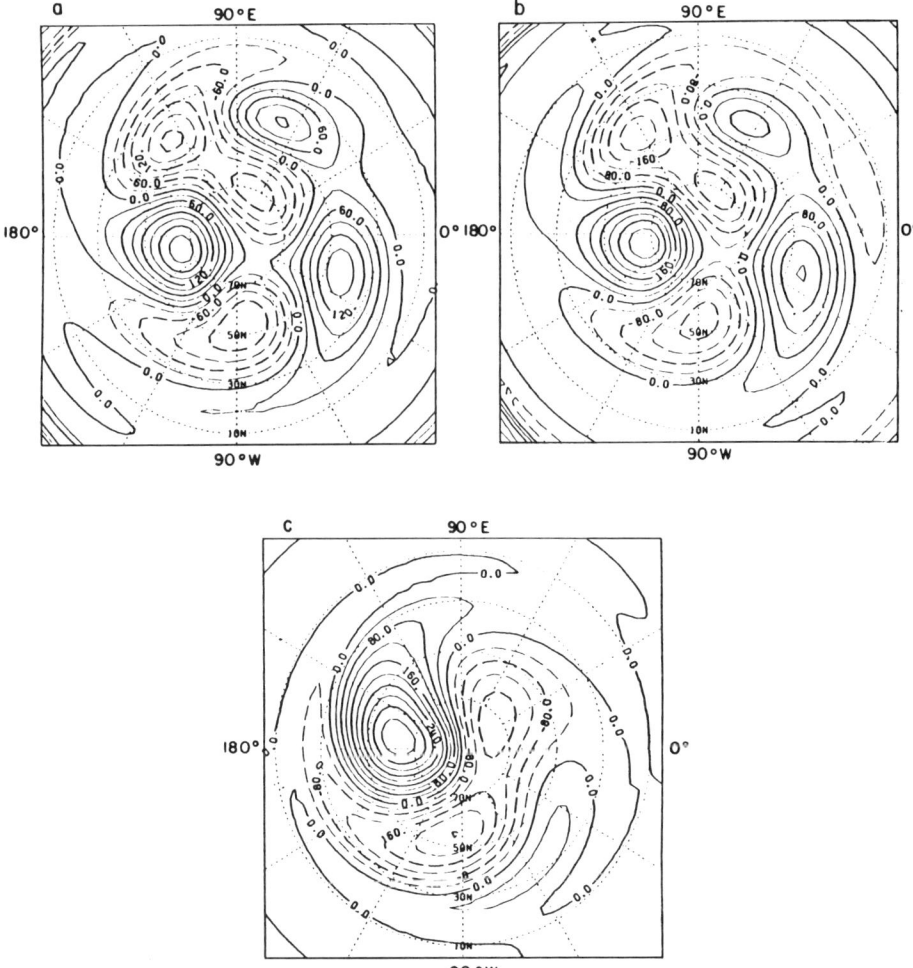

FIG. 10. Stationary-wave height field at various levels calculated from a high-resolution linear model with mountain and thermal forcing. (a) 6 km; (b) 10 km; (c) 30 km. (From Jacqmin and Lindzen, 1985.)

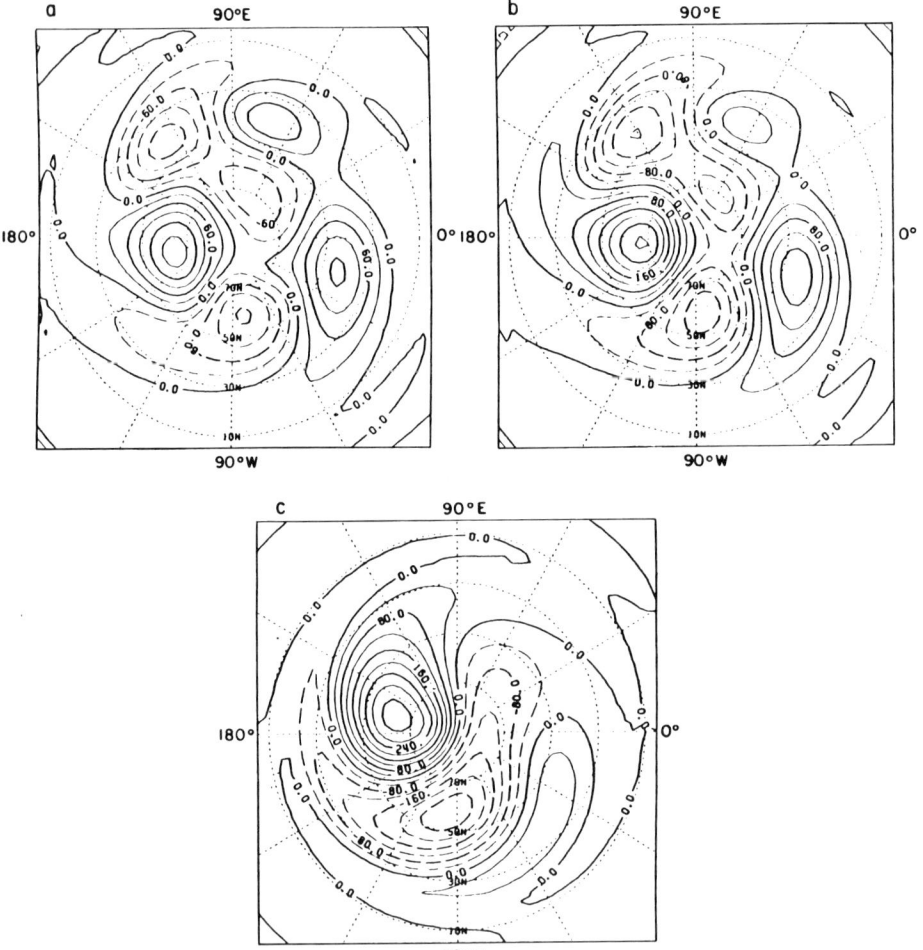

FIG. 11. Same as Fig. 10, but with mountain forcing only.

Fig. 12 includes thermal forcing from all latitudes. The contribution from tropical forcing alone is responsible for only about half of the midlatitude response shown in Fig. 12. These results are completely consistent with the observationally based analysis of Plumb (1985) cited earlier in this article. They are also compatible with results obtained by Nigam and shown in Fig. 9.

The question finally arises as to how to reconcile these results with the multitudinous results showing the important influence of tropical heating on Northern Hemisphere stationary waves (Hoskins, 1978; Hoskins and

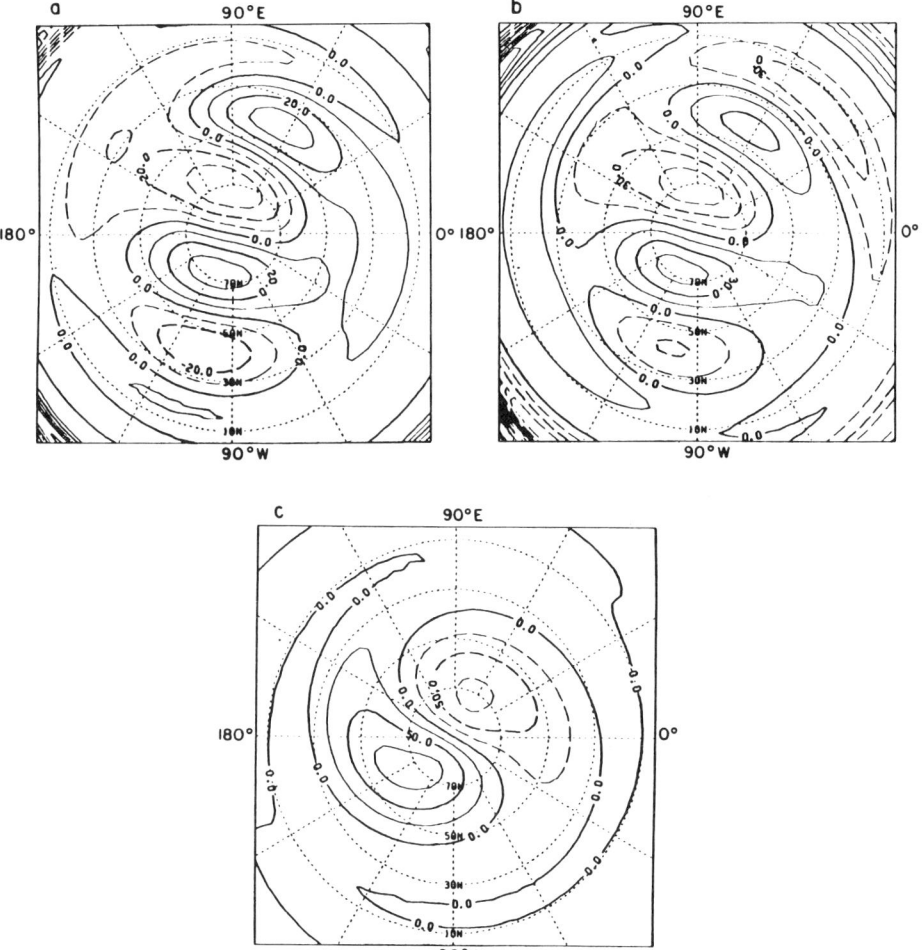

FIG. 12. Same as Fig. 10, but with thermal forcing only.

Karoly, 1981; Karoly and Hoskins, 1983; Egger, 1976; Opsteegh and van den Dool, 1980; Webster, 1981; Simmons, 1982; and many others). Most of these calculations, in fact, used oversimplified models (one- or two-level models) whose shortcomings are obvious. However, Simmons (1982), with a nine-level model in a spherical geometry, did not, in large measure, suffer from these shortcomings. His global response to tropical forcing is shown in Fig. 13. The midlatitude response at 500 mbar is somewhat large but certainly smaller than the climatological stationary-wave amplitudes (essentially those shown in Fig. 10). If Simmons had used more plausible damping,

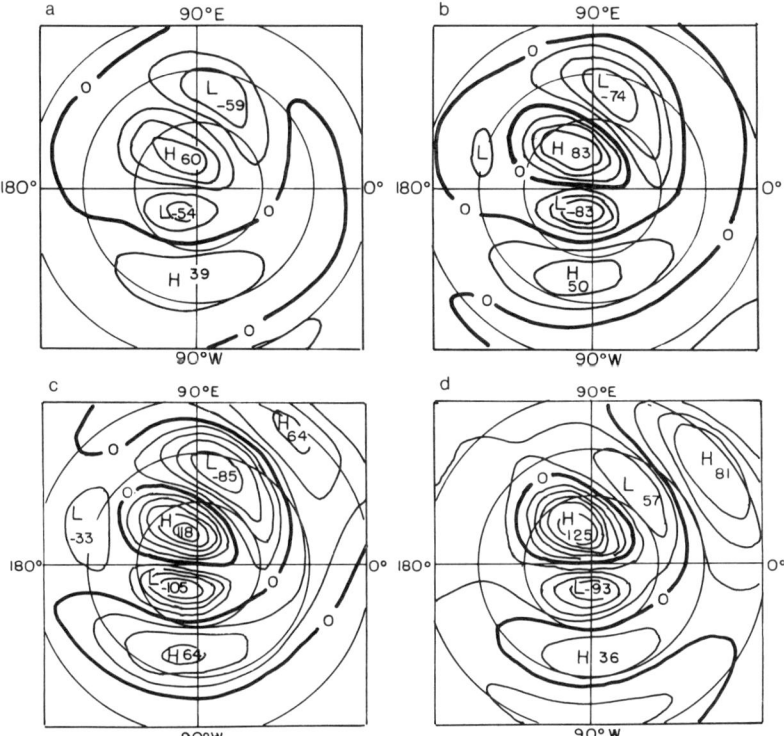

FIG. 13. Polar plots of the response to tropical forcing. (a) 700 mbar; (b) 500 mbar; (c) 300 mbar; (d) 100 mbar. (After Simmons, 1982.)

it is likely that the modest reduction in response would bring it into consistency with observed interannual variability.

This, however, is only part of the story. Jacqmin and Lindzen (1984) found that for their reference basic state, the response to an isolated tropical heat source was similar in magnitude to that obtained by Simmons, but the shape of the response was very different. However, for a moderately different (but still reasonable) distribution of zonal wind, they could, in fact, closely duplicate Simmons' result. This sensitivity to zonal wind suggests that it will be difficult to use tropical sea surface temperature as a long-term predictor of midlatitude stationary waves, simply because the zonal winds themselves are variable over relatively short periods.

Where does the above leave us? If Jacqmin and Lindzen (1984) are correct in identifying flow over topography (primarily the Himalayas and secondarily the Rockies) as the dominant forcing for Northern Hemisphere midlatitude stationary waves, then flow over these mountains should at least provide a short-term precursor for downstream stationary-wave anomalies.

Dole (in this volume) has recently found some weak evidence in support of this. As far as tropical anomalies go, they may not be of practical importance in midlatitude forecasting, but they are certain to be a prominent factor in determining stationary waves (i.e., the Walker circulation) in the tropics.

6. Free Rossby Waves and the Meaning of Persistence

In discussing unusual persistence, it is, of course, essential to know what one means by persistence. As we have already noted, persistence ought to mean *long lasting* compared to the time scales expected for transients of the same geographical scale. Historically, Rex (1950) noted that blocking tended to persist longer than the much smaller scaled synoptic disturbances whose period was as short as 2 days. Clearly, in view of the dissimilar scales, this comparison is not particularly relevant. The relevant transients are the planetary-scale free Rossby waves.

The periods of these waves have been calculated many times. Table III shows results calculated by Kasahara (1976) for basic states both at rest and with a barotropic mean flow corresponding to observed 500-mbar zonal flows in various seasons. We see that for zonal wavenumbers 1–3, the main symmetric meridional mode has a period 0 (5 days), the next antisymmetric mode a period 0 (10 days) and the next symmetric mode a period 0 (16 days). It has further been noted by Lindzen *et al.* (1984) that oscillations associated with the longer periods are dominant in the data. Using data from the First GARP Global Experiment (FGGE) year, Lindzen *et al.* (1984) found that 500-mbar data projected on individual Hough functions (the eigenfunctions appropriate to free Rossby waves) displayed almost precisely the predicted phase speeds. However, amplitudes varied significantly, with large amplitudes occurring episodically and with episodes seldom lasting longer than 2

TABLE III. Rossby Wave Periods[a]

Mode	Winter	Spring	Summer	Autumn
(1,2)	4.85	4.87	4.85	4.85
(1,3)	9.91	9.99	9.49	9.52
(1,4)	18.39	17.22	16.68	17.40
(2,3)	3.84	3.84	3.79	3.84
(2,4)	7.27	7.55	6.93	7.07
(2,5)	14.23	13.54	12.71	13.22
(3,4)	4.28	4.22	4.10	4.22
(3,5)	7.40	7.60	6.73	7.06
(3,6)	13.65	13.89	11.78	12.76

[a] Days, as calculated by Kasahara (1980) for climatological mean 500-mbar winds.

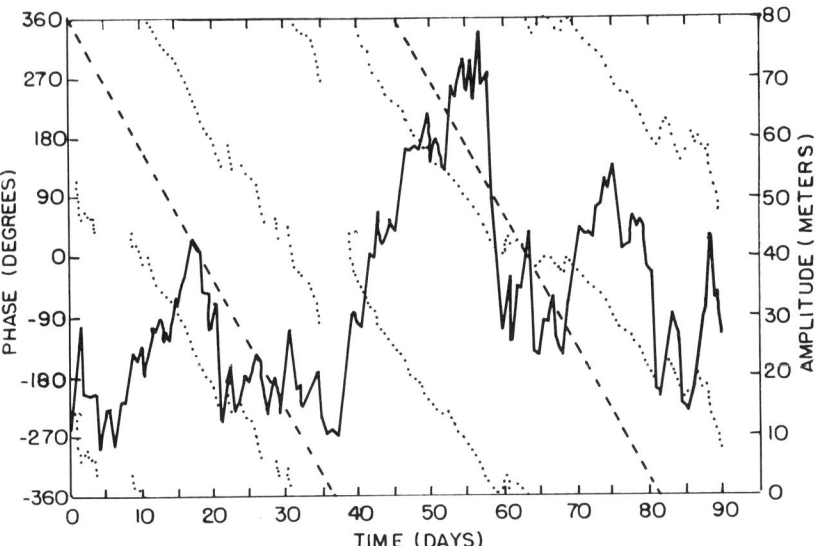

Fig. 14. Amplitude (solid lines) and phase (dotted lines) of the (1,4) Hough mode for December 1978–February 1979 from FGGE data. (From Lindzen et al., 1984.)

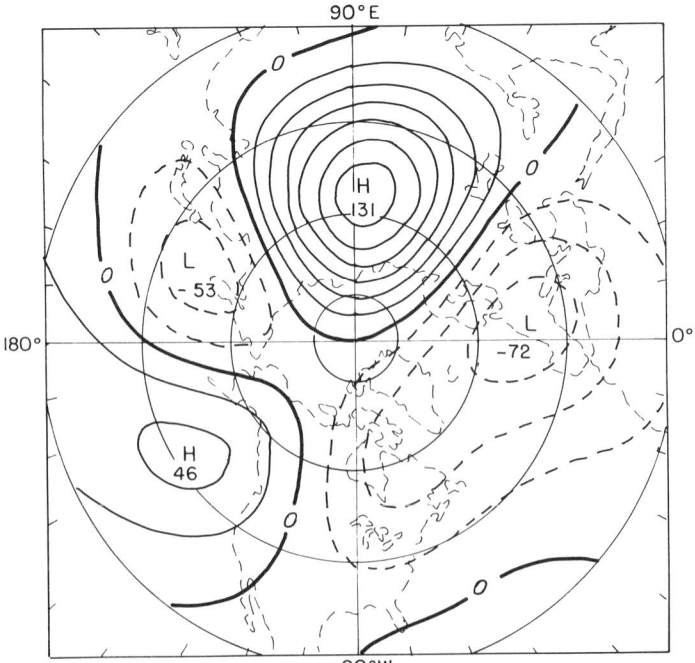

Fig. 15. Geopotential height field at 500 mbar due to nine main Hough modes, for 1200 GMT 12 January 1979. Contour interval is 20 m. (From Lindzen et al., 1984.)

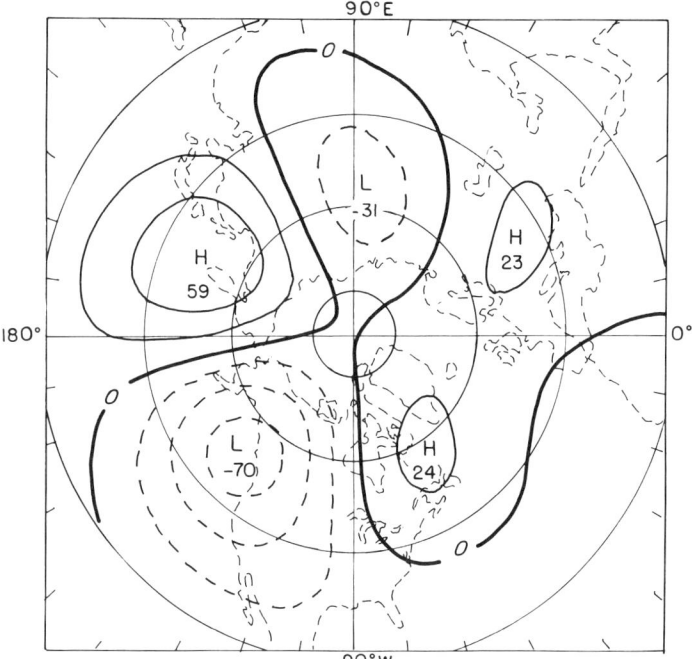

FIG. 16. Same as Fig. 15, but for 22 October 1979.

weeks. As an example, Fig. 14 shows the time series for the amplitude and phase of the 1,4 mode over a 3-month period. The particular choice of mode and period was completely arbitrary. Although individual modes rarely were found to exceed amplitudes of 60–80 m, the sum of the nine modes analyzed could contribute significantly more to the height field. Figure 15 shows this sum for a single day when at least one mode was particularly strong. We see anomalies amounting to as much as 137 m. Although anomalies in excess of 100 m due to these waves did not last in excess of 3 days, this is enough to contribute substantially to the overall picture. (Note from Fig. 2 that two-thirds of the anomalies do *not* last longer than 3 days.) Even on days when no modes were especially strong, the sum remains significant; this is seen in Fig. 16.

7. Concluding Remarks

The point of the present paper has been to suggest that anomalies of planetary waves are not, for the most part, unusually persistent, though some interannual variability exists. We argue that there is neither observational call nor physical basis for current models of multiple equilibria. We also show that realistic anomalies in tropical heating produce only modest anomalies [0 (50 m) at 500 mbar] in midlatitude stationary waves and that the distribution of even these anomalies depends strongly on the zonal wind distribution. This is not inconsistent with the modest amplitude of actual interannual variability, which is readily accounted for by some combination of anomalous tropical heating and variations of zonal flow over the Himalayas and Rockies. Finally we note that observed free Rossby waves contribute significantly to the large, transient planetary-wave anomalies, but by no means totally account for them.

Acknowledgments

The research in this paper was supported by the National Science Foundation under Grant 8342482-ATM and by NASA under Grant NAGW-525. Discussions with R. Dole and S. Nigam are gratefully acknowledged. This paper was completed while the author was a guest of the Physical Research Laboratory, Ahmedabad, India.

References

Charney, J. G. (1973). Planetary fluid dynamics. In "Dynamic Meteorology" (P. Morel, ed.), pp. 97–352. Reidel, Dordrecht.

Charney, J. G., and De Vore, J. G. (1979). Multiple flow equilibria in the atmosphere and blocking. *J. Atmos. Sci.* **36**, 1205–1216.

Dole, R. M. (1985). Life cycles of persistent anomalies. In "Proceedings of the 1984 Stanstead Seminar, July 9–13, 1984" (J. Derome, ed.). McGill University, Montreal.

Dole, R. S. and Gordon, M. D. (1983). Persistent anomalies of the extratropical Northern Hemisphere wintertime circulation: Geographical distribution and regional persistence characteristics. *Mon. Weather Rev.* **111**, 1567–1586.

Dorman, C. E., and Bourke, R. H. (1979). Precipitation over the Pacific Ocean 30°S to 60°N. *Mon. Weather Rev.* **107**, 751–774.

Egger, J. (1976). On the theory of steady perturbations in the troposphere. *Tellus* **28**, 381–389.

Frederiksen, J. S. (1979). The effects of long planetary waves on the regions of cyclogenesis: Linear theory. *J. Atmos. Sci.* **36**, 195–206.

Held, I. M. (1983). Stationary and quasi-stationary eddies in the extratropical troposphere: Theory. In "Large-scale Dynamical Processes in the Atmosphere" (B. J. Hoskins and R. P. Pearce, eds.), pp. 127–168. Academic Press, New York.

Horel, J. D., and Wallace, J. M. (1981). Planetary scale atmospheric phenomenon associated with the Southern Oscillation. *Mon. Weather Rev.* **109**, 813–829.

Hoskins, B. J. (1978). Horizontal wave propagation on the sphere. In "The General Circulation" (M. L. Blackmon, ed.), pp. 144–153. Summer Colloquium Notes, N.C.A.R.

Hoskins, B. J., and Karoly, D. J. (1981). The steady linear response of a spherical atmosphere to thermal and orographic forcing. *J. Atmos. Sci.* **38**, 1175–1196.

Jacqmin, D., and Lindzen, R. S. (1985). The causation and sensitivity of the northern winter planetary waves, *J. Atmos. Sci.*, **42**, 724–745.

Karoly, D. J., and Hoskins, B. J. (1982). Three dimensional propagation of planetary waves. *J. Meteorol. Soc. Jpn.* **60**, 109–123.

Kasahara, A. (1976). Normal modes of ultralong waves in the atmosphere. *Mon. Weather Rev.* **104**, 669–690.

Lau, K. M., and Phillips, T. J. (1986). Extratropical geopotential height fluctuation associated with tropical convection. *J. Atmos. Sci.,* in press.

Lindzen, R. S., and Hong, S. S. (1974). Effects of mean winds and horizontal temperature gradients on solar and lunar semidiurnal tides in the atmosphere. *J. Atmos. Sci.* **31**, 1421–1446.

Lindzen, R. S., Straus, D. M., and Katz, B. (1984). An observational study of large-scale atmospheric Rossby waves during FGGE. *J. Atmos. Sci.* **41**, 1320–1335.

McIntyre, M. E., and Palmer, T. N. (1984). The 'surf zone' in the stratosphere. *J. Atmos. Terr. Phys.* **46**, 825–849.

Namias, J. (1978). Multiple causes of the North American abnormal winter 1976–1977. *Mon. Weather Rev.* **106**, 279–295.

Niehaus, M. C. W. (1980). Instabilities of non-zonal baroclinic flows. *J. Atmos. Sci.* **37**, 1447–1460.

Nigam, S. (1983). On the structure and forcing of tropospheric stationary waves. Ph. D. thesis, Princeton University.

Opsteegh, J. D., and van den Dool, H. M. (1980). Seasonal differences in the stationary response of a linearized primitive equation model: Prospects for long range weather forecasting? *J. Atmos. Sci.* **37**, 2169–2185.

Plumb, R. A. (1985). On the three-dimensional propagation of stationary waves. *J. Atmos. Sci.* **42**, 217–229.

Rex, D. F. (1950). Blocking action in the middle troposphere and its effects on regional climate. II. The climatology of blocking action. *Tellus* **2**, 275–301.

Schutz, C., and Gates, N. L. (1972). Supplemental global climatic data, January. Rand Corp. Rep.

Simmons, A. J. (1982). The forcing of stationary wave motion by tropical diabatic heating. *Q. J. R. Meteorol. Soc.* **108**, 503–534.

Tung, K.-K. (1979). A theory of stationary long waves. III. Quasi-normal modes in a singular waveguide. *Mon. Weather Rev.* **107**, 751–774.

Tung, K.-K., and Lindzen, R. S. (1979). A theory of stationary long waves. I. A simple theory of blocking. *Mon. Weather Rev.* **107**, 714–734.

Tung, K.-K., and Rosenthal, A. J. (1985). The nonexistence of multiple equilibria in the atmosphere: Theoretical and observational considerations. *J. Atmos. Sci.* **42**, 2804–2819.

Van Loon, H., Jenne, R. L., and Labitzke, K. (1973). Zonal harmonic standing waves. *J. Geophys. Res.* **78**, 4463–4471.

Wallace, J. M., and Gutzler, D. S. (1981). Teleconnections in the geopotential height field during the Northern Hemisphere winter. *Mon. Weather Rev.* **109**, 784–812.

Webster, P. J. (1981). Mechanisms determining the atmospheric response to sea surface temperature anomalies. *J. Atmos. Sci.* **38**, 554–571.

Youngblut, C., and Sasamori, T. (1980). The nonlinear effects of transient and stationary eddies on the winter mean circulation. I. Diagnostic analysis. *J. Atmos. Sci.* **37**, 1944–1957.

Part IV

NUMERICAL EXPERIMENTS

INSTABILITY THEORY AND NONLINEAR EVOLUTION OF BLOCKS AND MATURE ANOMALIES

J. S. Frederiksen

Commonwealth Scientific and Industrial Research Organization
Division of Atmospheric Research
Aspendale, Victoria 3195
Australia

1. Introduction

The results of studies which indicate that blocking and teleconnection patterns may be generated through the three-dimensional instability of atmospheric flows are presented. The instability properties of climatological flow for the Northern Hemisphere are examined. It is found that three-dimensional instability theory produces very low-frequency teleconnection patterns such as the Pacific–North American and North Atlantic Oscillation patterns, as well as medium-frequency onset of blocking modes upstream of the mature patterns and higher frequency monopole cyclogenesis modes, which have maximum amplitudes in the observed storm tracks. Thus, the main observed modes of variability in the troposphere have analogs obtained from instability theory.

The instability solutions are compared with the structural changes that occur in the time evolution of observed anomalies such as blocks and with nonlinear simulations of the development of mature anomalies.

Blocking, which was noted as long ago as 1904 by Garriott (1904), is a primary example of anomalous circulation. It has subsequently stimulated numerous synoptic studies because of its profound effect upon the weather and climate (see, for example, Namias, 1947; Berggren et al., 1949; Rex, 1950; O'Connor, 1963). In the early tentative theories blocking was considered as a barotropic process (e.g., Yeh, 1949; Rossby, 1950; Rex, 1950; Thompson, 1957). In particular, Rossby (1950) and Rex (1950) suggested that block development was analogous to a hydraulic jump in open-channel flow.

In recent years there has been a resurgence of interest in studying anomalous circulations based on observations, theories, and numerical simulations. Many of these aspects are covered in this volume. Here, our purpose is to provide a relatively self-contained presentation of three-dimensional instability theory of the onset of blocking, large-scale teleconnection patterns, and cyclogenesis. As well we compare the results of this theory with studies based on observations and numerical simulations.

2. Three-Dimensional Instability Theory

2.1. Model Details

The two-layer quasi-geostrophic P-model that is used for this study was discussed in detail in Frederiksen (1982a) and was originally derived by Lorenz (1960). Here we briefly summarize the equations and describe how the instability problem is formulated. In dimensionless form the P-model equations take the form:

$$\partial \theta / \partial t = -J(\psi,\theta) + \sigma \nabla^2 \chi + F_\theta \tag{1a}$$

$$\partial \nabla^2 \psi / \partial t = -J(\psi, \nabla^2 \psi + 2\mu) - J(\tau, \nabla^2 \tau) + F_\psi \tag{1b}$$

$$\partial \nabla^2 \tau / \partial t = -J(\psi, \nabla^2 \tau) - J(\tau, \nabla^2 \psi + 2\mu) + \nabla \cdot 2\mu \nabla \chi + F_\tau \tag{1c}$$

$$\nabla^2 \theta = \nabla \cdot 2\mu \nabla \tau \tag{1d}$$

The original dimensional equations are made dimensionless by taking the earth's radius a and the inverse of the earth's angular velocity Ω^{-1} as length and time scales, and $a^2\Omega^2/bC_p$ as a temperature scale, where C_p is the specific heat of air at constant pressure and b (-0.124) is a dimensionless constant.

In Eq. (1), the average ψ and shear τ streamfunctions are taken to be related to the streamfunctions in the upper ψ^1 and lower ψ^3 layers through the relations

$$\psi = \tfrac{1}{2}(\psi^1 + \psi^3), \qquad \tau = \tfrac{1}{2}(\psi^1 - \psi^3) \tag{2}$$

Further, θ is the average potential temperature, χ is the velocity potential in the lower layer, and μ is $\sin \phi$, where ϕ is latitude. The two layers are frequently taken to have streamfunctions representative of 250 and 750 mbar, although other levels are also commonly taken as representative. Here we shall take the representative levels for the basic flow field in the upper and lower layers as being 300 and 850 mbar. The dimensionless static stability parameter $\bar{\sigma}$ is given by

$$\bar{\sigma} = bC_p \, \Delta\theta / 2a^2 \Omega^2 \tag{3}$$

where $\Delta\theta$ is the mean potential temperature difference between the upper and lower layers. The effects of diabatic heating, topography, dissipation, etc. are taken to be included formally in the operators F_ψ, F_τ, and F_θ.

2.2. Basic State

In this section, we use basic climatological Northern Hemisphere winter 300- and 850-mbar streamfunction fields derived from observations in the manner described by Blackmon (1976). The average over the eight winters 1963/1964 to 1970/1971 was used. The basic streamfunctions (denoted by a bar) were expanded in a general spherical harmonic decomposition.

$$\overline{\psi}^j(\lambda, \mu) = -\text{Re}\left[\sum_{\rho=-\infty}^{\infty} \sum_{\nu=|\rho|}^{\infty} \Psi^j_{\rho\nu} P^\rho_\nu(\mu) e^{i\rho\lambda}\right] \quad (4)$$

$$j = 1, 3$$

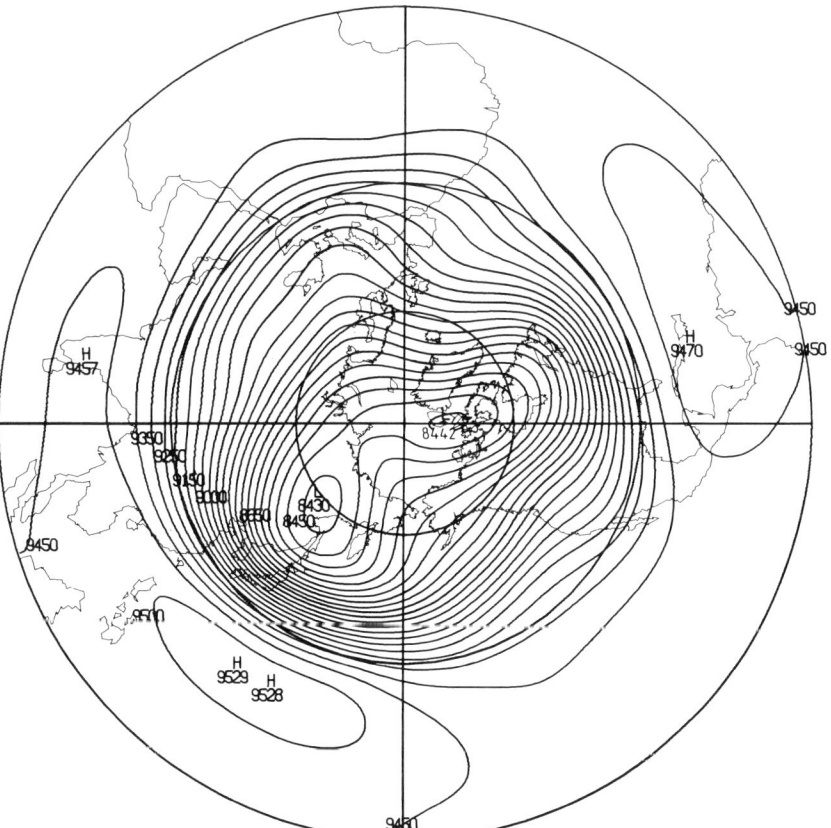

FIG. 1. Geopotential height (meters) of 300-mbar surface for Northern Hemisphere average winter.

at each level. Here λ is longitude, $P_\nu^\rho(\mu)$ are orthonormal Legendre functions, $\Psi_{\rho\nu}$ are spectral coefficients, ρ is zonal wavenumber, and ν is total wavenumber. A triangular truncation with truncation wavenumber 18 was used. Figure 1 shows the corresponding geopotential height for 300 mbar.

Once $\bar{\psi}$ has been specified through Eq. (4) the relations Eq. (1d) determines the basic potential temperature $\bar{\theta}$. The basic state is taken to be stationary and the operators F_ψ, F_τ, and F_θ are supposed to balance the basic state. For the sake of simplicity these operators are also supposed not to act on the perturbations, i.e., they are inhomogeneous. The implications of this prescription have been discussed in some detail in Frederiksen (1979).

In this article we concentrate on the results for one value of the static stability parameter, viz.,

Case 1. $\qquad \bar{\sigma} = 0.0067;$ $\quad$ i.e., $\Delta\bar{\theta} = 23$ K $\qquad$ (5a)

We also briefly mention some results for two other values of $\bar{\sigma}$ also considered in Frederiksen (1982a, 1983b), viz.,

Case 2a. $\qquad \bar{\sigma} = 0.008;$ $\quad$ i.e., $\Delta\bar{\theta} = 28$ K $\qquad$ (5b)

Case 3. $\qquad \bar{\sigma} = 0.01;$ $\quad$ i.e., $\Delta\bar{\theta} = 35$ K $\qquad$ (5c)

where $\Delta\bar{\theta}$ denotes the mean potential temperature difference between the two layers. In addition we shall consider a barotropic model for which the equation of motion is Eq. (1b) with τ set to zero. Our barotropic results are essentially a subset of those obtained by taking the static stability parameter to infinity in the baroclinic model.

2.3. Disturbances

For the disturbances we take general perturbations [which for convenience we denote by the original symbols in Eq. (1)] of the form

$$f(\lambda, \mu, t) = \text{Re}\left\{ \sum_{m=-\infty}^{\infty} \sum_{n=|m|}^{\infty} f_{mn} P_n^m(\mu) \times \exp[i(m\lambda - \omega t)]\right\} \qquad (6)$$

where f is any of the disturbance fields. Here m is the zonal wavenumber, n is the total wavenumber, and $\omega = \omega_r + i\omega_i$ is the complex angular frequency. In practice only a finite number of nonzero f_{mn} are retained in Eq. (6). Here a parallelogrammic truncation is used in which $|m| = 0, 1, \ldots, 15$; we consider only disturbance streamfunctions which are antisymmetric between the two hemispheres and take $n = |m| + 1, \ldots, |m| + 11$ for the streamfunctions.

The method of linearizing Eq. (1) about the basic states and solving the consequent eigenvalue–eigenvector equations for the complex angular frequency ω and the amplitude coefficients f_{mn} is described in the appendices of Frederiksen (1982b).

2.4. Growth Rates and Phase Frequencies

The nondimensional and dimensional growth rates (ω_i and ω_i^d) and phase frequencies (ω_r and ω_r^d) together with the periods are shown in Table I, for Case 1, and for all modes from the fastest growing down to the first nonpropagating but growing mode. We notice that there is quite a variety in the phase speeds of the modes, from very rapidly propagating modes with periods of 1.8 days to modes that are nonpropagating.

2.5. Disturbance Streamfunctions

To discuss individually each of the 22 modes in Table I, as well as the modes for Cases 2a and 3, would probably not be profitable. Rather, one can make a rough classification of the modes. Any such classification will serve only as a guide, and some modes will have characteristics of more than one group.

2.5.1. Monopole Cyclogenesis Modes (Class A). The typical member of this class is Mode 1 of Case 1 for which the instantaneous disturbance streamfunction[1] in the upper layer is shown in Fig. 2a, while the amplitude of the disturbance streamfunction in the upper layer is shown in Fig. 2b. The instantaneous disturbance has an essentially monopole structure (with latitude) except for a hint of secondary poles in the Atlantic in the upper layer. Both the instantaneous disturbance and the amplitude of the disturbance have largest values in the observed positions of the storm tracks.

The regions of preferential development may be seen to be slightly downstream and poleward of the jetstream maxima and the reason for this may be shown to be related to instability criteria as described in Frederiksen (1979). The location of these regions is in quite good agreement with bandpass-filtered observations (e.g., Blackmon, 1976, Fig. 5a), but the development off the east coast of North America is somewhat more pronounced than in the Pacific compared with the observations. This, however, depends on the particular mode examined as discussed below.

---

[1] It is the same as that shown in Fig. 2a of Frederiksen (1982a) except that the arbitrary phase and amplitude are different.

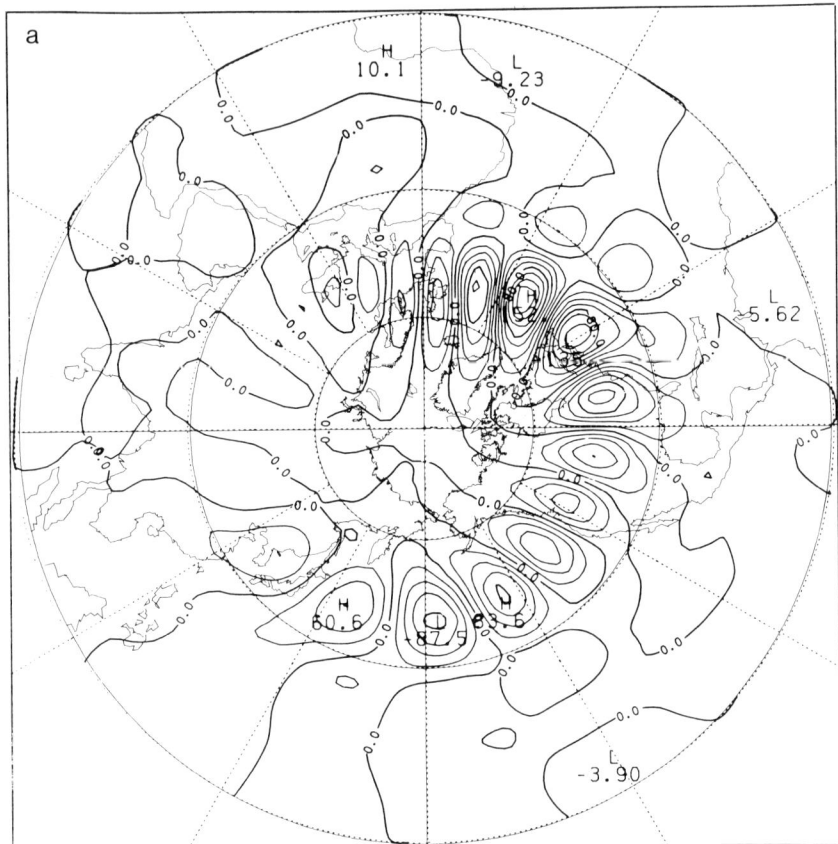

FIG. 2. (a) Disturbance streamfunction and (b) amplitude of disturbance streamfunction in upper layer for fastest growing mode in Case 1 (of class A).

Similar modes, for which the amplitude in the North Atlantic ocean is larger than in the Pacific, are Modes 2 and 4 of Case 1 as well as a number of modes of Cases 2a and 3. In each case, their periods are quite small. There are also monopole cyclogenesis modes which are similar to that shown in Fig. 9b of Frederiksen (1983a) for which the amplitudes in the Atlantic and Pacific are nearly equal. Mode 6 of Case 1 is an example. The modes with the very shortest periods, such as 8 and 14 of Case 1, have considerably larger amplitudes in the Pacific than in the Atlantic. In each case, the regions of preferential development occur in the same geographical locations as the observed storm tracks (Blackmon, 1976; Blackmon et al., 1977).

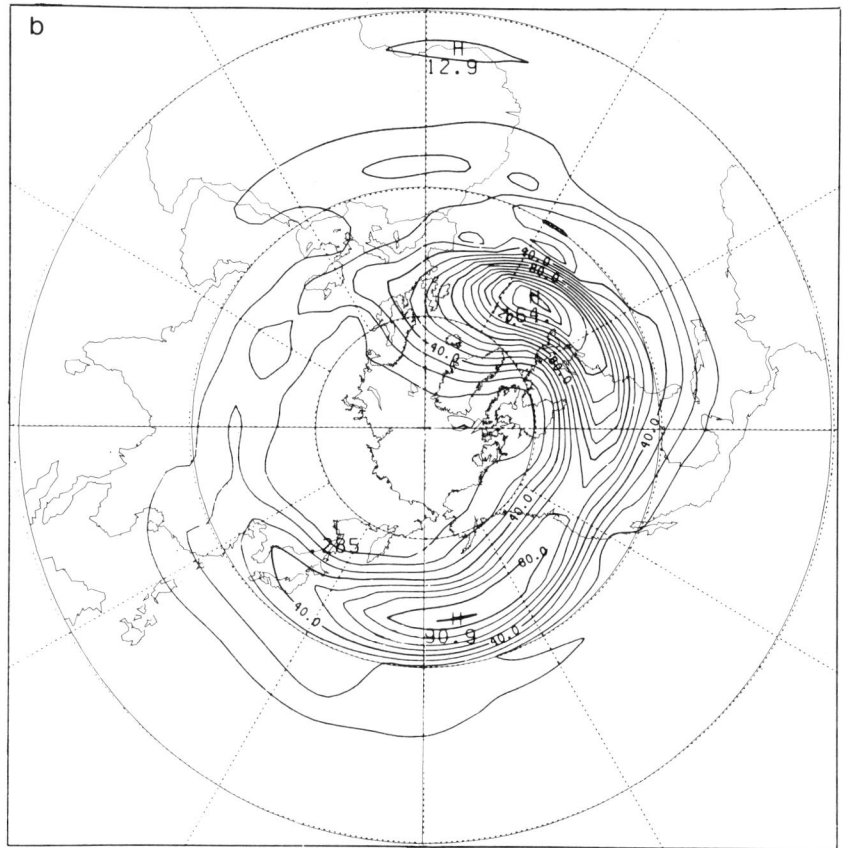

FIG. 2b.

2.5.2. Pacific Onset-of-Blocking Dipole Modes (Class B). Mode 3 of Case 1, for which the instantaneous disturbance streamfunction in the upper layer is shown in Fig. 3, is a typical member of this class; so is Mode 1 of Case 2a shown in Fig. 6 of Frederiksen (1982a). These modes have longer periods than the monopole cyclogenesis modes, and become the fastest growing modes when the static stability parameter is increased from 0.0067 to 0.008 so that the importance of baroclinic processes is reduced. They have dipole structures in the Pacific Ocean, with largest amplitudes upstream of the observed regions of mature blocks. Moreover, the zonal scale of the disturbances is larger than for the monopole cyclogenesis modes. The fact that the high–low vortex pairs have largest amplitude upstream of the main regions of the occurrence of mature blocks in the Pacific would seem to be consistent

TABLE I. PHASE FREQUENCIES, GROWTH RATES, AND PERIODS[a]

Mode	Class	$|\omega_r|$	$|\omega_r^d|$ (deg day^{-1})	T_r^d (days)	ω_i	ω_i^d (day^{-1})
1	A	0.30486	109.7	3.3	0.06796	0.4270
2	A	0.25510	91.8	3.9	0.06638	0.4171
3	B	0.13536	48.7	7.4	0.06246	0.3924
4	A	0.20759	74.7	4.8	0.06195	0.3892
5	B	0.16698	60.1	6.0	0.06125	0.3848
6	A	0.35928	129.3	2.8	0.05991	0.3764
7	B	0.10662	38.4	9.4	0.05978	0.3756
8	A	0.42264	152.2	2.4	0.05499	0.3455
9	C	0.071627	25.8	14.0	0.05217	0.3278
10	B	0.16149	58.1	6.2	0.05216	0.3277
11	C	0.12252	44.1	8.2	0.05090	0.3198
12	B	0.19731	71.0	5.1	0.04498	0.2826
13	F	0.08983	32.3	11.1	0.04128	0.2594
14	A	0.55167	198.6	1.8	0.03921	0.2464
15	F	0.03059	11.0	32.7	0.03593	0.2258
16	C	0.09676	34.8	10.3	0.03581	0.2250
17	F	0.05758	20.7	17.4	0.03454	0.2170
18	B	0.24884	89.6	4.0	0.03409	0.2142
19	E	0.02386	8.6	41.9	0.03358	0.2109
20	C	0.128288	46.2	7.8	0.03137	0.1971
21	C	0.081154	29.2	12.3	0.03136	0.1970
22	D	0.0	0.0	∞	0.02725	0.1712

[a] Data are for all growing modes, from the fastest to the first nonpropagating mode for Case 1.

with modes of this type being onset-of-blocking modes which initiate the blocking process. Evidence in support of this hypothesis will be presented in the following sections. As shown in Table I, Modes 5, 7, 10, 12, and 18 of Case 1 are also similar to the above modes. On the whole, most of these modes have periods somewhat longer than those of the monopole cyclogenesis modes but shorter than those of the class of Pacific and Atlantic onset-of-blocking dipole modes now to be considered.

2.5.3. Pacific and Atlantic Onset-of-Blocking Dipole Modes (Class C). The typical modes of this class are the ninth fastest growing mode for Case 1, for which the upper layer disturbance streamfunction is shown in Fig. 4, and Mode 1 of Case 3 shown in Fig. 7 of Frederiksen (1982a). They have longer periods than the Pacific onset-of-blocking dipole modes and have dipoles over a larger area of the globe, in particular in both the Pacific and Atlantic oceans. Modes 11 and 16 of Case 1 also are qualitatively quite similar to these patterns; Modes 20 and 21 of Case 1 also show many of the same characteristics.

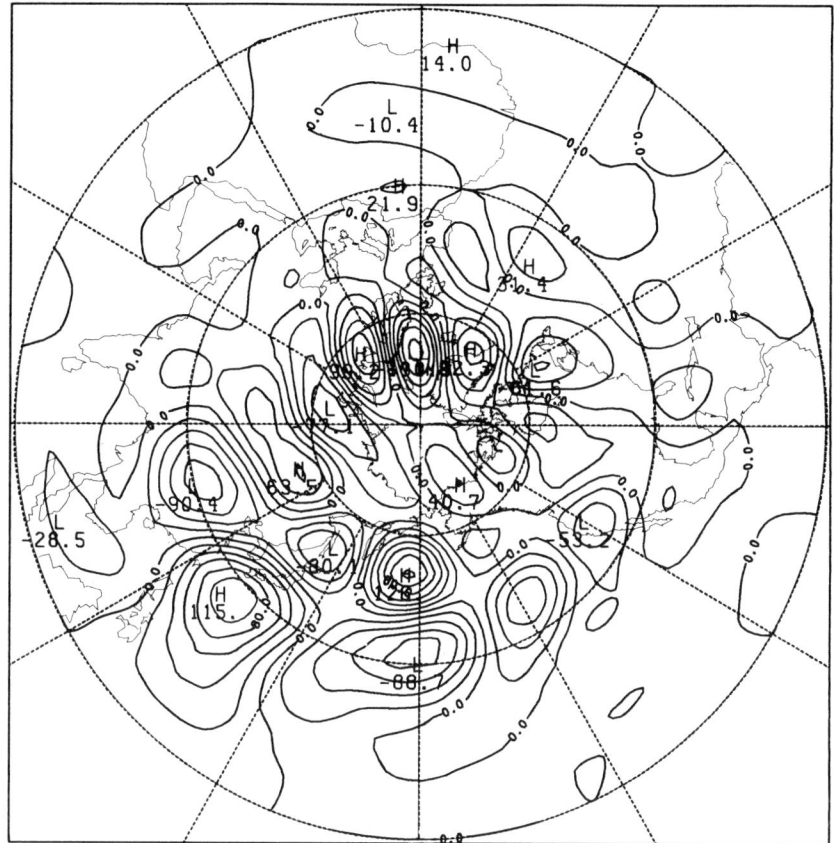

FIG. 3. As in Fig. 2a for third fastest growing mode in Case 1 (of class B).

2.5.4. Pacific–North American Mature Anomaly Pattern Modes (Class D). Next we examine the modes with the longest periods. Mode 22 of Case 1 has infinite period. Figure 5 shows the upper layer disturbance streamfunction for Mode 22 of Case 1. The amplitude of the disturbance is shown in Fig. 4 of Frederiksen (1983b) and is very similar to the diagram obtained by simply taking the absolute values of the disturbance streamfunction in Fig. 5.

In the extratropical region, particularly in the Pacific–North American region, this pattern is qualitatively quite similar to the Pacific anomaly pattern in Fig. 4.6f of Dole 1983 and in Figs. 16, 17c, and 18c of Wallace and Gutzler (1981). In both of these observational studies, the region between the equator and 20°N is excluded. There is, of course, a certain amount of arbitrariness in the definition of these patterns, as seen from Fig. 16 of Wallace and Gutzler and also by comparing the patterns of Wallace and

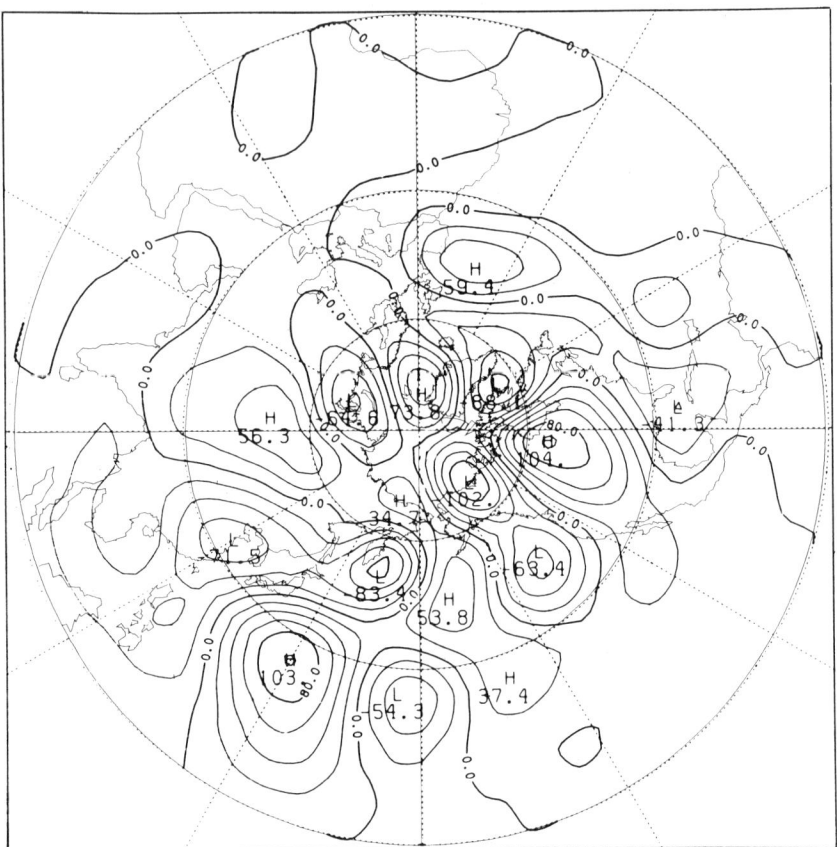

FIG. 4. As in Fig. 2a for ninth fastest growing mode (of class C).

Gutzler with those of Dole. Further, one would not expect a simple linear two-layer model to be able to reproduce all the details of the observations. Nevertheless, in the upper layer, the large-scale high – low – high structure (or vice versa for negative cases) from the Pacific across America closely resembles the observed 500-mbar patterns. Both Wallace and Gutzler and Dole also note that their 1000-mbar patterns are mainly confined to the key region in the Pacific – North American region, whereas the 500-mbar patterns are more extensive and complex. Again, the instability mode has a similar behavior, with more extensive wave trains in the upper layer and the lower layer streamfunction [Fig. 3 of Frederiksen (1983b)] more confined south of the Aleutians and over the west coast of North America.

For the barotropic model, none of the modes of classes A, B, and C occurs. However, with the 300-mbar Northern Hemisphere streamfunction as basic

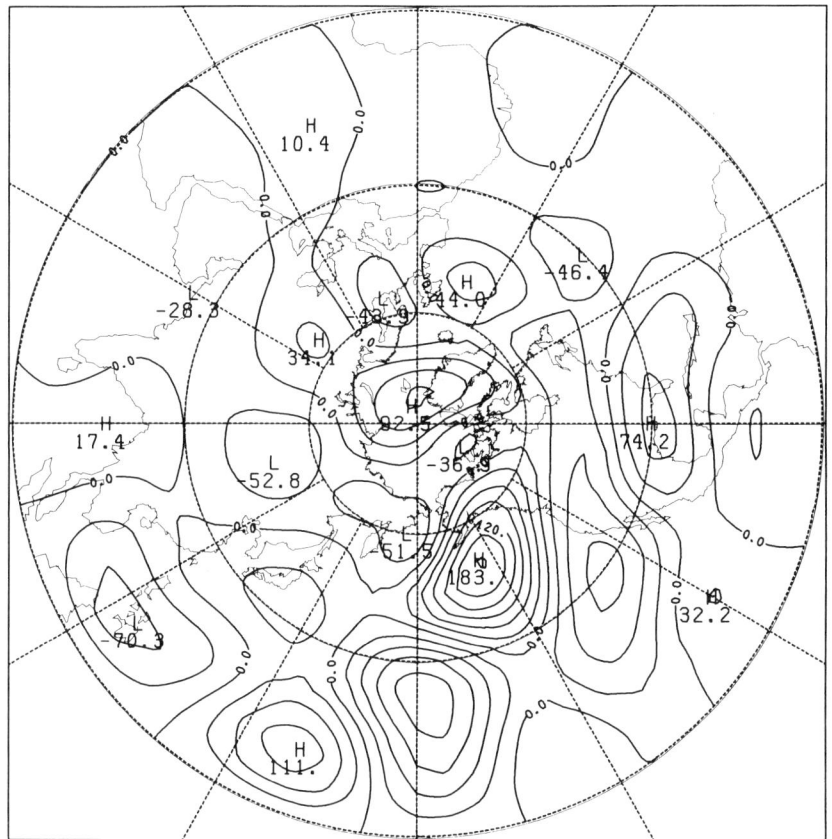

FIG. 5. As in Fig. 2a for twenty-second fastest growing mode (of class D).

state, the modes obtained with the barotropic model contain a mode like the Pacific–North American mature anomaly pattern. The fastest growing mode is a Pacific–North American mature anomaly mode; it is shown in Fig. 1 of Frederiksen (1983b) and a very similar mode was obtained in a barotropic model by Simmons *et al.* (1983).

In the key Pacific–North American region, Mode 22 of Case 1 is equivalent barotropic in contrast to the Pacific and Pacific and Atlantic onset-of-blocking dipole modes, which have a distinct westward tilt with height (not shown) indicating the importance of baroclinic processes.

2.5.5. North Atlantic Mature Anomaly Pattern Modes (Class E). The mode with the second longest period (of just over 40 days) is Mode 19 for Case 1; the upper layer disturbance streamfunction is shown in Fig. 6. Mode

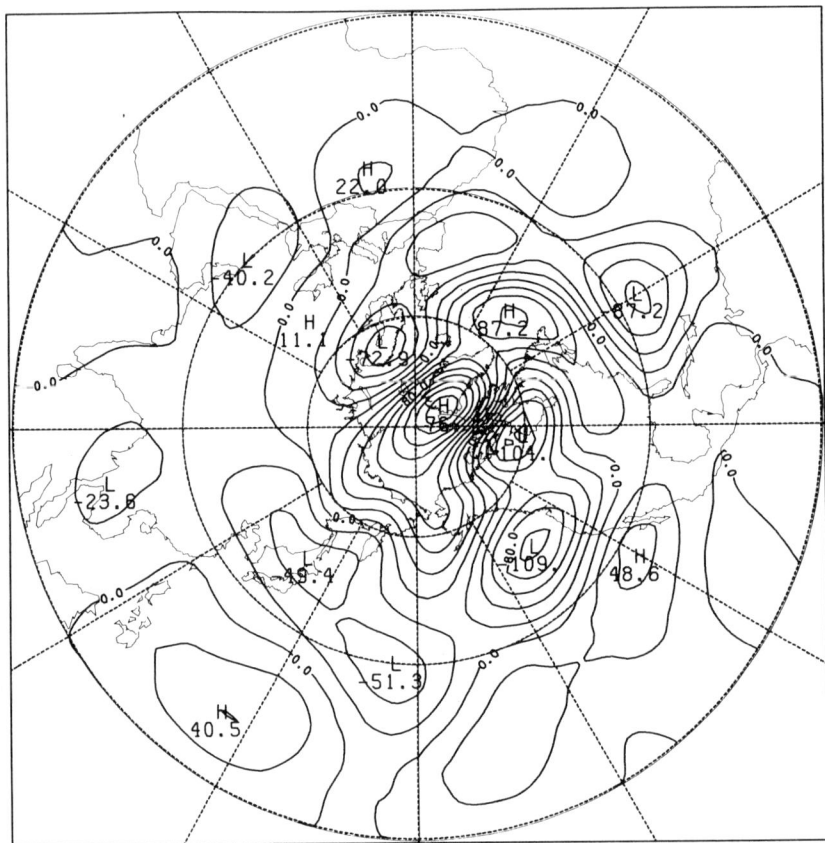

FIG. 6. As in Fig. 2a for nineteenth fastest growing mode (of class E).

2 for the barotropic model shown in Fig. 2 of Frederiksen (1983b) resembles that shown in Fig. 6 in the upper layer, although the similarity is perhaps not quite as good as for the corresponding Pacific–North American patterns. Mode 19 for Case 1 is essentially equivalent barotropic.

The pattern in Fig. 6 resembles most closely the North Atlantic Oscillation of Walker and Bliss (1932) and van Loon and Rogers (1978), the Atlantic teleconnection patterns of Wallace and Gutzler (1981), and the Atlantic anomaly pattern of Dole (1982). In making these comparisons, it should be noted, however, as discussed by Wallace and Gutzler (1981), that while these observed patterns show a large degree of consistency in the Greenland–North Atlantic area, the amplitude of the patterns in the polar region, as well as the presence or absence of a center of action in the North Pacific, depends on the particular dataset used. The North Atlantic mature anomaly instability mode appears to bear closest resemblance to the sea level pressure anomaly pattern in Fig. 11 of van Loon and Rogers (1978).

2.5.6. Intermediate Modes (Class F). Among the modes in Table I, there is also a group whose properties, in qualitative terms, are intermediate between the onset-of-blocking dipole modes and the mature anomaly pattern modes discussed above. Mode 15 of Case 1 has the third largest period (of ~ 30 days); Fig. 6 of Frederiksen (1983b) shows the disturbance streamfunction. As seen there, this mode appears to have characteristics of both of the major mature anomaly modes previously discussed. However, the scale of the centers making up the mode appears on the whole to be smaller.

Modes 17 and 13 of Case 1 have similar structures and periods of 17.4 and 11.1 days, respectively; Fig. 7 of Frederiksen (1983b) shows the perturbation streamfunction for Mode 17. These modes have properties and, in particular, scales intermediate between the mature anomaly pattern modes and the Pacific and Atlantic onset-of-blocking modes.

3. Time Evolution of Observed Mature Anomalies

At the time of the study of Frederiksen (1982a), there appeared to be no systematic statistical studies on the observed time evolution of mature anomalies such as blocks. Nevertheless, it seemed reasonable that the slower propagating modes with dipoles in the Pacific and Atlantic regions, called class B and C modes here, would initiate the process of the formation of mature anomalies such as blocks further downstream.

The thesis of Dole (1982), parts of which have been published in Dole (1983) and also in this volume, provided important insights into many aspects of observed mature anomalies, including their time evolution. Here we shall relate some of the instability solutions of Section 2 to Dole's observations.

In our discussion of Dole's studies, we shall concentrate on anomalies corresponding to the Pacific–North American pattern, since this is the case for which he gives the most extensive analysis. Similar results also hold for the North Atlantic (and northern Soviet Union) anomalies, which he also treats in lesser detail. Dole finds in his studies not only positive anomalies in the key region corresponding to blocking, but also negative anomalies. He notes that the two extreme stages of the index cycle (Namias, 1950) appear in his analyses as opposite phases of a single basic anomaly pattern, with high zonal index corresponding to negative anomaly and low zonal index corresponding to positive anomaly, such as caused by blocking.

Figure 4.6 of Dole (1983) shows the time evolution of composite analyses of 15 low-pass-filtered positive anomaly cases at 500 mbar in the Pacific. The sequence starts 4 days before (on day −4) the appearance (on day 0) of the essentially stationary large-scale positive anomaly in the key region in the north central Pacific and finishes 6 days thereafter. The corresponding sequences for unfiltered anomalies between days −3 and 0 are shown in Fig. 4.7 of Dole (1983). For the period leading up to the appearance of the positive anomaly in the key region, referred to in Frederiksen (1982a) as the onset-of-blocking period, Dole notes that the "sequence of development suggests that the initial rapid growth of the main centre is primarily associated with the propagating intensifying disturbance which originates in mid-latitudes near Japan." He also notes that "this disturbance continues to intensify as it becomes quasi-stationary over the key region." The dipole wave train that appears in Fig. 4.7 of Dole (1983) at day −3 is qualitatively very similar to that in our Fig. 3 and as well to that in Fig. 6a of Frederiksen (1982a).

Further evidence that modes of class B are involved in the initial period may be obtained from Fig. 5.10 of Dole (1982), which shows longitude–pressure cross sections at 45° and 20°N of the unfiltered Pacific composite anomalies at 1-day intervals between days −3 and 0.

The dipole nature of the developing anomaly is clearly evident particularly at day −3, and the zonal scale of the anomaly at day −3 is practically the same as shown in Fig. 6 of Frederiksen (1982a) and in our Fig. 3. The developing anomaly has a definite westward tilt with height, as do the class B modes. Dole notes: "This feature has pronounced westward tilts with height during this period, suggesting that a substantial baroclinic contribution is involved in its amplification."

Following day 0, the development of the 500-mbar height anomaly is as shown in Fig. 5.1d–f of Dole (1982), with the anomalies being largely equivalent barotropic during this period. Intensification of the centers occur with little phase propagation and by day 4 and especially at day 6 the Pacific–North American pattern is established.

The above observational and theoretical results suggest that the development of mature anomalies such as blocks may be thought of as approximately consisting of two stages, which, for the Pacific–North American pattern, are as follows:

1. A rapidly growing and relatively rapidly eastward-propagating dipole (class B) disturbance which tilts westward with height forms in the Pacific Ocean–East Asia region through the combined baroclinic–barotropic instability of the three-dimensional basic state. As the disturbance grows, the regions of largest amplitude propagate into the central Pacific and the disturbance increases its zonal scale, becoming quasi-stationary and essentially

equivalent barotropic through the operation of nonlinear effects. At this stage the mode has a structure reminiscent of class D modes.

2. The (nonlinear analog of A) class D mode amplifies without phase propagation and through the operation of largely equivalent barotropic effects to form the large-amplitude mature Pacific–North American anomaly pattern.

We have concentrated in this section on the composite results of Dole (1983), which present a very clear-cut example of the change in the horizontal and vertical structures of the anomaly pattern, starting upstream of the key region with the formation of westward-tilting smaller scale dipole disturbances. The importance of transients in the development of blocks has also been pointed out by many other authors, e.g., Lejenas (1977), Green (1977), Austin (1980), Youngblutt and Sasamori (1980), Tucker (1979), Illari and Marshall (1983), Colucci (1982), Hansen and Chen (1982), and Shutts (1983). Hansen and Chen, in particular, note that in their observational studies of blocking, intense upstream cyclogenesis preceded the growth of the blocking ridge. Further, "the nonlinear transfer of kinetic energy from baroclinically active cyclone-scale waves to the planetary-scale waves provided the major source of kinetic energy for the block." Lejenas (1977) notes the importance of the combination of barotropic and baroclinic instability during the formation of blocks. Barnett (1984), in an observational study, also finds support for instability normal modes playing important roles in the generation of anomaly patterns.

For the Southern Hemisphere, Frederiksen (1984, 1985) has carried out three-dimensional instability studies for climatological flows and for synoptic flows leading to the formation of blocks in the Australian–New Zealand area, as well as cyclogenesis in a number of regions. Instability theory appears to be able to pick out many of the important developments. In particular, dipole instability modes are produced in or slightly upstream of the blocking region. Lejenas (1984), in an observational study of Southern Hemisphere blocking, also finds support for dipole instability modes being responsible for the majority of blocking episodes while the additional influence of forcing controls the duration.

4. Nonlinear Simulation

Next, we relate the results of numerical simulations of evolving baroclinic waves to the properties of the instability solutions and to observations of developing anomalies. We shall primarily consider a study by Frederiksen and Puri (1985) of nonlinear baroclinic waves growing on an initial three-dimensional basic Northern Hemisphere flow in a multilevel primitive equation model.

4.1. Model Details

This study was carried out using the σ-coordinate spherical primitive equation spectral model developed by Bourke (1974) and Bourke et al. (1977), and used by Frederiksen (1981a,b) for studying life cycles of nonlinear baroclinic waves growing on initially zonally averaged flows. The equations for vorticity, divergence, thermodynamics, and continuity are described in Section 2 of Frederiksen (1981a). Since initial three-dimensional basic states are considered constant, forcing functions are included on the right-hand sides of Eqs. (2.1a)–(2.1d) of Frederiksen (1981a). The model also includes topography, so the term $-\nabla^2 \Phi_*$ is added to the right-hand side of the divergence equation, where Φ_* is the geopotential height of the Northern Hemisphere topography.

A five-level version of the model with five equispaced σ levels of $\sigma = 0.1$, 0.3, 0.5, 0.7, and 0.9 and a rhomboidal truncation scheme with a truncation number $J = 21$ are used. The model includes biharmonic diffusion and surface drag. For the basic state we take the monthly averaged flow field for January 1978, which has zonal variations that are qualitatively quite similar to those of the average winter field (shown in Fig. 1) in the upper troposphere. The streamfunction fields for the basic state are shown in Fig. 1 of Frederiksen and Puri (1985). The forcing functions mentioned above are chosen in such a way that, in the absence of drag and diffusion, the flow field for January 1978 is an *exact* steady-state solution of the equations of motion. However, if the basic state is perturbed, the perturbation will extract energy from it and grow.

For the small-amplitude perturbation to the basic state, the fastest growing linear instability mode for the January 1978 basic state obtained from a three-dimensional instability calculation in a quasi-geostrophic model (Frederiksen, 1983a) is used. The disturbance amplitude was chosen to be sufficiently large, so that time-stepping errors are comparatively small, but small enough so that the different interesting stages of evolution can be studied. The disturbance kinetic energy per unit mass is 6.03×10^{-2} m^2 sec^{-2}, and the disturbance streamfunction at $\sigma = 0.7$ is shown in Fig. 7. We see that the disturbance is a monopole cyclogenesis mode with largest amplitudes in the storm tracks in the Atlantic and Pacific Oceans. That is, it is similar to Mode 1 of Case 1 in Fig. 2, but is somewhat more localized.

The model was integrated for a period of 10 days from the perturbed initial conditions and the evolution of the disturbance was examined. Here the

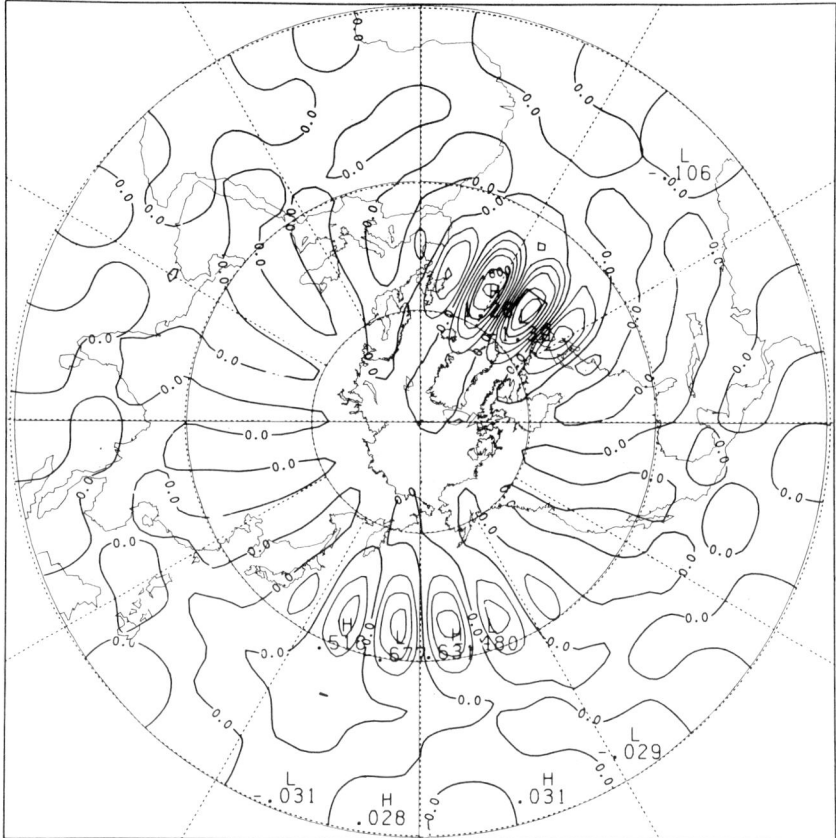

Fig. 7. Disturbance streamfunction at $\sigma = 0.7$ used to perturb basic state in primitive-equation model.

disturbance at a given time is the difference between the flow field at a given time and the basic January 1978 flow, which (in the absence of time-stepping errors and drag and diffusion) would remain constant with the specified forcings.

4.2. Disturbance Kinetic Energy Spectra

The developments of the disturbance kinetic energy zonal wavenumber m spectra are shown in Fig. 8 for the 10-day period. We see that there is a gradual and continual change in the energy spectra, until by day 10 the disturbance energy preferentially populates the larger scales. In fact, errors

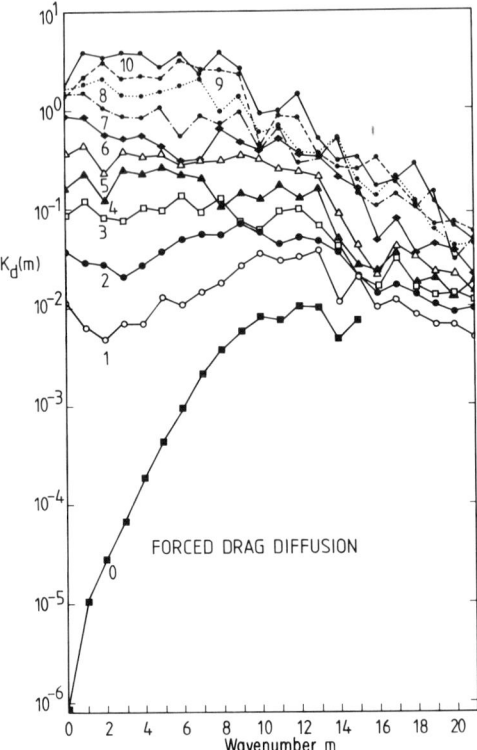

FIG. 8. Disturbance kinetic energy spectra as function of zonal wavenumber m and day for a 10-day period.

tend to predominate in the largest scales after sufficient time, irrespective of whether the errors are generated by model differences (Frederiksen and Sawford, 1980, Figs. 9 and 10, and references therein; Bengtsson and Lange, 1982) or by differences in initial conditions as in Fig. 8.

4.3. Disturbance Streamfunctions

The disturbance streamfunctions at days 2, 4, 6, and 10 and at $\sigma = 0.7$ are shown in Fig. 9. Figure 10 shows the disturbance streamfunction at $\sigma = 0.1$ on days 6 and 10. Although it is not shown explicitly in these diagrams, we note that the disturbance starts as a rather shallow structure in the Atlantic and Pacific Oceans, with largest amplitude at the lowest level, and during the integration penetrates increasingly into the upper troposphere. In addition, there are a number of new areas of development that occur during the 10-day period. The shift in the maximum of the disturbance amplitude is in fact qualitatively similar to that found in corresponding nonlinear experiments of disturbances growing on zonally averaged basic flows (Gall, 1976; Simmons and Hoskins, 1978; Frederiksen, 1981a,b).

Next we consider in more detail the synoptic sequence of events shown in Figs. 9 and 10, concentrating mainly on the two levels shown there. We mention, however, that at $\sigma = 0.9$ more of the smallest scale features are visible, while only the largest scale features tend to penetrate to $\sigma = 0.1$. On day 1 (not shown) there are the first indications of slight poleward movements of the anticyclones and equatorward movements of the cyclones in both the Pacific and Atlantic sectors. Further, there are the beginnings of development of a cyclone over the Himalayas, with an associated weaker

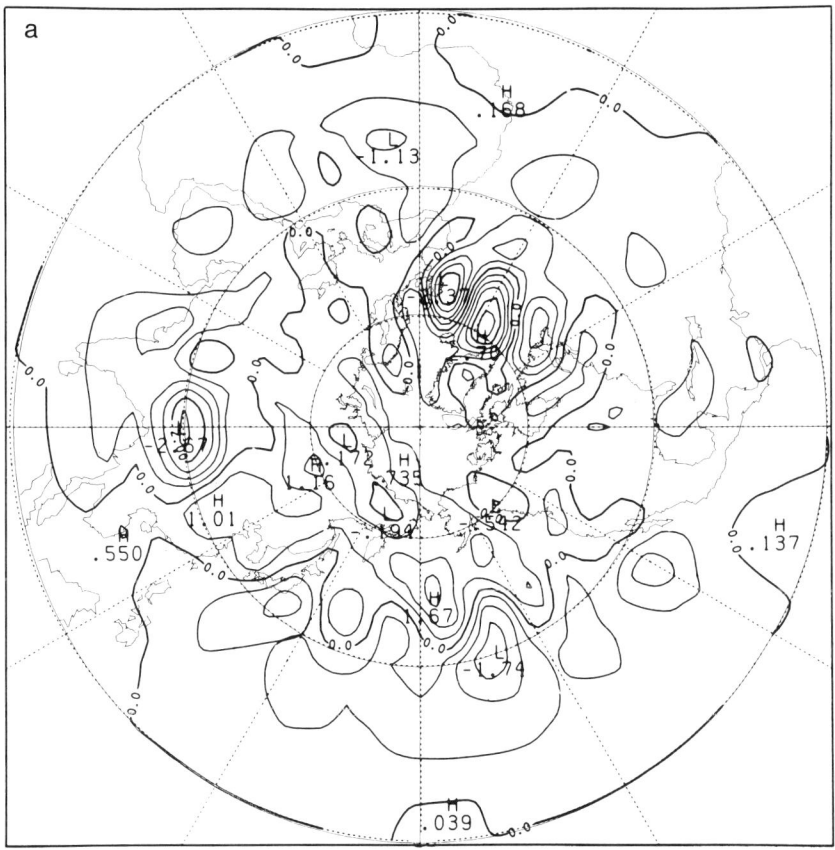

FIG. 9. Disturbance streamfunction on (a) day 2, (b) day 4, (c) day 6, and (d) day 10 at $\sigma = 0.7$.

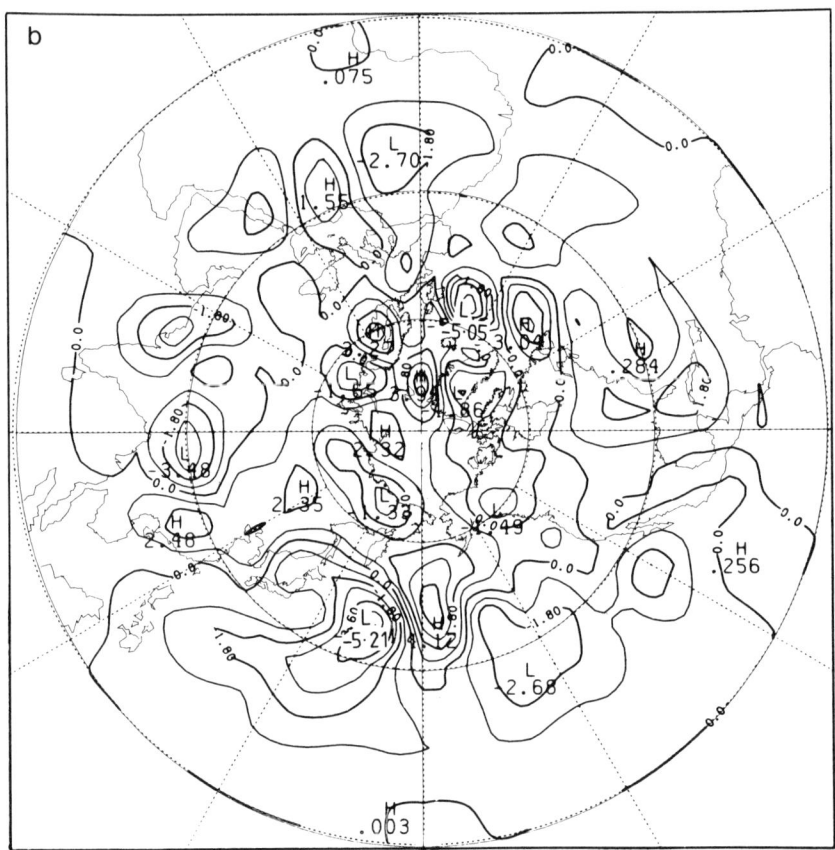

FIG. 9b. See legend on p. 295.

anticyclone. These features are even more pronounced on day 2, as seen from Fig. 9. In fact, at $\sigma = 0.3$ (not shown), the splitting has yielded an anticyclone over Greenland with lows both upstream and downstream.

The developments that occur in the early stages appear to have the following causes. First, in the Pacific and Atlantic sectors one is seeing the beginnings of development of dipoles or multipoles. In general terms, these may be regarded as the nonlinear analog of modes like the Pacific, and in particular Pacific and Atlantic, onset-of-blocking dipole modes shown in Figs. 6 and 7 of Frederiksen (1982a) and in our Figs. 3 and 4. Second, in the Tibetan and Greenland areas the large-scale flow changes during the first few days largely through the presence of drag and nonlinear interactions. These changes in turn mean that the new (unbalanced) large-scale flow impinges on the topography to produce new disturbances. These disturbances then project onto (or

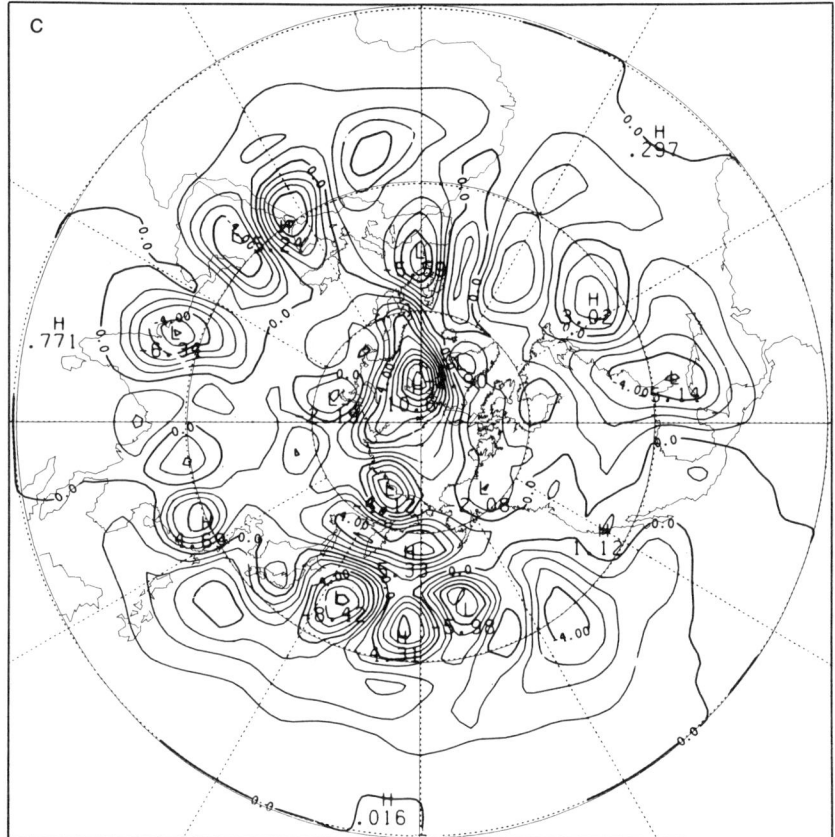

FIG. 9c. See legend on p. 295.

may be expanded in terms of) growing, neutral, and decaying modes, with the growing modes playing the most important role, as may be seen from the amplification that occurs during the integration period. Among the (nonlinear analog of) growing instability modes that would make contributions to the disturbances are modes like the Pacific and Atlantic onset-of-blocking dipole mode in Fig. 7 of Frederiksen (1982a), the intermediate mode in Fig. 7 of Frederiksen (1983b), and mature modes like the North Atlantic Oscillation mode in Fig. 5 of Frederiksen (1983b).

On day 4 the disturbance still has a significant westward tilt with height in most regions (not shown), indicating that baroclinic processes are making a substantial contribution in the developments. By day 4 we see the beginnings of wave trains from the North Atlantic–Greenland sea region across the northern Soviet Union into northeast Asia and from North Africa across

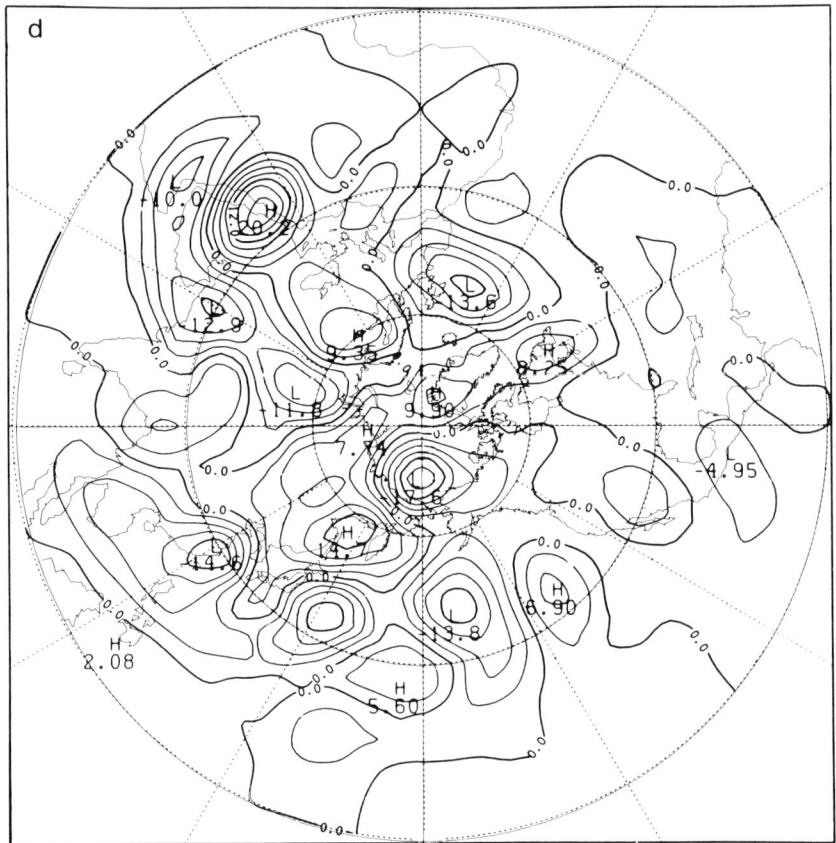

FIG. 9d. See legend on p. 295.

Arabia into central Asia. A further wave train is beginning to form across the United States and goes into the Atlantic Ocean. The signature of the onset-of-blocking dipole modes is also seen in the momentum fluxes (Fig. 9 of Frederiksen and Puri, 1985), which, particularly at higher levels, have multiple centers of poleward and equatorward fluxes in the Atlantic and Pacific, leading to convergence and divergence zones. This is a characteristic of the onset-of-blocking modes as shown in Fig. 9 of Frederiksen (1982a) and in Fig. 10c of Frederiksen (1984). It is interesting to note that Ji and Tibaldi (1983) also find that during the early stages of block development their observed latitudinal distribution of momentum flux also shows convergence and divergence zones, indicating the buildup of separate branches of the jet.

By day 6 there has been a considerable intensification of the anticyclone in

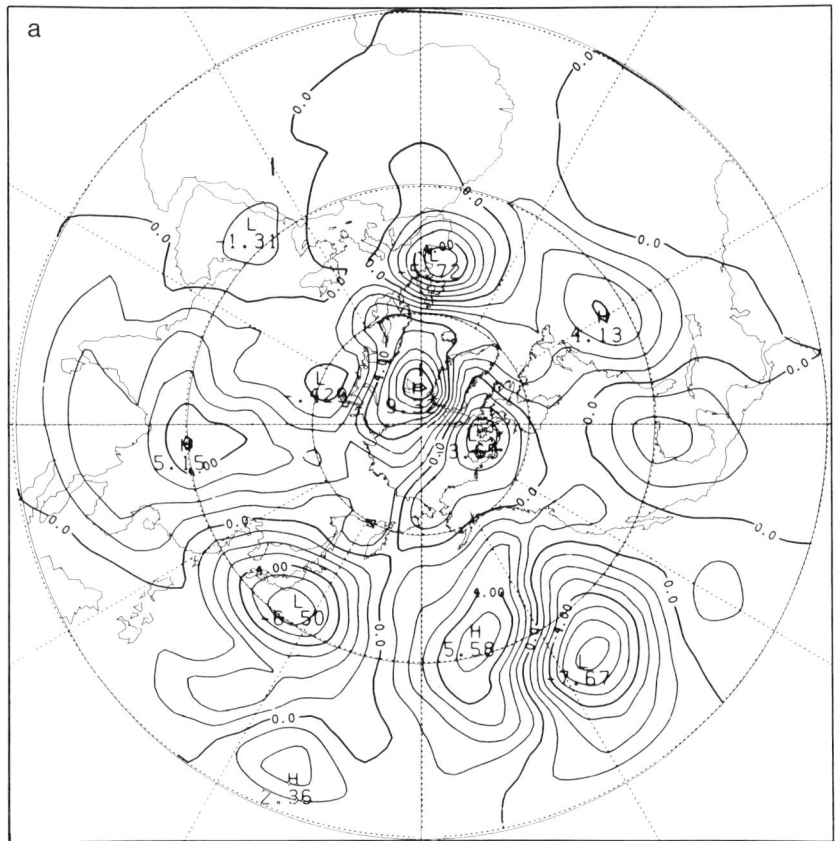

FIG. 10. Disturbance streamfunction on (a) day 6 and (b) day 10 at $\sigma = 0.1$.

the North Atlantic–Greenland sea area at all levels. We note the high meridional wavenumber of the disturbance in the Pacific Ocean, particularly at $\sigma = 0.7$. There has also been marked intensification of the disturbance in the North Africa–Arabia and North America–West Atlantic area, with some downstream propagation between days 4 and 6. At this stage, the wave train emanating from the Tibetian region has effectively merged with the Pacific wave train. In the North Atlantic–Arctic region we see the emergence of the North Atlantic Oscillation pattern. This is disguised to some extent at the lower levels by the presence of small-scale features and is best seen at the very highest model level of $\sigma = 0.1$ shown in Fig. 10a. In the North Atlantic–Arctic region there is quite reasonable agreement between Fig. 10a and the North Atlantic mature anomaly instability mode in Fig. 6, which was found to be quite similar to the North Atlantic Oscillation of van Loon and Rogers

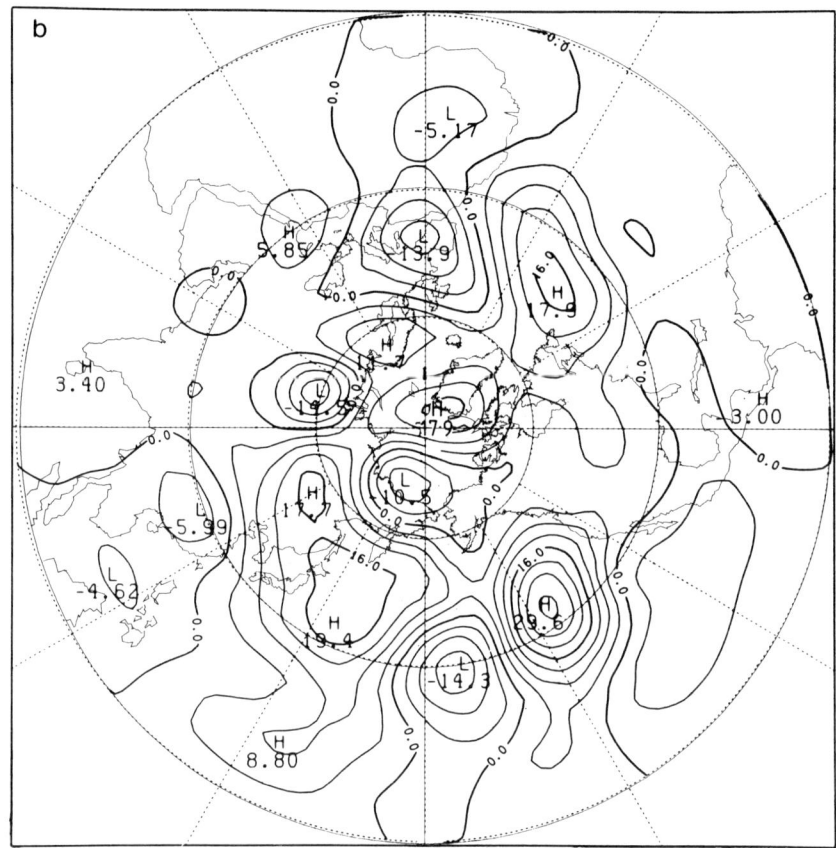

FIG. 10b. See legend on p. 299.

(1978). Their sea level pressure pattern is, however, a larger scale pattern due to averaging over many years, which removes the smaller scale features and resembles more the $\sigma = 0.1$ disturbance streamfunction in Fig. 10a rather than the disturbance at lower levels. There is perhaps some anticlockwise rotation in Fig. 10a compared with the observations.

There is a reasonable resemblance between the disturbance in the North Atlantic–Arctic region and the North Atlantic Oscillation pattern obtained from GCM experiments shown in Fig. 14 and, in particular, in Fig. 22 of Lau (1981). We also note that the North Atlantic Oscillation pattern of van Loon and Rogers (1978) and Lau (1981) has features in common with some of the patterns described by other authors. For example, it has common features with the Atlantic mature anomaly of Dole (1982; Fig. 5.7) of appropriate sign. We note that, as discussed by van Loon and Madden (1981), many of the well-known teleconnection patterns are contained within the comprehensive Southern Oscillation pattern (Walker and Bliss, 1932; Troup, 1965; Trenberth, 1976; Trenberth and Paolino, 1981; van Loon and Rogers, 1981).

ACKNOWLEDGMENTS

It is a pleasure to thank Roberto Benzi for organizing, and IBM Italy for sponsoring, the workshop on "Global scale anomalous circulation in the atmosphere and blocking" at which this work was presented.

REFERENCES

Austin, J. F. (1980). The blocking of middle latitude westerly winds by planetary waves. *Q.J.R. Meteorol. Soc.* **106**, 327–350.

Barnett, T. P. (1984). Variations in near-global sea level pressure. *J. Atmos. Sci.* **42**, 478–501.

Bengtsson, L., and Lange, A. (1982). Results of the WMO/CAS MWP data study and intercomparison project for forecasts for the Northern Hemisphere in 1979-1980. World Meteorological Organization, Geneva.

Berggren, R., Bolin, B., and Rossby, C. G. (1949). An aerological study of zonal motion, its perturbations and break-down. *Tellus* **1**, 14–37.

Blackmon, M. L. (1976). A climatological spectral study of the 500 mb geopotential height of the Northern Hemisphere. *J. Atmos. Sci.* **33**, 1607–1623.

Blackmon, M. L., Wallace, J. M., Lau, N., and Mullen, S. L. (1977). An observational study of the Northern Hemisphere wintertime circulation. *J. Atmos. Sci.* **34**, 1040–1053.

Bourke, W. (1974). An efficient, one-level primitive-equation spectral model. *Mon. Weather Rev.* **100**, 683–689.

Bourke, W., McAvaney, B., Puri, K., and Thurling, R. (1977). Global modeling of atmospheric flow by spectral methods. *Methods Comput. Phys.* **17**, 267–324.

Colucci, S. J. (1982). Diagnostic studies of atmospheric blocking patterns. Ph. D. thesis, State University of New York at Albany.

Dole, R. M. (1982). Persistent anomalies of the extratropical Northern Hemisphere wintertime circulation. Ph. D. thesis, Massachusetts Institute of Technology.

Dole, R. M. (1983). Persistent anomalies of the extratropical Northern Hemisphere wintertime circulation. *In* "Large-Scale Dynamical Processes in the Atmosphere" (B. J. Hoskins and R. P. Pearce, eds.). Academic Press, New York.

Frederiksen, J. S. (1979). The effect of long planetary waves on the regions of cyclogenesis: Linear theory. *J. Atmos. Sci.* **36**, 195–204.

Frederiksen, J. S. (1981a). Growth and vacillation cycles of disturbances in Southern Hemisphere flows. *J. Atmos. Sci.* **38**, 1360–1375.

Frederiksen, J. S. (1981b). Scale selection and energy spectra of disturbances in Southern Hemisphere flows. *J. Atmos. Sci.* **38**, 2573–2584.

Frederiksen, J. S. (1982a). A unified three-dimensional instability theory of the onset of blocking and cyclogenesis. *J. Atmos. Sci.* **39**, 969–987.

Frederiksen, J. S. (1982b). Instability of the three-dimensional distorted stratospheric polar vortex at the onset of the sudden warming. *J. Atmos. Sci.* **39**, 2313–2329.

Frederiksen, J. S. (1983a). Disturbances and eddy fluxes in Northern Hemisphere flows: Instability of three-dimensional January and July flows. *J. Atmos. Sci.* **40**, 836–855.

Frederiksen, J. S. (1983b). A unified three-dimensional instability theory of the onset of blocking and cyclogenesis. II. Teleconnection patterns. *J. Atmos. Sci.* **40**, 2593–2609.

Frederiksen, J. S. (1984). The onset of blocking and cyclogenesis in Southern Hemisphere synoptic flows: Linear theory. *J. Atmos. Sci.* **41**, 1116–1131.

Frederiksen, J. S. (1985). The geographical locations of Southern Hemisphere storm tracks: Linear theory. *J. Atmos. Sci.* **42**, 710–723.

Frederiksen, J. S., and Puri, K. (1985). Nonlinear instability and error growth in Northern Hemisphere three-dimensional flows: Cyclogenesis, onset-of-blocking and mature anomalies. *J. Atmos. Sci.* **42**, 1374–1397.

Frederiksen, J. S., and Sawford, B. L. (1980). Statistical dynamics of two-dimensional inviscid flow on a sphere. *J. Atmos. Sci.* **37**, 717–732.

Gall, R. (1976). A comparison of linear baroclinic instability theory with the eddy statistics of a general circulation model. *J. Atmos. Sci.* **33**, 349–373.

Garriott, E. B. (1904). Long range forecasts. *U.S. Weather Bur. Bull.* **35**.

Green, J. S. A. (1977). The weather during July, 1976: Some dynamical considerations on the drought. *Weather* **32**, 120–126.

Hansen, A. P., and Chen, T. C. (1982). A spectral energetics study of atmospheric blocking. *Mon. Weather Rev.* **110**, 1146–1165.

Illari, L., and Marshall, J. C. (1983). On the interpretation of eddy fluxes during a blocking episode. *J. Atmos. Sci.* **40**, 2232–2242.

Ji, L. R., and Tibaldi, S. (1983). Numerical simulation of a case of blocking: The effects of orography and land-sea contrast. *Mon. Weather Rev.* **111**, 2068–2086.

Lau, N. (1981). A diagnostic study of recurrent meteorological anomalies appearing in a 15-year simulation with a GFDL general circulation model. *Mon. Weather Rev.* **109**, 2287–2311.

Lejenäs, H. (1977). On the breakdown of the westerlies. *Atmosphere* **15**, 89–113.

Lejenäs, H. (1984). Characteristics of Southern Hemisphere blocking as determined from a time series of observational data. *Q.J.R. Meteorol. Soc.* **110**, 967–979.

Lorenz, E. N. (1960). Energy and numerical weather prediction. *Tellus* **12**, 364–373.

Namias, J. (1947). Characteristics of the general circulation over the northern hemisphere during the abnormal winter 1946-1947. *Mon. Weather Rev.* **75**, 145–152.

Namias, J. (1950). The index cycle and its role in the general circulation. *J. Meteorol.* **7**, 130–139.

O'Connor, J. (1963). The weather and circulation of January 1963. *Mon. Weather Rev.* **91**, 209–218.

Rex, D. (1950). Blocking action in the middle troposphere and its effect upon regional climate. *Tellus* **2**, 196–211.

Rossby, C. G. (1950). On the dynamics of certain types of blocking waves. *J. Chin. Geophys. Soc.* **2**, 1–13.

Shutts, G. J. (1983). The propagation of eddies in diffluent jetstreams: Eddy vorticity forcing of 'blocking' flow fields. *Q.J.R. Meteorol. Soc.* **109**, 737–761.

Simmons, A. J., and Hoskins, B. J. (1978). The life cycles of some nonlinear baroclinic waves. *J. Atmos. Sci.* **35**, 414–432.

Simmons, A. J., Wallace, J. M., and Branstator, G. W. (1983). Barotropic wave propagation and instability, and atmospheric teleconnection patterns. *J. Atmos. Sci.* **40**, 1363–1392.

Thompson, P. D. (1957). A heuristic theory of large-scale turbulence and long period velocity variations in barotropic flow. *Tellus* **9**, 69–91.

Trenberth, K. E. (1976). Spatial and temporal variations of the Southern Oscillation. *Q.J.R. Meteorol. Soc.* **102**, 639–653.

Trenberth, K. E., and Paolino, D. A. (1981). Characteristic patterns of variability of sea level pressure in the Northern Hemisphere. *Mon. Weather Rev.* **109**, 1169–1189.

Troup, A. J. (1965). The Southern Oscillation. *Q.J.R. Meteorol. Soc.* **91**, 490–506.

Tucker, G. B. (1979). Transient synoptic systems as mechanisms for the meridional transport: An observational study in the Southern Hemisphere. *Q.J.R. Meteorol. Soc.* **105**, 657–672.

van Loon, H., and Madden, R. A. (1981). The Southern Oscillaton. Part I: Global associations with pressure and temperature in the northern winter. *Mon. Weather Rev.* **109**, 1150–1162.

van Loon, H., and Rogers, J. C. (1978). The seesaw in winter temperatures between Greenland and Northern Europe. Part I: General description. *Mon. Weather Rev.* **106**, 296–310.

van Loon, H., and Rogers, J. C. (1981). The Southern Oscillation, Part II: Associations with changes in the middle troposphere in the northern winter. *Mon. Weather Rev.* **109**, 1163–1168.

Walker, G. T., and Bliss, E. W. (1932). World weather. *Mem. R. Meteorol. Soc.* **4**, 53–84.

Wallace, J. M., and Gutzler, D. S. (1981). Teleconnections in the geopotential height field during the Northern Hemisphere winter. *Mon. Weather Rev.* **109**, 784–812.

Yeh, T. C. (1949). On energy dispersion in the atmosphere. *J. Meteorol.* **6**, 1–16.

Youngblutt, C. E., and Sasamori, T. (1980). The nonlinear effects of transient and stationary eddies on the winter mean circulation. Part I: The diagnostic analysis. *J. Atmos. Sci.* **37**, 1944–1957.

NUMERICAL PREDICTION: SOME RESULTS FROM OPERATIONAL FORECASTING AT ECMWF

A. J. SIMMONS

*European Centre for Medium Range Weather Forecasts
Shinfield Park, Reading, Berkshire RG2 9AX
England*

1. INTRODUCTION

Prediction of atmospheric behavior using comprehensive numerical models may be considered as falling into three categories. One is the "deterministic" prediction of the instantaneous state of the atmosphere for as many days or weeks ahead as may be possible. Such predictions are generally referred to as short-range or medium-range weather forecasts. The other two categories form what can be regarded as climate predictions of the *first* and *second* kinds (Lorenz, 1975). The first kind is the prediction of statistical properties of the atmosphere, for example temporal or spatial means, for time ranges beyond the limit of deterministic predictability. Initial conditions in the atmosphere (and in the ocean and at the earth's land surface) may be in general as important in this case as they are in deterministic prediction, although different aspects of the initial state may be crucial. Climate prediction of the second kind concerns the prediction of the long-term impact of prescribed changes in the boundary conditions, composition, or external forcing of the atmosphere. Statistics which are essentially independent of any experimental initial conditions are sought.

Results relating to the subject of large-scale anomalies of the general circulation have been obtained from all three categories of prediction, although in this contribution emphasis will be placed on practical experience with deterministic weather forecasting, for which a considerable body of data exists. Long-range prediction by numerical methods is at a much less developed stage than short- and medium-range prediction, although encouraging first results have been obtained from specific attempts at prediction (e.g., Miyakoda *et al.*, 1983) and from study of the potential for skillful prediction (Shukla, 1981). In addition, climate studies too numerous to be referenced individually have examined the long-term response to anomalous boundary forcing, for example anomalous distributions of sea surface temperature, and have thereby also contributed toward our understanding of the predictability of anomalies on a monthly time scale and beyond.

Routine operational forecasting for a period up to 10 days ahead, using a global prediction system, began at the European Centre for Medium Range Weather Forecasts (ECMWF) on 1 August 1979. Archiving of both initial analyses and forecasts began 1 month later. Thus at the time of writing there exists an almost 5-year record of forecast results which can be examined to shed light on a number of questions concerning atmospheric predictability on a time range of 10 days and beyond. The record may be used, for example, to define present levels of forecast accuracy, including its geographical and temporal variability, and dependence on the nature of the prevailing synoptic situation. Trends in predictive skill can be identified, and estimates made of the scope for further improvements. Information relevant to developing a capability for the accurate prediction of longer term anomalies may be gathered from the treatment of such anomalies over the 10-day forecast period, and the general evolution of forecast error can be suggestive of some of the causes of this error, which may aid the improvement of numerical models used for all types of atmospheric prediction.

Some results from studies of the above type will be presented in this paper. Operational experience from the first 2 years of medium-range weather prediction at ECMWF has been summarized by Bengtsson and Simmons (1983), and much of what is reported here is in agreement with the earlier results. Although the discussion of forecast skill will not be restricted to cases of large-scale anomalies and blocking, particular attention will be devoted to these features.

The following section contains a short description of the ECMWF forecasting system, including the developments that have taken place over the past 5 years. Methods of assessment of the forecasts are then discussed. There follows a section in which the current overall level of forecast accuracy is summarized, and some specific discussion of the prediction of blocking and cutoff lows is given in Section 5. Section 6 then deals with the development of forecast skill, both that actually achieved in recent years and that expected in the future. The seventh section deals with the ability of the forecast model to maintain anomalous features of the general circulation present in the initial conditions over a monthly time scale, and the overall question of systematic errors is briefly discussed in Section 8. This is followed by some concluding remarks.

2. THE ECMWF FORECASTING SYSTEM

2.1. The Data Assimilation

The ECMWF data assimilation system uses a three-dimensional multivariate analysis scheme, described by Lorenc (1981), to produce initial analyses of geopotential and wind fields for the forecast model, and a correction method for the analysis of humidity. Analyses are produced every 6 hr, with the forecast model providing the necessary first guess for the analysis, which in turn provides the initial state for the 6-hr forecasting step required to produce the first guess for the following analysis. Observations are analyzed at 15 standard pressure levels from 1000 to 10 mbar, and first-guess data are interpolated (or extrapolated) to these levels from the terrain-following coordinate surfaces of the forecast model. In the version of the data assimilation system first introduced into operational use, complete pressure-level fields from the analysis were similarly interpolated back to the model levels, prior to nonlinear normal-mode initialization (Temperton and Williamson, 1981; Williamson and Temperton, 1981).

Although the basic nature of the data assimilation has not changed over 5 years of operational implementation, there have been numerous revisions. Perhaps the most significant of these have been the introduction late in 1980 of the interpolation from pressure to model levels of the difference between the analyzed and first-guess fields, rather than the full analyzed fields, and a recent substantial change in data control and selection and in optimal interpolation statistics. The former change was such as to preserve the model boundary-layer structure through the analysis, since the lowest analyzed pressure levels of 1000 and 850 mbar are insufficient to define such structure. The more recent change is the result of several years experience with the operation of the original scheme, drawing on the general accumulation of statistics of the performance of the scheme. Many of the other revisions have been individually minor ones concerned with the use of data, but changes worthy of note (particularly from the viewpoint of tropical prediction) are the introduction of a distinction between temperature and virtual temperature in November 1980, a revised vertical interpolation of humidity the following November, use of sea surface temperature analyses produced by the National Meteorological Center (in the United States) in July 1982, and the introduction 2 months later of diabatic forcing of lower frequency gravity-wave modes in the initialization. An analysis of soil moisture and snow depth began in November 1983. In addition to these changes, the data analysis has also benefitted from the general development of the forecast model.

2.2. *The Forecast Model*

A major change in the numerical formulation of the forecast model has taken place over the 5-year period considered here. The version chosen for the first phase of operational forecasting used a potential-enstrophy-conserving finite-difference scheme for a staggered (C) grid, a σ coordinate for the vertical representation, and a semi-implicit time scheme for the treatment of gravity-wave terms. A resolution of 1.875° in latitude and longitude, with 15 levels in the vertical, was adopted. Details have been given by Burridge and Haseler (1977) and Burridge (1979). Changes in the horizontal diffusion and the introduction of a more realistic representation of the orography and coastlines took place within the first 2 years of operational use.

The major change took place in April 1983. The new version uses a spectral formulation in the horizontal, with triangular truncation at total wavenumber 63, a vertical coordinate (with 16-level resolution) that is terrain following at low levels but which reduces to pressure in the stratosphere, and a modified, more efficient, time-stepping scheme. The operational change to this version was accompanied by a change to a higher "envelope" orography in the model. Papers relating to these changes include those of Girard and Jarraud (1982), Simmons and Burridge (1981), Simmons and Jarraud (1984), Simmons and Strüfing (1983), and Wallace et al. (1983). In the light of operational experience, minor adjustments of the orography, horizontal diffusion, and time scheme have subsequently been made.

The parameterization schemes described by Tiedtke et al. (1979) have been used with both versions of the operational model. They include a convection scheme following Kuo (1974), a stability-dependent representation of boundary and free-atmospheric turbulent fluxes (Louis, 1979), and a radiation scheme (Geleyn and Hollingsworth, 1979) that includes interaction with model-generated clouds. A number of minor adjustments of the parameterizations have taken place during operational use, but the most noteworthy change was the introduction of a diurnal radiative cycle in May 1984. Other more substantial changes are at an advanced stage of testing and will be mentioned briefly in Section 9.

3. METHODS OF ASSESSMENT

For much of this paper attention will be concentrated on the results of objective verifications of forecasts carried out at ECMWF, using methods discussed by Arpe (1980) and Nieminen (1983). Results will be presented mostly for the anomaly correlation of the height field, the correlation between observed and predicted deviations from climatology of the height field at one or more levels in the atmosphere, evaluated over a certain region of the globe. This has generally been found to give a reliable indication of overall forecast skill, although as with any method of assessment care must be taken in interpreting results from very limited numbers of forecasts. Nieminen (1983) illustrates how within a particular season a good agreement is found between use of anomaly correlation and standard deviation of forecast error as measures of forecast accuracy.

Forecasts are also routinely subjected to synoptic assessment, particularly over the European area by the member states of ECMWF. In Fig. 1, results of one such assessment, of day $5\frac{1}{2}$ forecasts of 500-mbar height from January 1981 to July 1982, are compared with an assessment using anomaly correlations. The subjective marking was carried out for forecasts for the neighborhood of Scandinavia by the Swedish Meteorological and Hydrological Institute, while anomaly correlations were computed for a larger European area. The results indicate that an average subjective classification of the day $5\frac{1}{2}$ forecasts is "useful" and that this corresponds approximately to a 0.6 value for the anomaly correlation coefficient, in agreement with the previous use of 0.6 as a limit of useful predictability (e.g., Hollingsworth *et al.,* 1980). Keeping in mind the different verification areas, it also appears from Fig. 1 that the subjective and objective assessments are in broad agreement regarding differences in forecast quality from month to month.

Despite such agreement it must nevertheless be recognized that judgment of the quality of particular forecasts, or spells of forecasts, can depend sensitively on the place for which the forecast is to be made, on the prevailing

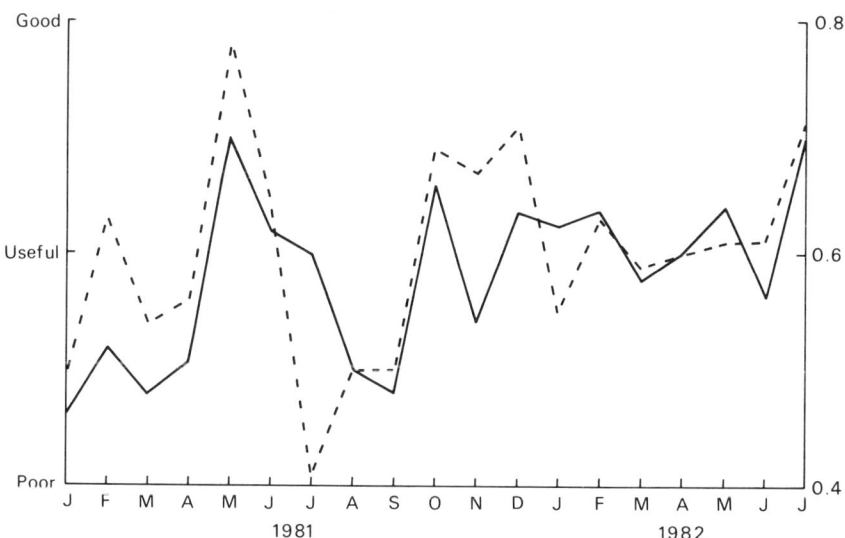

FIG. 1. Monthly-mean assessments of day $5\frac{1}{2}$ 500-mbar height forecasts. The solid line denotes results from a subjective assessment for the Scandinavian region, with averages computed from individual scores given on a scale of 1 to 5, with 4 representing a "good" forecast, 3 a "useful" forecast, and 2 a "poor" forecast. The dashed line denotes means of anomaly correlations computed over a European area (12°W–42°E; 36°N–72°N).

synoptic situation, and on the nature of the forecast error. For example, medium-range predictions for some locations may be particularly sensitive to small phase errors in quasi-stationary situations involving meridional flow, whereas a useful medium-range outlook of weather type could be given in predominantly zonal situations for which there might be phase errors in the prediction of individual transient disturbances. Conversely, the forecasts may be rated badly in zonal situations at locations where systematic error in the latitude of the jetstream causes persistent biases in the track of cyclones. General statements concerning forecast skill must therefore be interpreted with caution.

4. The Accuracy of Forecasts in the Medium Range

Anomaly correlations calculated daily using height fields at standard pressure levels from 1000 to 200 mbar for the extratropical Northern Hemisphere, and averaged over the year of 1983, are shown in the upper plot of Fig. 2 for the 10-day forecast range. Representing the mean over a large number of forecasts, the plot shows a monotonic decline in forecast accuracy with increasing range, with the correlation coefficient of 0.6 reached after about 5.5 days. A figure of 5.8 days is found for the 500-mbar level alone, while accuracy is somewhat less at 1000 mbar, the 0.6 level being reached at day 5.3.

The lower plot of Fig. 2 shows corresponding correlations computed separately for three groups of zonal wavenumber. Accuracy is evidently highest for the longest zonal scales, and there is a rapid growth of error in the shorter synoptic scales, here represented by zonal wavenumbers 10–20. Data for the winter of 1980/1981 analyzed by E. Kalnay (personal communication) shows that viewed in terms of the error variance of total wavenumber n, rather than zonal wavenumber, error in scales with $n > 15$ on average reaches saturation within the first 5 days of the forecast. Synoptic analysis reveals both cases in which the large-scale pattern is well predicted despite error in the forecast of shorter transient disturbances, and cases in which the erroneous forecast of a small-scale feature is followed by deterioration of the forecast over a much larger area.

Anomaly correlations as in the upper plot of Fig. 2, but averaged only over the winter months of January, February, and December or the summer months of June, July, and August, are shown in the upper panel of Fig. 3. Although seasonal differences are probably exaggerated by an unusually good performance in January 1983 (which will be discussed later), these and other results (e.g., Bengtsson and Simmons, 1983) reveal a generally more accurate performance, as measured by anomaly correlation, in winter. In absolute terms, the growth of error variance is smaller in summer than in

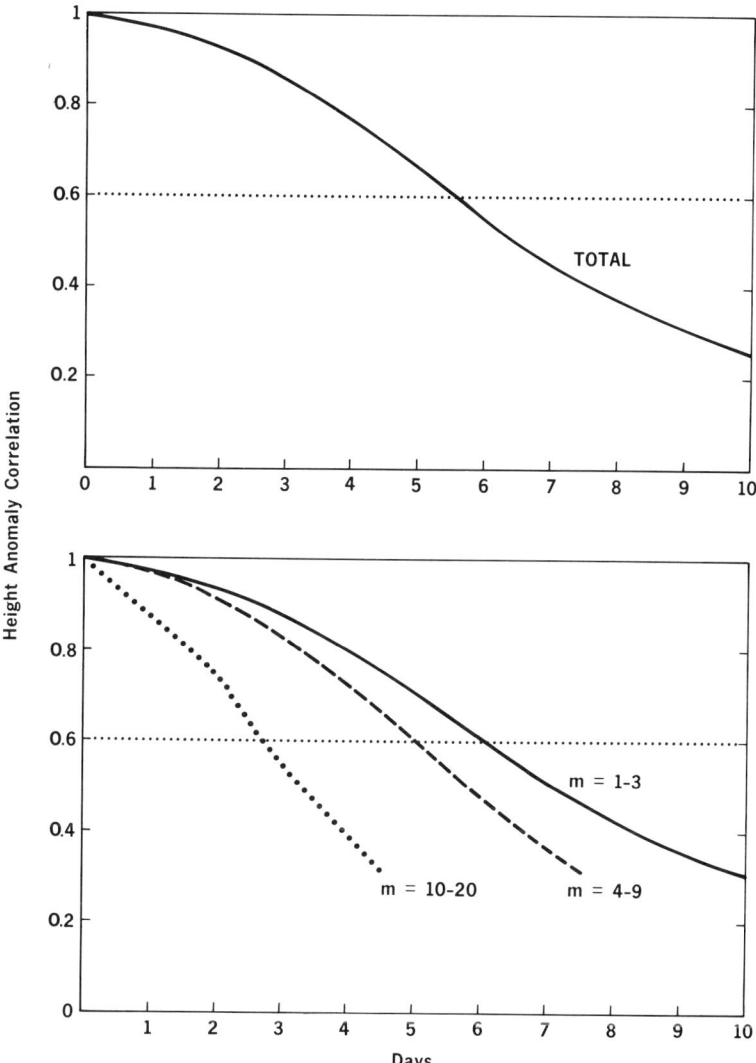

FIG. 2. Anomaly correlations of 1000- to 200-mbar height for the area from 20°N to 82.5°N averaged over all operational forecasts carried out in 1983. Values are plotted as a function of forecast range. The upper panel is for the total fields, and the lower panel shows results for different groups of zonal wavenumber m.

winter, but when normalized by a measure of the actual atmospheric variability, for example the error variance of a persistence forecast (as shown for individual months in a later figure), the relative inaccuracy of the summer forecasts again becomes evident. A smaller absolute rate of error growth in summer is consistent with smaller growth rates for baroclinic instability in

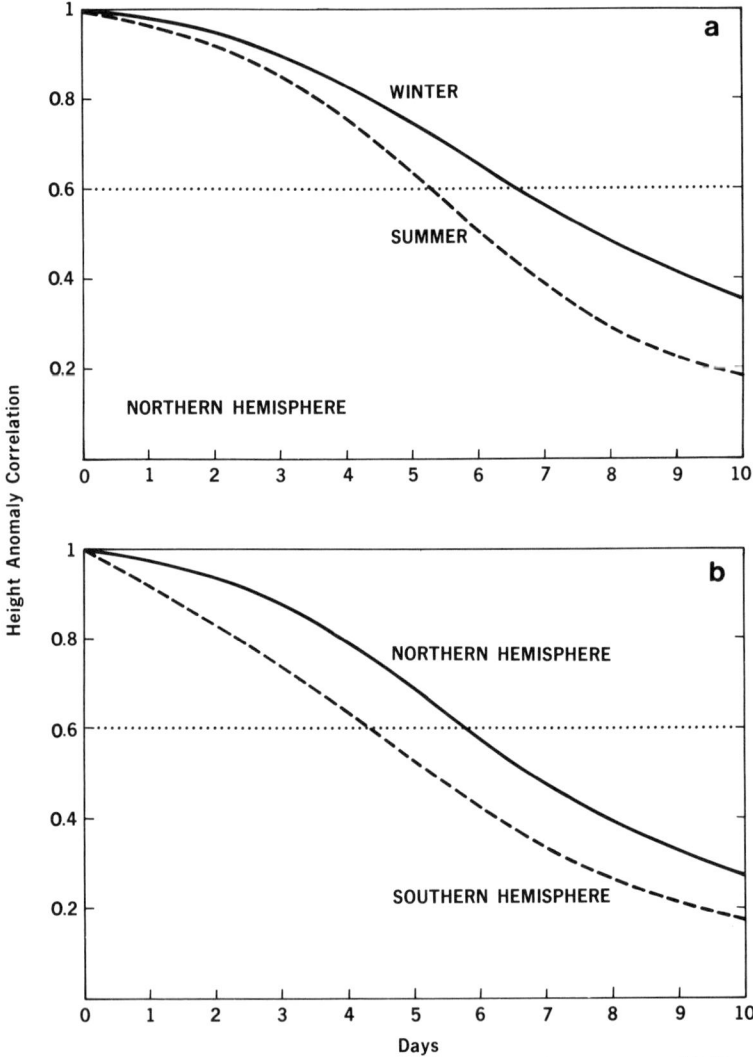

FIG. 3. Height anomaly correlations for 1983 as functions of forecast range. (a) Means for January, February, and December (marked winter) and June, July, and August (marked summer) for the 1000- to 200-mbar levels of the extratropical Northern Hemisphere. (b) Annual means for the extratropical Northern and Southern Hemispheres for the 500-mbar level.

the weaker summer flow; relative inaccuracy may arise from a tendency for shorter scales to be more prevalent in summer and a greater sensitivity of summer circulation systems to processes requiring parameterization in the forecast model.

The accuracy of forecasts for the extratropical Northern and Southern

Hemispheres, again for the year 1983, is also compared in Fig. 3. A clear difference in predictability is found, amounting to about $1\frac{1}{2}$ days at a correlation of 0.6. Although some differences between the hemispheres could arise from interhemispheric differences in the nature of the general circulation, the rapid initial growth of error shown in Fig. 3 for the Southern Hemisphere is indicative of a much larger error in the determination of the initial state for this hemisphere. This is not surprising in view of the sparsity of some types of observational data in the Southern Hemisphere, and generally more accurate forecasts for this region have been reported using the enhanced data coverage of the First GARP Global Experiment (FGGE) year (Bengtsson et al., 1982).

Verification statistics have also been accumulated routinely for smaller areas of the globe. In Table I, predictability, as measured by the time at which the monthly-mean anomaly correlation of 500-mbar height reaches the value of 0.6, is shown for January and July of each operational year for three regions of the extratropical Northern Hemisphere. Such regional results show much more variability from month to month and year to year than those for the hemisphere as a whole, and general conclusions are difficult to draw. On the basis of the 5-year record, the winter forecasts are not found to be clearly worse over Europe than over the other continental regions, in contrast to the indication discussed by Bengtsson and Simmons (1983) on the basis of the first 2 years of operational prediction. An overall tendency for more accurate summer predictions at 500 mbar over Europe is found from examination of Table I and from results for other months, although the more northern latitude of this region should be borne in mind. More substantial differences are found for the anomaly correlation of 1000-mbar height, with

TABLE I. PREDICTABILITY FOR THREE REGIONS OF THE EXTRATROPICAL NORTHERN HEMISPHERE

Month/year	Forecast ranges[a]		
	Europe ($36°-72°$N; $12°$W$-42°$E)	East Asia ($24°-60°$N; $102°-150°$E)	North America ($24°-60°$N; $120°-72°$W)
Jan 1980	4.4	5.4	5.8
July 1980	5.5	5.7	4.0
Jan 1981	4.8	6.6	7.0
July 1981	4.1	5.2	3.9
Jan 1982	5.2	5.9	4.6
July 1982	6.4	6.1	4.7
Jan 1983	8.6	5.7	5.6
July 1983	5.3	3.8	5.2
Jan 1984	5.8	6.4	4.8
July 1984	5.7	5.5	5.4

[a] Days at which monthly-mean anomaly correlations of 500-mbar height reach a value of 0.6 over different continental regions of the Northern Hemisphere.

the 0.6 correlation coefficient reached on average about 1 day later in the forecast period over Europe than over North America. However, the extent of these differences is not confirmed by examination of another measure of skill, the standard deviation of forecast error, normalized by persistence, for the 850-mbar wind. A definitive specification of regional differences in forecast accuracy over the extratropical Northern Hemisphere is thus difficult to give on the basis of such objective verification.

Forecasting for the tropics and immediate subtropics undoubtedly poses particular problems. There is not only in general a severe deficiency in data coverage, but also strong sensitivity to aspects of the data assimilation and parameterization schemes of the forecasting system. In addition, the strong persistence of some regional circulations emphasizes difficulties. As an example, the upper panel of Fig. 4 shows the absolute correlation of the 850-mbar vector wind for a region covering India and part of Southeast Asia for July 1983. Overall, the numerical forecasts barely improve on a persistence forecast for the summer monsoon flow, although individual cases of quite accurate forecasts of transient behavior over several days can be found. The verification for Northwest Africa also shown in Fig. 4 indicates a much better performance relative to persistence, although the absolute correlation of the forecast evolves similarly for the two regions. Some further discussion of tropical forecast accuracy is given in Section 6.

The accuracy of forecasts for a given month or season can vary substantially from year to year even on a hemispheric domain. Extreme cases for the extratropical Northern Hemisphere, comparing anomaly correlations for January 1982 and 1983 and for July 1983 and 1984, are presented in Fig. 5. Although changes in the forecasting system over the two intervening 1-year periods may have contributed to the differences, particularly between 1983 and 1984, the tests of the various changes that were made indicate that most of the differences seen in Fig. 5 are not due to development of the model or data assimilation. Rather, differences in predictability of up to 2 days at the 0.6 level of anomaly correlation appear to be mostly a consequence of differences in the circulation patterns from month to month. The extent to which this is due to a fundamentally higher predictability, due to a greater sensitivity to data coverage or analysis techniques, or due to a greater sensitivity to systematic model deficiencies, in certain synoptic situations is unclear.

The results shown in Fig. 5 are confirmed by corresponding plots of the standard deviation of forecast error. However, it is also found that standard deviations for persistence forecasts show some of the same variability. In Fig. 6 we plot a measure of forecast accuracy based on the standard deviation of the model forecast normalized by the standard deviation of persistence, for the same months as in Fig. 5. For the two Januaries there appears little

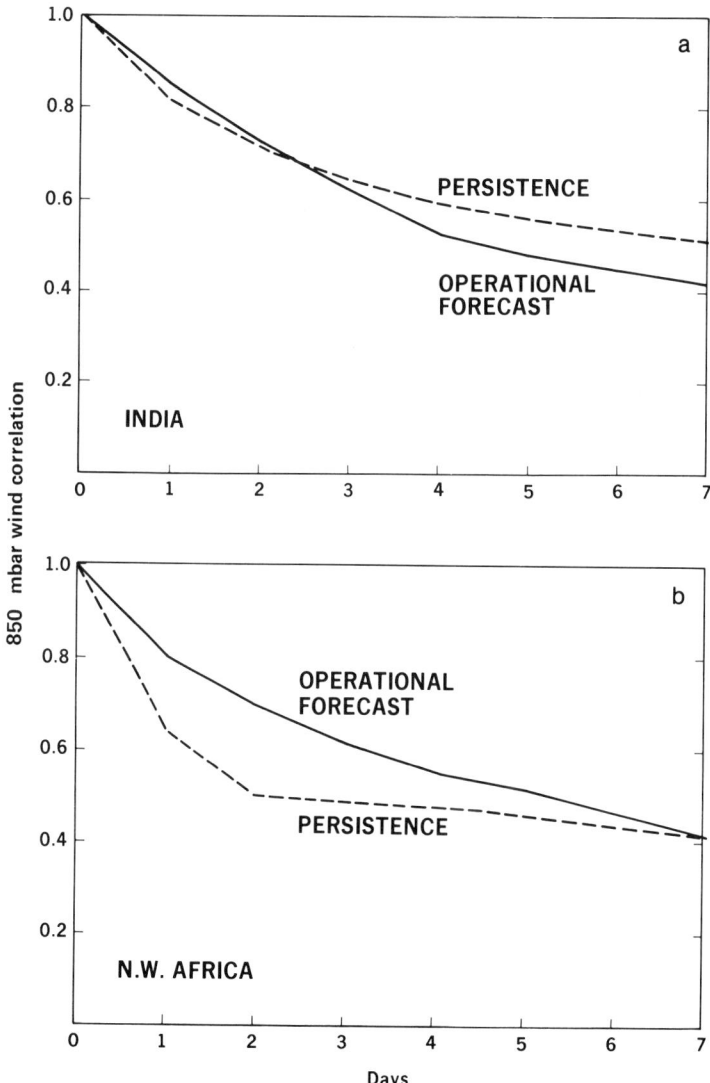

FIG. 4. Monthly means of the absolute correlation of 850-mbar vector wind for July 1983 as functions of forecast range. Results are shown for both operational numerical forecasts and persistence forecasts for two regions (6°–33°N, 72°–102°E, a; 12°–36°N, 21°W–12°E, b) spanning the tropics and subtropics.

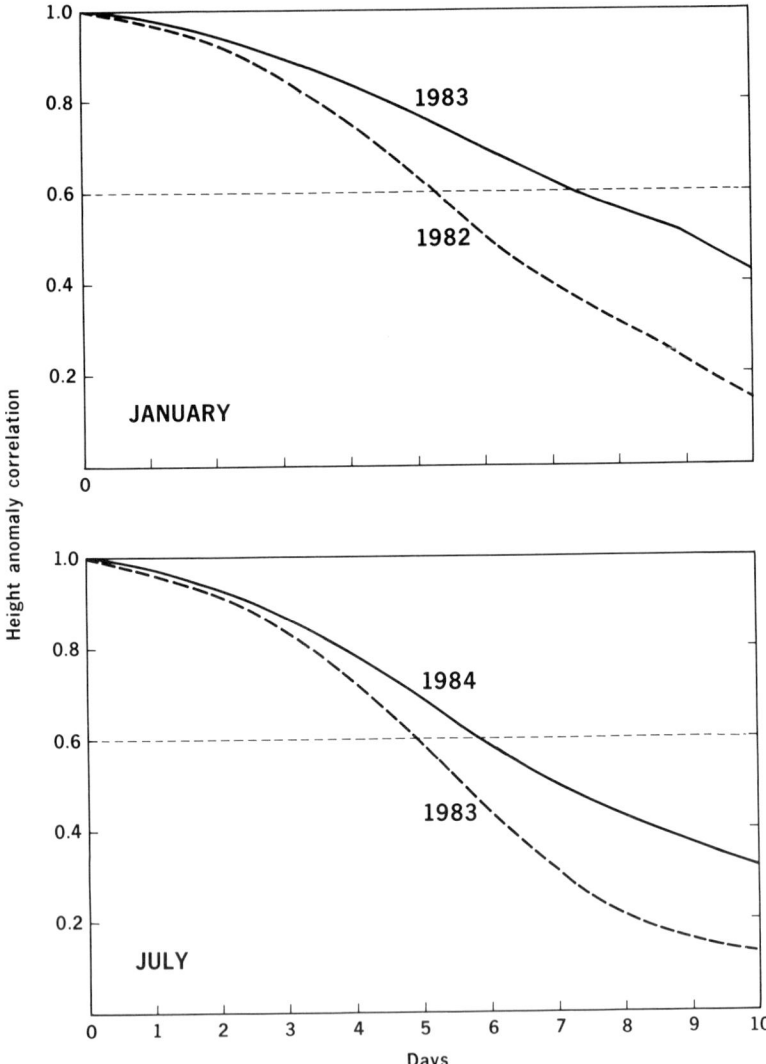

FIG. 5. Monthly-mean anomaly correlations of height for 1000- to 200-mbar levels and the extratropical Northern Hemisphere as functions of forecast range for January 1982 and 1983, and July 1983 and 1984.

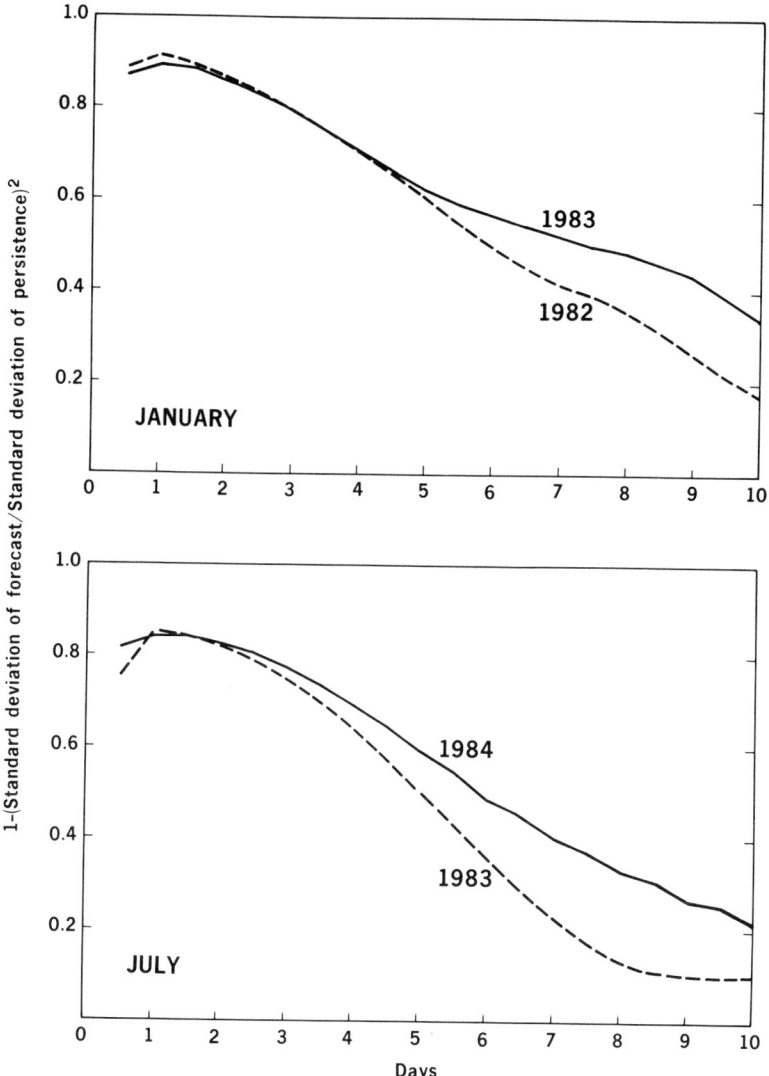

FIG. 6. As Fig. 5, but for a measure of forecast accuracy based on standard deviation normalized by persistence.

difference in forecast quality out to day 4, according to this measure, although the superiority of 1983 still becomes evident in the second half of the forecast range. July 1984 appears clearly better than July 1983 beyond day 2. Thus although differences are not quite so pronounced for a measure normalized by the skill of a persistence forecast, they are nevertheless substantial in the medium range.

5. The Prediction of Blocking and Cutoff Lows

Spells of above- and below-average predictability are also found within a monthly time scale. Examples have been discussed by Bengtsson and Simmons (1983), and here we present a more recent case, November 1983, as a basis for choice of specific examples of forecasts involving blocking and the formation of cutoff lows.

Figure 7 shows height anomaly correlations for 3-, 7-, and 10-day forecasts performed from initial dates within November 1983. Particularly evident in the 7-day forecasts is a slow variation in forecast skill over the course of the month, with below-average performance at the beginning of the month and in the final week, and an accurate spell during the second and third weeks. This variation is discernible in the 3-day forecasts, and more so at day 10.

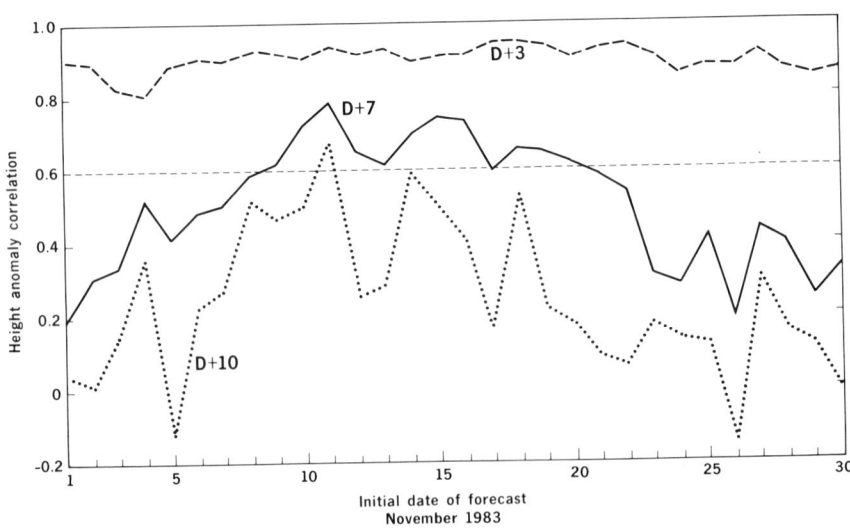

FIG. 7. Anomaly correlations of height for 1000–200 mbar and the extratropical Northern Hemisphere for 3-, 7-, and 10-day forecasts performed from initial dates within the month of November 1983.

Small daily variations in the accuracy of individual forecasts at day 7 are seen to be amplified at day 10.

To illustrate the range of accuracy of large-scale forecasts within this month, 5-day mean maps of 500-mbar height from analyses and two forecasts are shown in Fig. 8. The forecast from 11 November shown in the left-hand column is classified as the most accurate of the month according to the anomaly correlations at days 7 and 10 shown in Fig. 7. The right-hand forecast plot shows the result from the forecast from 26 November, which ranks as the poorest on the hemispheric scale.

The forecast from 11 November is successful in its representation of most large-scale features present over the final 5 days of the forecast period. It has captured the retrogression of the block initially located over western Europe, the enhancement of the closed low to the northeast, and the decay of the low over northern Canada. Conversely, major error is seen in the forecast from 26 November. In particular, its evolution has produced a predominant zonal wavenumber 2 pattern in the high-latitude height field rather than the strong wavenumber 3 pattern which occurred in reality. The forecast is successful in its formation of a cutoff low in the European sector, but for practical application suffers from an important error in the position of this low, and in the eastward extension of the zonal jet over the Atlantic.

It is of interest from the viewpoint of systematic model errors (to be discussed further in Section 8) to examine in more detail the development of the latter features of the forecast from 26 November. Figure 9 shows the initial 500-mbar height analysis for the European/Atlantic region, together with the corresponding maps for the 5-day forecast and its verifying analysis. At the 5-day range a pronounced ridge extends over western Europe, and the cutoff low has just formed to the southeast in both the forecast and reality, although it is already seen to be located further east in the forecast. To the west, the forecast exhibits a characteristic error in that the trough over the eastern Atlantic has a phase lag at southern latitudes, and fails to produce a weak cutoff west of the Iberian Peninsula.

Figure 10 shows the 6 day forecast and verifying analysis. As in idealized studies of mature baroclinic waves (e.g., Simmons and Hoskins, 1978) the tilted eastern Atlantic trough in the forecast has decayed and enhanced westerly flow in this sector, whereas in reality a weak cutoff remains, with high pressure over southern England. In addition, the major Mediterranean cutoff has drifted slightly eastward in the forecast, rather than becoming established over the south of Italy.

Grønaas (1983) has carried out an extensive study of forecasts for the European/Atlantic area for the years 1980 and 1981. He identified 20 spells lasting 7 days or longer in which the anomaly correlation coefficient of 500-mbar height for the European area indicated above- or below-average

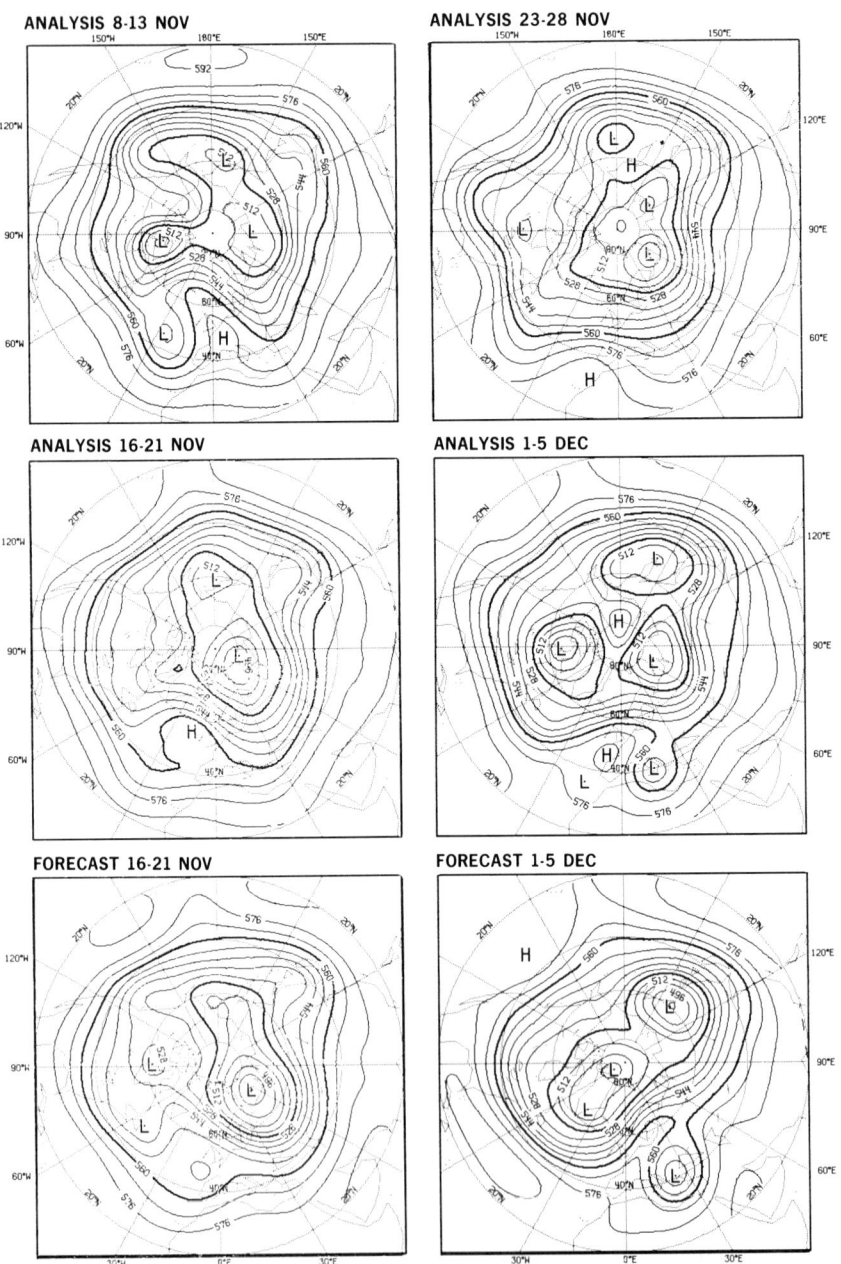

FIG. 8. Mean 500-mbar height analyses for the periods 8–13 November (upper left), 16–21 November (middle left), 23–28 November (upper right), and 1–5 December (middle right). Corresponding means for 16–21 November and 1–5 December are also shown for the forecasts from 11 and 26 November, respectively. The contour interval is 8 dam.

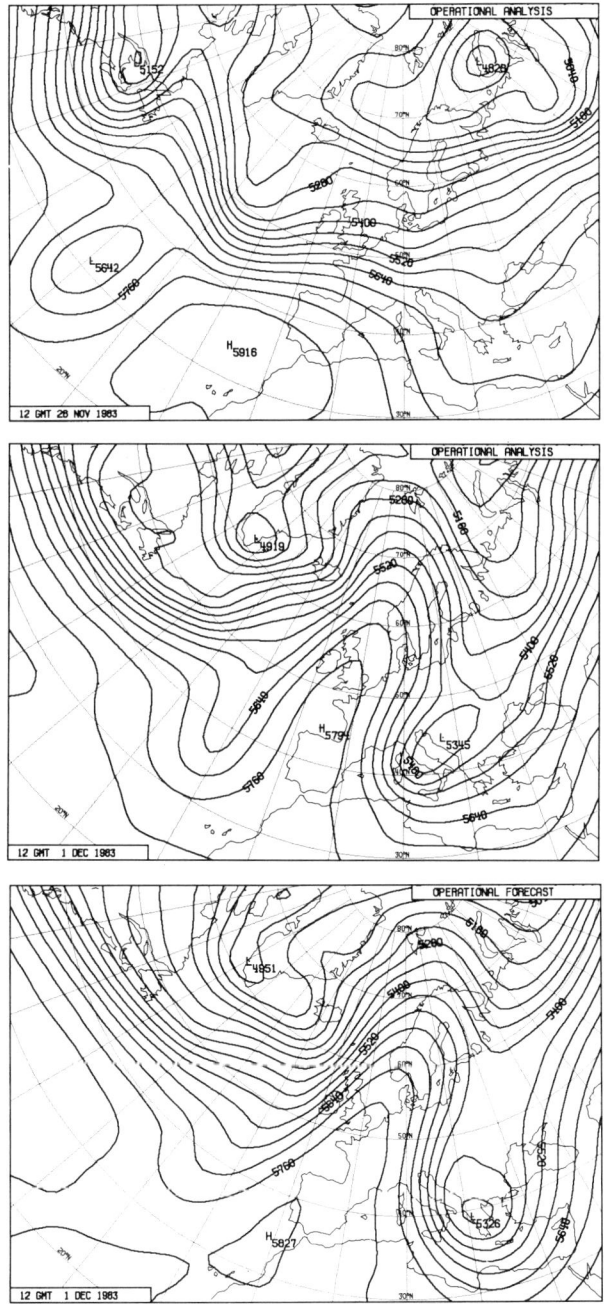

FIG. 9. The 500-mbar height analysis for 12 GMT 26 November (upper) and 1 December (middle), and the 500-mbar height field forecast for 1 December (lower) from initial conditions for 26 November. The contour interval is 6 dam.

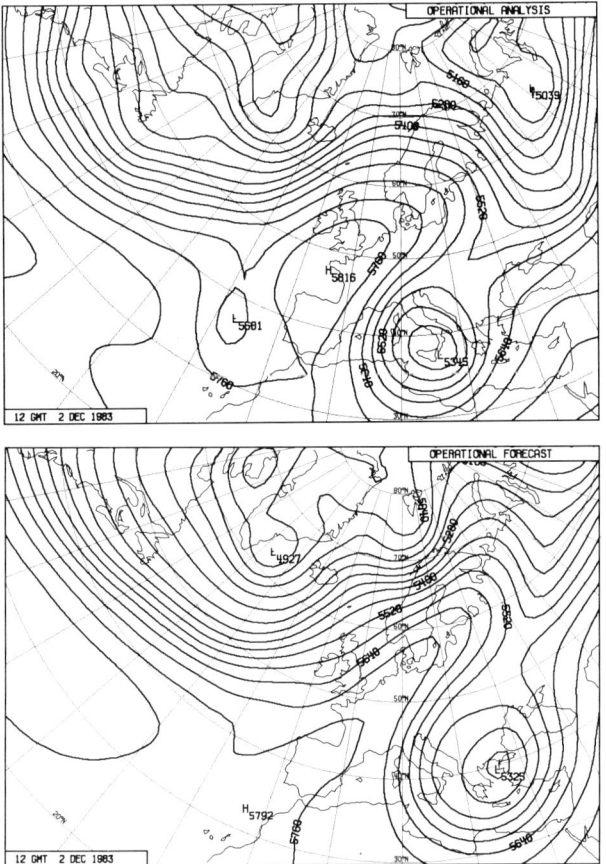

Fig. 10. The 500-mbar height field for 2 December (upper) and the corresponding 6-day forecast (lower) from 26 November.

forecast quality. These spells were typed synoptically, and it was discovered that the best scores were found for situations characterized by some form of blocking, with poorest results in prevailing zonal situations. Mean scores for day 7 in blocking spells were comparable with those for day 5 in cases of zonal flow.

More specifically, Grønaas (1983) reports above-average forecast scores when persistent large-scale synoptic features such as blocking and cutoff lows are present in the initial analyses, or predicted within the first 3 days of the forecast. The poorer performance in zonal situations reflects not only phase errors of traveling disturbances, but also errors (which have a systematic element) in the cyclone tracks and in the rate of filling of cyclones. The systematic errors appear to inhibit the formation of blocking later in the forecast period and give rise to a tendency for the cyclonic activity on the western side of a ridge or block to break down that feature. An indication of this has already been discussed with respect to the forecast from 26 November 1983.

Situations in which the development of blocking is accurately predicted may also be used to examine the mechanisms and interactions involved, and the features of the forecasting system which are of crucial importance. This can be achieved by controlled numerical experimentation, and case studies examining a range of sensitivities to such features as orography, model resolution, and parameterizations have been reported by Bengtsson (1981), Ji and Tibaldi (1983), Tibaldi and Buzzi (1983), and Tibaldi and Ji (1983).

6. Developments in Predictive Skill

Miyakoda *et al.* (1972) suggested that the results of their first comprehensive trial of 2-week predictions, using a hemispheric model, might be taken as a benchmark for future comparisons. Their ensemble mean anomaly correlations of 500-mbar height for the extratropical Northern Hemisphere, based on 12 forecasts from January cases taken from the years 1964 to 1969, are presented in Fig. 11, together with corresponding results from the operational ECMWF forecasts for the December to February period for the years 1982/1983 and 1983/1984. Included in the ECMWF results is the month of January 1983, which we have already shown to have been one of high predictability with the then-operational gridpoint model, but results with the new spectral model and envelope orography for January 1984 are only slightly poorer than achieved in the operational forecasts for the previous January, and Fig. 11 thus appears to be reasonably representative of current ECMWF skill for the winter months.

Although a number of other qualifying remarks could be made about the comparison presented in Fig. 11, it is nevertheless clear that a substantial improvement has taken place over the past 10–15 years in our ability to predict at least the larger scales of motion over the extratropical Northern Hemisphere. The improvement is confirmed by comparison of limited-area verifications of operational forecasts over Europe (L. Bengtsson, personal communication). Figure 11 shows a large increase in accuracy in the very early part of the forecast, suggesting an important contribution from a more accurate specification of the initial state. Overall, the more recent forecasts reach the 0.8 level of anomaly correlation almost 2.5 days later, and they also exhibit a somewhat smaller error growth beyond this time, as measured either by anomaly correlation or by standard deviation. Differences between

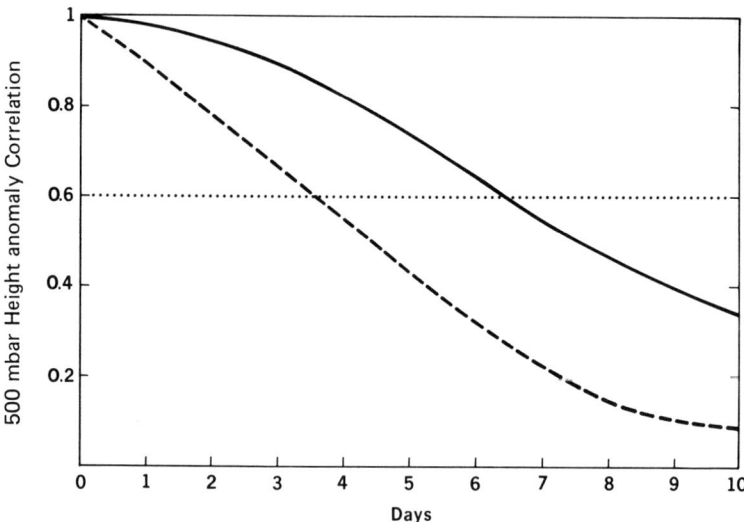

FIG. 11. Mean 500-mbar height anomaly correlations as functions of forecast range. The solid line denotes average results from ECMWF operational forecasts for the winters of 1983 and 1984, and the dashed curve shows the mean of 12 forecasts from January cases chosen from the years 1964–1969, as reported by Miyakoda *et al.* (1972).

the two sets of forecasts amount to 2.8 days at the 0.6 level and 3.7 days at the 0.4 level of anomaly correlation. The mean standard deviation of the 500-mbar height error reaches a value of about 130 m by day 10 of the ECMWF forecasts; this level of error was reached soon after day 6 in the experiments of Miyakoda *et al.*

Improvements in predictive skill can also be identified within the shorter period over which operational forecasting has been carried out at ECMWF. This may be seen in Fig. 12, which shows times at which monthly-mean anomaly correlations of 500- and 1000-mbar height reach the value of 0.6 for the two extratropical hemispheres, for the period from January 1980 to July 1984. Annual running means are also plotted. Results for both hemispheres exhibit generally higher correlations at 500 than at 1000 mbar, and the variability from month to month that has already been noted for the Northern Hemisphere. They indicate overall improvements of about $\frac{1}{2}$ day in the "useful predictability" attained by the forecasting system, with somewhat larger improvements at 1000 than at 500 mbar. Such advances are of course not unique to the ECMWF system; Lange and Hellsten (1984) present results showing distinct improvements to have been achieved at the 3-day range by a number of operational forecasting centers.

Corresponding results for the tropical belt are presented in Fig. 13, in this

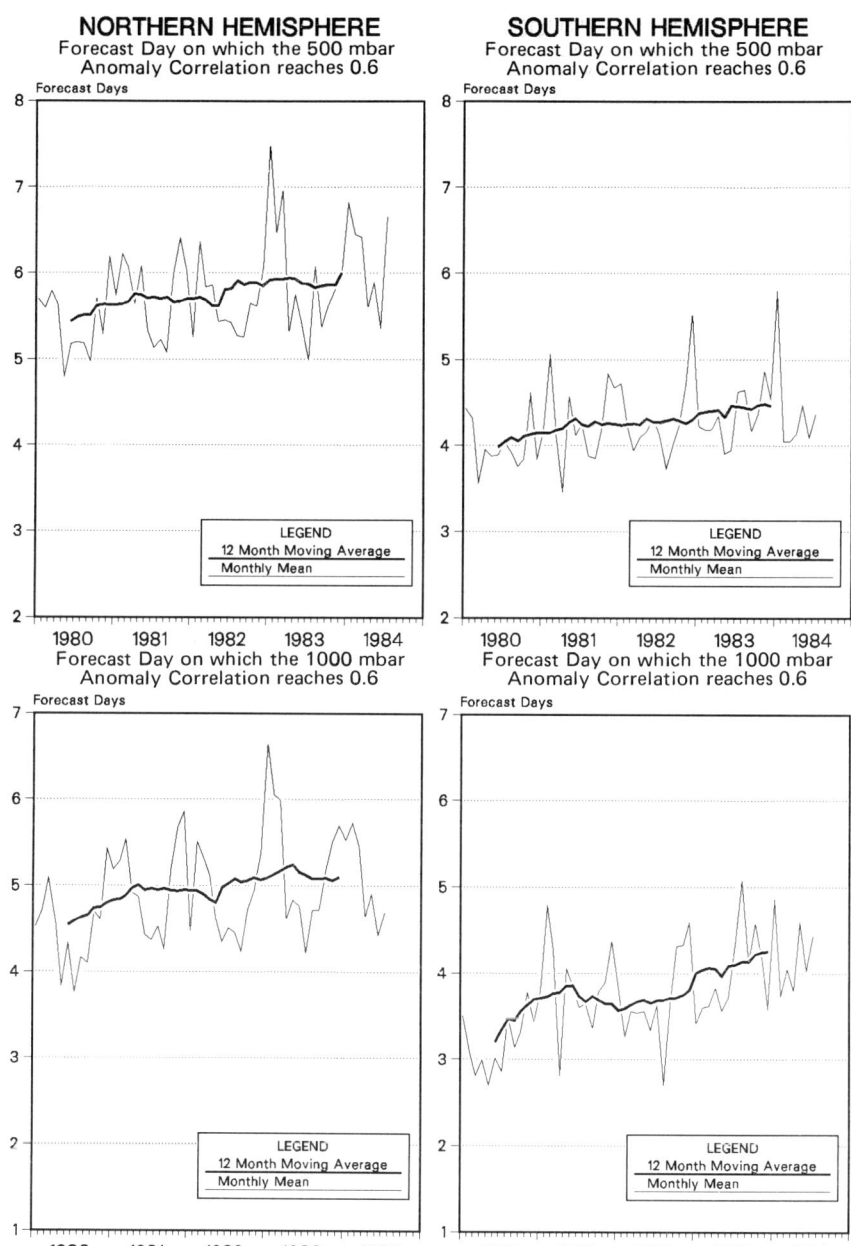

FIG. 12. The forecast ranges (days) at which monthly-mean height anomaly correlations reach values of 0.6. Results are shown for the extratropical Northern (left) and Southern (right) Hemispheres, and for the 500-mbar (upper) and 1000-mbar (lower) levels. Both individual monthly values and 12-month running means are plotted.

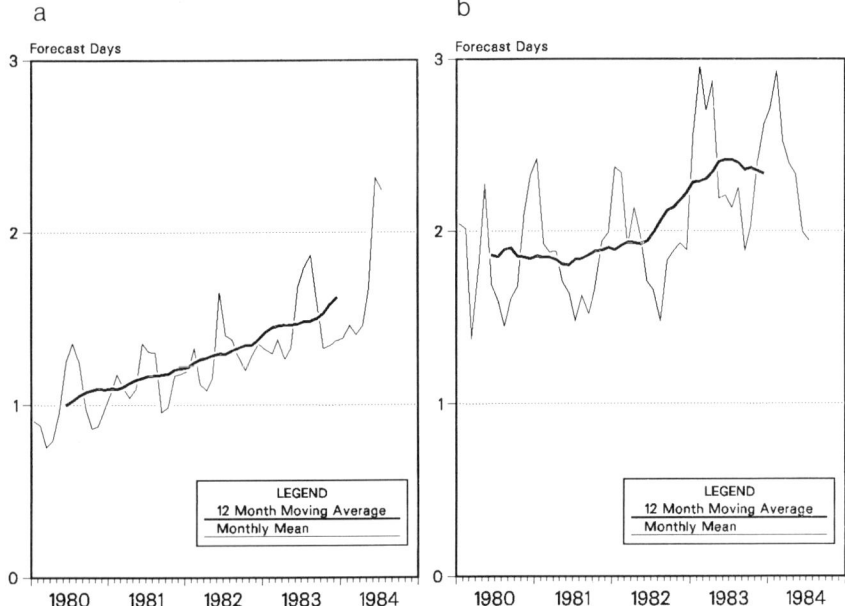

FIG. 13. The forecast ranges (days) at which monthly-mean absolute correlations of 850-mbar (a) and 200-mbar (b) vector winds evaluated for the tropical belt fall below a value of 0.75.

case using the forecast range at which the absolute correlation of vector wind falls below 0.75 to indicate the improvements brought about in the forecasting system. The level of accuracy of wind forecasts achieved initially at the 1-day range is seen to be reached now at about day $1\frac{1}{2}$ at 850 mbar, and improvement is also evident at 200 mbar. This indication is consistent with a reduction in root-mean-square errors, which at 850 mbar have fallen (in the annual mean) from a maximum of 4.9 m sec^{-1} to a value of 4.2 m sec^{-1} for the 48-hr forecasts and from 3.7 to 3.2 m sec^{-1} at the 24-hr range. Falls at 200 mbar have been from 11.8 to 9.8 m sec^{-1} at the 3-day range and from 10.5 to 8.5 m sec^{-1} for the 2-day forecasts. Although the overall level of forecast skill in the tropics remains far from satisfactory, it is encouraging that the performance of the forecasting system has shown itself to be sensitive to the various changes that have been made in the data assimilation and forecast model.

An indication of some of the potential for improving extratropical forecasts further has been derived by Lorenz (1982) from analysis of the performance of the ECMWF forecasting system. For a particular point in time, the current analysis of the atmosphere and the 1-day forecast from the day before provide two estimates of the actual atmospheric state. Thus the growth (as time t increases) of differences between a particular forecast for

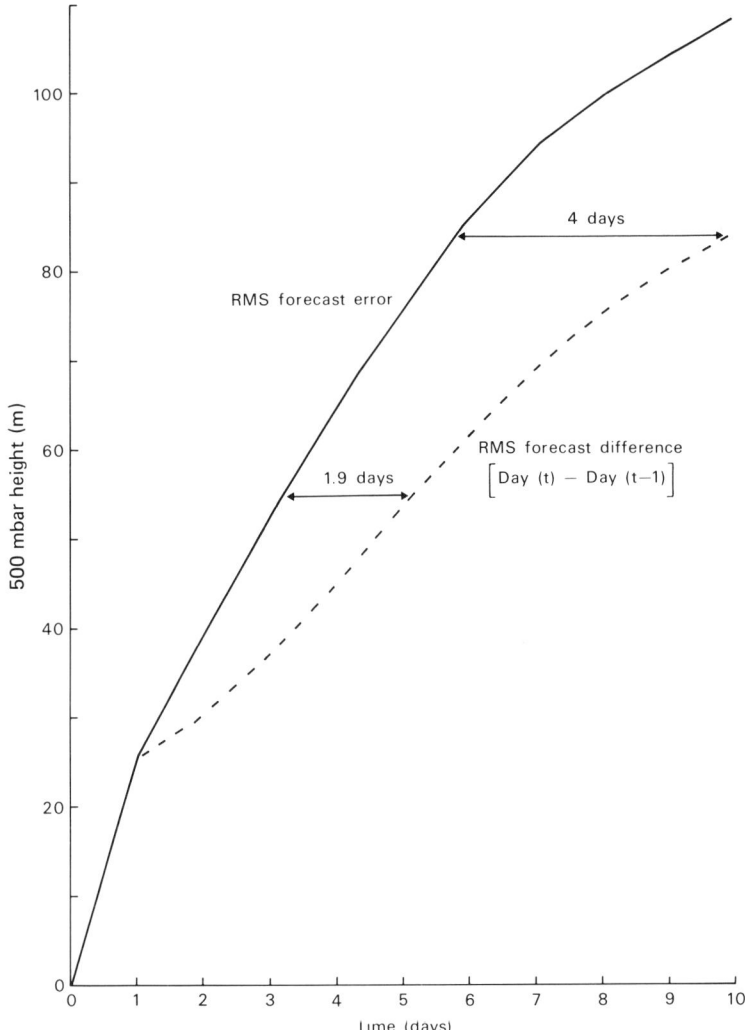

FIG. 14. The global root-mean-square error of 500-mbar height (solid line) for all forecasts from 1 December 1980 to 10 March 1981, and the corresponding root-mean-square difference between forecasts for a particular time starting from initial analyses 1 day apart. Adapted from the work of Lorenz (1982).

day t and the previous day's forecast for day $(t + 1)$ shows how initial, relatively small differences amplify in time. If this growth in the model occurs at a rate similar to that of real differences in the atmosphere, then it indicates a limit to the accuracy with which a forecast can be made, assuming the 1-day forecast error to remain unchanged. Specific results such as shown in Fig. 14 suggest a maximum possible improvement of about 2 days at the level of predictability currently reached at day 3, and about 4 days at the level now reached at day 6. Further improvements would result from increased accuracy of the 1-day forecast and Lorenz argues that halving the 1-day root-mean-square error should add 2 days to the range of predictability. To put such a reduction in context, present ECMWF results for winter indicate a 1-day error which is 40% of the value reported by Miyakoda *et al.* (1972). Thus an average limit of deterministic predictability of the order of 2 weeks or more is indicated, a result in general agreement with earlier estimates discussed by Charney *et al.* (1966) and Smagorinsky (1969).

7. The Representation of Monthly-Mean Anomalies

Some results of relevance to the longer-range prediction of large-scale anomalies may be obtained by examining the extent to which observed monthly-mean deviations from climatology are present in the day 10 forecast fields. For example, Bengtsson and Simmons (1983) illustrate how the mean of all ECMWF forecasts for 10 days ahead produced daily during July 1980 succeeded in representing the temperature anomaly associated with a major heat wave and drought over southeastern North America. Such an anomaly was, of course, present in the initial conditions for the predictions, but the forecast model (which used a number of climatological initial surface conditions) was evidently capable of maintaining the strength of the anomaly over at least a 10-day period.

A brief examination of the treatment of Northern Hemispheric 500-mbar height anomalies during the winter months of 1980/1981 and 1981/1982 has been carried out and some objective verification of day 10 forecasts is presented in Fig. 15. Since mean anomalies computed for a period of about 30 days change little when the averaging period is shifted by 10 days, persistence forecasts score highly in Fig. 15, and the challenge for the numerical model is in the first place to maintain initial anomalies over the 10-day forecast period, even though the ultimate requirement for long-range prediction models is to forecast the change in anomaly pattern from one month (or season) to the next. The ECMWF forecasting system is evidently partially successful in maintaining anomalies, achieving an average correlation of about 0.5 when monthly means of day 10 forecast fields are verified. The lower mean of about 0.25 obtained from averaging the day 10 correlations of individual forecasts reflects the inaccurate prediction by day 10 of transient disturbances which do not directly influence the verification of monthly-mean fields.

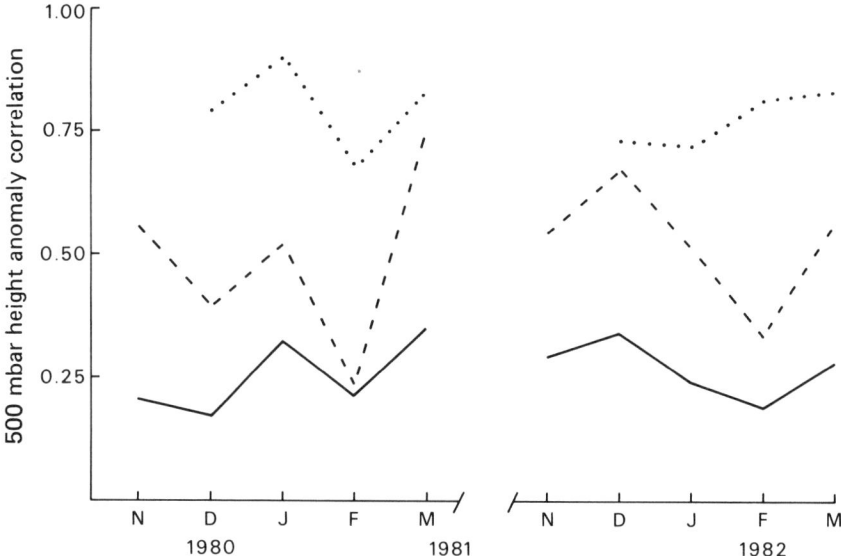

FIG. 15. Monthly-mean 500-mbar height anomaly correlations for the extratropical Northern Hemisphere. (——) Mean of anomaly correlations of day 10 forecasts; (---) anomaly correlation of mean of day 10 forecasts; (· · ·) anomaly correlation of mean of day 10 persistence forecasts.

Inspection of individual cases indicates both anomalies which are well represented throughout the forecast period and anomalies which virtually disappear over 10 days of model integration. As an example, Fig. 16 presents analyzed and day 10 forecast anomalies for January 1981. It is clear that the anomaly resembling the Pacific–North American teleconnection pattern discussed by Wallace and Gutzler (1981) has been well maintained by the forecasts, whereas the anomalously high pressure to the west of Europe has disappeared by day 10. In this case there may be an element of coincidence as the anomalously low pressure which occurred in reality over the Pacific matches a systematic tendency of the ECMWF model to produce lower than normal pressure in this region. However, the simulated pattern over North America does not correspond particularly with a systematic error pattern. Bengtsson and Simmons (1983) have noted the successful representation of low-level temperature anomalies for this case, and the sensitivity of the mean error implied by Fig. 16 to the representation of orography in the forecast model is discussed by S. Tibaldi in this volume.

Another interesting example, for November 1981, is illustrated in Fig. 17. The predominant anomaly during this month has the appearance of a wave train over Europe and Asia, and the amplitude of this wave train is substan-

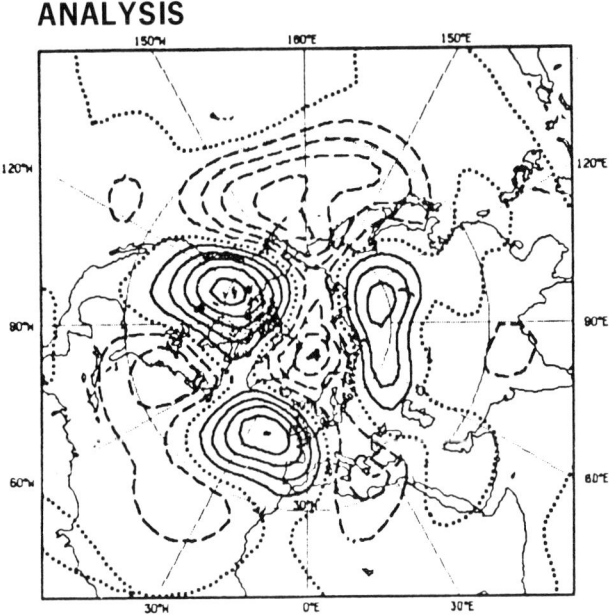

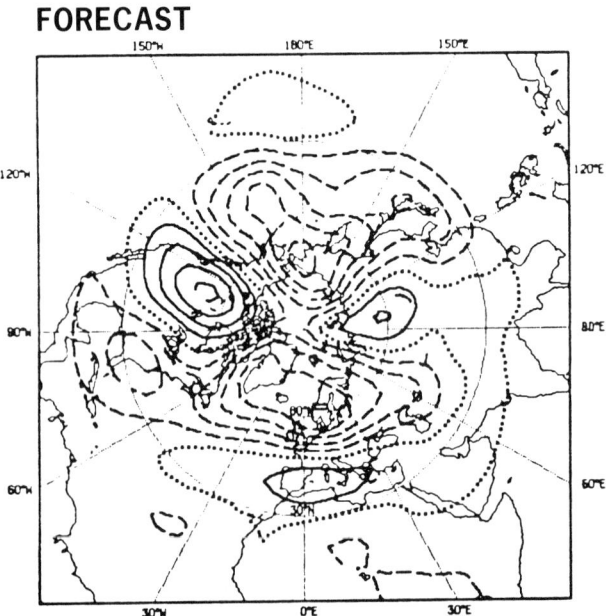

FIG. 16. The mean deviation of the 500-mbar height from climatology for January 1981 (upper), and the corresponding anomaly computed from all day 10 forecasts verifying within this month (lower). The zero contour is dotted, and negative contours dashed. The contour interval is 4 dam.

ANALYSIS

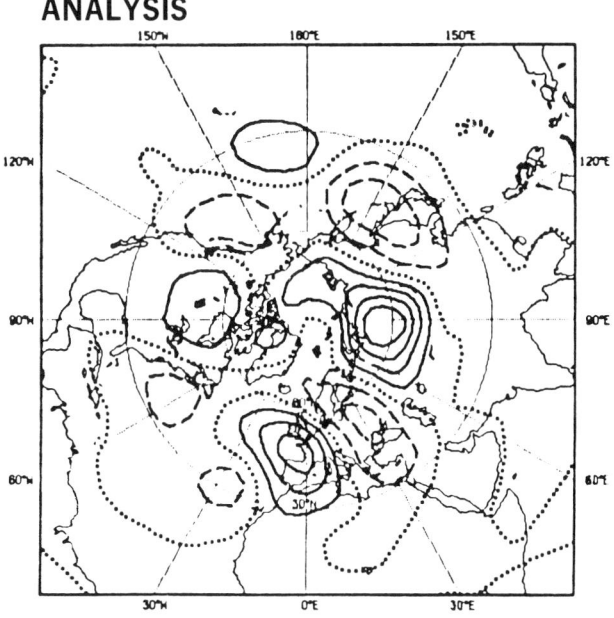

FORECAST

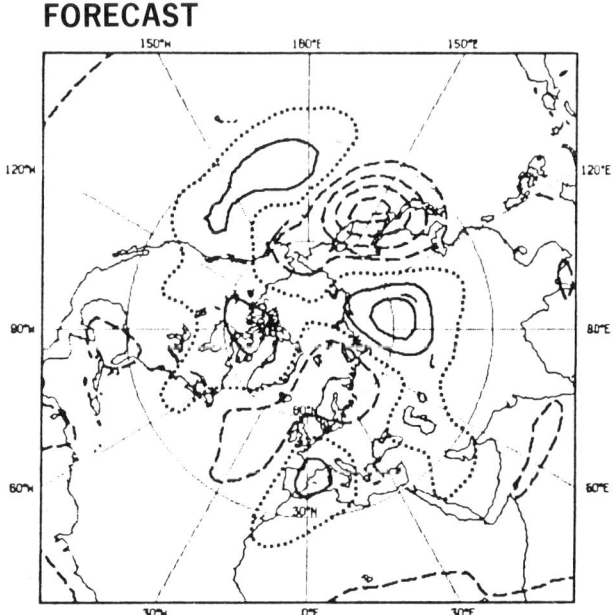

FIG. 17. As in Fig. 16, but for November 1981.

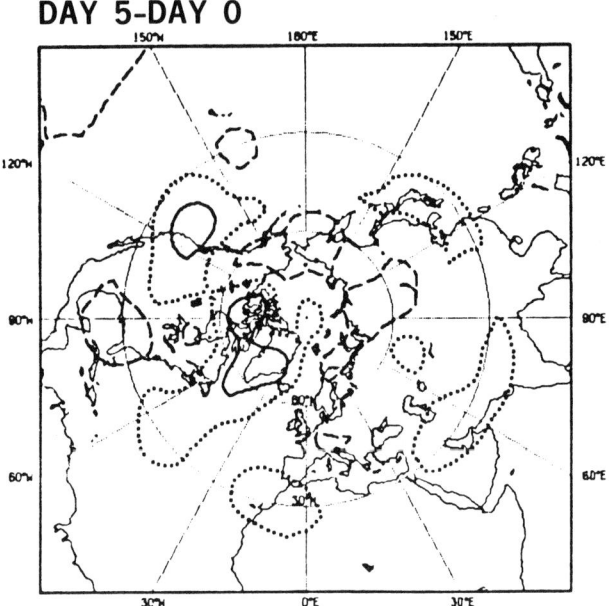

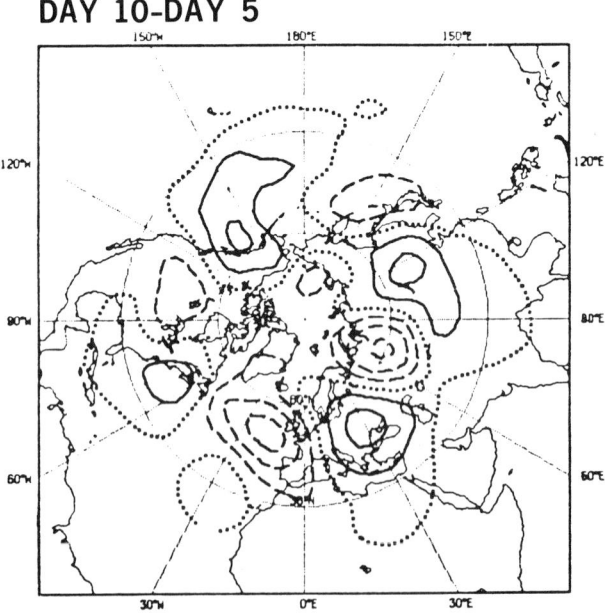

FIG. 18. The mean 500-mbar height error of day 5 forecasts verifying in November 1981 and the mean difference between day 5 and day 10 forecasts verifying within this month. Contouring is as in Figs. 16 and 17.

tially weakened over all but its easternward limit in the mean of the day 10 forecasts. A wave train of smaller amplitude extending over North America from the Pacific to the Atlantic can also be seen to decay over the forecast period. Figure 18 shows that almost all of the weakening of the monthly-mean anomaly pattern occurs in the second half of the forecast period. Further diagnosis would be required to determine the reason for this result, but it could, for example, be that the model misrepresents a localized remote forcing or propagation which does not influence middle latitudes until day 5 or later. In any case, it would appear from these examples that the archived results from operational medium-range prediction form a valuable data source which may be diagnosed to further the understanding and numerical modeling of short-term climatic anomalies.

8. Systematic Model Errors

Examination of monthly-mean maps such as discussed in the preceding section shows a tendency for certain errors to recur. These "systematic errors" characteristically grow in amplitude throughout the forecast period, and their general similarity to errors in the model climatology revealed by integration over extended periods indicates that their growth represents a gradual drift from the climate of the atmosphere (as represented by the average of many initial states) to the climate of the model. The rate of this drift is found to vary from case to case, but the overall nature of the errors appears to be independent of the initial data. The errors can be recognized in substantial distortions of the flow pattern within individual forecasts toward the end of the 10-day range, and for longer range prediction they may act to destroy predictability by, for example, causing an erroneous response to anomalous surface forcing which cannot be corrected in a simple statistical way.

A particularly important error of the ECMWF forecast model, and indeed of a number of other prediction and climate models (as discussed, for example, by Wallace *et al.,* 1983, or Simmons and Bengtsson, 1984), occurs in the representation of the large-scale quasi-stationary wave patterns of the extra-tropical hemispheres. At the surface, centers of erroneously low pressure are characteristically found over the eastern Atlantic and Pacific Oceans, with too zonal a flow over North America and Europe. Stationary wave patterns in the middle and upper troposphere are typically weakened in both hemispheres. Over Europe this appears as a southward shift of the jetstream over the northern part of the continent, with a related erroneous track of transient disturbances, and an eastward shift of the mean trough over the eastern Mediterranean. A manifestation of the latter within an individual forecast has been discussed earlier in relation to Figs. 9 and 10. Overall, the error pattern has an "equivalent barotropic" structure suggestive of inadequate orographic or remote thermal forcing of the stationary waves, although a contribution from systematic deficiencies in the behavior of mature baroclinic eddies cannot be ruled out. The interaction between transient eddies and the time-averaged nonzonal flow has been the subject of several recent diagnostic studies (e.g., Hoskins *et al.*, 1983; Illari and Marshall, 1983; Lau and Holopainen, 1984), and this interaction is almost certainly biased in the ECMWF model by a tendency to exaggerate the meridional phase tilts of baroclinic waves (Arpe, 1983). Again, the example shown in Figs. 9 and 10 illustrates this deficiency in the forecast trough over the eastern Atlantic, which appears to contribute to a spurious eastward extension of the zonal Atlantic jet.

One specific area of investigation has involved examination of the prescription of the model orography. Diagnostic and barotropic model studies reported by Wallace *et al.* (1983) have suggested that use of a grid-square mean orography significantly underestimates the orographic forcing of the synoptic-scale and larger scale flow in the ECMWF model. Prediction experiments, some of which are described by Wallace *et al.*, have been carried out using a series of "envelope" orographies formed by adding to the mean orography multiples of the standard deviation of the actual orography over the grid square, this being computed from a very high-resolution data set. Some significant improvements in the accuracy of medium-range winter forecasts have been found, and the growth of some systematic forecast errors has been substantially reduced. Further discussion is given by S. Tibaldi in this volume.

Dependence of these errors on other aspects of the model formulation has also been identified. In extended integrations they are found, in part at least, to be less at lower horizontal resolution (Cubasch, 1981), as reported earlier for another model by Manabe *et al.* (1979). However, M. Kanamitsu (personal communication) has found larger mean errors for T21 than for T42 or T63 spectral resolutions in a sample of 10 winter forecasts averaged over the 8- to 10-day forecast range. Errors have also been found to be sensitive to the parameterization of boundary-layer turbulence and shallow convection, and some sensitivity to the representation of the stratosphere has been found in longer integrations. The latter is in agreement with experience elsewhere (e.g., Mechoso *et al.*, 1982; Boville, 1984), but it should be noted that the impact on the 10-day time range has not been found to be large at the vertical resolution used operationally (Simmons and Strüfing, 1983).[1]

---

[1] *Note added in proof:* Subsequent experiments have shown a larger impact of increased stratospheric resolution. This results principally from a better assimilation of stratospheric data.

9. Concluding Remarks

A summary has been given of the predictive skill attained by the ECMWF forecasting system. We have largely utilized an extensive body of results from routine objective verification, although conclusions are generally borne out by subjective synoptic assessment. Some aspects of the prediction of blocking have been discussed, and a superficial investigation of the treatment of anomalies on the monthly-mean time scale by the forecast model has been reported. Indications of significant developments in forecast accuracy have been presented.

Forecasts for the extratropical Northern Hemisphere are generally of good quality for 3 to 4 days ahead, with useful indications of weather type given on average for a further 2 or 3 days. Substantial variations in accuracy occur on a time scale of weeks and months, and determination of the reason for this is of importance, as it may lead either to an ability to give guidance as to the expected accuracy of a particular forecast, or to ways of improving less accurate forecasts. Predictability is typically poorer by about $1\frac{1}{2}$ days in the more data-sparse Southern Hemisphere, and more substantially poorer in the tropical belt. Both data assimilation and numerical modeling for the tropics pose particular problems, but objective verification indicates that significant advances are being made.

A stage has been reached in the development of the ECMWF forecasting system at which a number of major changes are anticipated. An enhancement of computing power has made possible a forthcoming increase in model resolution, and several changes are expected in the parameterization, with the introduction of a representation of shallow convection and new treatments of long-wave radiation and the specification of clouds. Changes in the parameterization of deep convection may also be introduced. In addition, the data analysis is being recoded to allow more flexibility, particularly with respect to the specification of structure functions, the use of data, and the various interpolations carried out. We are unable to give a reliable estimate of the net improvement in forecast accuracy that will result from these changes and await the monitoring of future operational performance with interest.

The treatment of blocking in the forecasts appears to pose no outstanding problem, and cases of high predictability involving blocking have been noted. Such problems as do occur appear to be related to general deficiencies of the forecast model. Some success in the maintenance of pronounced large-scale anomalies on the monthly time-scale has been discussed, but examples of the failure to represent such features have also been presented. Systematic errors can seriously influence predictions later in the forecast period, and become more pronounced in integrations carried out beyond the

10-day range. Reduction of these errors is of importance for the improvement of medium-range forecasting, and must be a crucial element in the development of a capability for the longer range prediction of anomalies.

ACKNOWLEDGMENTS

Many staff members and visiting scientists of ECMWF have contributed to the work reported here in some way or another, and gratitude is expressed to them.

REFERENCES

Arpe, K. (1980). Confidence limits for verification and energetic studies. ECMWF Tech. Rep. No. 18.
Arpe, K. (1983). Diagnostic evaluation of analysis and forecasts: Climate of the model. *Proc. ECMWF Semin. Interpret. Num. Weather Predict. Products* 99-140.
Bengtsson, L. (1981). Numerical prediction of atmospheric blocking—A case study. *Tellus* 33, 19-42.
Bengtsson, L., and Simmons, A. J. (1983). Medium range weather prediction—Operational experience at ECMWF. *In* "Large-Scale Dynamical Processes in the Atmosphere" (B. J. Hoskins and R. P. Pearce, eds.), pp. 337-363. Academic Press, New York.
Bengtsson, L., Kanamitsu, M., Kållberg, P., and Uppala, S. (1982). FGGE research activities at ECMWF. *Bull. Am. Meteorol. Soc.* 63, 277-303.
Boville, B. A. (1984). The influence of the polar night jet on the tropospheric circulation in a GCM. *J. Atmos. Sci.* 41, 1132-1142.
Burridge, D. M. (1979). Some aspects of large scale numerical modelling of the atmosphere. *Proc. ECMWF Semin. Dyn. Meteorol. Num. Weather Predict.* 2, 1-78.
Burridge, D. M., and Haseler, J. (1977). A model for medium range weather forecasts—Adiabatic formulation. ECMWF Tech. Rep. No. 4.
Charney, J. G., Fleagle, R. G., Lally, V. E., Riehl, H., and Wark, D. Q. (1966). The feasibility of a global observation and analysis experiment. *Bull. Am. Meteorol. Soc.* 47, 200-220.
Cubasch, U. (1981). The performance of the ECMWF model in 50-day integrations. ECMWF Tech. Memo No. 32.
Geleyn, J.-F., and Hollingsworth, A. (1979). An economical analytical method for the computation of the interaction between scattering and line absorption of radiation. *Beitr. Phys. Atmos.* 52, 1-16.
Girard, C., and Jarraud, M. (1982). Short and medium range forecast differences between a spectral and grid point model. An extensive quasi-operational comparison. ECMWF Tech. Rep. No. 32.
Grønaas, S. (1983). Systematic errors and forecast quality of ECMWF forecasts in different large-scale flow patterns. *Proc. ECMWF Semin. Interpret. Num. Weather Predict. Products* 161-206.
Hollingsworth, A., Arpe, K., Tiedtke, M., Capaldo, M., and Savijärvi, H. (1980). The performance of a medium-range forecast model in winter—Impact of physical parameterizations. *Mon. Weather Rev.* 108, 1736-1773.

Hoskins, B. J., James, I., and White, G. H. (1983). The shape, propagation and mean-flow interaction of large-scale weather systems. *J. Atmos. Sci.* **40,** 1595–612.

Illari, L., and Marshall, J. C. (1983). On the interpretation of eddy fluxes during a blocking episode. *J. Atmos. Sci.* **40,** 2232–2242.

Ji, L. R., and Tibaldi, S. (1983). Numerical simulations of a case of blocking. The effects of orography and land-sea contrast. *Mon. Weather Rev.* **111,** 2068–2086.

Kuo, H. L. (1974). Further studies of the influence of cumulus convection on large-scale flow. *J. Atmos. Sci.* **31,** 1232–1240.

Lange, A., and Hellsten, E. (1984). Results of the WMO/CAS NWP Data Study and Intercomparison Project for forecasts for the Northern Hemisphere in 1983. WMO Geneva.

Lau, N.-C., and Holopainen, E. O. (1984). Transient eddy forcing of the time-mean flow as identified by geopotential tendencies. *J. Atmos. Sci.* **41,** 313–328.

Lorenc, A. C. (1981). A global three-dimensional multivariate statistical interpolation scheme. *Mon. Weather Rev.* **109,** 701–721.

Lorenz, E. N. (1975). Climate Predictability, GARP Publication Series No. 16, pp. 132–136. WMO, Geneva.

Lorenz, E. N. (1982). Atmospheric predictability experiments with a large numerical model. *Tellus* **34,** 505–513.

Louis, J.-F. (1979). A parametric model of vertical eddy fluxes in the atmosphere. *Boundary-Layer Meteorol.* **17,** 187–202.

Manabe, S., Hahn, D. G., and Holloway, J. L. (1979). Climate simulation with GFDL spectral models of the atmosphere: Effect of spectral truncation. GARP Publication Series No. 22, 41–94. WMO, Geneva.

Mechoso, C. R., Suarez, M. J., Yamazaki, K., Spahr, J. A., and Arakawa, A. (1982). A study of the sensitivity of numerical forecasts to an upper boundary in the lower stratosphere. *Mon. Weather Rev.* **110,** 1984–93.

Miyakoda, K., Hembree, G. D., Stricker, R. F., and Shulman, I. (1972). Cumulative results of extended forecast experiments. I: Model performance for winter cases. *Mon. Weather Rev.* **100,** 836–55.

Miyakoda, K., Gordon, T., Caverly, R., Stern, W., Sirutis, J., and Bourke, W. (1983). Simulation of a blocking event in January 1977. *Mon. Weather Rev.* **111,** 846–869.

Nieminen, R. (1983). Operational verification of ECMWF forecast fields and results for 1980–1981. ECMWF Tech. Rep. No. 36.

Shukla, J. (1981). Dynamical predictability of monthly means. *J. Atmos. Sci.* **38,** 2547–2572.

Simmons, A. J., and Bengtsson, L. (1984). Atmospheric general circulation models. Their design and use for climate studies. *In* "The Global Climate" (J. T. Houghton, ed.), pp. 37–62. Cambridge University Press, London.

Simmons, A. J., and Burridge, D. M. (1981). An energy and angular momentum conserving vertical finite difference scheme and hybrid vertical coordinates. *Mon. Weather Rev.* **109,** 758–766.

Simmons, A. J., and Hoskins, B. J. (1978). The life cycles of some nonlinear baroclinic waves. *J. Atmos. Sci.* **35,** 414–432.

Simmons, A. J., and Jarraud, M. (1984). The design and performance of the new ECMWF operational model. *Proc. ECMWF Semin. Num. Meth. Weather Predict.,* 113–164.

Simmons, A. J., and Strüfing, R. (1983). Numerical forecasts of stratospheric warming events using a model with a hybrid vertical coordinate *Q.J.R. Meteorol. Soc.* **109,** 81–111.

Smagorinsky, J. (1969). Problems and promises of deterministic extended range forecasting. *Bull. Am. Meteorol. Soc.* **50,** 286-311.

Temperton, C., and Williamson, D. L. (1981). Normal mode initialisation for a multi-level grid-point model, Part I: Linear aspects. *Mon. Weather Rev.* **109,** 729–743.

Tibaldi, S., and Buzzi, A. (1983). Effects of orography on Mediterranean lee cyclogenesis and its relationship to European blocking. *Tellus* **35A,** 269-286.

Tibaldi, S., and Ji, L. R. (1983). On the effect of model resolution on numerical simulation of blocking. *Tellus* **35A,** 28-38.

Tiedtke, M., Geleyn, J.-F., Hollingsworth, A., and Louis, J.-F. (1979). ECMWF model, parameterization of subgrid scale processes. ECMWF Tech. Rep. No. 10.

Wallace, J. M., and Gutzler, D. S. (1981). Teleconnections in the geopotential height field during the northern hemisphere winter. *Mon. Weather Rev.* **109,** 784-812.

Wallace, J. M., Tibaldi, S., and Simmons, A. J. (1983). Reduction of systematic forecast errors in the ECMWF model through the introduction of an envelope orography. *Q.J.R. Meteorol. Soc.* **109,** 683-717.

Williamson, D. L., and Temperton, C. (1981). Normal mode initialisation for a multi-level grid-point model, Part II: Nonlinear aspects. *Mon. Weather Rev.* **109,** 744-757.

ENVELOPE OROGRAPHY AND MAINTENANCE OF THE QUASI-STATIONARY CIRCULATION IN THE ECMWF GLOBAL MODELS

Stefano Tibaldi

European Centre for Medium Range Weather Forecasts
Shinfield Park, Reading, Berkshire RG2 9AX
England

1. Introduction

It is a well-known fact that general circulation models (GCMs) are still far from simulating satisfactorily both the mean values and the variability of observed atmospheric parameters. These departures of model-produced meteorological fields from their observed counterparts have been referred to "errors in the model's climatology." Such models are often used to forecast actual weather and this is done by integrating them forward in time from observed initial conditions. It is not unreasonable to think of those initial conditions (e.g., the day 0 forecast) as one realization of the climate of the real earth's atmosphere. During the model integration, a progressive drift therefore takes place from that single realization of the earth's atmosphere's climatology toward the model's own climatic state. The ensemble means of forecast error, for increasing validity time (day 1, day 2, and so on) document, therefore, such a drift; these ensemble means of forecast error are often referred to as the model's systematic error (SE).

There is an ever increasing wealth of literature documenting the characteristics of such errors for different observational global (or sometimes hemispheric) forecasting models, e.g., Leary (1971), Saunders and Gyakum (1980), Hollingsworth *et al.* (1980), Derome (1981), Bengtsson and Lange (1981), Wallace and Woessner (1981), Silberberg and Bosart (1982), Tiedtke (1983), Wallace *et al.* (1983), Bettge (1983), Arpe (1983), Tibaldi (1984), Chouinard (1984), Sumi and Kanamitsu (1984), Arpe and Klinker (1984), Heckley (1984), Lange and Hellsten (1983, 1984), and Arpe *et al.* (1985).

In some of these cited works, notably Hollingsworth *et al.* (1980), Derome (1981), and Wallace *et al.* (1983) (hereafter referred to as WTS, after Wallace, Tibaldi, and Simmons), the model's climate drift, or the onset of the SE, was mostly described as the progressive surfacing of the model's inability to maintain the observed climatological amplitude of the ultralong planetary-scale stationary waves. In WTS, in particular, the hypothesis was put forward that this was to be interpreted as the model's own response to an erroneous forcing and that an inadequate representation of small-scale (possibly subgrid) orographic features would be responsible for such an erroneous boundary forcing. Numerical experiments with an enhanced ("envelope") orography in the ECMWF global gridpoint model seemed also to lend support to this idea. In Section 2 we will briefly review WTS's work and some of the main problems left open by it. The main purpose of this article is to present another, larger set of experiments of this type, performed with the same global gridpoint model, and, in so doing, to lend support to some of the conclusions of WTS; this will be done in Section 3. Section 4 will be devoted to the results of a further experiment designed to clarify the relative role of different scales of orographic forcing in producing the results shown previously. In Section 5 we will discuss the mechanisms by which such an enhanced orographic forcing can influence the onset of the SE in the ECMWF global models' midlatitudes and, ultimately, their operational performance. In Section 6 we will examine some effects of the "envelope" orography on the tropical circulation. A summary of the work and a list of the main conclusions are presented in Section 7.

2. Orographic Forcing and the Systematic Error of the ECMWF Gridpoint Model: The Envelope Orography

Figure 1 shows Northern Hemispheric maps of ensemble mean (systematic) error for the 500-mbar geopotential height for days 1, 4, 7, and 10 superimposed on the observed 500-mbar height mean field. The sequence, therefore, describes the progressive climate drift of the ECMWF operational model; the ensemble mean was constructed using 100 daily analyses and corresponding day 1, 4, 7, and 10 forecasts verifying on those days. The period spans from 1 December 1980 to 10 March 1981 and is, therefore, characteristic of winter conditions only and was originally chosen by Wallace *et al.* (1983) because, during that period, the SE of the ECMWF operational model was very large, even compared to other winters. It should be remembered that the amplitude of the SE during winter is normally already much larger than during summer.

Although the maps shown in Fig. 1 have been already discussed in some depth by WTS, we will briefly summarize here the main characteristics of the evolution of the SE that can be recognized in them:

1. The error evolves from a "small-scale" configuration to a "large-scale" pattern.

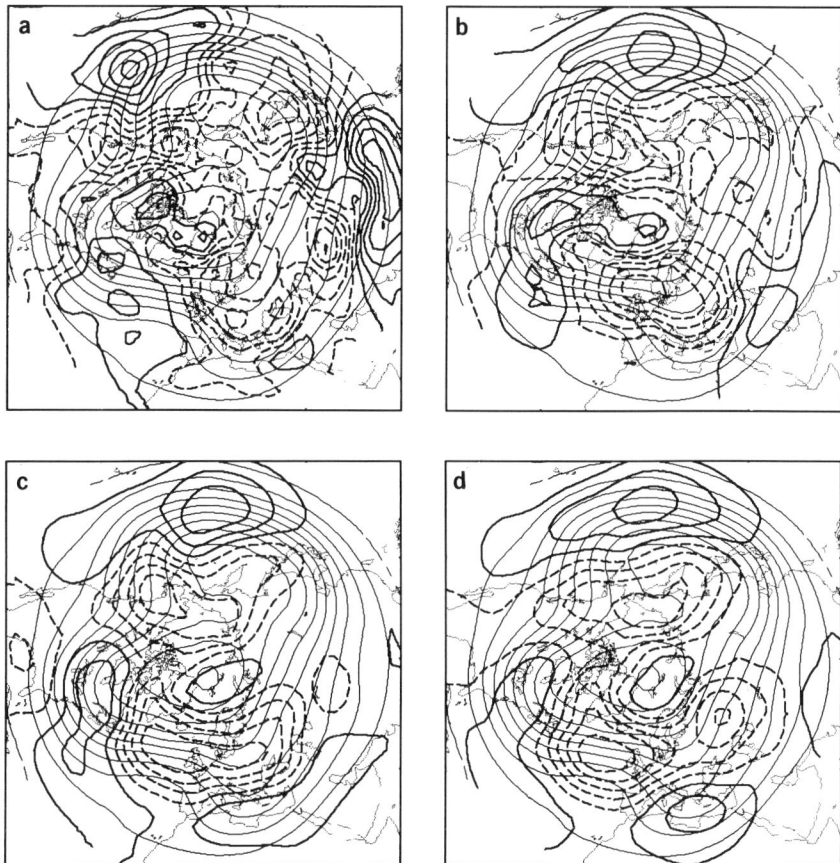

FIG. 1. Ensemble mean forecast error fields for ECMWF operational forecasts of 500-mbar height for the 100-day period, 1 December 1980 to 10 March 1981 inclusive. (a) Day 1 forecasts, contour interval 5 m; (b) day 4 forecasts, contour interval 16 m; (c) day 7 forecasts, contour interval 30 m; (d) day 10 forecasts, contour interval 30 m. The background field (lighter contours) is the mean 500-mbar height field based on ECMWF operational analyses for the same period, contour interval 80 m. Negative contours denoted by dashed lines. (From Wallace et al., 1983.)

2. The smaller scale day 1 error pattern can be related, to a degree, to the underlying orographic features, with important (but not unique) maxima in direct correspondence to the Rockies, the Himalayas, and the Euro-Carpatian mountains.

3. The larger scale day 10 error pattern is distinctly correlated with the planetary-scale wave pattern, with negative errors always occurring over ridges and positive errors over troughs; this suggested to WTS an inadequate

representation, in the model's atmosphere, of those forcings that are mostly responsible for the maintenance of the ultralong planetary waves (e.g., orography).

4. There is more area covered by negative error than there is by positive error, indicating the onset of a progressive bias of the model's atmosphere. Since the same effect is weaker at 1000 mbar and stronger at 300 mbar (see Fig. 2), this is the result of a progressive cooling of the model's troposphere.

5. Negative error centers occupy, on average, more northerly positions, while positive error centers lie on more southerly positions, indicating (in the geostrophic assumption) that an error in the mean westerlies also develops, with the model's atmosphere becoming, on average, too westerly after 10 days. This "zonalization" of the model's flow is one of the most prominent features common to a large number of GCMs used for operational forecasting.

Looking again at Fig. 2, which shows the 10-day geopotential height SE at 1000 and 300 mbar, we also note that:

6. The SEs are mostly equivalent barotropic in character, since the phase of their large-scale features changes little with height.

The work by WTS proceeded to verify that the day 10 SE could be interpreted as the model's own response to the constant erroneous forcing represented by the day 1 error growth rate; this was done using a barotropic model with an *ad hoc* forcing to maintain the observed mean circulation. To this

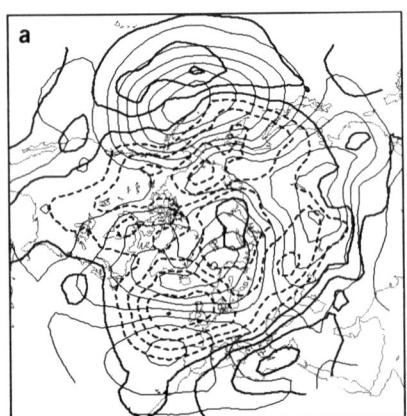
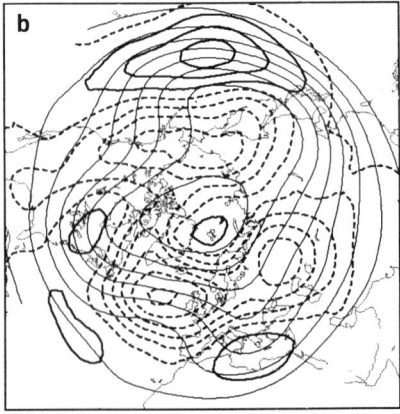

FIG. 2. Ensemble mean day 10 forecast error fields for the same 100-day period as Fig. 1; (a) 1000-mbar height, contour interval 20 m; (b) 300-mbar height, contour interval 40 m; superimposed are the correspondent mean observed fields for the same period (lighter contours) based on ECMWF operational analyses, contour intervals 30 and 160 m, respectively. Negative contours denoted by dashed lines. (From Wallace *et al.*, 1983.)

main forcing an additional forcing was superimposed, proportional to the day 1 error growth rate. The strong resemblance between the model's day 10 response to such an additional forcing and the observed day 10 error added additional confidence to the "erroneous forcing" hypothesis made by WTS.

A number of 10-day forecasts were then attempted using an enhanced ("envelope") orography[1] that showed, even within the limits of the small sample, a remarkably reduced SE, more notably around day 7. The reader was then left with a note of caution regarding the fact that the observed decrease of the SE in the later stages of the forecast ensemble seemed to occur despite much less convincing changes in the day 1 SE pattern, showing no apparent reduction or even a small increase, at least in some areas. The limits imposed on the applicability of the WTS result to other situations by the smallness of the sample of forecasts is obvious, and this was also pointed out.

The possibility of incorporating some sort of envelope orography in the ECMWF operational forecasting required experimentation based on a much larger sample of cases. It was then decided that January 1981, with the largest ever systematic error in the Northern Hemisphere, was an ideal test period. In the next section we will describe in some detail this set of experiments. Some preliminary results were reported in Tibaldi (1984).

3. The January 1981 Set of Experiments

We will now describe a larger set of experiments of the same type as described in WTS, that is, "mean" against "envelope" orography 10-day forecasts. In order to maintain the experiment as unbiased as possible and eliminate all additional causes of different behavior, the global observed data for the whole of January 1981 were reassimilated twice with the ECMWF operational data assimilation system, once using a mean orography in the assimilating model and once using a $2\mu_h$ envelope orography. Then 2×31 10-day forecasts were run from the 1200 GMT analysis of every day with the two different orographies, and these two ensembles of forecasts were compared. It should be pointed out that, in the envelope ensemble, a correction was made to the soil temperature, soil moisture, and snow cover fields to allow for the vertical displacement of the ground surface, to minimize the spurious thermal effects coming from an enhanced orography and to focus on the mechanical effects alone as much as possible.

---

[1] The gridpoint "envelope" orography (h_E) (Wallace *et al.*, 1983) was constructed as the mean terrain height defined on the gridbox (h_M) plus a given proportion of the standard deviation of such terrain height μ_h as computed from a dataset of much higher spatial resolution than the grid of the model, namely the U.S. Navy global orography dataset. In the WTS experiments the proportion of standard deviation added to the mean was 2, e.g., $h_E = h_M + 2\mu_h$.

Figures 3 and 4 show the impact of the orography on the day 4, day 7, and day 10 SE of geopotential height at 500 and 1000 mbar, respectively. It is evident that the impact is beneficial, since the onset of the SE is slowed down and its final (day 10) value is reduced both at the 500- and 1000-mbar levels. Also noteworthy features of the envelope SE pattern are the now fairly equidistributed positive and negative error areas (reduced overall tropospheric cooling) and a greatly reduced tendency to concentrate the negative error centers to the north and the positive to the south (reduced zonalization of the model's troposphere). Both these qualitative impressions are quantitatively confirmed by observing the latitude–height cross-sections of ensemble mean zonally averaged temperature errors shown in Fig. 5. Both the midlatitude and the tropical troposphere experience a greatly reduced cooling, and the large excessive warming of the day 7 tropical stratosphere is also reduced by up to 70%. The tropospheric zonal mean of zonal wind SEs are also greatly reduced (Fig. 6), with the excess of westerlies in the Northern Hemisphere subtropical midlatitudes ($35°-55°N$) being halved for the whole validity period of the medium-range forecast. The negative error maximum in $\bar{u}$ located in the tropical troposphere and lower stratosphere is also reduced throughout the same period (Fig. 6). The upper tropospheric wind error is, on the contrary, somewhat enhanced.

These general ameliorations in the mean mass and wind fields are well reflected in the objective forecast skill scores. Figure 7 shows the geopotential height anomaly correlation coefficient averaged over the Northern Hemisphere troposphere and its breakdown over different spectral bands. These diagrams show that the usefulness of the model's forecasts is increased by approximately 8 hr, if measured by the time it takes the correlation coefficient to reach the 60% value. The improvement is most noticeable in the zonal part of the flow and in the ultralong waves, wavenumber (WN) band 1–3, while long waves (WN 4–9) suffer an equivalent setback (but they explain much less variance of the total signal and therefore weigh less in the total correlation coefficient). Synoptic-scale waves (WN 10–20) show very little sensitivity in such a skill score.

Figure 8a shows box diagrams for the energetics of the entire month of January 1981 for the analyzed data (top), and average day 1 to day 10 forecast data with the mean orography (middle) and with the envelope orography (bottom). Despite an increase from an excess of 1% to an excess of 6% of total available potential energy (A), the use of the envelope orography reduces the error on the baroclinic conversion from around 167% to a more tolerable 83%. The barotropic zonal-to-eddy conversion term (CK) is also sensitive to the change on orography, going from a 28% overestimation (in absolute value) to a 12% underestimation. This last change comes mostly from the barotropic nonlinear interactions between the zonal flow and the

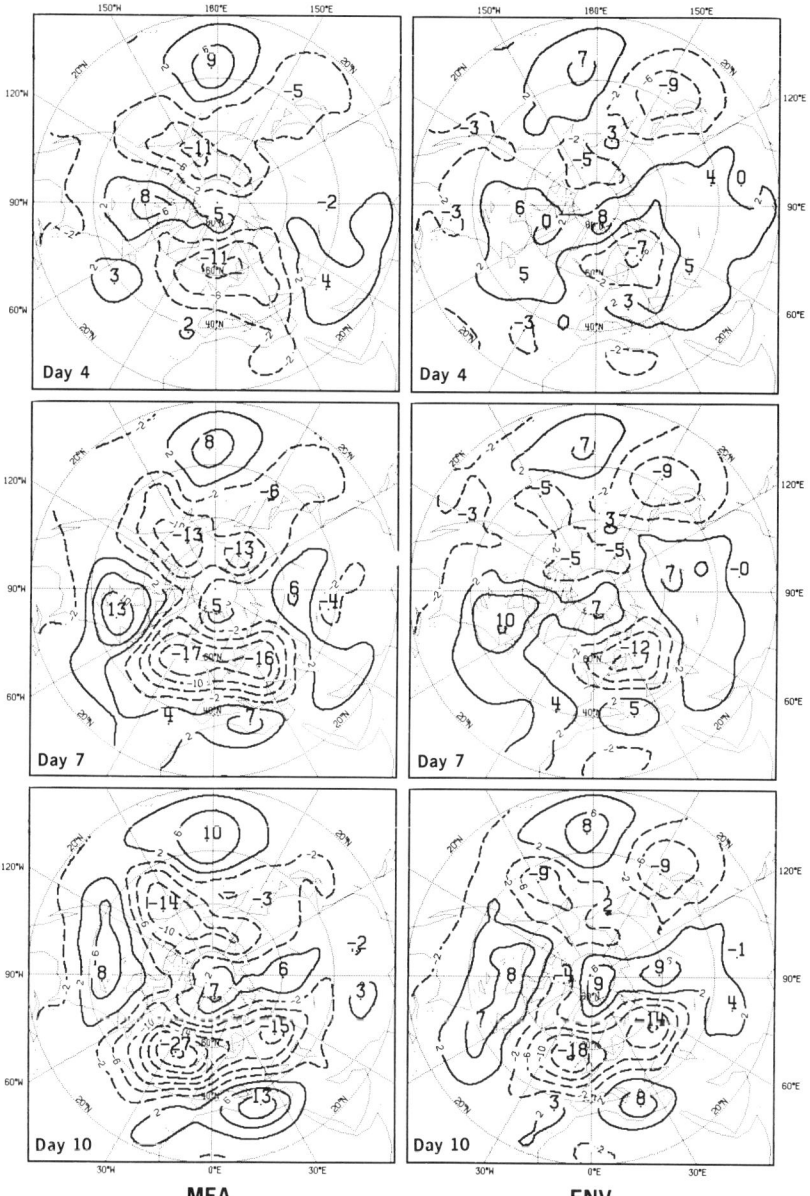

FIG. 3. Ensemble 500-mbar geopotential height forecast error fields for the 31 cases of January 1981 with the operational mean-type orography (MEA, left) and the envelope orography (ENV, right). From top to bottom, day 4, day 7, and day 10. Dashed contours represent negative systematic error (SE), full contours positive SE: isolines drawn every 4 dam.

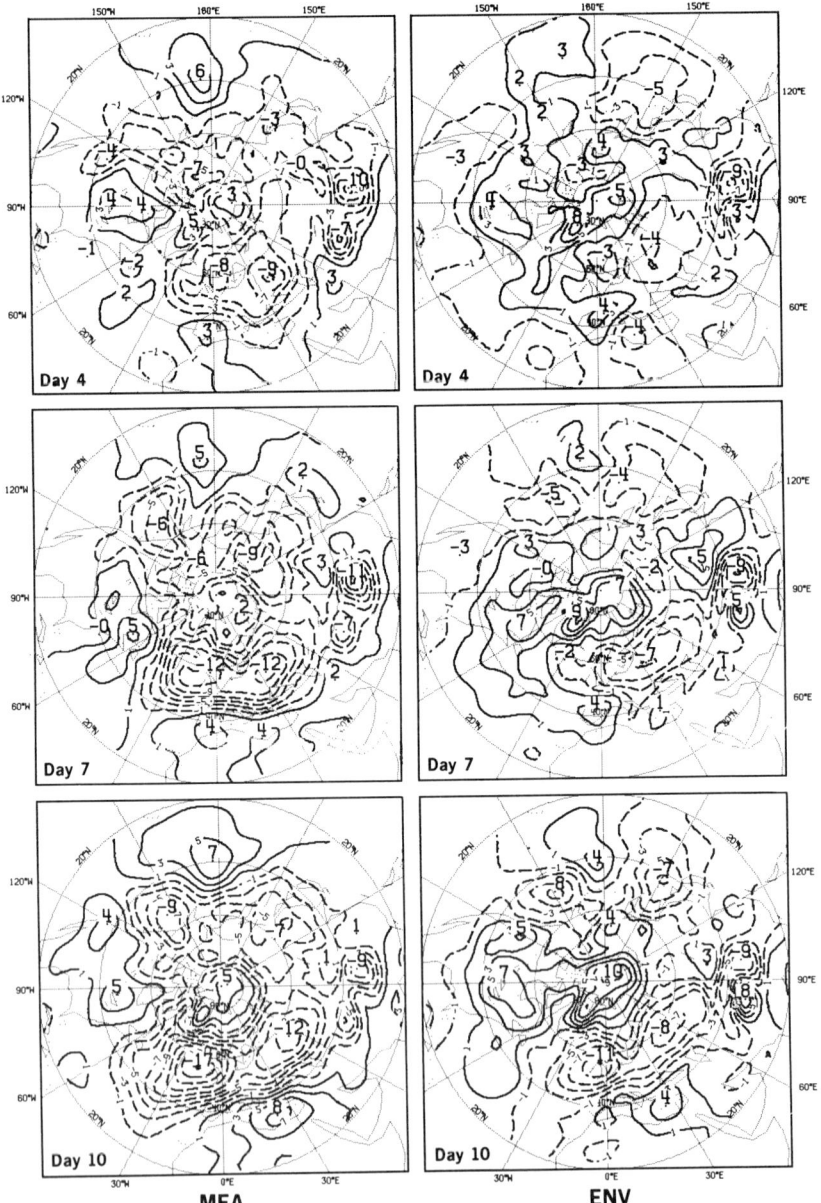

FIG. 4. Ensemble 1000-mbar geopotential height forecast error fields for the 31 cases of January 1981 with the operational mean-type orography (MEA, left) and the envelope orography (ENV, right). From top to bottom, day 4, day 7, and day 10. Dashed contours represent negative SE, full contours positive SE: isolines drawn every 2 dam.

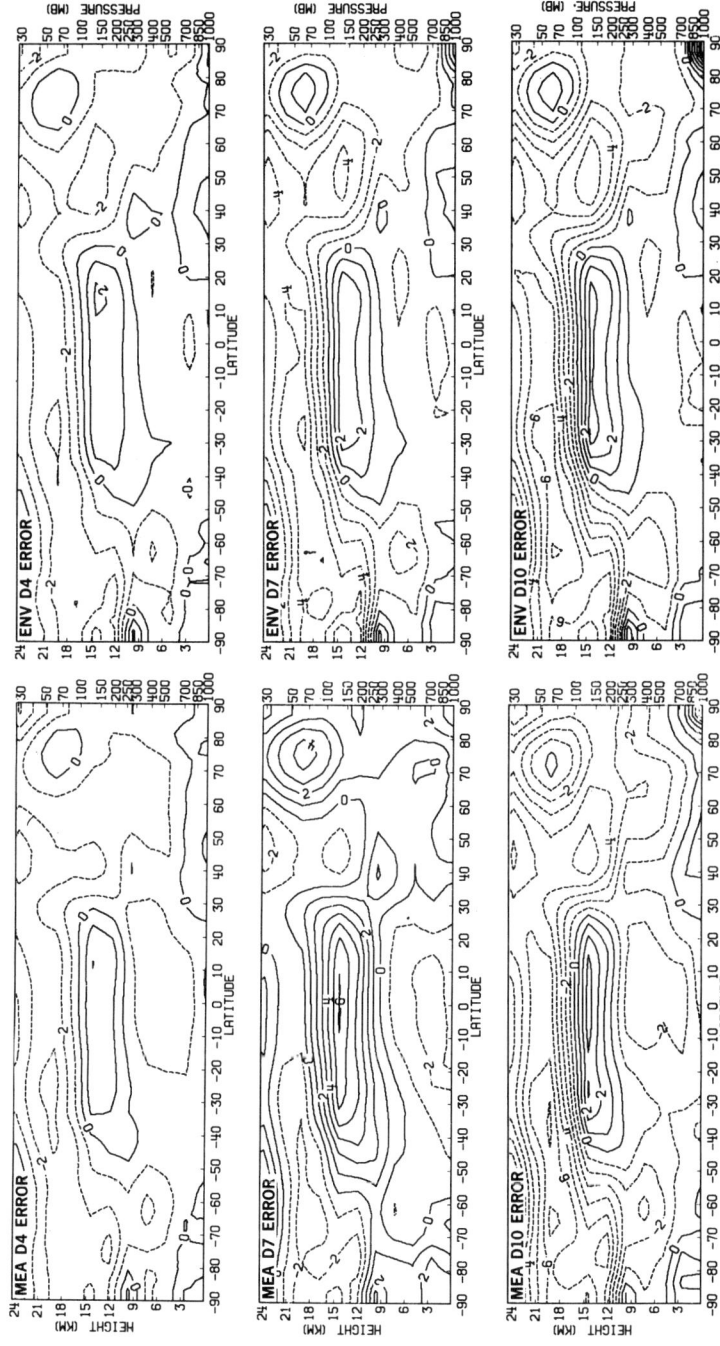

FIG. 5. Height–latitude cross sections of zonal-mean temperature errors, averaged over the ensemble of 31 forecasts of January 1981. Left panels: mean temperature error for the mean orography ensemble (°K) for day 4, day 7, and day 10 forecasts (top to bottom), respectively; right panels: the same for the envelope orography ensemble.

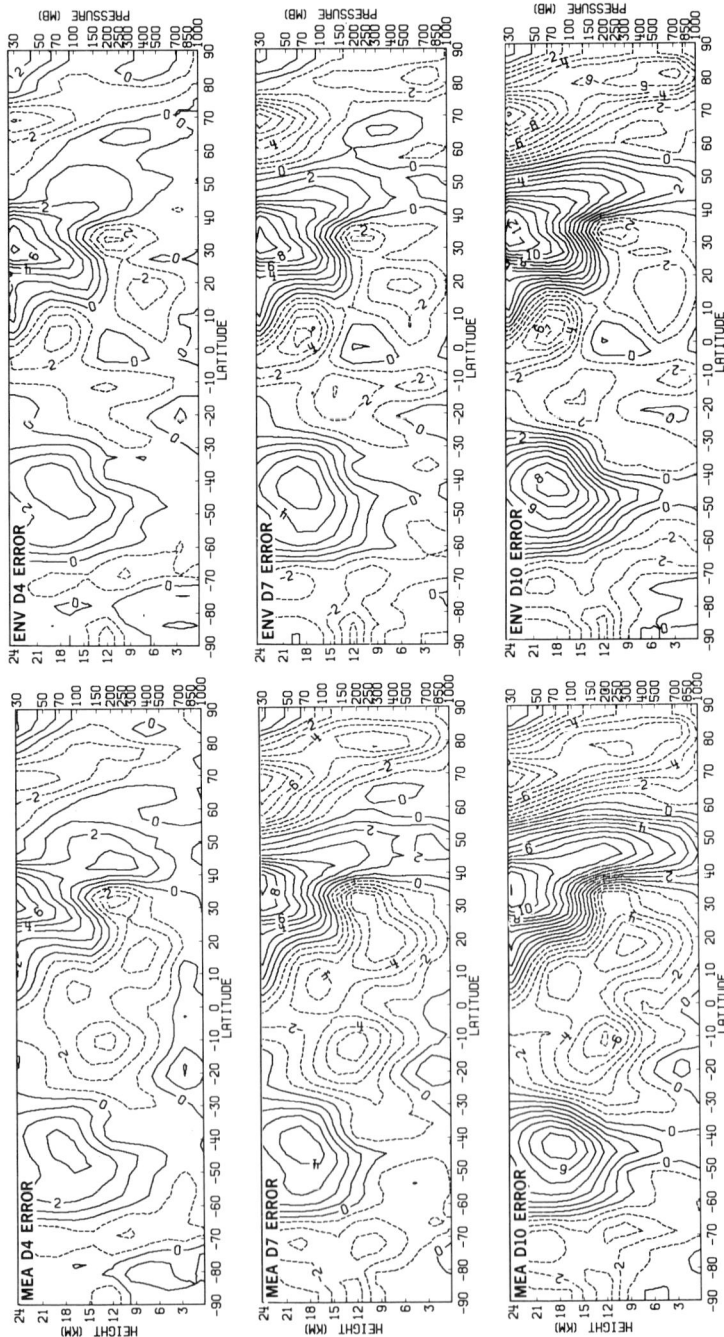

FIG. 6. Height–latitude cross sections of zonal mean of zonal wind errors, averaged over the ensemble of 31 forecasts of January 1981. Left panels: mean zonal wind error for the mean orography ensemble (meters per second) for the day 4, day 7, and day 10 forecasts (top to bottom), respectively; right panels: the same for the envelope orography ensemble.

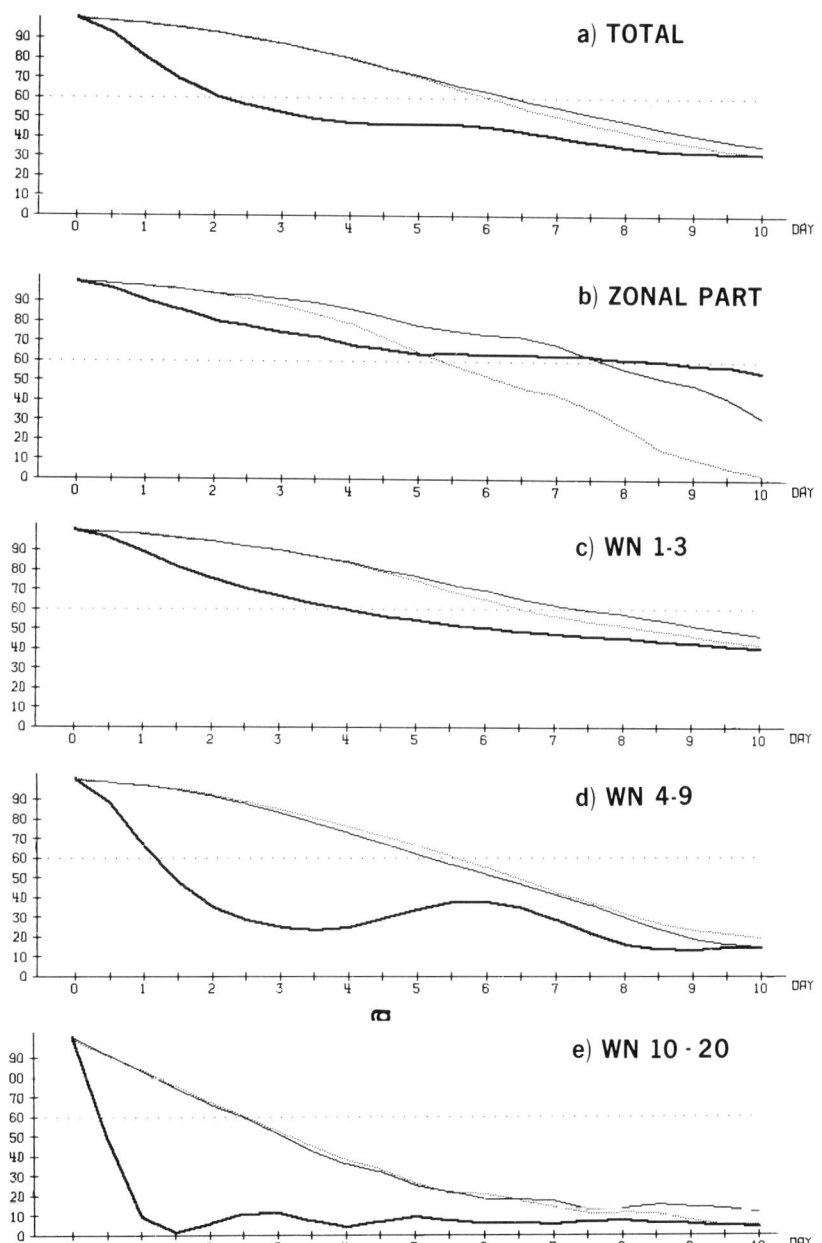

FIG. 7. (a) Vertically averaged (200–1000 mbar) correlation coefficient between geopotential forecast height anomalies and the corresponding verification as a function of forecast time averaged over the ensemble of 31 forecasts of January 1981 and computed for the area extending from 20° to 82.5°N. Heavier curves indicate persistence, thin solid curves are the envelope forecasts, and dashed curves are the corresponding mean orography forecasts. (b–e) A spectral breakdown in selected wavenumber (WN) bands of the total anomaly correlation shown in (a).

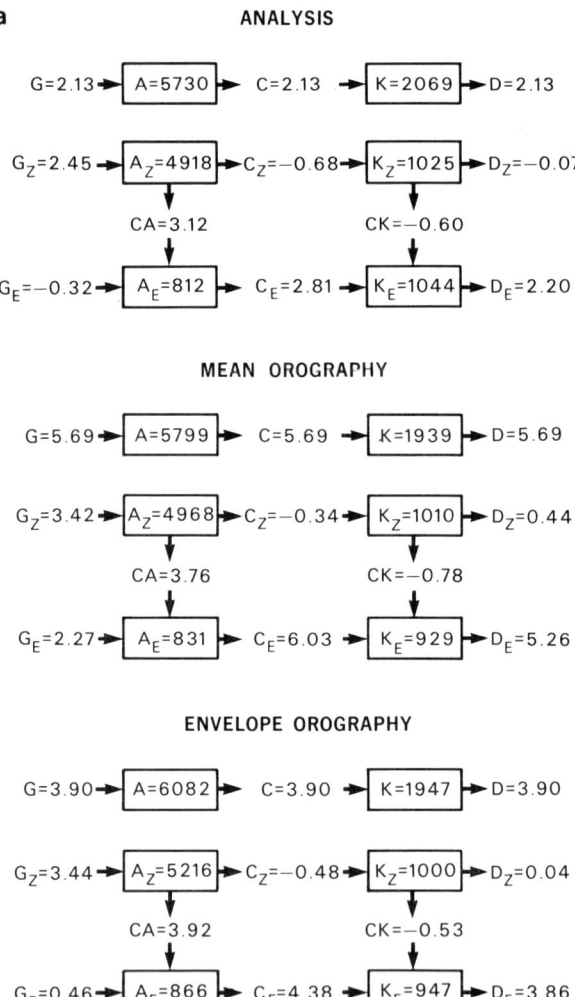

FIG. 8. Box diagrams (Lorenz, 1955) of energy conversion terms for the two ensembles of 31 10-day forecasts and for the correspondent analyses [panel (a)]. Subscript Z indicates zonal quantities, while E indicates eddy quantities. No subscript means totals. All other symbols and conventions are as usual. Panel (b) shows the spectral breakdown of the nonlinear barotropic kinetic energy conversion term CK between the eddies and the mean zonal flow. All energies are expressed in kilojoules per square meter and all conversions in watts per square meter. All terms are integrated between 1000 and 50 mbar and over the entire Northern Hemisphere.

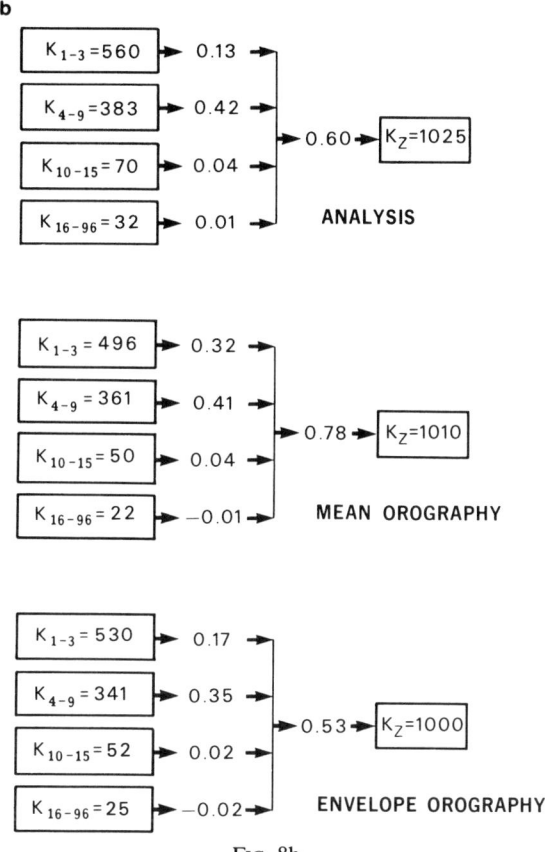

FIG. 8b.

ultralong eddies, grossly exaggerated by the mean orography forecast ensemble (0.32 against an observed value of 0.13) and much better represented by the envelope orography ensemble (0.17) (see Fig. 8b). This improvement is, unfortunately, achieved at the expense of a reequilibration of all other barotropic conversions, with the wavenumber band WN 4–9 conversion being, now, too low (0.35 against an observed 0.42). The overall spectral distribution of CK seems, however, more plausible in the envelope than in the mean ensemble, and the same can be said of the residual generation and dissipation terms (G and D) as shown in Fig. 8a. Figure 9 shows that most of this spurious transfer of kinetic energy from the eddies to the mean zonal flow is, in the model, taking place at the jet-maximum level (~200 mbar) with a positive ($C_Z \rightarrow C_E$) maximum around 40°N and a negative ($C_E \rightarrow C_Z$) maximum around 25°N. The latitudinal integral of CK is shown, as a function only of

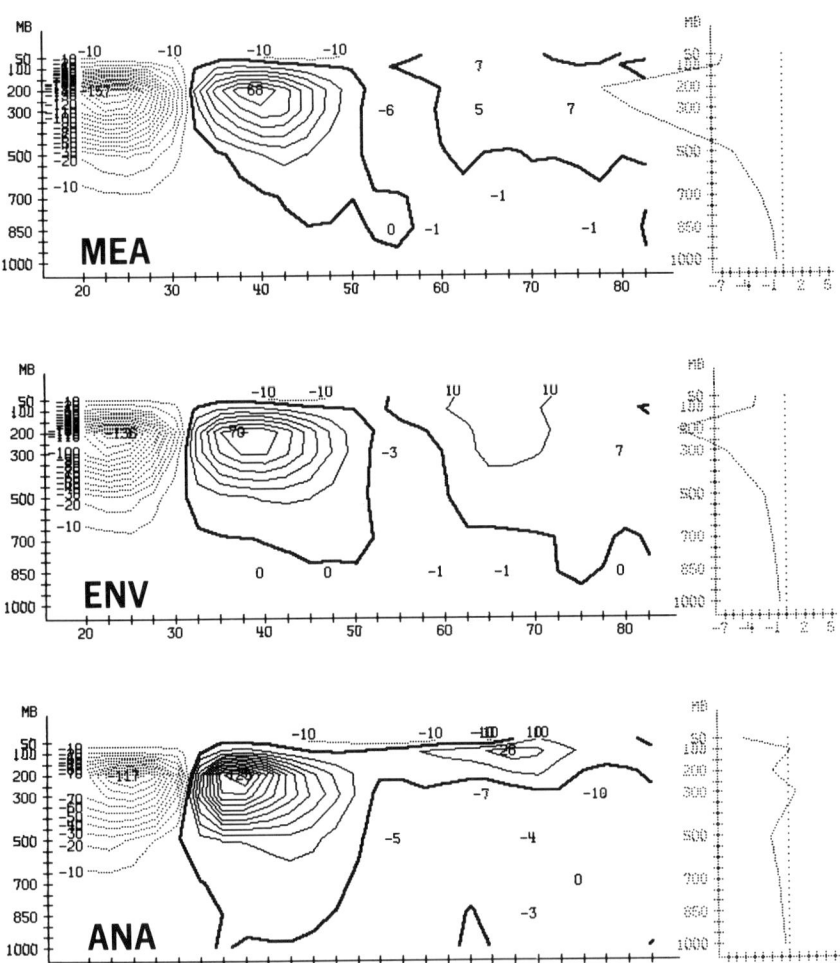

FIG. 9. Latitude–height cross-sections of barotropic conversion of kinetic energy from the zonal flow to the eddies (CK) averaged over the 31 cases and over the second 5 days of the forecast period. All wavenumbers from 1 to 20 are included. On the right the integrals over latitude of the same quantity are shown as a function of height only. Units are 10^{-1} W m^{-2} bar^{-1} throughout. MEA, Mean orography; ENV, envelope orography; ANA, analyzed value.

the vertical coordinate p in the three smaller diagrams on the right of Fig. 9. It is again evident that the introduction of the enhanced orography largely decreases the transfer of kinetic energy to the mean zonal flow previously taking place in the upper troposphere and lower stratosphere.

Concerning the effect on the planetary-scale waves, Fig. 10 shows the eddy kinetic energy at several levels in the model's troposphere from 300 to 1000

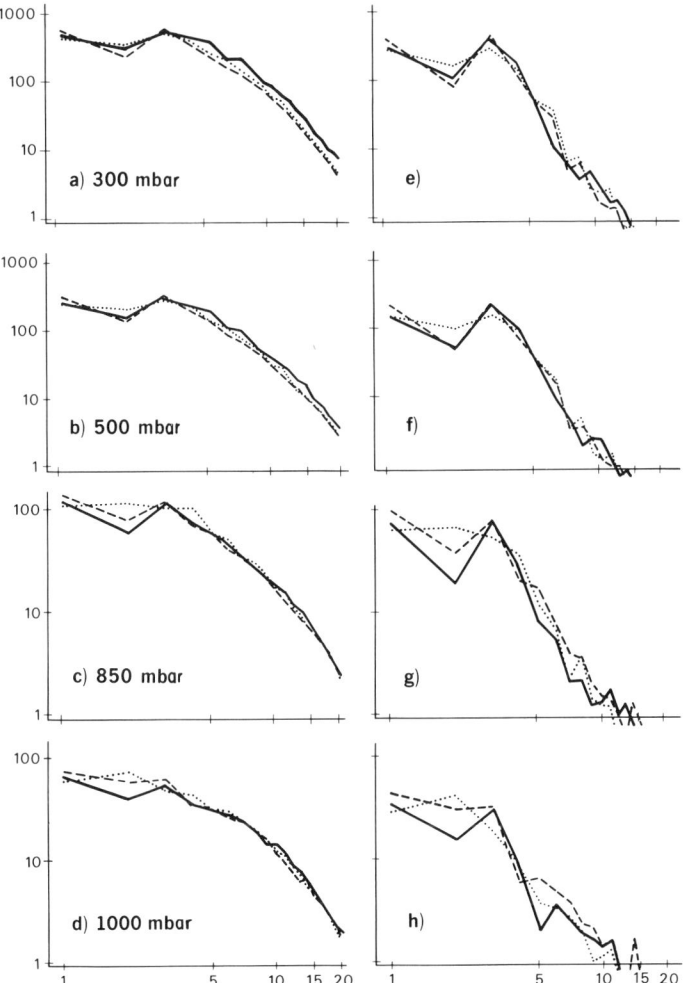

FIG. 10. Midlatitude kinetic energy spectra (in units of kJ m^{-2} bar^{-1}) at various levels. Usual ensemble average over 31 cases and the last 5 days of forecast period. Solid lines, ANA; dashed lines, ENV; dotted lines, MEA. Mean between 40° and 60°N. Left panels (a–d), total energy; right panels (e–h), stationary waves energy only.

mbar, both for stationary plus transient (left) and stationary waves only (right). It is evident from the figure that the largest response is in the first few wavenumbers, with the net effect of the envelope orography consisting of displacing the emphasis from the WN 2 response to the WN 3 response (as observed), more intensely so where the error seemed to be at its largest, namely in the lower part of the troposphere, and with emphasis on the stationary part of the waves. We can then summarize the message contained in the diagrams of Fig. 10 by saying that the envelope orography contributes to largely (but not completely) restore the observed spectral distribution of kinetic energy among the planetary-scale wavenumbers, with this effect somehow reducing in amplitude with height and being more evident for the stationary part of the eddy flow. The somehow vertically evanescent nature of the "envelope" effect on the kinetic energy spectra is in some contrast with the widely accepted view that orographic mechanical forcing is a dominating effect in the upper troposphere and in the stratosphere, but less important, compared to thermal forcing, in explaining the stationary features of the lower troposphere (e.g., Manabe and Terpstra, 1974). Figure 11 contains the isotropic amplitude spectra for both orographies and shows that, in going from a mean to an envelope orography, the spectral amplitude has been

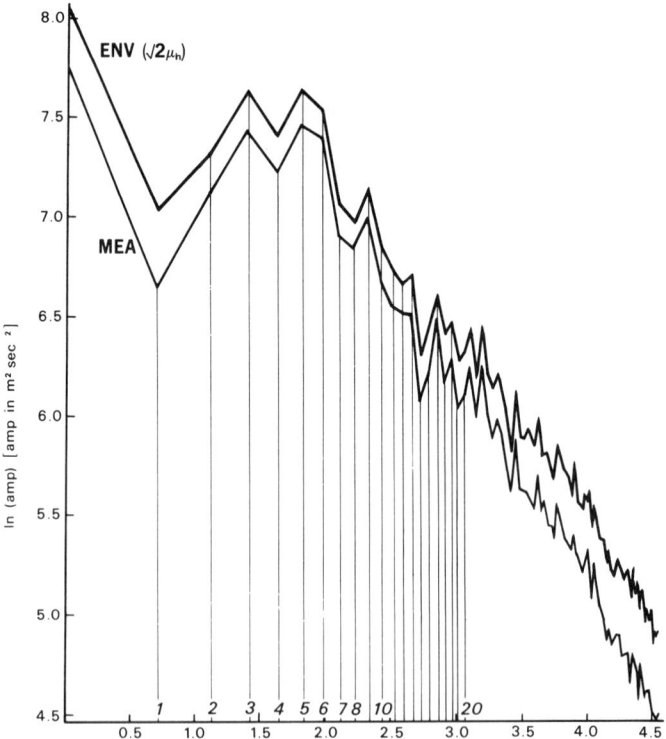

FIG. 11. Amplitude spectra of mean and $\sqrt{2}\mu_h$ envelope orographies as a function of the isotropic two-dimensional wavenumber (Legendre polynomial index n). On the x axis the natural logarithm of $(n + 1)$, on the y axis the natural logarithm of the amplitude of surface geopotential (m^2 sec^{-2}).

increased almost identically for almost all wavenumbers up to WN 16, with a somewhat larger proportional increase for WN 1 and for wavenumbers larger than 16. This, in a linear framework, hardly justifies the response implied by the diagrams of Fig. 10, where we showed an increased WN 3 response and a largely decreased WN 2 response.[2] A picture has therefore emerged of a "selective" model response (in wavenumber space and location in the vertical) to a generally increased orographic forcing, with a large, necessarily indirect, impact on the zonal mean mass and wind fields.

At this stage it is interesting to try to diagnose whether this behavior of the eddies and of the zonal flow in the envelope ensemble of forecasts is the consequence of a rapid (linear) response of the model atmosphere's eddy flow to an enhanced large-scale eddy orographic forcing, eventually modifying the zonal flow via nonlinear wave–mean flow interactions, or the zonal flow is affected directly by the small scales of the envelope orography and it, in turn, affects the response of the quasi-stationary planetary waves. The results of an experiment designed to shed light on precisely this question is discussed in Section 4.

4. The Experiments with the Blended Orographies

In order to address the question raised at the end of the previous section, it was decided to construct two spectrally 'blended' orographies in the following way: the envelope increment, that is, the proportion of the subgrid terrain height variance that is added to the gridbox mean orography, was decomposed in spherical harmonics, using the same base functions used by the ECMWF spectral model. Two orographies were then resynthesized, blending the spherical harmonic coefficients in the total wavenumber band 0–3 from the mean (envelope) orography with the coefficients in the wavenumber band 4 to truncation from the envelope (mean) orography. We will call the first of these two LSMO, for large-scale mean orography (implying that the small scales come from the envelope orography), and the second LSEO, for large-scale envelope orography (implying that the small scales come from the mean orography). Now we have four orographies at our disposal: MEA, LSMO, LSEO, and ENV.

---

[2] We should point out that although the spectra of Fig. 10 (zonal wavenumber spectra) and of Fig. 11 (isotropic wavenumber spectra) are not directly comparable, since we are concentrating here on the planetary scales, the argument is little affected. The correspondent "m" orography spectrum (not shown) would show a relative change similar to the one shown by Fig. 11 in going from a mean to an envelope orography. The "n" spectra were preferred in this case because of their easier interpretation in terms of the orography "blending" experiments described in Section 4.

If the behavior of the forecasting model, measured looking in particular at the WN 1–3 response, using, for example, LSEO, is similar to its behavior using ENV and different from the one obtained using MEA, while, conversely, the behavior using LSMO is similar to the behavior using MEA, we will infer that the "envelope" effect on the ultralong waves comes from direct forcing by the correspondent wavenumbers and that the mean zonal flow is then influenced via wave–mean flow interaction. If, however, the contrary is true (LSEO behaves like MEA and LSMO like ENV), we will infer that the smaller scales of the orography have a direct (beneficial) effect on the evolution of the mean zonal flow and that this improved zonal flow affects, in turn, the response of the planetary waves.

These experiments had to be carried out, for practical reasons, with the ECMWF T63 spectral model (Simmons and Jarraud, 1984) rather than with the gridpoint model. There are, however, a number of experiments showing that the impact of the envelope orography on the ECMWF T63 spectral model is very similar to the one we have documented in the previous section on the gridpoint model (e.g., Jarraud and Simmons, 1984). Another difference that should be mentioned here and that is also reported and explained in Jarraud and Simmons (1984) is that the envelope orography used with the spectral model is of the type $h_E = h_M + \sqrt{2}\mu_h$ and not $h_E = h_M + 2\mu_h$ as it was in the case for the earlier experiments using the gridpoint model. This difference has its basis in ECMWFs operational needs (we should point out here that the T63 spectral model using a $\sqrt{2}\mu_h$ envelope orography replaced in operations at ECMWF the N48 gridpoint model with mean orography in April 1983) and was carried over to the experiments described here only for reasons of convenience. All available experimentation with the T63 ECMWF spectral model seems to indicate that whatever strong signal is emerging from this set of experiments can be transported to the gridpoint model environment, and vice versa.

Figure 12 shows the Northern Hemispheric tropospheric anomaly correlation coefficient of geopotential height for all waves (a, b, and c) and for the wavenumber band WN 1–3 only (d, e, and f) for the case chosen for this experiment. The initial conditions were 30 January 1982 at 12 GMT and all forecasts were conducted up to 10 days. These particular initial conditions were chosen for a variety of practical reasons, the main ones being that the MEA and ENV integrations were already available and that this was the case study that had shown possibly the greatest sensitivity to the change in orography, among those used to test the impact of the envelope orography on the spectral model before their joint operational implementation in April 1983. Let us first concentrate on panels a, b, d, and e. The important conclusion that emerges is that, of the two alternatives described earlier in this section, the second is by far the most plausible one, since LSEO behaves almost

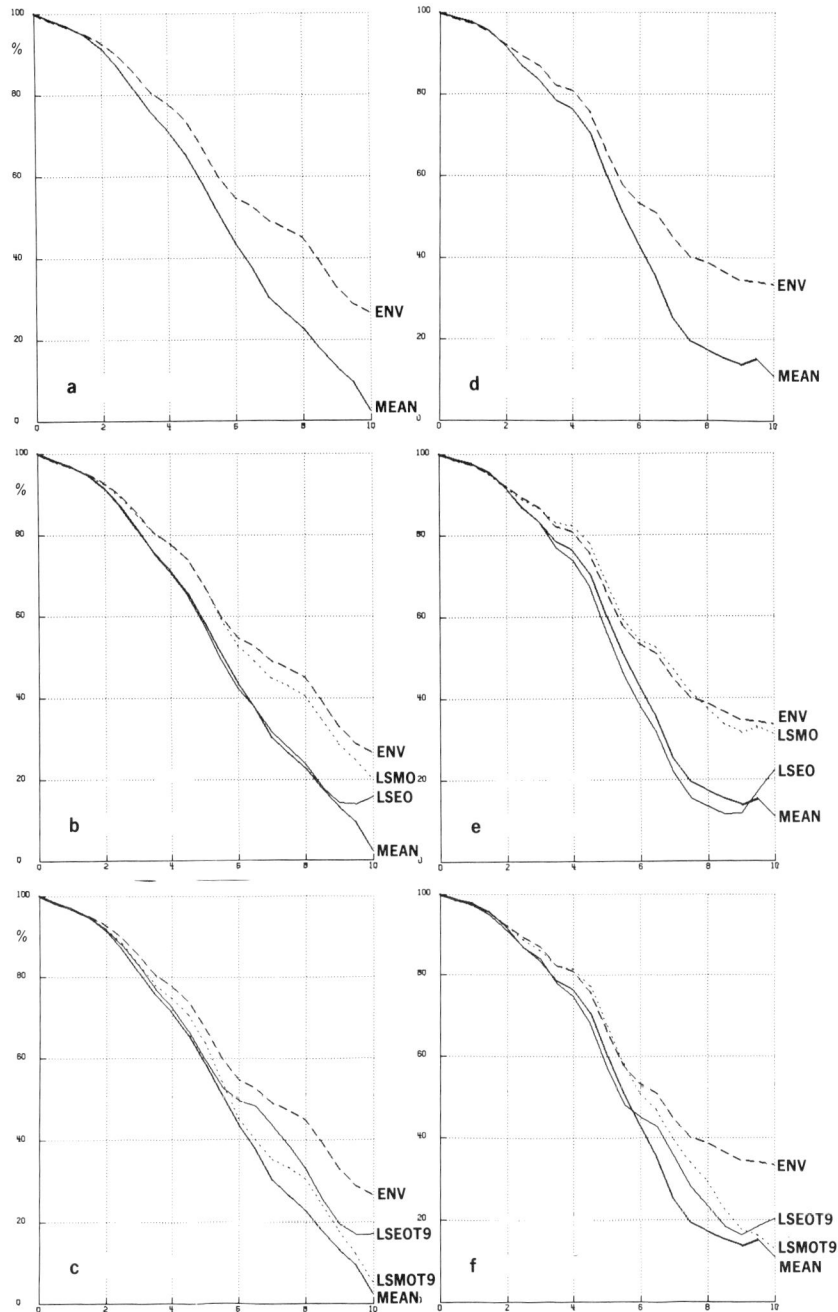

FIG. 12. Mean tropospheric anomaly correlation coefficient of geopotential height as a function of forecast time. (a–c), Full fields; (d–f), wavenumbers 1–3 only. ENV, Envelope orography model; MEA, mean orography model; LSEO, large-scale envelope orography model; LSMO, large-scale mean orography model. All experiments were performed using the T63 ECMWF spectral model. The initial data for all experiments was 30 January 1982 12 GMT.

identically to MEA, and LSMO to ENV. This shows, therefore, that the envelope effect is due to the smaller scales of the orography. Before concluding from this that the smaller scales of the orography exert their beneficial impact by directly affecting the zonal flow (and it, therefore, in turn affecting the response of the planetary-scale waves), we should mention that the behavior of the two integrations that use the blended orographies (LSMO and LSEO) evidenced so neatly by the objective scores of Fig. 12 is not confined to this particular indicator (Northern Hemisphere tropospheric geopotential height correlation coefficient). Mean cross-sections of zonal wind error for the last 5 days of the 10-day integrations show that, for the lower tropospheric belt 50°–60°N, LSEO and MEA have a mean zonal wind in excess of the observed of almost 3 m sec^{-1}. For LSMO and ENV this error is reduced to around 1 m sec^{-1}.

Panels c and f of Fig. 12 show the results from further 10-day integrations using yet two different blended orographies. They have been produced using exactly the same procedure as LSEO and LSMO but with the wavenumber cut off at WN 9 instead of WN 3. Orography LSEOT9 (LSMOT9) is, therefore, resynthesized with WN 0–9 from the $\sqrt{2}\mu_h$ envelope (mean) orography and all other waves from the mean ($\sqrt{2}\mu_h$ envelope) orography. This was done in order to isolate more efficiently the spectral zone where the envelope increment is directly felt and the spectral zone where we are measuring the impact (WN 1–3). Panel f shows that, at least up to day 6, an envelope orography that is "envelope" only in the very short scales still has the same effect on the ultralong planetary scales, and this effect shows clearly even in the earliest part of the forecast (days 2 to 3), when an influence of the smaller scales of motion through progressive up-the-spectrum nonlinear interactions could not yet be felt by the planetary, ultralong scales since the time elapsed is much shorter than the eddy turnover time (Lorenz, 1969; Leith, 1971).

We conclude therefore that the smaller scales of the orography have an effect on the mean zonal flow by reducing its westerly strength in the lower tropospheric midlatitudes. Simple linear theory of stationary Rossby waves (e.g., Held, 1983) could then suggest that shorter stationary wavelengths are favored by the weakening of the mean (westerly) zonal wind since $k_s = \sqrt{(\beta/\bar{u})}$, where k_s is the stationary Rossby wavenumber, β is the latitudinal variation of the Coriolis parameter, and $\bar{u}$ is the strength of the mean zonal wind. Despite the crudity of this argument, this is quite consistent with the decrease of WN 2 response and the corresponding increase of WN 3 response brought about by the introduction of the envelope orography, as shown in Fig. 10.

The next section will be devoted to the description of some further diagnostic work aiming to investigate the physical mechanism by which the (smaller scales of the) orography can influence the zonal flow on comparatively shorter time scales.

5. Mountain Torque and Zonal Flow

We have so far concluded that there must be a direct and effective link between the specification of orography and the strength of the tropospheric zonal flow. Let us remember that the mountain torque is, together with the frictional torque, the only possible source or sink of total angular momentum for the global atmosphere:

$$\frac{\partial}{\partial t}\left[\int_m M\, dm\right] = \int_A (T_F + T_M)\, dA \qquad (1)$$

where

$$M = M_\Omega + M_r = \Omega r^2 \cos\phi + ur\cos\phi$$

is the total angular momentum for unit mass, m is the total mass of the atmosphere, A is the total surface of the earth, and T_F and T_M are the local frictional and mountain zonal torques per unit area.

The mountain torque would therefore seem as a very good candidate to hold the responsibility for an efficient, dynamical coupling between the atmosphere and the underlying solid earth. Figure 13 shows the mean analyzed mountain torque for the month of January 1981 as a function of latitude (integrated over longitude) for the mean orography data assimilation cycle (a) and for the envelope orography data assimilation cycle (b), while Fig. 14 shows the same for the ensembles of the day 7 forecasts. The numerical computation was performed using the N48 gridpoint model (analysis or forecast) field of surface pressure p_* and surface terrain height h and refers, therefore, to the $2\mu_h$ envelope orography. The mean zonal torque per unit area τ_M computed over a (thin) latitude belt of width $\Delta\theta$ was computed using the relationship

$$\tau_M = \frac{-1}{2\pi r^2 \cos\theta\, \Delta\theta} \int_{\Delta\theta} \int_0^{2\pi} r^2 \cos\theta\, p_* \frac{\partial h}{\partial \lambda}\, d\theta\, d\lambda \simeq \frac{-1}{2\pi} \sum_{1\ i}^{N} p_{*i}(h_{i+1} - h_{i-1})$$

where θ and λ are, respectively, latitude and longitude, r is the radius of the earth, the latitudinal integral is taken over one gridbox ($\Delta\theta = 1.875°$), and the longitudinal integral has been replaced by a sum over a longitudinal index $i = 1, 2, \ldots, N; N = 192$. This method of computing the mountain torque has well known problems associated with truncation errors in regions of very steep topography (see Manabe and Terpstra, 1974; Wahr and Oort,

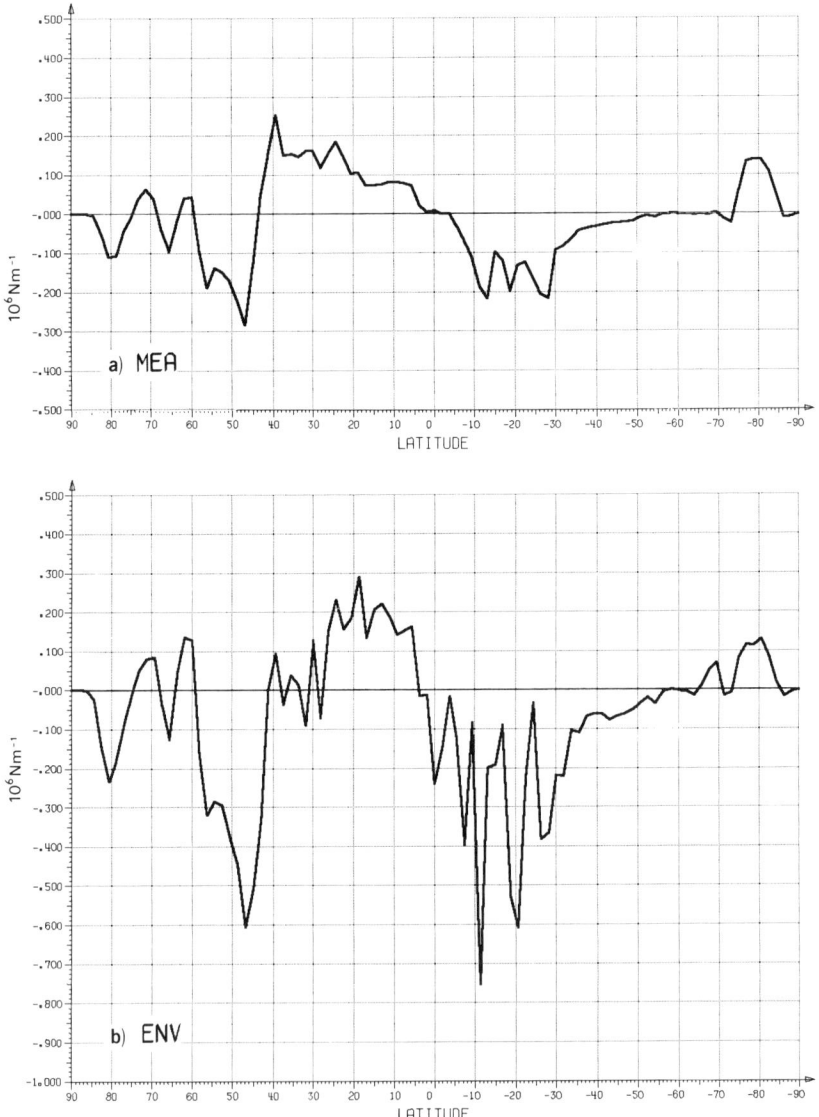

FIG. 13. Analyzed mountain torque per unit area averaged for the January 1981 period (ensemble of 31 analyses) as a function of latitude. (a) MEA, Mean orography used in the assimilating model; (b) ENV, envelope orography ($2\mu_h$) used in the assimilating model.

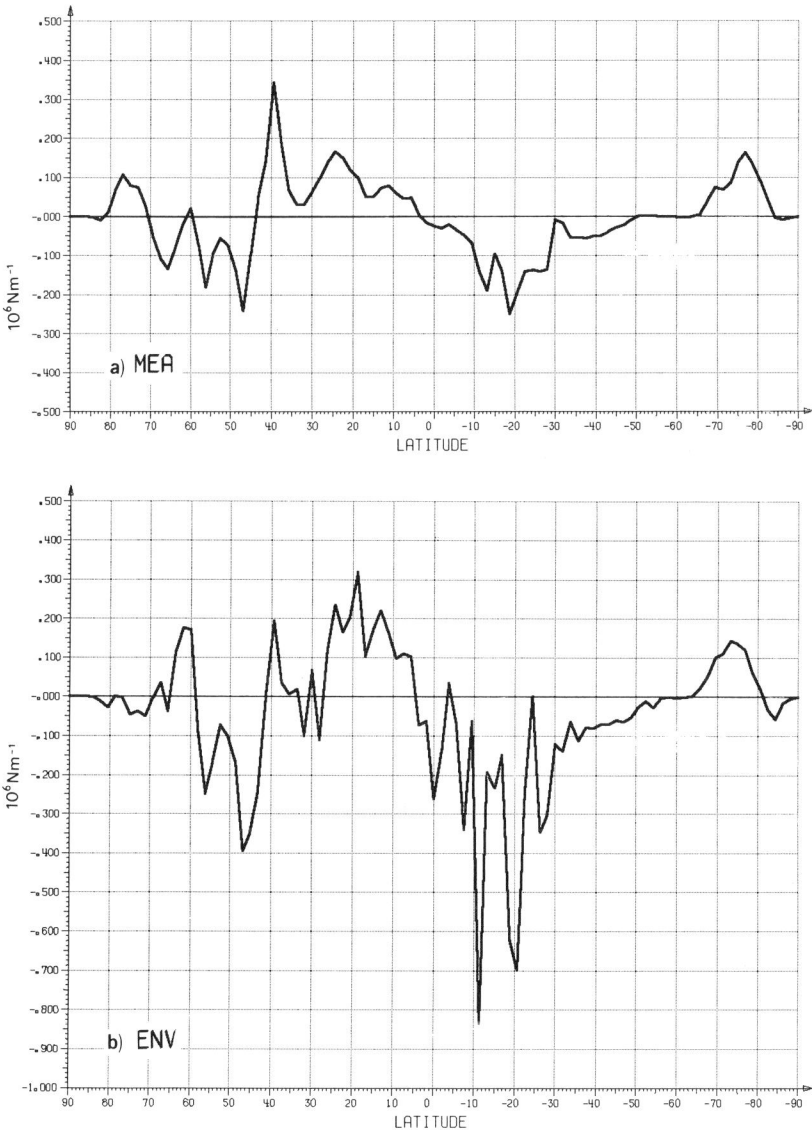

FIG. 14. Analyzed mountain torque per unit area averaged for the January 1981 period (ensemble of all day 7 forecasts) as a function of latitude. (a) MEA, Mean orography model; (b) ENV, envelope orography ($2\mu_h$) model.

1984), but was employed for convenience and in the light of the comparatively high resolution of the ECMWF gridpoint model. These problems seem to become excessive only in the region of the Andes and only in the case of the much steeper envelope orography. Since we will concentrate here on the Northern Hemisphere, we are confident that the technique employed is adequate.

Figure 13 shows the comparison between the torques produced by the mean and by the envelope orography in the ensemble of 31 initial conditions for the January 1981 set of experiments. Figure 14 shows the same for the ensemble of the day 7 forecasts. In the Northern Hemisphere midlatitudes (30°–60°N) the higher negative torque exerted by mountains (mostly the Rockies) is, using the envelope orography, such as to slow down the lower tropospheric westerlies (approximately between 100 and 500 mbar) and is, therefore, consistent with the observed reduction of the model's excessive westerlies shown in Fig. 6. The increase of the positive torque taking place in the Northern Hemisphere subtropical and tropical belt (0°–30°N) is also quite consistent with the exaggerated easterlies of the mean orography model, also reduced by the introduction of the envelope orography.

Figure 15 shows the time evolution during the January 1981 10-day integrations of the ensemble mean latitudinal integral of the mountain torque between approximately 30° and 60°N. The mean and envelope orography

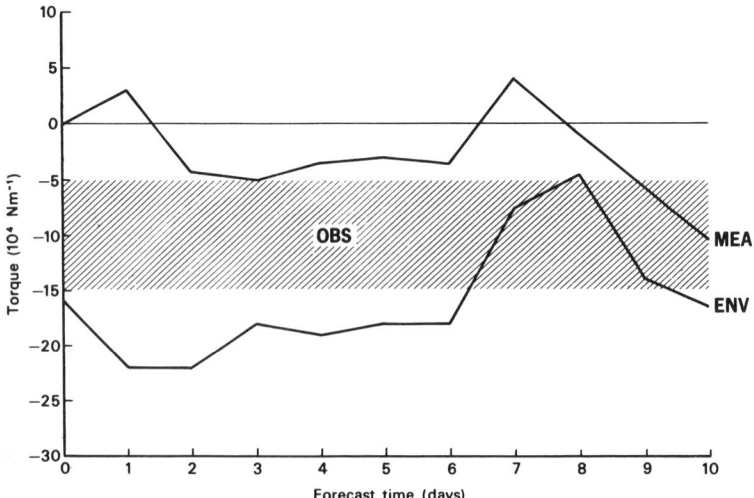

FIG. 15. Mountain torque per unit area averaged for the 31 forecasts of January 1981 and over the latitudinal belt 28°–60°N as a function of forecast time. MEA, Mean orography model; ENV, envelope ($2\mu_h$) orography model; OBS, appropriate observed winter values (e.g., Oort and Bowman, 1974).

ensembles are compared with estimates of the real torque derived from observations (Oort and Bowman, 1974). It is quite clearly shown that the ensemble of mean orography forecasts underestimates the mountain torque, while the $2\mu_h$ envelope orography overestimates it for the first 6 days. Both ensembles tend toward more reasonable (and closer) values toward the end of the forecast period, suggesting that a new dynamical balance has been attained and that the surface pressure pattern has adjusted to give again a reasonable total torque. We know, however, from Fig. 6, that these new balances are attained for different values of the mean zonal wind for that latitude belt.

If we add to the results of this diagnosis the fact that the smaller scales of the orography (and of the pressure field) are likely to dominate the generation of mountain torque, we can conclude that the changes brought about in the Northern Hemisphere midlatitudinal mountain torque are quite consistent with the reduced model errors in the zonal mean of zonal wind, indicating this as a plausible physical mechanism to account for an important part of the observed impact of the envelope orography.

6. The Effects of the Envelope Orography in the Tropical Regions

Figures 5 and 6 show that the effects of the envelope orography on the Southern Hemisphere's midlatitudes are, as it was reasonable to expect, smaller in amplitude but essentially of the same nature of those experienced by the Northern Hemisphere. The impact, however, is largest on the day 7 upper tropospheric and lower stratospheric tropical temperature SE. The zonal-wind SE is also strongly affected in this portion of the model's atmosphere but, unlike for temperature for which the benefit has vanished by day 10, the impact seems to last well into the whole forecast range explored. It becomes, therefore, natural to investigate the impact of the envelope orography or the mean meridional circulation and, in particular, on the Hadley cell.

Figure 16 shows the mean meridional circulation for the two January 1981 ensembles at days 1, 4, 7, and 10. The most conspicuous difference is a considerable reduction of the spin-up time of the model's tropical atmosphere. The mean orography ensemble needs 8–10 days to reach the value of 140 ton sec^{-1} for the total Hadley cell mass transport, while the envelope ensemble shows an "overshoot" value of 180 ton sec^{-1} at day 1 followed by a lower value of 130 ton sec^{-1}, then settles down at values around 160 ton sec^{-1} during the second 5 days of the forecast. The existence of a spin-up phase that is an integral part of the model's systematic error is a well known

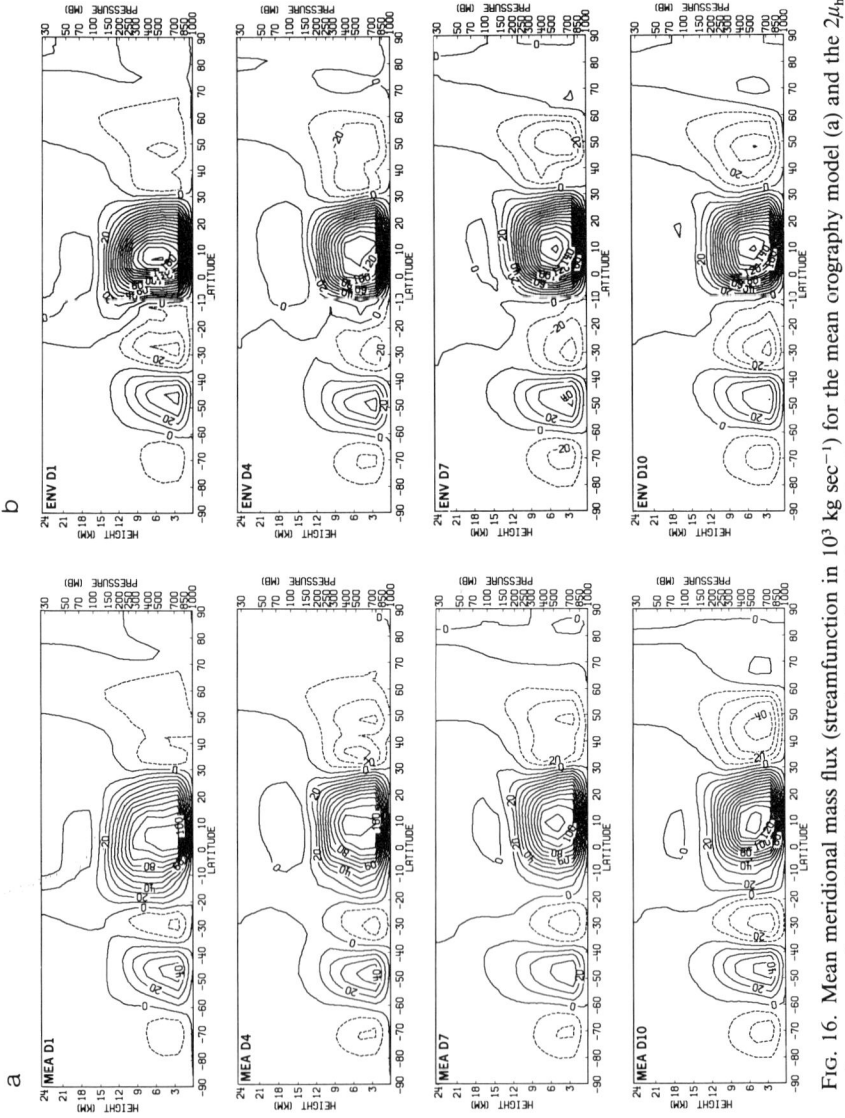

FIG. 16. Mean meridional mass flux (streamfunction in 10^3 kg sec^{-1}) for the mean orography model (a) and the $2\mu_h$ envelope orography model (b). From top to bottom: day 1, day 4, day 7, and day 10.

fact (e.g., Heckley, 1984) and has been mostly, but not uniquely, attributed to the lack of divergent motion in the tropical regions brought about by the use of nonlinear normal mode initialization (albeit diabatic) as an integral part of the ECMWF's forecasting system (Williamson and Temperton, 1981). Figure 16 shows that, not only has the envelope orography ensemble a shorter spin-up (130 against 100 ton sec^{-1} at day 4, 170 against 120 ton sec^{-1} at day 7, and 160 against 140 ton sec^{-1} at day 10), and therefore reaches sooner a regime phase, but also the final (day 10) value of the total Hadley cell mass flux is 15% higher (40% higher at day 7, however, indicating a degree of overshoot in the envelope spin-up process). It is well known (e.g., Palmen and Newton, 1969; Oort and Peixoto, 1983) that the tropical Hadley cell is driven both directly, by convection, and indirectly, by frictional and mountain torques. We have shown in the previous section that the envelope orography causes, in Northern Hemispheric *tropical* regions and during approximately the first 7 days of model integration, a considerable increase in the mountain torque, in agreement with the observed reduction of the model's excessive *easterlies*. This is also quite consistent with an enhanced Hadley circulation. It has been shown (e.g., Kidson *et al.*, 1969) that the inertial torque exerted on the upper and lower branches of the Hadley cell ($f\bar{v}$) is balanced, respectively, by the convergence of eddy flux of zonal momentum [$(\partial/\partial y)\overline{u^*v^*}$] aloft and by the generation terms (frictional and mountain torques) associated with the lower boundary.

Although we have not evaluated directly the frictional torque, the general reduction of the low-level tropical easterlies induced by the enhanced orography must imply that the frictional torque has not increased in those latitudes (if not decreased). The problem, however, remains to explain whether this intensification of the Hadley circulation was caused by increased direct forcing (enhanced convection) or by indirect forcing (enhanced imbalance between positive mountain torque and negative frictional torque). We will attempt to give a qualitative answer to this question by examining the changes in the precipitation field.

Figure 17 shows the zonally averaged total 24-hr precipitation of the two January 1981 ensembles compared with Jaeger's climatology (Jaeger, 1976, 1983) for days 1, 2, 4, 7, and 10. Together with documenting the known deficiencies of the ECMWF operational rainfall forecasts (Heckley, 1984; Molteni and Tibaldi, 1985), these diagrams show a substantial increase, maintained throughout the 10-day forecast period, of the tropical precipitation caused by the envelope orography and coupled with a concentration of such precipitation in a latitudinal band narrower than in Jaeger's climatology. Although this increase is excessive, and mostly in its equatorial maxima, it goes in the right direction, since the model underestimates the tropical convective activity and, therefore, the associated direct meridional circula-

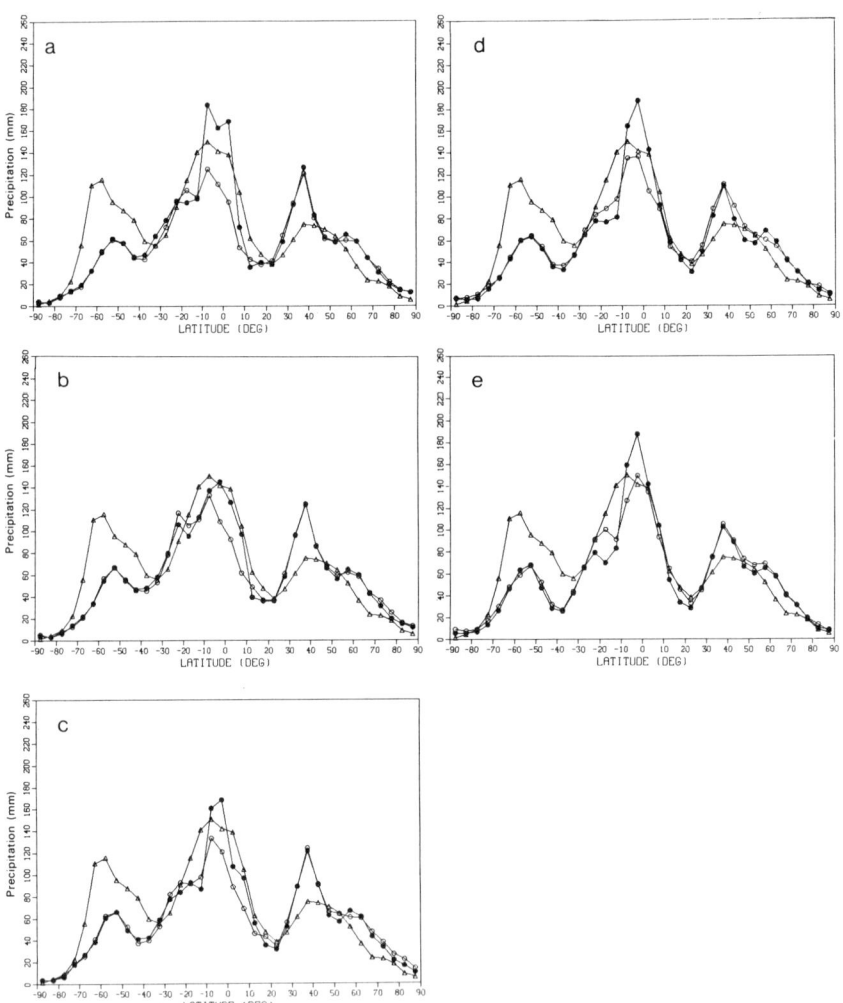

FIG. 17. Latitudinal cross sections of zonally averaged total 24-hr precipitation (millimeters). ○, Mean orography; ●, envelope ($2\mu_h$) orography; △, climatological estimates for January (Jaeger, 1976). (a–e), Day 1, day 2, day 4, day 7, and day 10, respectively.

tion. The excess of the precipitation increase is also consistent with the overshooting effect on the Hadley circulation shown by Fig. 16 and mentioned above.

Figure 18 concentrates on day 7 results and shows the two separate totals over land points and over sea points. Here the picture is less clear: Intertropical Convergence Zone (ITCZ) precipitation increases strongly over sea areas

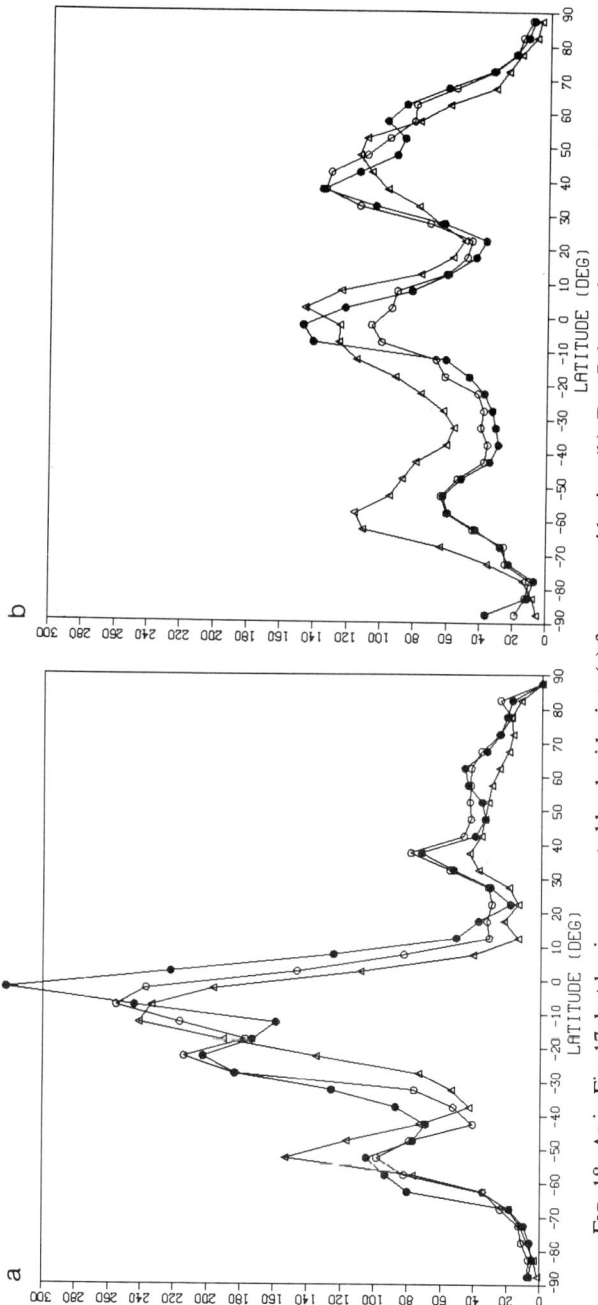

FIG. 18. As in Fig. 17, but having separated land gridpoints (a) from sea gridpoints (b). Day 7 data only are presented.

as well as over land areas, and this, although not in disagreement with the observed climatology, is somewhat difficult to understand in terms of an increased and steeper orography that should, conceivably, have had an effect only on land points. Figure 19, however, provides the explanation for this apparent contradiction by allowing us to examine the geographical distribution of the rainfall for the two ensembles (at day 7) and compare them to Jaeger's climatology. It is evident from this comparison that the enhanced orography has the net effect of enhancing and concentrating the tropical rainfall on all orographic reliefs in the intertropical region. This takes place, to some minor degree, even at the expense of the marine ITCZ rainfall; note the weakening of the rainfall belt associated with both the Atlantic and the Pacific portions of the ITCZ. Noteworthy are also the excesses of precipitation over the Indonesian Island, with maxima peaking at 2000 mm day^{-1} around 150°E–8°S. These maxima, certainly excessive, are caused by an undesirable interaction between the steeper envelope orography and the modeling of precipitation (both convective and large scale) in the ECMWF gridpoint model (see Tibaldi, 1982). They occur, to a lesser extent, also in the spectral version of ECMWF's GCM and they are mostly due to the concurrence of two facts: the first is that the Kuo-type convection scheme (as employed in both the gridpoint and the spectral model) is prone to excessive feedback mechanisms involving the strong low-level mass convergence induced by intense deep convection and the sensitivity of the scheme itself to low-level moisture convergence. A typical example of this are the so called gridpoint storms that were (and, to a point, still are) typical occurrences in daily operational 10-day forecasts. The second is that horizontal diffusion of temperature and moisture on σ surfaces tends, in the presence of very steep mountains, to warm and moisten mountain tops and cool and dry valleys and foothills. Tibaldi (1982) showed that this is often enough to trigger, in the model, long-lasting and very deep convection towers producing just such excessive precipitation maxima. This can be only partially circumvented using an extra correction term in computing the horizontal gradient operator, as a zero-order approximation to the diffusion operator being applied on pressure surfaces; doing this strictly would almost completely prevent the spurious warming/moistening of mountain tops, but, unfortunately, would also be prohibitively expensive in terms of computer time. Once the convective cell has been initiated, for example, by a tropical island represented by only a few gridpoints with very high terrain elevation (e.g., the Indonesian Islands), a large amount of convective rain falls both on the island land gridpoints and on the neighboring sea gridpoints, explaining the apparent inconsistency between the large-scale *reduction* of maritime ITCZ precipitation mentioned above and panel (b) of Fig. 18, showing an *increase* in total precipitation over sea. A careful inspection of Fig. 19 reveals that the only

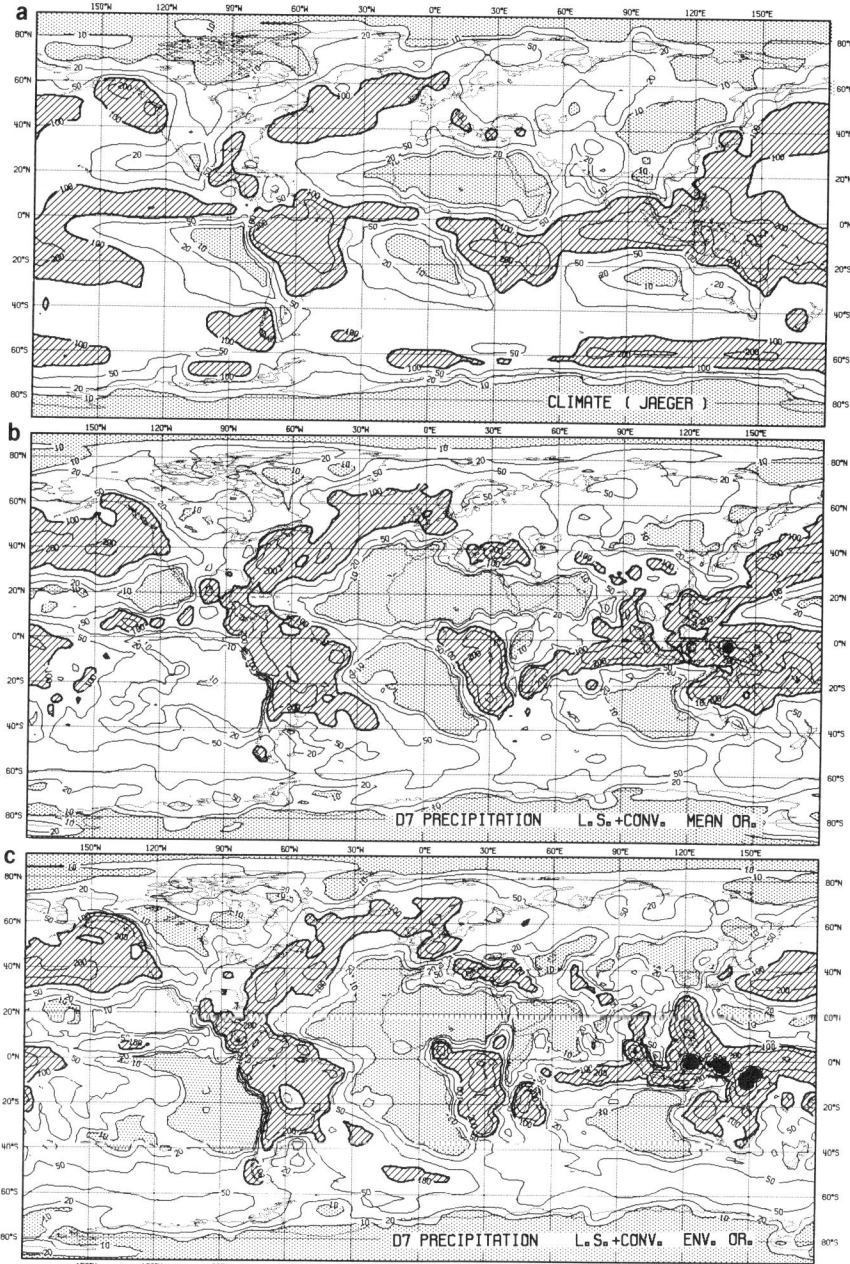

FIG. 19. Global maps of model day 7 total 24-hr precipitation compared to Jaeger's January climate: (a), climate; (b), mean orography model; (c), envelope ($2\mu_h$) orography model.

notable increases of tropical precipitation occurred in very limited "spots" centered around high mountain peaks, while everywhere else (including non-steep-orography land areas, e.g., South America) the precipitation decreases.

We conclude, therefore, that although the impact of the envelope orography on the zonally averaged tropical circulation is positive, showing a more vigorous Hadley cell and a much faster spin-up, much of this change is brought about by a spurious increase of (mostly convective) precipitation produced by an undesirable mishandling of steep orography by the horizontal diffusion algorithm coupled with the Kuo-type convection scheme operating in the model. The obvious inadequacy of Jaeger's climatology as a term of comparison, both for its own deficiencies (Molteni and Tibaldi, 1985) and because of sampling consideration (January interannual variability), make it impossible for us to estimate quantitatively the extent of such an effect, but, qualitatively, the existence of a "compensating error" effect can be in very little doubt. The extent to which this effect is propagating into midlatitudes is also an open question.

7. Summary and Conclusions

Setting out from the conclusions of the work by Wallace *et al.* (1983) (WTS) and from some of the questions left open by them, we have, first of all, attempted to give a more solid statistical basis, and therefore credibility, to their results concerning the beneficial impact of an enhanced (envelope) orography on the onset and magnitude of the systematic error of the ECMWF gridpoint GCM. The global data for the month of January 1981, notable for the anomalously large amplitude of the monthly-mean error of the operational ECMWF forecast model, have been reprocessed twice in a quasi-operational fashion using a mean and a $2\mu_h$ envelope orography in the data assimilation and forecast model. The two ensembles of 31 10-day forecasts have constituted the main dataset analyzed and diagnosed in this work. The large decrease (up to 50%) of the systematic error of the 1000- and 500-mbar geopotential height associated with the use of the envelope orography has been confirmed. This change manifests itself as a restoration of the correct spectral distribution of kinetic energy among the quasi-stationary ultralong waves, with WN 3 becoming, as observed, the dominant feature of the Northern Hemisphere mean tropospheric flow. At the same time, a large and positive impact on the zonally averaged temperature and zonal wind error was also observed, acting to reduce the excessive midlatitudinal tropospheric westerlies, this latter feature being common to several GCMs whose systematic errors have been documented in the literature. Careful diagnosis of the energy cycle in the two forecast ensembles showed much improved barotropic conversions between the eddies and the mean zonal flow, quite consistent with the improvements in the individual quantities.

The question, at this stage, was posed as to whether this entire pattern of behavior of the model employing the enhanced orography was the consequence of a direct response of the model's planetary waves to the modified large-scale eddy orographic forcing, with the waves eventually influencing the zonal flow through nonlinear wave–mean flow interactions, or whether the model's zonal flow was being directly affected by the (small scales of the) envelope orography and it then, in turn, produced a modified response of the low-frequency planetary waves of the model. An experiment using "spectrally blended" orographies was designed and its results showed that an orography that is "envelope" only in the smaller spatial scales has an effect on the ultralong waves of the model's atmosphere that is almost identical to the one produced by a complete envelope orography (envelope on all scales). Since such an efficient and rapid coupling between different scales of motion is likely to take place via the zonal flow, and the smaller scales of the orography can influence heavily the zonal flow via the mountain torque, we proceeded to diagnose this quantity in the two forecast ensembles and how it evolves during the 10 days of the integrations. We found that, in comparison with observed climatological estimates, the mean orography model underestimates heavily, during the first 4 to 7 days of integration, the (negative) mountain torque in the Northern Hemispheric midlatitudes. The $2\mu_h$ envelope orography model, on the contrary, slightly overestimates it, in very good agreement with their respective behavior in terms of zonal wind SE. We therefore conclude that the behavior of the model's midlatitudes is quite consistent with the idea that the smaller scales of the envelope orography exert an increased (negative) torque on the atmosphere, conducing to a reduced tropospheric mean zonal (westerly) wind. These modified westerlies in turn, and not the direct enhanced forcing of the planetary scales of motion, are to be held responsible for the improved response of the low-frequency ultralong waves always present in the envelope orography model integrations.

The effects of the envelope orography on the mean meridional circulation and on the direct forcing of the Hadley cell were then diagnosed. We found that the "envelope" integrations present, on average, a much reduced spin-up problem with an enhanced Hadley circulation, in agreement with observations. Unfortunately, however, this overall improvement of the zonally averaged indicators, including the tropical temperature SE, mostly came from an unrealistic increase of convective (and, to a smaller extent, even large scale) precipitation over steep mountain peaks. This effect is to be ascribed to an undesirable consequence of the algorithm to diffuse horizontally temperature and moisture, and to its interaction with the parameterization of moist convection in the model, and is physically quite unrealistic. In fact, although the zonally averaged total precipitation improves somehow as a result of this, the realism of its spatial distribution and its partitioning between land points and sea points suffers at the same time. Although it was not possible to quantify the extent to which this undesirable effect contributed to the apparent improvement of the midlatitudes of the model, it seems appropriate to conclude this work with a word of caution regarding some possible "compensating error" effects due to the envelope orography. Very careful numerical experimentation and diagnostic work should always be carried out prior to the acceptance of any comparatively new parameterization module in a complicated and deeply nonlinear system like a general circulation model.

Acknowledgments

A large number of colleagues from the Research and Operations Department of ECMWF and from outside ECMWF have contributed in various degrees to this project. B. Norris assisted in running the data assimilations for the January 1981 period; R. Strüfing assisted in running the gridpoint model; M. Jarraud helped with the running of the spectral model and the creation of the "blended" orographies; K. Arpe, E. Oriol, G. X. Wu, and H. Z. Liu assisted with diagnostic computations and F. Molteni with producing the precipitation maps. Discussions with R. Benzi, A. Sutera, and A. Speranza were essential to generate and clarify many of the ideas and experiments described in this work. K. Arpe, A. Simmons, and F. Mesinger carefully read successive versions of the manuscript. I am grateful to all of them.

Figures 1 and 2 were reproduced with the kind permission of the Royal Meteorological Society.

References

Arpe, K. (1983). Diagnostic evaluation of analysis and forecasts: Climate of the model. *ECMWF Semin. Proc. Interpret. Num. Weather Predict. Products, 13–17 Sept. 1982* 99–140.

Arpe, K., and Klinker, E. (1984). Systematic errors of the ECMWF operational spectral model. Part 1: Midlatitudes. ECMWF SAC Paper. Available from ECMWF.

Arpe, K., Hollingsworth, A., Tracton, M. S., Lorenc, A. C., Uppala, S., and Kallberg, P. (1985). The response of numerical weather prediction systems to FGGE level II-b data. Part II: Forecast verifications and implications for predictability. *Q. J. R. Meteorol. Soc.* **111**, 67–102.

Bengtsson, L., and Lange, A. (1981). Results of the WMO/CAS numerical weather prediction data study and intercomparison project for forecasts for the Northern Hemisphere in 1979-80. Short and Medium Range Weather Prediction Research Publication Series No. 1. WMO, Geneva.

Bettge, W. T. (1983). A systematic error comparison between the ECMWF and NMC prediction models. *Mon. Weather Rev.* **111**, 2385–2389.

Chouinard, C. (1984). A study of the Systematic Errors of the Canadian Meteorological Centre's Spectral Model. *Atmos.-Ocean* **22**, 226-243.

Derome, J. (1981). On the average errors of an ensemble of forecasts. *Atmos.-Ocean* **19**, 103-127.

Heckley, W. A. (1984). Systematic errors of the ECMWF operational spectral model. Part II: Tropics. ECMWF SAC Paper. Available from ECMWF.

Held, I. M. (1983). Stationary and quasi-stationary eddies in the extratropical troposphere: Theory. *In* "Large-Scale Dynamical Processes in the Atmosphere" (B. J. Hoskins and R. P. Pearce, eds.), pp. 127-168. Academic Press, New York.

Hollingsworth, A., Arpe, K., Tiedtke, M., Capaldo, M., and Savijärvi, H. (1980). The performance of a medium-range forecast model in winter. Impact of physical parameterizations. *Mon. Weather Rev.* **108**, 1736-1773.

Jaeger, L. (1976). Monatskarten des Niederschlags für die ganze Erde. *Ber. Dtsch. Wetterd.* **18** (139).

Jaeger, L. (1983). Monthly and areal patterns of mean global precipitation. *In* "Variations in the Global Water Budget" (A. Street-Percott, M. Berand, and R. Ratcliffe, eds.), pp. 129-140. Reidel, Dordrecht.

Jarraud, M., and Simmons, A. (1984). Development of the new forecast model: Numerical aspects. ECMWF 1984 SAC paper. Available from ECMWF.

Kidson, J. W., Vincent, D. G., and Newell, R. E. (1969). Observational studies of the general circulation of the Tropics: Long term mean values. *Q. J. R. Meteorol. Soc.* **95**, 258-287.

Lange, A., and Hellsten, E. (1983). Results of the WMO/CAS NWP Data Study and Intercomparison project for forecasts for the Northern Hemisphere in 1981/82. Short and Medium Range Weather Prediction Research Publication Series No. 2. WMO, Geneva.

Lange, A., and Hellsten, E. (1984). Results of the WMO/CAS NWP Data Study and Intercomparison project for forecasts for the Northern Hemisphere in 1983. Short and Medium Range Weather Prediction Research Publication Series No. 7. WMO, Geneva.

Leary, C. (1971). Systematic errors in operational National Meteorological Center primitive-equation surface prognoses. *Mon. Weather Rev.* **99**, 409-413.

Leith, C. E. (1971). Atmospheric predictability and two-dimensional turbulence. *J. Atmos. Sci.* **28**, 145-161.

Lorenz, E. N. (1955). Available potential energy and the maintenance of the general circulation. *Tellus* **7**, 157-167.

Lorenz, E. N. (1969). The predictability of a flow which possesses many scales of motion. *Tellus* **21**, 289-307.

Manabe, S., and Terpstra, T. B. (1974). The effects of mountains on the general circulation of the atmosphere as identified by numerical experiments. *J. Atmos. Sci.* **31**, 3-42.

Molteni, F., and Tibaldi, S. (1985). Climatology and systematic error of rainfall forecasts at ECMWF. ECMWF Tech. Rep. No. 51 (91 pp.).

Oort, A. H., and Bowman, H. D. (1974). A study of the mountain torque and its interannual variations in the northern hemisphere. *J. Atmos. Sci.* **31**, 1974-1982.

Oort, A., and Peixoto, J. P. (1983). Global angular momentum and energy balance requirements from observations. *Adv. Geophys.* **25**, 355-490.

Palmèn, E., and Newton, C. W. (1969). "Atmospheric Circulation Systems." Academic Press, New York.

Saunders, F., and Gyakum, J. R. (1980). Synoptic-dynamic climatology of the "bomb." *Mon. Weather Rev.* **108**, 1589-1606.

Silberberg, S. R., and Bosart, L. F. (1982). An analysis of systematic cyclone errors in the NMC LFM-II model during the 1978-79 cool season. *Mon. Weather Rev.* **110**, 254-271.

Simmons, A., and Jarraud, M. (1984). The design and performance of the new ECMWF Operational Model Proceedings of the ECMWF 1983 Seminars on 'Numerical Methods for Weather Prediction,' Vol. 2, pp. 113-164. Available from ECMWF.

Sumi, A., and Kanamitsu, M. (1984). A study of systematic errors in a numerical weather prediction model. Part I: General aspects of the systematic errors and their relation with the transient eddies. *J. Meteorol. Soc. Jpn.* **62**, 234-250.

Tibaldi, S. (1982). The production of a high resolution global orography and associated climatological surface fields for operational use at ECMWF. *Riv. Meteorol. Aeronaut.* **42**, 285-308.

Tibaldi, S. (1984). On the relationship between the systematic error of the ECMWF forecast model and orographic forcing. Predictability of fluid motions (La Jolla Institute, 1983). *AIP Conf. Proc.* **106**, 397-418.

Tiedtke, M. (1983). Winter and summer simulations with the ECMWF model. Proceedings of the ECMWF Workshop on 'intercomparisons of large-scale models used for extended range forecasts,' pp. 263-314. Available from ECMWF.

Wahr, J. M., and Oort, A. H. (1984). Friction and mountain torque estimates from global atmospheric data. *J. Atmos. Sci.* **41**, 190-209.

Wallace, J. M., and Woessner, J. K. (1981). An analysis of forecast errors in the NMC hemispheric primitive equation model. *Mon. Weather Rev.* **109**, 2444-2449.

Wallace, J. M., Tibaldi, S., and Simmons, A. (1983). Reduction of systematic forecast error in the ECMWF model through the introduction of an envelope orography. *Q. J. R. Meteorol. Soc.* **109**, 683-717.

Williamson, D. L., and Temperton, C. (1981). Normal mode initialisation for a multi-level grid-point model. Part II: Nonlinear aspects. *Mon. Weather Rev.* **109**, 744-757.

NUMERICAL FORECASTS OF TROPOSPHERIC AND STRATOSPHERIC EVENTS DURING THE WINTER OF 1979: SENSITIVITY TO THE MODEL'S HORIZONTAL RESOLUTION AND VERTICAL EXTENT

CARLOS R. MECHOSO, MAX J. SUAREZ,* KOJI YAMAZAKI,† AKIO KITOH,† AND AKIO ARAKAWA

Department of Atmospheric Sciences
University of California, Los Angeles
Los Angeles, California 90024

1. INTRODUCTION

In this paper we report the results of several medium-range (10-day) experimental forecasts with the University of California, Los Angeles (UCLA) general circulation model (GCM). Our primary interest in these experiments is in the impact of increased horizontal resolution and vertical extent on the accuracy of the forecasts. It is a pervasive idea in numerical weather prediction that increased horizontal resolution generally results in more accurate forecasts (e.g., Bengtsson, 1981). On the other hand, the impact of the artificial upper boundary is a recognized but relatively unexplored problem. Here we note that having the model top at zero pressure is, in practice, equivalent to having the model top at some small finite nonzero pressure as a result of inevitable finite-difference errors (Lindzen *et al.,* 1968).

Lambert (1980) investigated the sensitivity of short-range forecasts to increased vertical resolution and the addition of stratospheric levels. He compared forecasts with versions of the Canadian Atmospheric Environ-

---

* Present address: Laboratory for Atmospheres, National Aeronautics and Space Administration, Goddard Space Flight Center, Greenbelt, Maryland 20771.

† Present address: Meteorological Research Institute, Yatabe-machi, Tsukuba-gun, Ibaraki-ken 305, Japan.

ment Service model which have 5, 11, and 15 levels. The first two versions have the top level at approximately 100 mbar, while the last version has the top at 10 mbar and is identical to the 11-level version below 100 mbar. He found that the differences between these forecasts were generally very small up to 48 hr. Derome (1981) analyzed the averaged fields for an ensemble of seven 10-day forecasts with a 15-level version of the European Centre for Medium Range Weather Forecasts (ECMWF) model. He found that errors in geopotential height increase with time at all levels while having maximum amplitudes at the uppermost levels. He also detected downward propagation of the geopotential error growth throughout the entire vertical domain in about 2 days. Simmons and Strüfing (1983) examined the sensitivity of tropospheric forecasts to stratospheric resolution using different versions of the ECMWF model. They compared forecasts with versions which have 14, 16, and 18 levels and top levels at 50, 25, and 10 mbar, respectively. They found that the forecasts for 500 mbar up to 10 days are largely insensitive to changes in the position of the top level. This conclusion was based on skill scores, which are integrated quantities. However, when they looked for impact on specific phenomena, they found some sensitivity of the planetary-wave structures at high latitudes.

In Mechoso *et al.* (1982) we studied the impact of an upper boundary in the lower stratosphere on tropospheric forecasts using two versions of the UCLA GCM which have 9 and 15 layers and tops at 51.8 and 1 mbar, respectively. We took a 15-day period from a long-term integration with the 15-layer version as the "control" and performed a "forecast" for this period with the 9-layer version. We found that significant "errors" in the geopotential field rapidly appeared in the uppermost model layers, and later propagated to all other layers. In particular, the ultralong waves in the forecast showed a strong tendency to quickly become equivalent barotropic due to reflection at the lowered upper boundary. Such a behavior of the ultralong waves progressively affected shorter waves. At 500 mbar, significant errors in the ultralong waves appeared within 5 days, while errors in the total field became large after 10 days. From this experiment we expect that a lowered position of the upper boundary can contaminate actual tropospheric forecasts after a few days, and that this contamination will become more apparent as models improve and the period of useful prediction is lengthened.

Our strategy in this paper consists of performing 10-day actual forecasts with four versions of the UCLA GCM that differ in horizontal resolution and/or vertical extent, and of analyzing in detail the accuracy of these forecasts. We have selected four cases from the Northern Hemisphere winter of 1979. Two of these cases are from January 1979, when there was a minor stratospheric warming; and the other two cases are from February 1979, when there was a major stratospheric warming.

We begin in Section 2 by describing some selected observed features of the tropospheric and stratospheric circulations during January and February of 1979. The UCLA GCM is briefly described in Section 3. Tropospheric and stratospheric forecasts are presented in Sections 4 and 5, respectively. In Section 6 we discuss the influence of the upper boundary on tropospheric forecasts.

2. Selected Features of the Atmospheric Circulation during the Northern Hemisphere Winter of 1979

In this section we describe some selected observed features of the tropospheric and stratospheric circulations of the Northern Hemisphere during the period 12 January to 27 February 1979. The data are identical to those used by Mechoso et al. (1985). They consist of the First GARP Global Experiment (FGGE) Level III-b analysis by ECMWF below 10 mbar, and United States National Meteorological Center (NMC) global height and temperature fields above that level. Winds above 10 mbar and at latitudes higher than 20° were obtained by using the geostrophic relations, while for

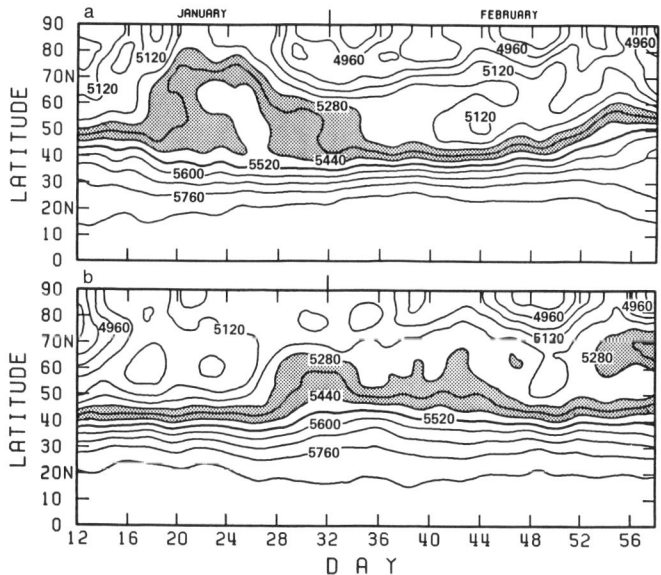

FIG. 1. Longitudinally averaged geopotential field Z (meters) at 500 mbar (a) between 90°W and 0° and (b) between 135°E and 135°W during the period 12 January 1979 to 27 February 1979.

lower latitudes they were obtained by linear interpolation between 20°N and 20°S.

We begin with the circulation at 500 mbar and consider two sectors of the Northern Hemisphere: the Atlantic Ocean sector (90°W to 0°), and the Pacific Ocean sector (135°E to 135°W). Figure 1 shows that there were episodes with persistent ridges at 500 mbar between about 40° and 70°N in both sectors. For the Atlantic Ocean sector, the major episode was roughly during 20–31 January. For the Pacific Ocean sector, it was after 20 February. These episodes were accompanied by splitting of the westerly flow into two branches separated by easterly flow, as shown in Fig. 2. They were also accompanied by strong zonally averaged northward eddy heat flux at high latitudes in the lower stratosphere, as shown in Fig. 3. This is revealing since the zonally averaged northward eddy heat flux is proportional to the vertical component of the Eliassen–Palm flux vector, which represents upward energy flux in the quasi-geostrophic framework. Therefore, both major episodes were associated with intense tropospheric forcing on the stratosphere. Figure 3 also shows that zonal wavenumber 1 was the principal contributor to the forcing during the January maximum, while zonal wavenumber 2 was dominant during the February maximum.

The stratospheric events of the Northern Hemisphere winter of 1979 have

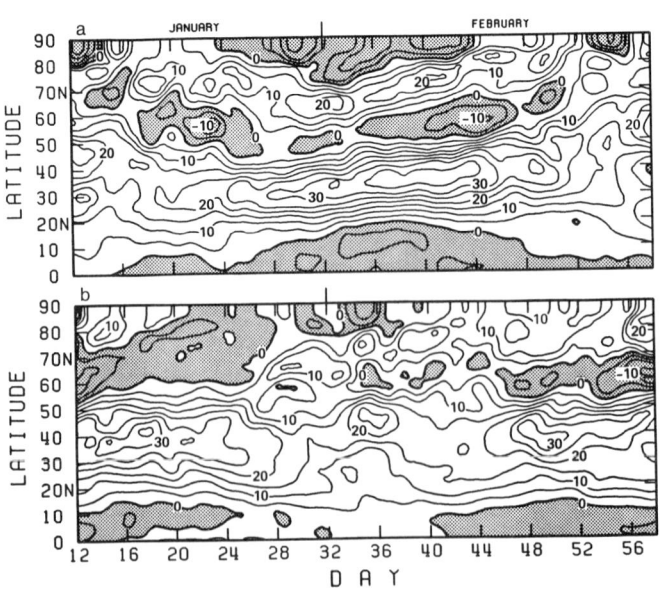

FIG. 2. As in Fig. 1, except for zonal wind u (m sec^{-1}).

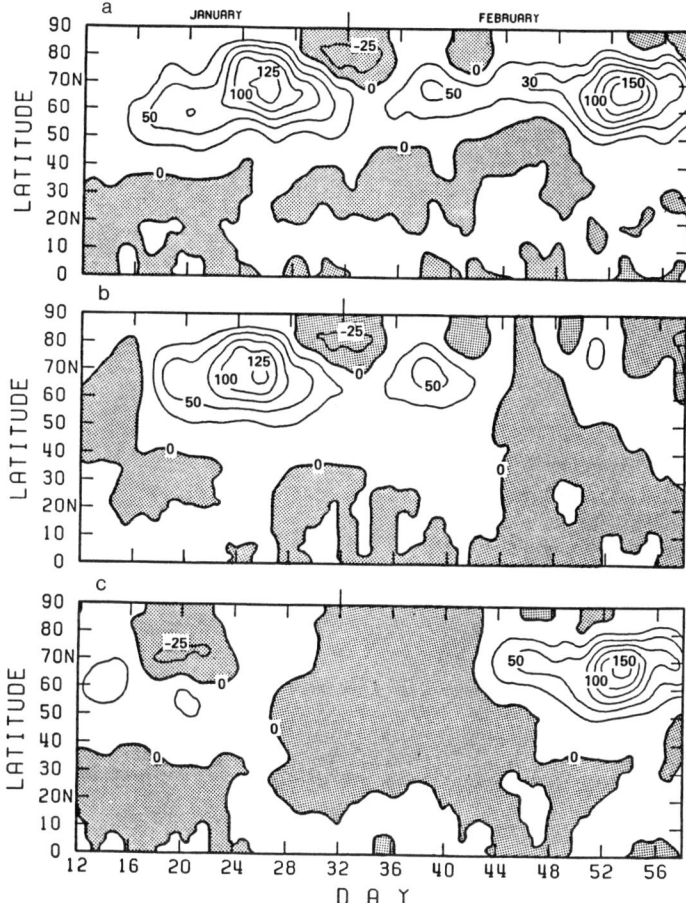

FIG. 3. Zonal mean northward eddy heat flux $[v^*T^*]$ (K m sec^{-1}) at 50 mbar for (a) all waves, (b) wavenumber 1, and (c) wavenumber 2 during the period 12 January 1979 to 27 February 1979.

been described in several studies (e.g., Quiroz, 1979; Palmer, 1981; McIntyre, 1982). At 10 mbar the typical wintertime circulation consisting of a circumpolar vortex changed during a minor warming in mid-January to a pattern consisting of a cyclonic vortex and an Aleutian high. This pattern persisted with slight variations in phase until it changed again during a major warming in February to another pattern consisting of an anticyclonic vortex over the polar region and two well-separated cyclonic vortices. The major warming reversed the direction of the zonal flow in deep stratospheric layers.

3. Description of the Model

The UCLA GCM is described mainly in Suarez *et al.* (1983). Briefly, the model is based on the primitive equations in a modified σ-coordinate system for which the lower boundary, the top of the planetary boundary layer (PBL), and isobaric surfaces above a prescribed pressure level (100 mbar), including the model top, are coordinate surfaces. The primary prognostic variables of the model are the surface pressure, the horizontal velocity, and the potential temperature. In addition, the model predicts the water vapor and ozone mixing ratios and the PBL depth. Formulation of the diabatic processes include parameterizations of cumulus convection (Arakawa and Schubert, 1974) and of boundary-layer processes (Suarez *et al.*, 1983) and a radiation calculation (Katayama, 1972; Schlesinger, 1976) with both seasonal and diurnal changes.

In this study we use the 9- and 15-layer versions of the model with low and high resolutions (4° of latitude by 5° of longitude and 2.4° of latitude by 3° of longitude, respectively). The 15-layer version has the top at 1 mbar and 6 layers above 51.8 mbar (with interfaces at 26.8, 13.9, 7.20, 3.73, and 1.93 mbar), resulting in a stratospheric resolution of about 5 km. The 9-layer version has the top at 51.8 mbar and is identical to the 15-layer version below that level except that ozone is prescribed.

For the primary prognostic variables, the initial conditions of the forecasts were obtained from the dataset described in Section 2. The surface pressure was obtained from the observed sea level pressure and geopotential fields. The horizontal velocity was then obtained by interpolation of the observed values. The geopotentials at the interfaces of the model's layers were also obtained by interpolation from the observed values. This particular interpolation procedure is designed such that the model's pressure gradient force is a proper interpolation of that on the pressure surfaces. The potential temperature was then obtained from those geopotentials using the model's hydrostatic equation. No further initialization procedures were performed for these variables.

To obtain initial conditions for the ozone mixing ratio, the PBL depth, and the water vapor mixing ratio with some degree of consistency with the model, we performed an "update cycle," which consists of a 5-day integration preceding each forecast period. For this integration, the initial PBL depth was 50 mbar everywhere, the water vapor mixing ratio was obtained from the dataset described in Section 2 below 300 mbar and was set equal to 2.5×10^{-6} above that level, and the initial ozone mixing ratio was zonally uniform and taken from Schlesinger and Mintz (1979). During the update cycle, the primary prognostic variables were updated at regular time intervals. Below 10 mbar, the updating interval was 6 hr during the FGGE Special

Observing Period, when data are available with that frequency, and 12 hr outside of that period. Above 10 mbar, the interval was 24 hr, since the NMC data are available only at 1200 GMT.

4. Tropospheric Forecasts

4.1. Forecast from 16 January

The geopotential field at 500 mbar for 16 January is shown in Fig. 4. Notable features are two cutoff lows, one at about 40°N and 25°E and another at about 30°N and 30°W, and the ridges at higher latitudes over western and eastern Europe. The North American region shows a mature cyclone over the northeastern part of the continent and a prominent trough west of the Rocky Mountains. The Pacific Ocean region is dominated by the Aleutian Low while the Asian region exhibits a high-latitude cyclone and a middle-latitude trough.

The 5-day low-resolution 9- and 15-layer forecasts and the high-resolution 9-layer forecast are shown in Fig. 5. All three forecasts predict the merging of

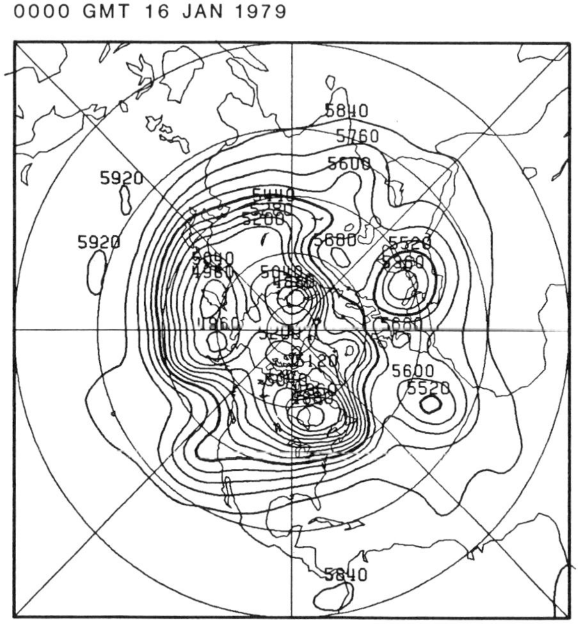

FIG. 4. Geopotential field (meters) at 500 mbar for 16 January.

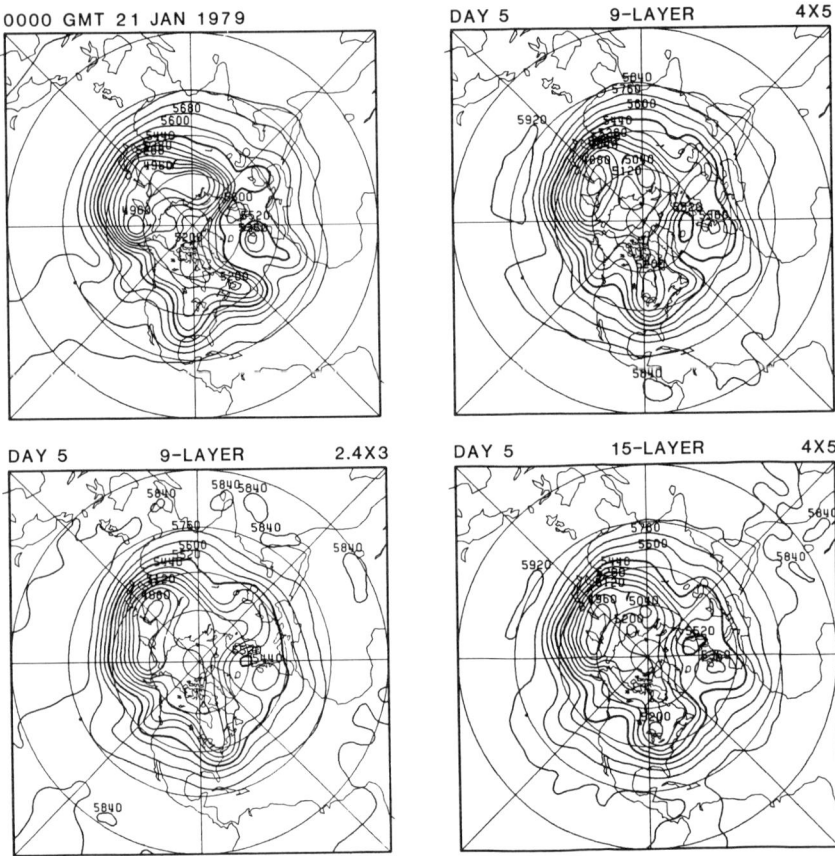

FIG. 5. Geopotential field (meters) at 500 mbar for 21 January (top left), and forecasts from 16 January with the low-resolution 9-layer version (top right), high-resolution 9-layer version (bottom left), and low-resolution 15-layer version (bottom right).

the ridges over northern Europe, the merging of the two cutoff lows, and the associated split of the westerly jet around 40°W. However, they fail to predict the deepening of the low off the east coast of North America, and placed the trough over the continent too far east.

The 5-day means from 21 to 26 January of the geopotential field at the 100-, 300-, 500-, and 700-mbar levels are shown in Fig. 6. These figures illustrate the vertical structure observed during the second half of the forecast period. The high over Greenland (the major episode over the Atlantic Ocean sector described in Section 2) and the low over the central North Atlantic Ocean have little tilt with height. On the other hand, the Aleutian Low has a

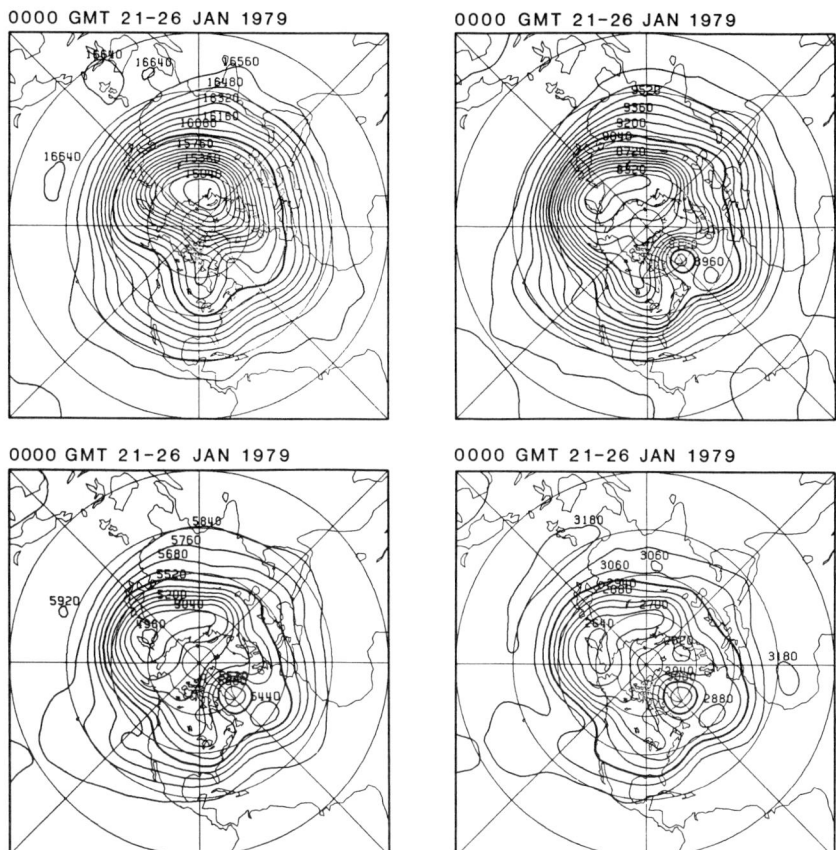

FIG. 6. Five-day means from 21 January to 26 January of the geopotential field at 100 (top left), 300 (top right), 500 (bottom left), and 700 mbar (bottom right).

pronounced westward tilt with height, particularly in the upper troposphere. The corresponding low-resolution 9-layer forecast, shown in Fig. 7, successfully predicts the equivalent barotropic structure over the North Atlantic Ocean, although the amplitudes are weaker than those observed. The predicted Aleutian Low also has an equivalent barotropic structure but here it disagrees with the observed. The low-resolution 15-layer forecast shown in Fig. 8, on the other hand, is more accurate in predicting the baroclinic structure of the Aleutian Low although it is less accurate over the North Atlantic Ocean. The predicted flow at 100 mbar by both versions is generally stronger than observed at midlatitudes, but weaker than observed over the polar region.

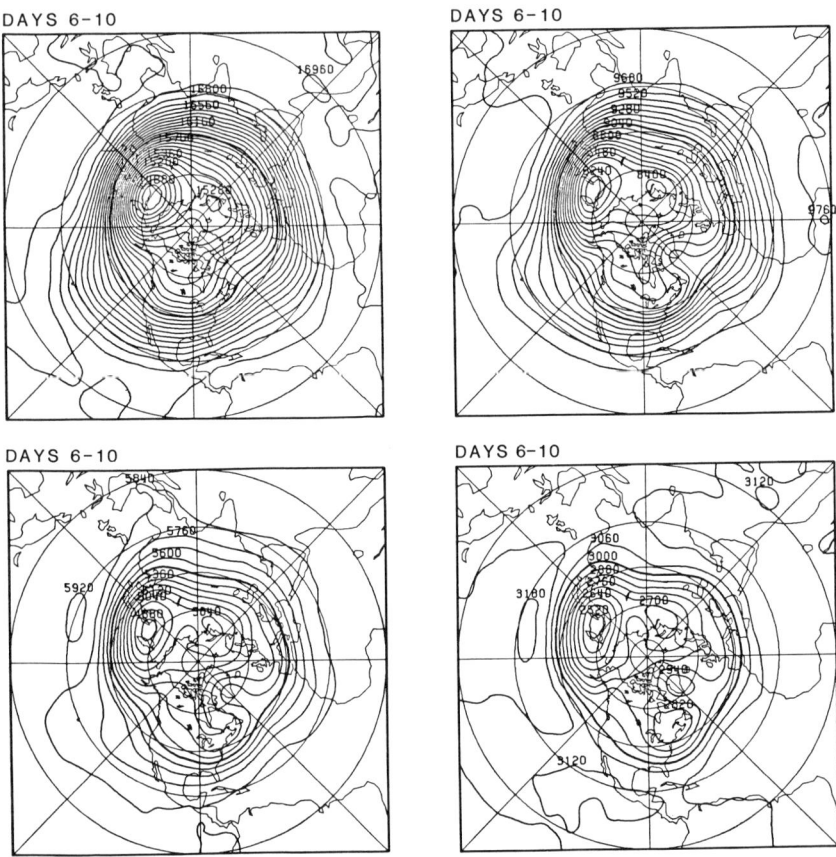

FIG. 7. As in Fig. 6, except for the low-resolution 9-layer forecast from 16 January.

4.2. Forecast from 21 January

The initial condition for the geopotential field at 500 mbar is in Fig. 5. The 5-day forecasts with the four versions of the model are shown in Fig. 9. The omega-shaped pattern over the North Atlantic Ocean, consisting of a high over the southern tip of Greenland flanked by two lows at lower latitudes, is visible in all forecasts, especially in those with the high resolution. On the other hand, the pronounced trough over the Rocky Mountains is underpredicted in all forecasts. The forecasts for high latitudes show some lack of detail. The accuracy of the forecasts decreases considerably after day 5.

A closer look at the evolution of zonal wavenumber 1 in geopotential height at 500 mbar, 60°N, is obtained from the harmonic dial shown in Fig.

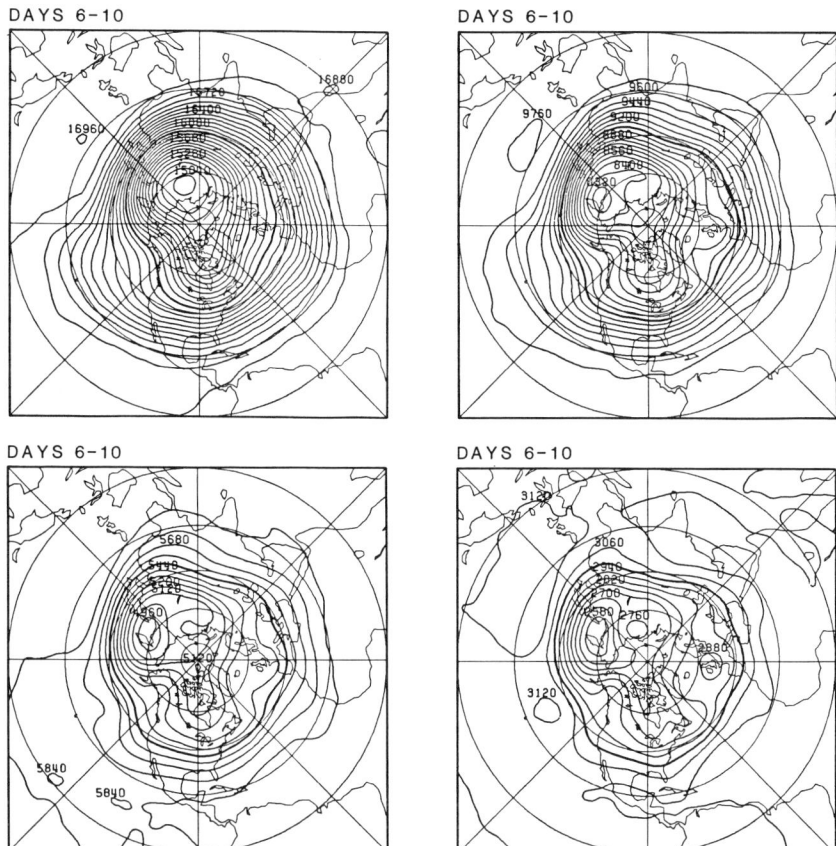

FIG. 8. As in Fig. 7, except for the low-resolution 15-layer forecast.

10. In this plot the distance from the origin represents the amplitude in meters, the right half of the horizontal axis corresponds to a ridge over the Greenwich Meridian, and a counterclockwise progression corresponds to an eastward translation of the ridge. Squares and circles on the lines indicate values at 1-day intervals. Wavenumber 1 decayed during this period while slowly moving westward (thick solid line). The forecasts show retrogression for the first few days, with the largest westward translation obtained from the low-resolution 9-layer forecast (dashed-dotted line), and the smallest from the high-resolution 15-layer forecast (dashed line). The predicted amplitude of wavenumber 1 decreases during the first half of all forecasts, but remains considerably higher than observed during the second half except for the low-resolution 15-layer forecast (thin solid line).

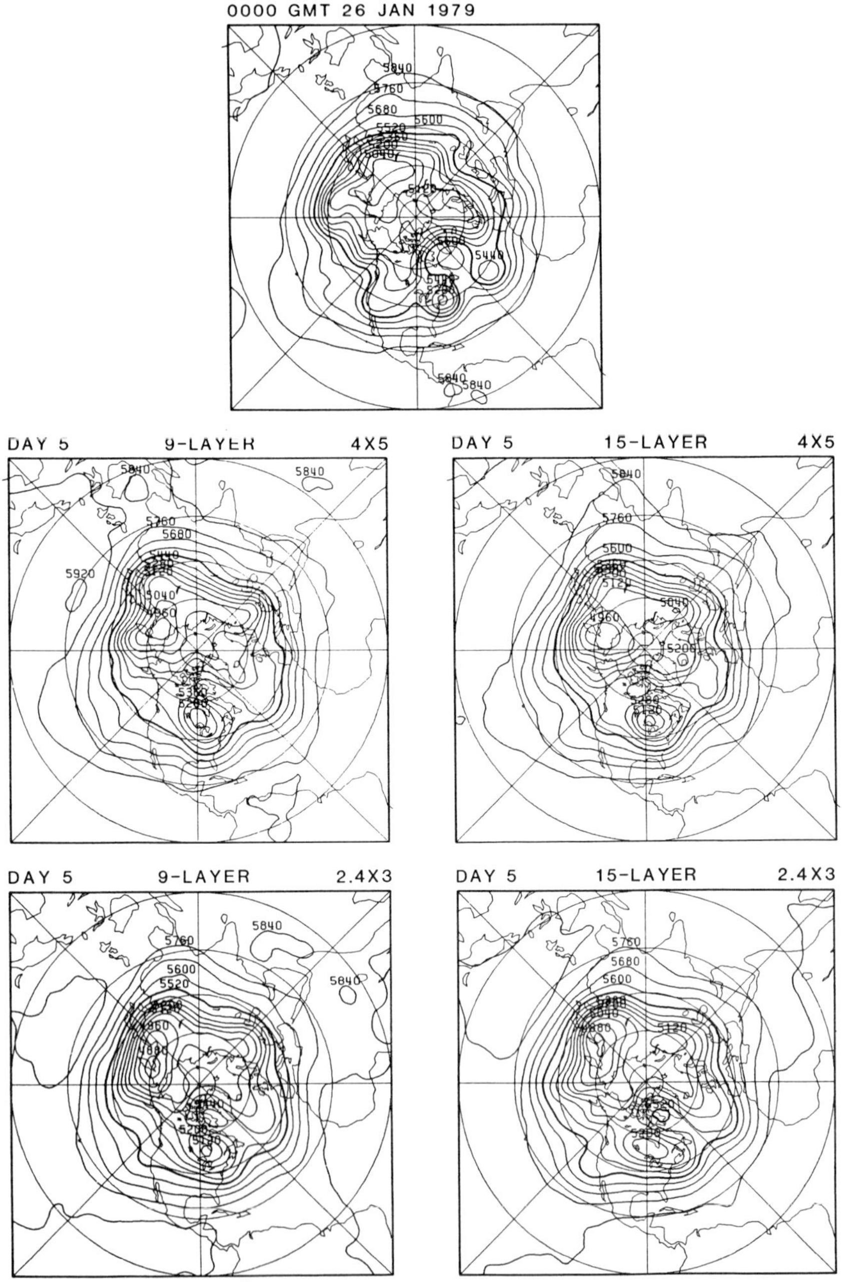

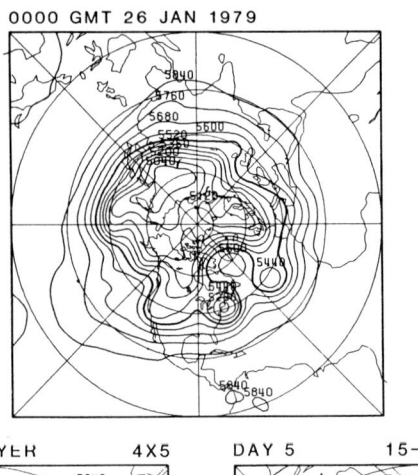

FIG. 9. Geopotential field (meters) at 500 mbar for 26 January (top), and forecasts from 21 January with the low-resolution 9-layer version (middle left), low-resolution 15-layer version (middle right), high-resolution 9-layer version (bottom left), and high-resolution 15-layer version (bottom right).

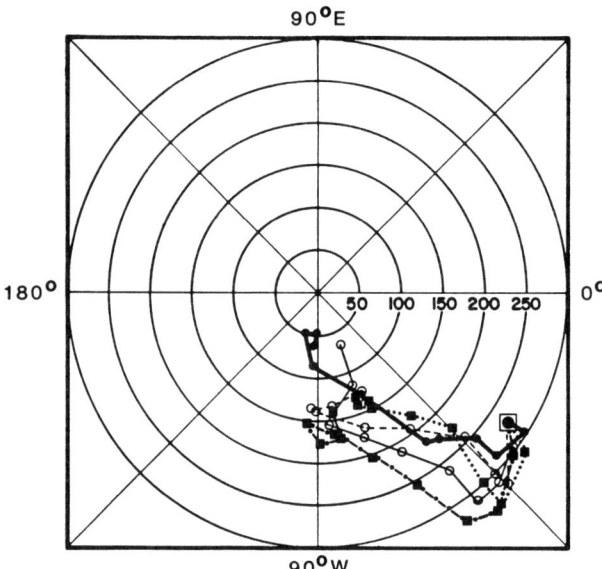

FIG. 10. Harmonic dial for zonal wavenumber 1 in observed 500-mbar geopotential height at 60°N during the period 21 January to 31 January (thick solid lines) and forecasts with the low-resolution 9-layer version (dashed-dotted lines), low-resolution 15-layer version (thin solid lines), high-resolution 9-layer version (dotted lines), and high-resolution 15-layer version (dashed lines). Other notations are explained in the text.

Whatever success the forecasts from 21 January have is largely due to the reasonably good forecast of the decay of zonal wavenumber 1, since the forecast of remaining ultralong waves are poor after 2 days. This is seen clearly in Fig. 11, which shows the histories at 500 mbar, 60°N, of the superposition of waves 2–4 in geopotential field. The observed field shows an eastward-propagating wave train which seems to originate early in the period east of the Rocky Mountains. Such a wave train results in an intensification of the omega shaped pattern over the North Atlantic Ocean and seems to be related to an intensification of a ridge west of North America. (Inspection of other levels reveals that this wavetrain has little tilt with height.) All forecasts performed poorly in this respect.

4.3. Forecast from 15 February

The initial geopotential field at 500 mbar is shown in Fig. 12. There are pronounced ridges over the eastern North Atlantic Ocean and over the Bering Strait. There is also an intense, cyclonic circulation at very high

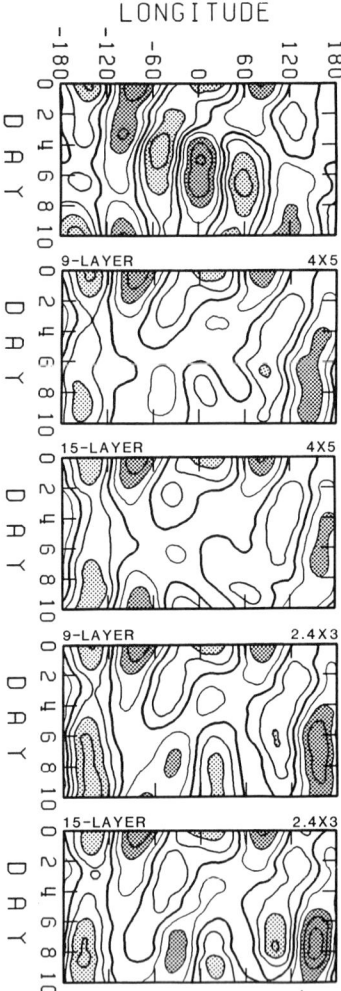

FIG. 11. Contribution to the geopotential field (meters) at 60°N, 500 mbar by wavenumbers 2–4 during the period 21 January to 31 January. From top to bottom: observed, low-resolution 9-layer version, low-resolution 15-layer version, high-resolution 9-layer version, and high-resolution 15-layer version. The contour interval is 60 m. Values greater than 120 m are lightly shaded, and values less than −120 m are heavily shaded.

0000 GMT 15 FEB 1979

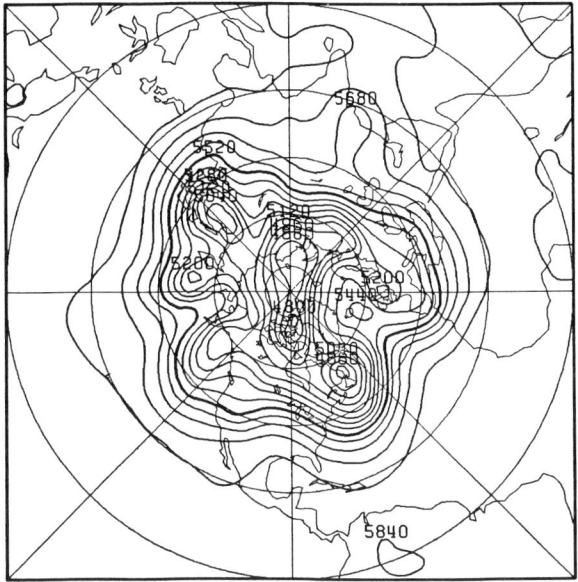

FIG. 12. Geopotential field (meters) at 500 mbar for 15 February.

latitudes with a clear wavenumber 2 pattern. The intense flows at very high latitudes over the prime meridian and the dateline remain 5 days after, as shown in Fig. 13. The ridge over the eastern North Atlantic Ocean is more intense than on 15 February and extends over northern Europe, poleward of a trough over the Mediterranean Sea. The ridge over the North Pacific Ocean (the major episode over the Pacific Ocean sector discussed in Section 2) is also more intense than on 15 February. There is a deep low over northeastern North America and a cyclone over northeastern Asia. Inspection of Fig. 13 reveals that, in general, the high-resolution forecasts are considerably more accurate than the low-resolution forecasts, particularly over the North Atlantic and Pacific Oceans. All forecasts show a much weaker flow over the polar region than observed.

The 5-day means from 20 February to 25 February of the observed and predicted geopotential field at 500 mbar, shown in Fig. 14, illustrate the higher accuracy of the high-resolution forecasts in the prediction of quasi-stationary features of the circulation during the second half of the period. The ridges over northwestern Europe and over the Bering Sea and the lows over northern North America and Siberia are well predicted, reflecting a good forecast of the evolution of wavenumber 2, which was very important

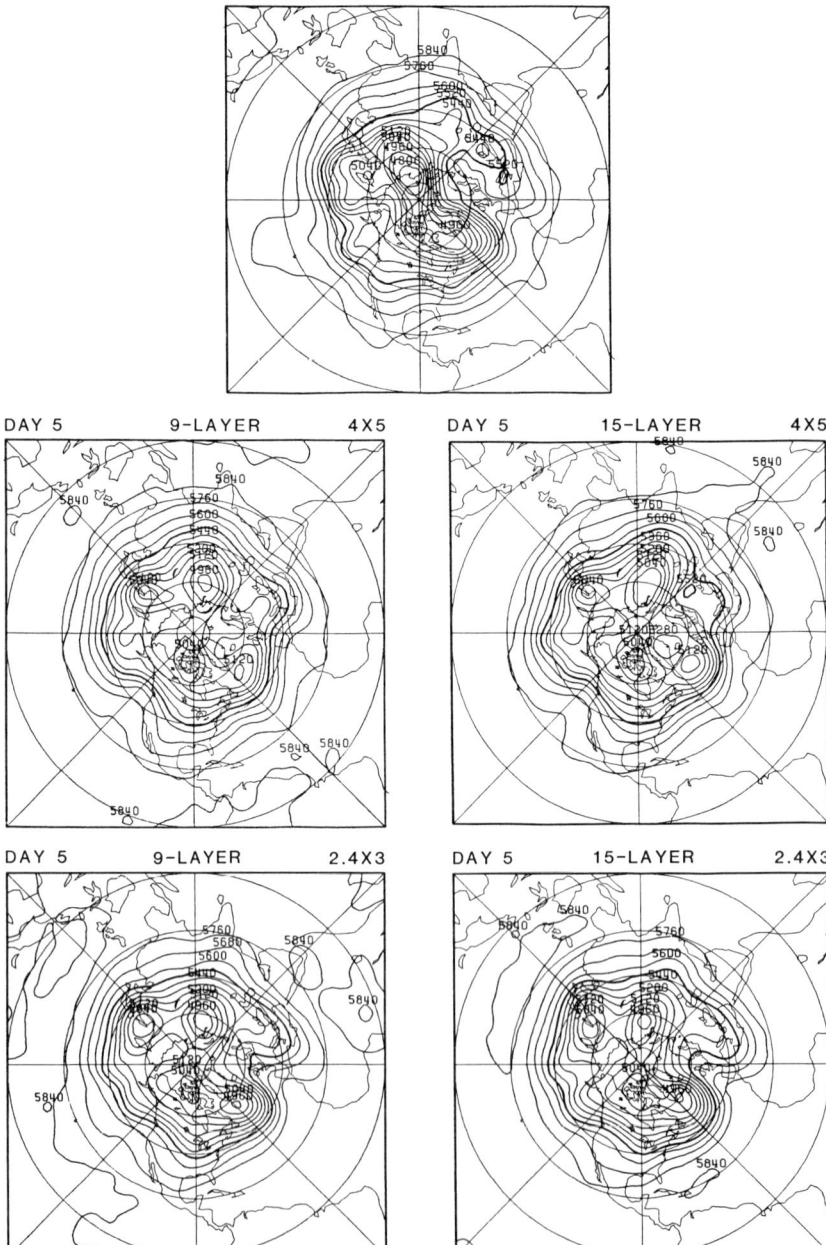

FIG. 13. As in Fig. 9, except for 20 February and forecasts from 15 February.

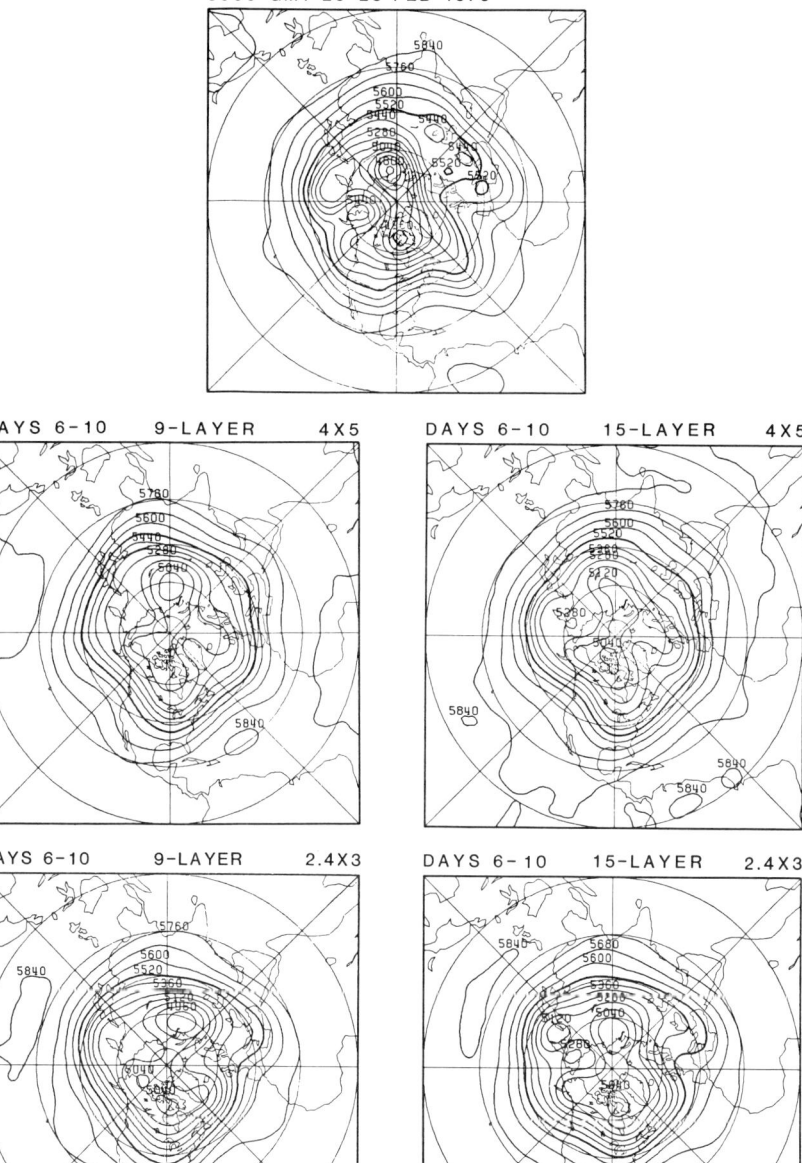

FIG. 14. As in Fig. 13, except for the 5-day means from 20 February to 25 February.

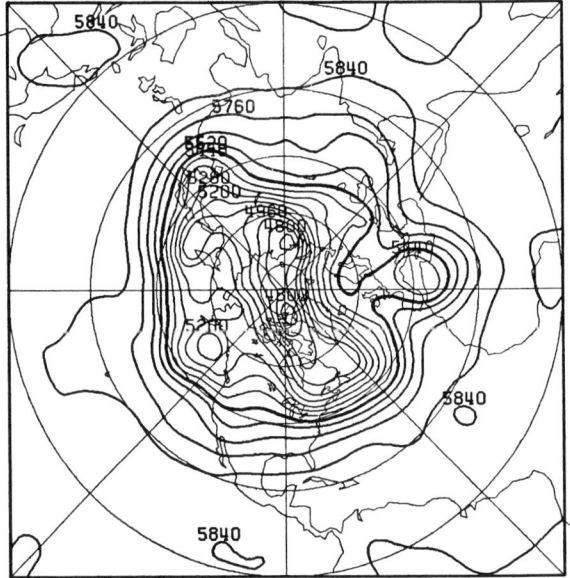

FIG. 15. Geopotential field (meters) at 500 mbar for 17 February.

during that period. On the other hand, the predicted midlatitude circulation over North America shows a trough (ridge) over the eastern (western) part of the continent, in disagreement with the observed situation.

4.4. Forecast from 17 February

The geopotential field at 500 mbar for 17 February is shown in Fig. 15, and the 5-day forecasts for that level are shown in Fig. 16. The ridge over the eastern North Atlantic is predicted with generally weaker amplitudes. A common failure of all forecasts, particularly of those with the low-resolution versions, is on the northern Pacific Ocean region where the ridge over the Bering Strait is greatly underpredicted. The accuracy of all forecasts decreases notably after day 5.

5. STRATOSPHERIC FORECASTS

These are discussed in detail in Mechoso *et al.* (1985). We summarize some of the results in what follows.

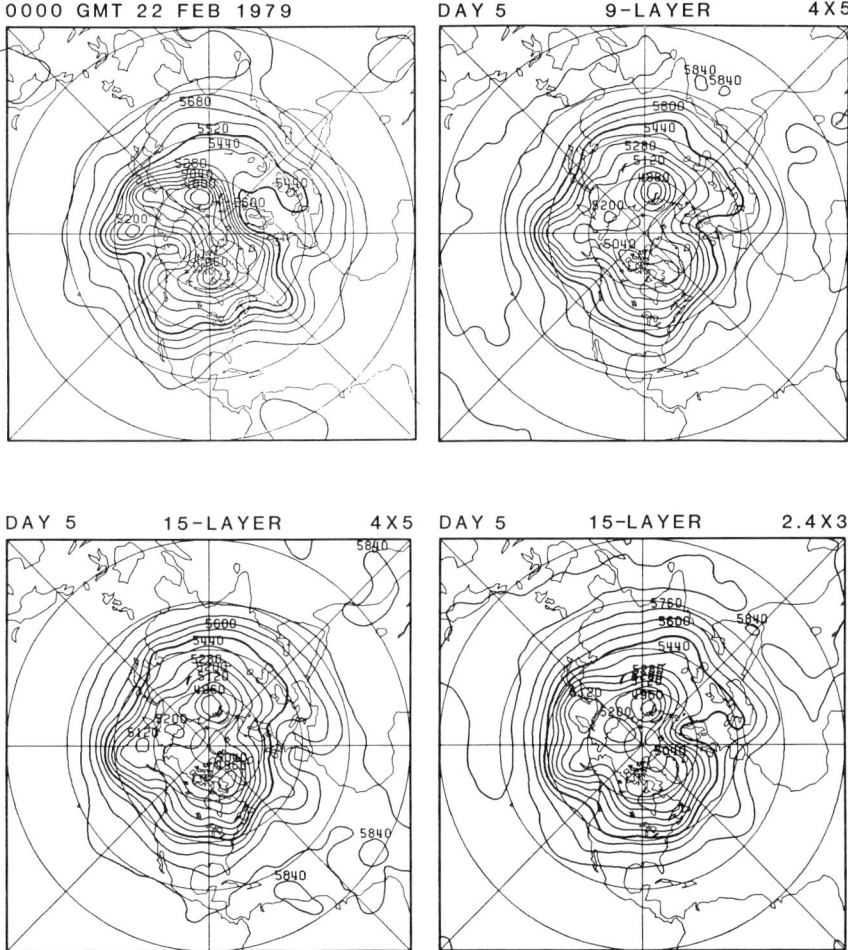

FIG. 16. Geopotential field (meters) at 500 mbar for 22 February (top left), and forecasts from 17 February with the low-resolution 9-layer version (top right), low-resolution 15-layer version (bottom left), and high-resolution 15-layer version (bottom right).

5.1. Forecasts from 16 January and 21 January

The 10-mbar geopotential field for 16 January is shown in Fig. 17. The 5- and 10-day low-resolution 15-layer forecasts from 16 January are shown in the same figure. The transition that occurred at this level, from a circumpolar cyclonic vortex to a two-vortex system, consisting of a cyclonic vortex

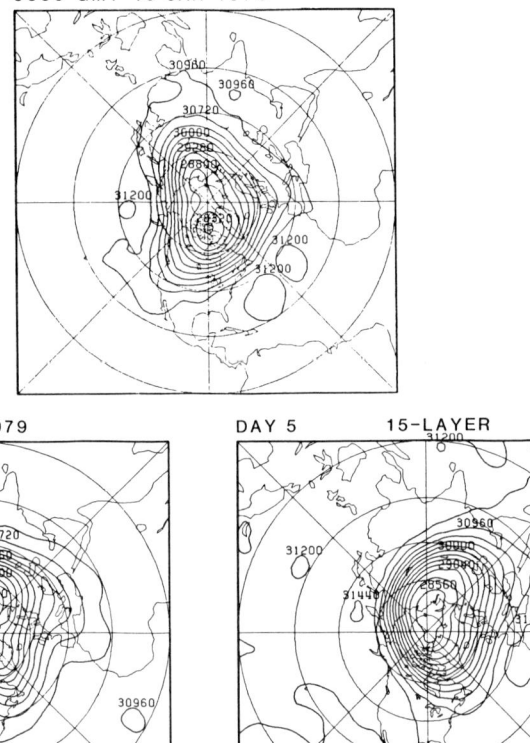
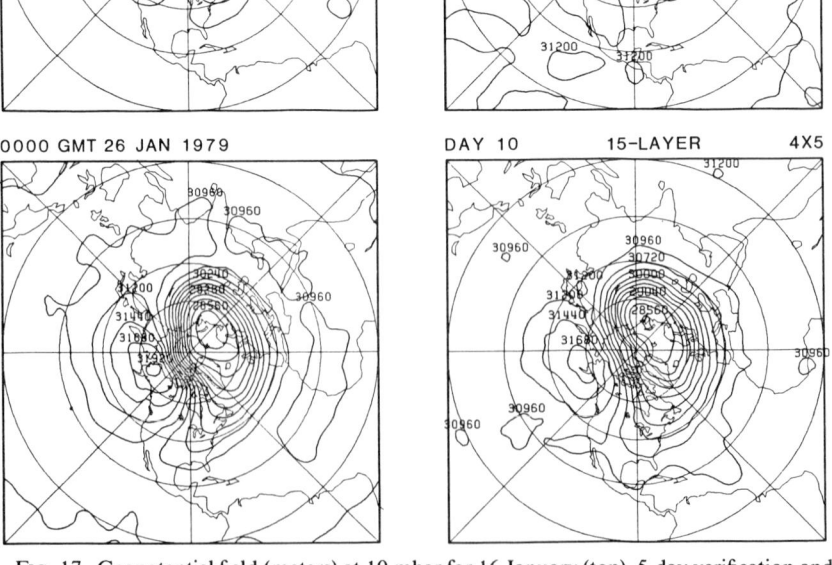

Fig. 17. Geopotential field (meters) at 10 mbar for 16 January (top), 5-day verification and forecast (middle row), and 10-day verification and forecast (bottom row) with the low-resolution 15-layer version.

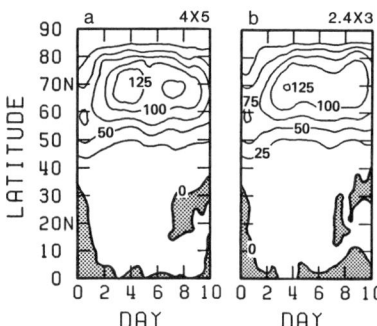

FIG. 18. Zonal-mean northward eddy heat flux v^*T^* (K m sec^{-1}) at 50 mbar for the forecasts from 21 January with (a) the low-resolution 15-layer version and (b) high-resolution 15-layer version.

with strong cross-polar flow and an Aleutian high, is accurately predicted. The 10-mbar forecasts from 21 January (not shown; see Mechoso et al., 1985) are less accurate than the forecast from 16 January. Nevertheless, the forecasts from 21 January are significantly accurate in the lower stratosphere. For example, the high values of the zonally averaged northward eddy heat flux at 50 mbar with both horizontal resolutions (see Fig. 18) compare favorably with those observed (see Fig. 3). Furthermore, we have confirmed that zonal wavenumber 1 is the principal contributor to the eddy heat flux at this level during this period, as observed.

5.2. Forecasts from 17 February and 15 February

Figures 19 and 20 show that the high-resolution forecast from 17 February is very successful at both 10 and 100 mbar. The breakdown of the two-vortex pattern and the development of an anticyclonic circulation over the polar region and two well-separated cyclonic vortices are accurately predicted. As discussed in Mechoso et al. (1985), the observed sequential amplification of zonal wavenumbers 1 and 2 are well predicted. The low-resolution forecast is somewhat less successful, particularly during the second half of the period. The analysis presented in Mechoso et al. (1985) indicates that such a poorer performance is associated with a poorer forecast of wave propagation in the upper troposphere resulting from errors in the predicted zonal-mean flow there.

The considerable success of the forecasts from 15 February for 500 mbar discussed in Section 4.3 does not extend to the lower stratosphere. Figure 21 shows the zonally averaged northward eddy heat flux at 50 mbar with both

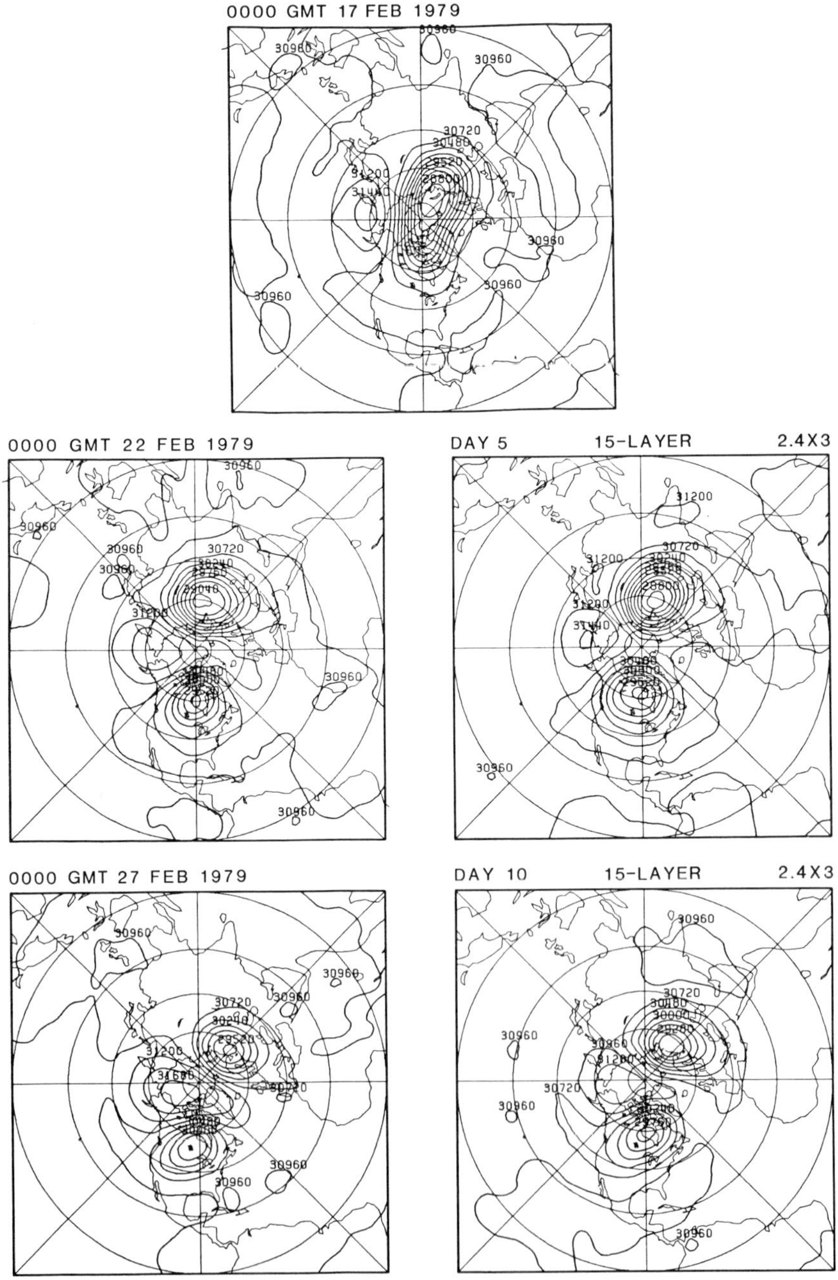

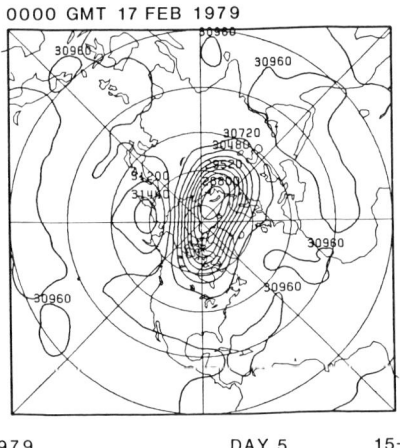

FIG. 19. As in Fig. 17, except for initial condition and forecasts from 17 February with the high-resolution 15-layer version.

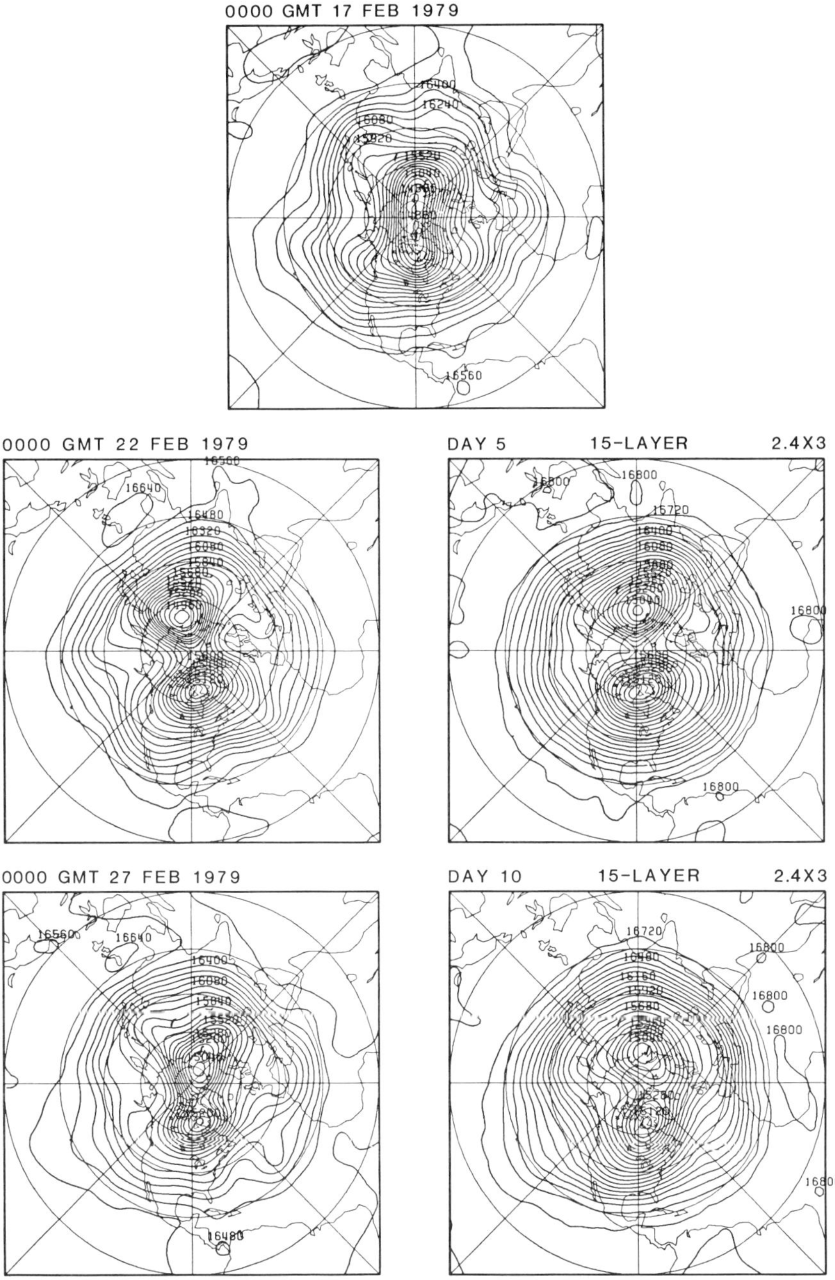

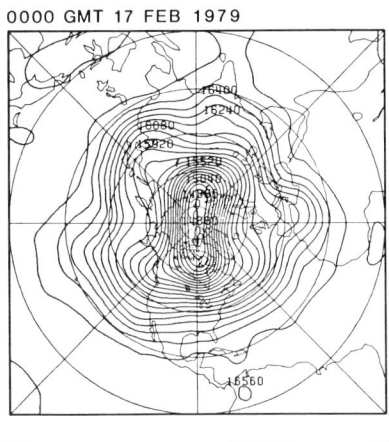

FIG. 20. As in Fig. 19, except at 100 mbar.

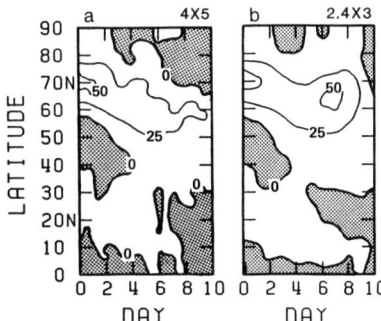

FIG. 21. As in Fig. 18, except for the forecasts from 15 February.

horizontal resolutions. Comparison with Fig. 3 shows that the flux is seriously underpredicted, revealing deficiencies in the predicted tropospheric forcing to the stratosphere. Associated with this feature, the forecasts from 15 February show only a tendency to predicting the observed breakdown of the cyclonic vortex at 10 mbar, failing in completing the breakdown (see Fig. 22). Again, this failure is attributed in Mechoso et al. (1985) to errors in the predicted zonal-mean upper tropospheric flow. These errors have small magnitudes during the first few days of forecast, but crucially alter the characteristics of upward wave propagation.

6. Impact of the Upper Boundary on Tropospheric Forecasts

To investigate the impact of the model's artificial upper boundary on the tropospheric forecasts, we analyze the differences between those with the 9- and 15-layer versions of the model. For illustration we have chosen the cases from 21 January and 15 February. The 15-layer forecasts have smaller root-mean-square (rms) errors with the low resolution in the 21 January case and with the high resolution in the 15 February case. In what follows, therefore, we concentrate on the low-resolution forecasts for the former case and on the high-resolution forecasts for the latter case.

6.1. Root-Mean-Square Error Fields

We begin by examining time–height plots of rms errors in geopotential height for the 9- and 15-layer forecasts at a selected latitude. Figures 23 and 24 show such plots at 60°N for the low-resolution 9- and 15-layer forecasts from 21 January, and high-resolution 9- and 15-layer forecasts from 15

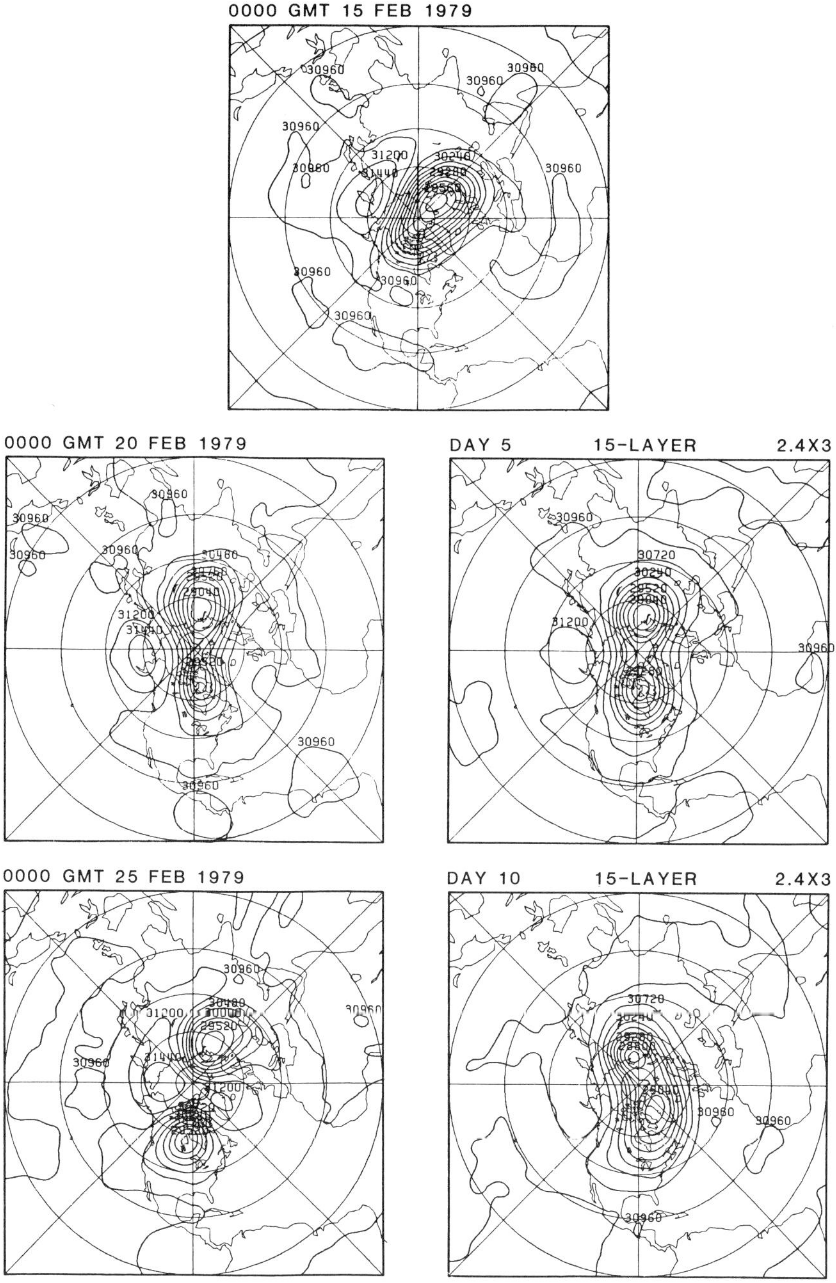

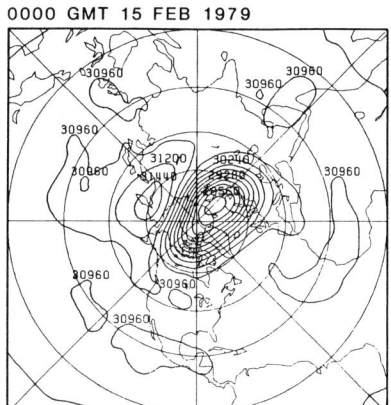

FIG. 22. As in Fig. 19, except from 15 February.

February, respectively. Figures 23a and 24a show that the 9-layer forecast errors develop rapidly above 200 mbar and more slowly at lower levels. Figures 23b and 24b show that the 15-layer forecast errors also develop rapidly at the uppermost levels, and that they have a local minimum at around 70 mbar. This level is close to the location of the top of the 9-layer model and, consequently, to the site of maximum errors in the 9-layer forecasts. The 15-layer forecast errors for both cases show local maxima at around day 7 at 250 mbar. These maxima are associated with an increase in amplitude of the long waves (zonal wavenumbers 3 and 4) that was observed during the second half of both forecast periods in the upper troposphere. The 9-layer forecast errors do not exhibit such distinct features since they have merged with the region of large errors close to the model's top.

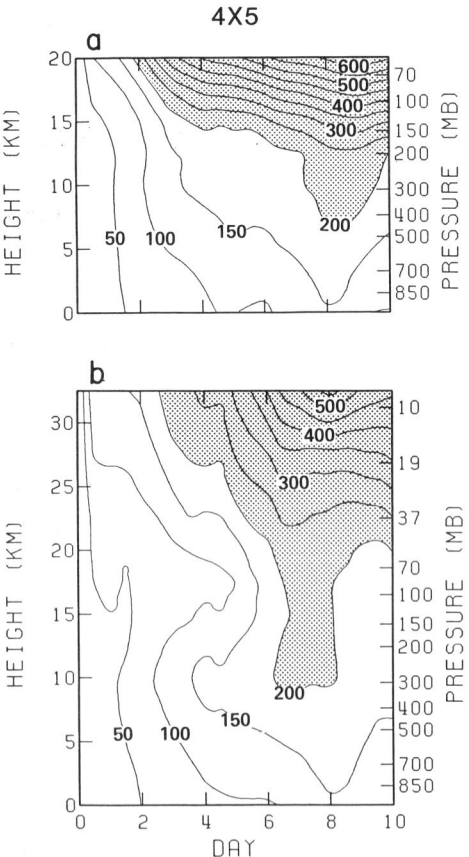

FIG. 23. Root-mean square of the height error at 60°N for forecasts from 21 January with (a) the low-resolution 9-layer version and (b) low-resolution 15-layer version.

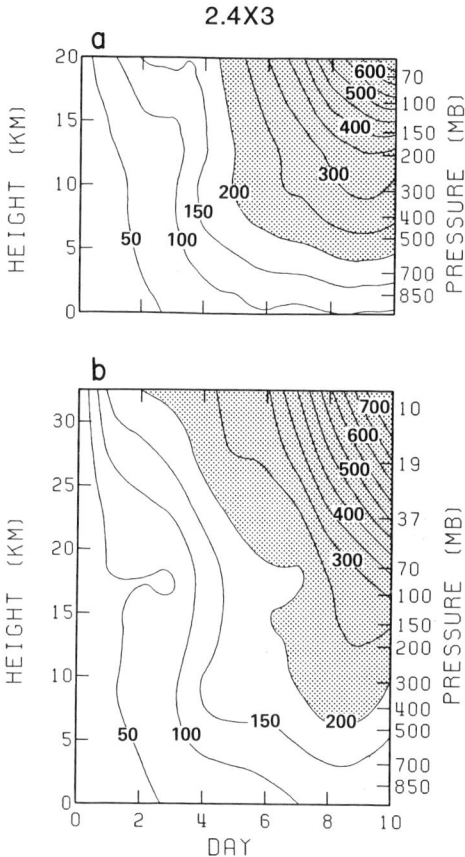

FIG. 24. As in Fig. 23, except from 15 February with high-resolution versions.

The errors depicted in Figs. 23 and 24 are due to a complex combination of factors, such as the lack of horizontal and/or vertical resolutions, deficiencies in the parameterizations of subgrid scale processes, and errors in the initial conditions. In addition, the artificial upper boundary contributes to the errors. Figures 23 and 24 clearly show that this is the case for the uppermost model layers and suggest downward propagation into the middle and lower troposphere, contaminating the forecast in that region. We address this problem in what follows.

The rms of the differences in geopotential height between the 9- and 15-layer forecasts selected above are shown in Figs. 25 and 26. The patterns here are similar to those in Fig. 8 of Mechoso *et al.* (1982). They are also similar to those in Figs. 23a and 24a herein, although the values are some-

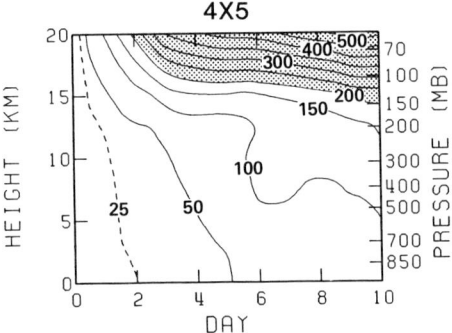

FIG. 25. Root-mean square of the height difference at 60°N between forecasts from 21 January with the low-resolution 9-layer and 15-layer versions.

what smaller in Figs. 25 and 26. These results confirm that, as expected, the lowered upper boundary has a considerable impact in the 9-layer forecasts for the lower stratosphere and upper troposphere. No conclusive results on such an impact can be reached from these plots, however, for the middle and lower troposphere. We have to look in more detail to the geographic distributions of the errors.

6.2. Error Distributions at 500 mbar

Let A_9 be a variable in the 9-layer forecast, A_{15} be the corresponding variable in the 15-layer forecast, and A be the observed value of that variable.

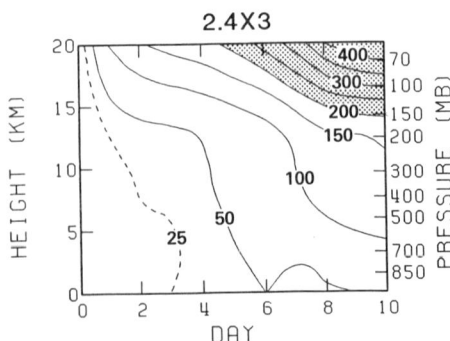

FIG. 26. As in Fig. 25, except between forecasts from 15 February with high-resolution versions.

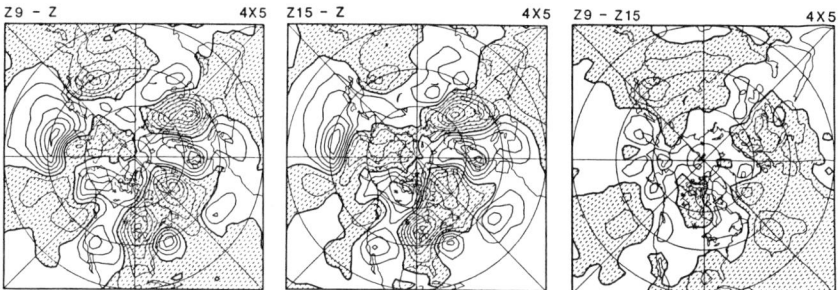

FIG. 27. The terms in Eq. (1) for geopotential height at 500 mbar for the 5-day forecast from 21 January with low-resolution versions. Term (I), left; term (II), middle; and term (III), right. The contour interval is 60 m and negative values are shaded.

Then the following relation for the 9-layer forecast error holds:

$$\underset{(\mathrm{I})}{(A_9 - A)} = \underset{(\mathrm{II})}{(A_{15} - A)} + \underset{(\mathrm{III})}{(A_9 - A_{15})} \quad (1)$$

Term (II) is the 15-layer forecast error and term (III) is the error due to the lowered upper boundary.

The terms in Eq. (1) are plotted for geopotential height at 500 mbar for the 5-, 7-, and 10-day forecasts from 21 January in Figs. 27, 28, and 29, respectively. At day 5, term (III) is small everywhere compared to terms (I) and (II). Term (III), however, gradually increases in magnitude with time, and at day 10 it becomes comparable to or even exceeds term (II) at several locations. The terms in Eq. (1) have a similar pattern of evolution for the 15 February case as shown in Figs. 30, 31, and 32, although the magnitude of term (III) remains considerably smaller than the magnitude of terms (I) and (II).

The above results illustrate that the error due to the lowered upper bound-

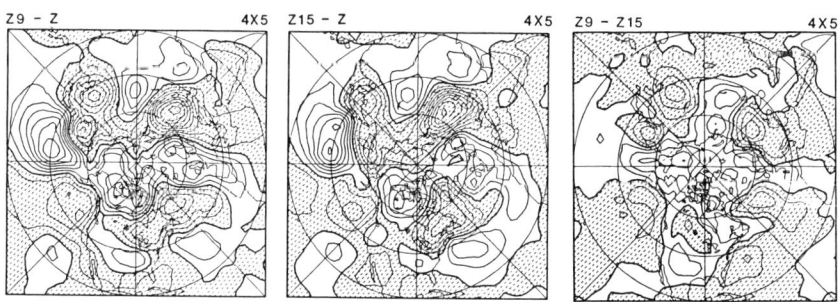

FIG. 28. As in Fig. 27, except for the 7-day forecast.

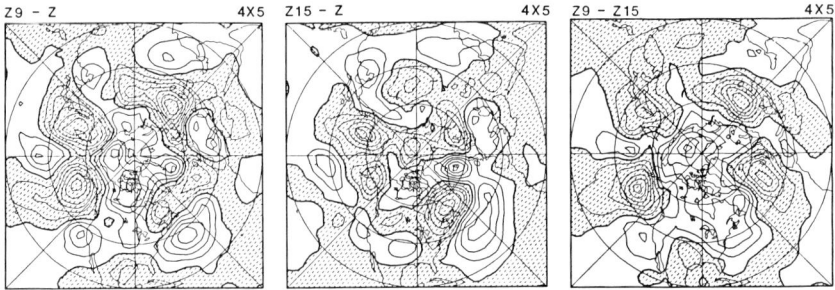

FIG. 29. As in Fig. 27, except for the 10-day forecast.

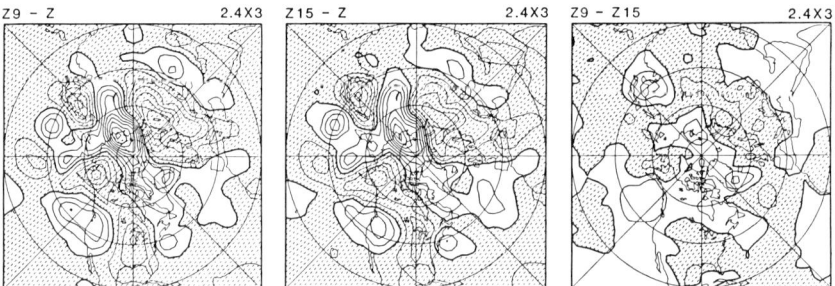

FIG. 30. As in Fig. 27, except for the 5-day forecast from 15 February with high-resolution versions.

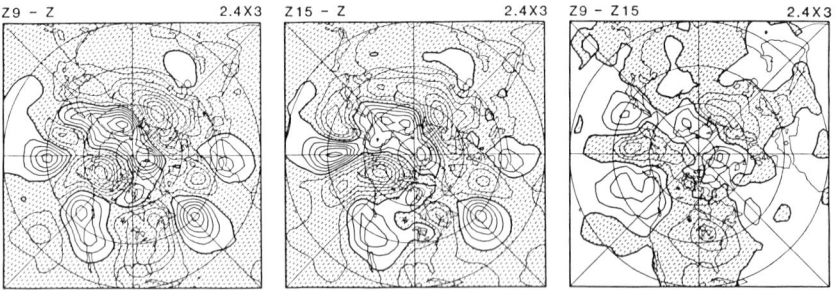

FIG. 31. As in Fig. 30, except for the 7-day forecast.

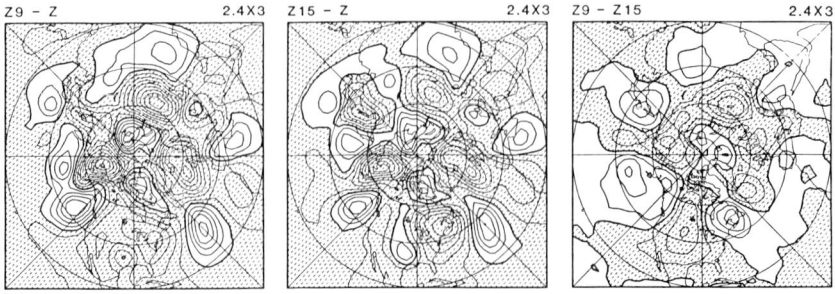

FIG. 32. As in Fig. 30, except for the 10-day forecast.

ary can become comparable to the errors due to other sources, at least for some cases. The large difference between the 15- and 9-layer forecasts for the 21 January case and the small difference for the 15 February case will be interpreted in the following discussions.

6.3. Ultralong Wave Forecasts

Figures 33 and 34 show the observed and predicted amplitude and phase of zonal wavenumber 1 in geopotential height at 60°N for the 5- and 10-day

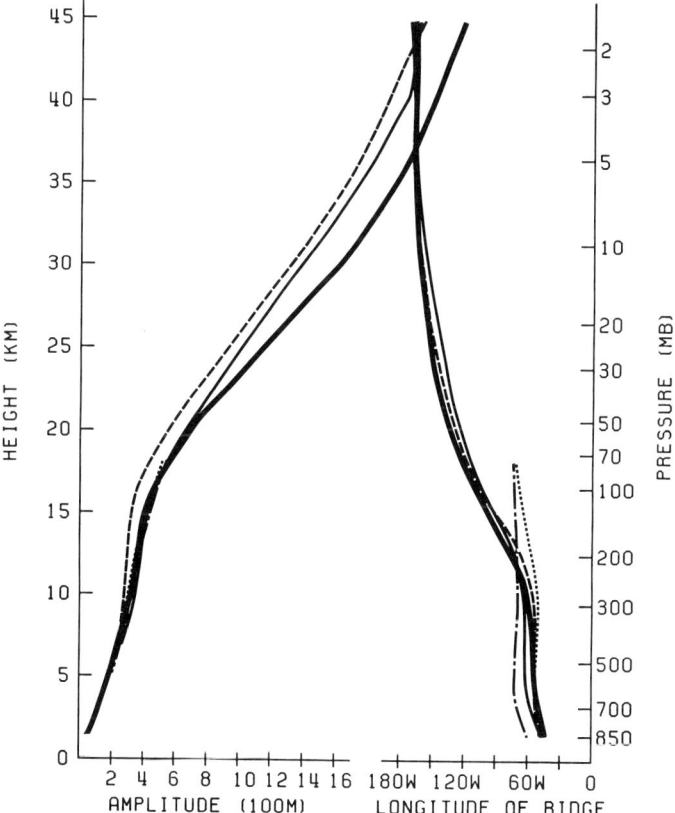

FIG. 33. Amplitude and phase (position of the ridge) of zonal wavenumber 1 in geopotential height at 60°N on 26 January (thick-solid lines) and 5-day forecasts with the low-resolution 9-layer version (dashed-dotted line), low-resolution 15-layer version (thin solid lines), high-resolution 9-layer version (dotted lines), and high-resolution 15-layer version (dashed lines).

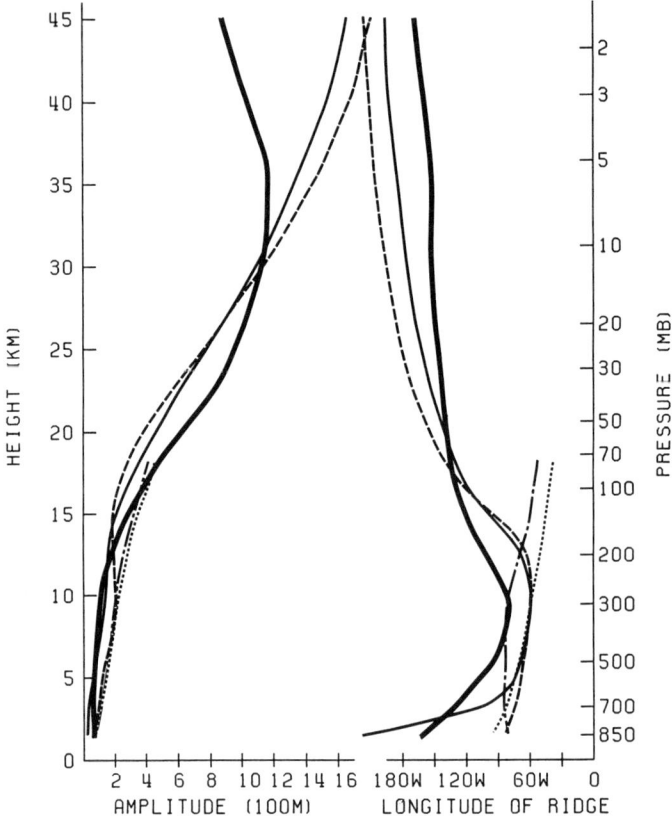

FIG. 34. As in Fig. 33, except for 31 January and 10-day forecasts.

forecasts from 21 January, respectively. At day 5, the wave amplitude is generally well predicted in the troposphere by all forecasts and underpredicted in the stratosphere by both 15-layer forecasts. The 9- and 15-layer forecasts for the wave phase exhibit significant differences: the 9-layer forecasts show a tendency toward reducing the vertical tilt with height, particularly above 300 mbar, while the 15-layer forecasts are quite accurate. At day 10, the wave amplitude is underpredicted in the lower stratosphere but overpredicted in the upper stratosphere by both 15-layer forecasts. The forecast errors for wave phase are generally larger than at day 5. However, the tendency of the 9-layer forecasts toward reducing the vertical tilt with height remains apparent. The significant retrogression of the wave below 500 mbar during the second half of the forecast period is qualitatively predicted by the low-resolution 15-layer forecast and ignored by all other versions.

The ultralong wave with zonal wavenumber 2 was predominant during the period of the forecast from 15 February. Figures 35 and 36 show the observed and predicted amplitude and phase of zonal wavenumber 2 at 60°N for that case. At day 5, the wave amplitude is underpredicted by all forecasts in the troposphere and lower stratosphere. The high-resolution forecasts seem to be the most accurate there. Contrary to the previous case, there are no major differences between the 9- and 15-layer forecasts for the wave phase. All of them show the same tendency toward reducing the vertical tilt with height in the middle and upper troposphere and in the lower stratosphere. At day 10, the errors are larger than at day 5 and the high-resolution versions show superior performance in predicting the wave amplitude.

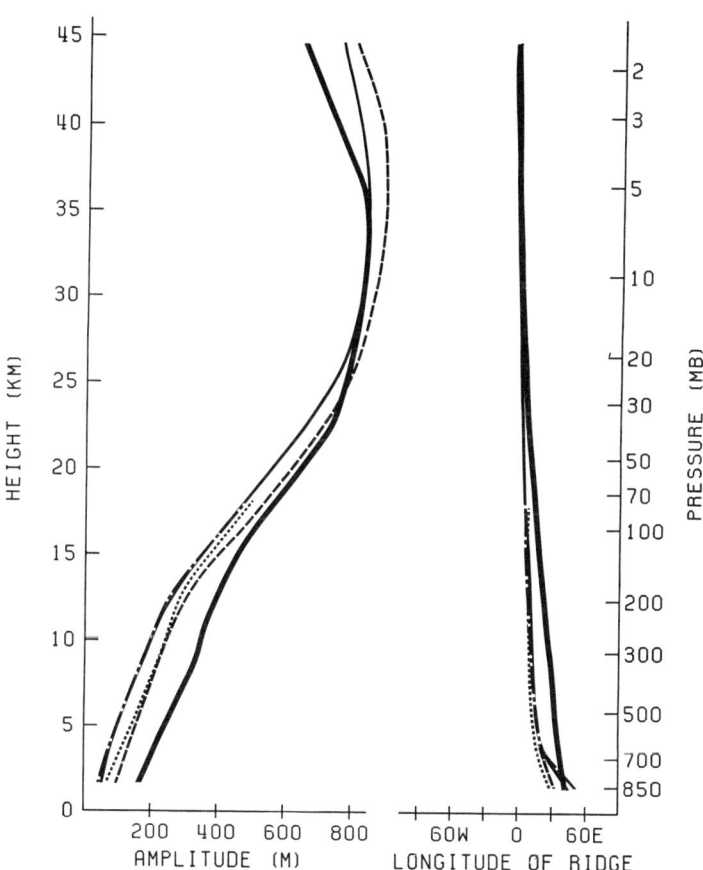

FIG. 35. As in Fig. 33, except for 20 February and forecasts from 15 February of zonal wavenumber 2.

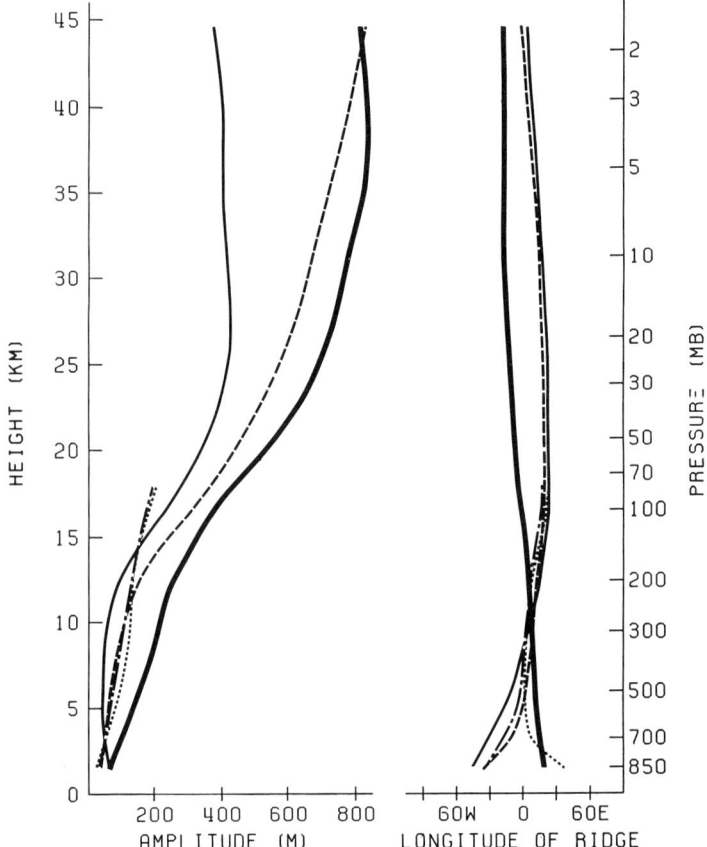

FIG. 36. As in Fig. 35, except for 25 February and 10-day forecasts.

6.4. On the Different Impact of the Lowered Upper Boundary for the Forecasts from 21 January and 15 February

We have shown that the 9- and 15-layer forecasts at 500 mbar are significantly different at day 10 in the 21 January case, while they are similar in the 15 February case. To interpret this, we first analyze the differences and similarities between the forecasts in the framework of the theory on the vertical propagation of planetary waves. According to the theory (Matsuno, 1970), the propagation characteristics of a linear, steady, planetary wave with zonal wavenumber k are given by the values of its refractive index squared, Q_k, defined by

$$Q_k = \frac{a\partial[q]/\partial\phi}{[u]} - \frac{k^2}{\cos^2\phi} - \frac{a^2 f^2}{4N^2 H^2} \quad (2)$$

Here u is the zonal component of the horizontal velocity, q is the quasi-geostrophic potential vorticity, ϕ is latitude, f is the Coriolis parameter, N is the Brunt–Väisälä frequency, H is the scale height, and a is the radius of the earth. Butchart *et al.* (1982) showed that higher (negative) values of Q_k increase (prevent) propagation of the planetary wave with zonal wavenumber k even in unsteady cases.

The Q_1 field for the 2-day, low-resolution 15-layer forecast from 21 January and the corresponding observed field are shown in Fig. 37. The broad regions with positive values in both fields indicate the possibility of vertical propagation for wavenumber 1. The upward propagation of this wave is predicted by the 15-layer forecast as indicated by the successful forecast of the minor stratospheric warming that occurred during this period. On the other hand, the structure of the wave in the 9-layer forecast is affected in the lower stratosphere according to the constraints imposed by the model's top, and those effects can propagate downward, reaching eventually the middle and lower troposphere.

The Q_2 field for the 2-day, high-resolution 15-layer forecast from 15 Feb-

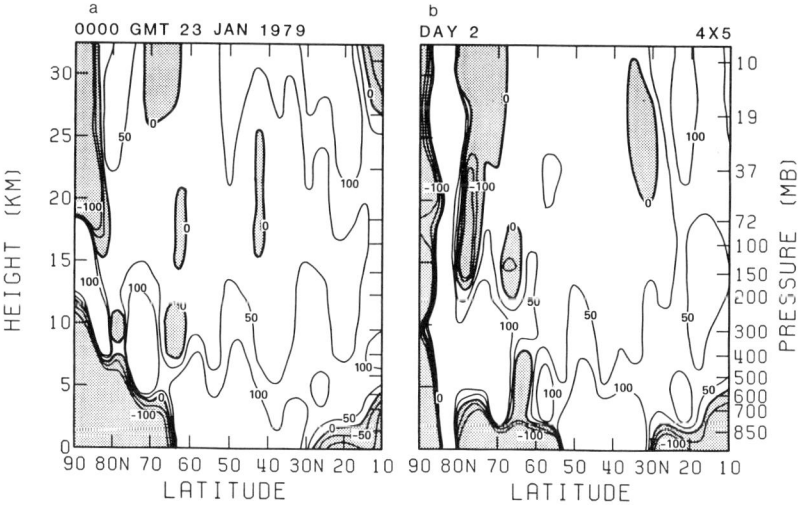

FIG. 37. Refractive index squared Q_1 for (a) 23 January and (b) 2-day forecast from 21 January with the low-resolution 15-layer version. Values greater than 100 or less than -100 are not contoured. For this figure $H = 7$ km.

ruary and the corresponding observed field, shown in Fig. 38, suggest different situations. The observed field exhibits a region of weak negative values centered at about 57°N and 500 mbar, and negative values at polar latitudes. The predicted field shows a broader region of negative values with large magnitudes centered at about 62°N and 400 mbar, and negative values at high latitude. These differences in Q_2 are associated with small errors in the prediction of the zonal wind and have a crucial effect on the characteristics of upward wave propagation, as indicated by the failure in predicting the major stratospheric warming that occurred during this period. If, for a 15-layer forecast, planetary waves forced near the surface cannot propagate through the upper troposphere, the forecast should not differ from another performed with a model that has its top in the lower stratosphere.

In Section 6.2, we have shown that the 15- and 9-layer forecasts for 500 mbar from 21 January are significantly different at day 10, while those from 15 February are similar even at day 10. This can now be interpreted as a consequence of the differences and similarities in the forecasts of ultralong waves during the early part of the forecast periods. In the "forecast" experiment performed by Mechoso *et al.* (1982), the error due to a lowered upper boundary appears first in ultralong waves (wavenumber 1 in that case) in the uppermost layers, and then propagates downward and progressively spreads to shorter waves. It is reasonable to assume that similar processes happened in the forecasts from 21 January. In the forecasts from 15 February, on the other hand, wavenumber 1 is not dominant and the vertical propagation of

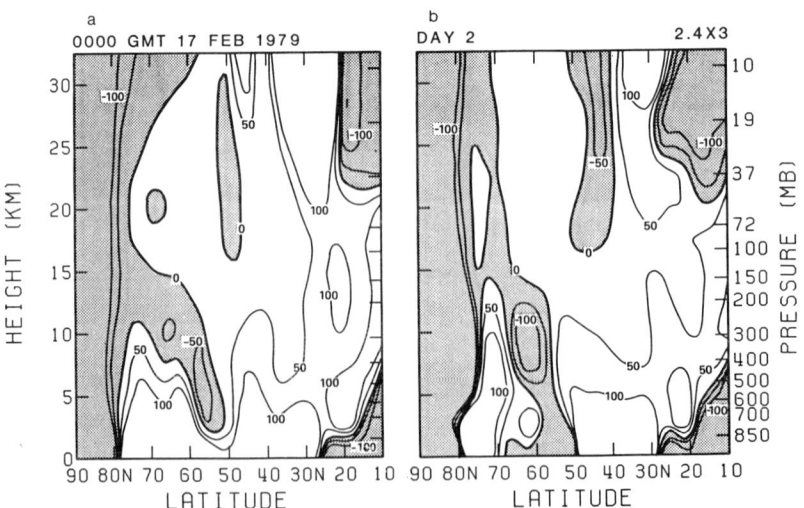

FIG. 38. As in Fig. 37, except for Q_2 for (a) 17 February and (b) 2-day forecast from 15 February with the high-resolution 15-layer version.

wavenumber 2 is underpredicted even in the 15-layer forecast. As a consequence, the 15- and 9-layer forecasts for 500 mbar remain similar.

7. Conclusions

We have compared 10-day forecasts with four versions of the UCLA GCM. The versions differ in horizontal resolution (the low-resolution is 4° of latitude by 5° of longitude, while the high-resolution is 2.4° of latitude by 3° of longitude) or in vertical extent (the 9-layer has the top at 51.8 mbar, while the 15-layer has the top at 1 mbar). In this article we have considered four cases with starting dates 16 and 21 January 1979 and 15 and 17 February 1979. These cases are characterized by episodes of tropospheric blocking and stratospheric sudden warmings in the Northern Hemisphere.

We have shown 5-day forecasts for 500 mbar which are generally accurate, particularly in midlatitudes. We have presented time means for the second 5-day period for selected forecasts which show the model's skill in predicting quasi-stationary features of the flow. We have shown very accurate 5- and 10-day forecasts for 10 mbar of stratospheric warming events during the winter of 1979. We conclude that the UCLA GCM has the ability to predict a variety of atmospheric phenomena at 5 days and beyond.

The impact of increasing the horizontal resolution (from 4° of latitude by 5° of longitude to 2.4° of latitude to 3° of longitude) on the accuracy of the forecasts at 500 mbar was not the same in all cases. The forecast from 16 January is quite insensitive to the increase in resolution. Also, there is no clear indication of impact in the 21 January and 17 February cases. On the other hand, the high-resolution forecasts from 15 February are more accurate than those with the low-resolution version, particularly in midlatitudes. The most visible impact of the increase in horizontal resolution was found in the stratosphere. In particular, the forecast of the stratospheric major warming of February 1979 was drastically improved when the horizontal resolution was increased. This case is discussed in detail by Mechoso et al. (1985), who argue that the reasons for improvement are not the more accurate numerics in the stratosphere, but the more accurate forecast of the zonal-mean wind in the upper troposphere. The slightly higher errors in the zonal-mean wind with the low-resolution version in that region can have a large impact on the characteristics of upward wave propagation. We conclude that increased horizontal resolution can significantly improve both tropospheric and stratospheric forecasts in some cases.

The impact of the upper boundary on forecasts was analyzed by studying rms forecast errors and by comparing fields predicted with the 9-layer forecast of the model, which has the top at 51.8 mbar, to those predicted with the

15-layer version, which has the top at 1 mbar. The 9- and 15-layer versions are identical below 51.8 mbar. As expected, the differences between 9- and 15-layer forecasts increased rapidly at the uppermost levels of the 9-layer version in all cases. However, the differences for 500 mbar were case dependent. They reached significant values in the 21 January case, and remained generally small in the 15 February case. We interpreted this different behavior on the framework of the theory on vertical propagation of planetary waves. In the first case, the 15-layer version predicts correctly the upward propagation of wavenumber 1, as indicated by the accurate forecast of the minor stratospheric warming that occurred during the period. The wave propagation through the lower stratosphere is altered in the 9-layer forecasts by the presence of the model's top. Furthermore, the effects of the top can propagate downward, eventually reaching the middle and lower troposphere. In the second case, the 15-layer version does not predict correctly the upward propagation of wavenumber 2, as indicated by the failure in predicting the major warming that occurred during the period. The poor forecast of wave propagation is associated with small errors in the predicted zonal wind in the upper troposphere. If upward propagation is not well predicted in the lower stratosphere by the 15-layer version because of the particular configuration of the flow, and in the 9-layer version for the effects of the model's top, then the 15- and 9-layer forecasts should be similar in the middle and lower troposphere. We conclude that, although the impact of the model's top on forecasts is largest at the uppermost model layers in all cases, it can be significant in the middle and lower troposphere in some cases.

The limited number of cases does not allow us to estimate the overall gain in predictability achieved by adding 6 more layers above the top of the 9-layer model. Such a gain might be small for this model, since the 15-layer version might lose its useful predictability before the differences between 15- and 9-layer forecasts became significant. In addition, the predictability of some specific phenomena might be artificially increased by the lowered model top. However, as models become more accurate, the presence of the unavoidable errors due to artificial boundaries will become more apparent.

Acknowledgments

Mr. Joseph Spahr provided invaluable programming help. We are grateful to Ms. Julia Lueken and Mr. David Silverfarb for technical assistance in preparing the manuscript. This research was supported by the Office of Naval Research, through the Naval Environmental Prediction Research Facility under Contract N00014-80-K-0947, and jointly by the National Science Foundation and the National Oceanic and Atmospheric Administration under Grant ATM 8218215 and the Naval Environmental Prediction Research Facility, Monterey, California, under Program Element 62759N, Project WF59-551, "Meteorological Models and Predictions."

REFERENCES

Arakawa, A., and Schubert, W. R. (1974). Interaction of a cumulus cloud ensemble with the large-scale environment, Part I. *J. Atmos. Sci.* **31,** 674-701.

Bengtsson, L. (1981). Numerical prediction of atmospheric blocking: A case study. *Tellus* **33,** 19-42.

Butchart, N., Clough, S. A., Palmer, T. N., and Trevelyan, P. J. (1982). Simulations of an observed stratospheric warming with quasigeostrophic refractive index as a model diagnostic. *Q. J. R. Meteorol. Soc.* **108,** 475-502.

Derome, J. (1981). On the average errors of an ensemble of forecasts. *Atmos.-Ocean* **19,** 103-127.

Katayama, A. (1972). A simplified scheme for computing radiative transfer in the troposphere. "Numerical Simulation of Weather and Climate." Tech. Rep. No. 6, Dept. of Meteorology, University of California, Los Angeles.

Lambert, S. J. (1980). The sensitivity of tropospheric numerical weather forecasts to increased vertical resolution and the incorporation of stratospheric data. *Atmos.-Ocean* **18,** 53-64.

Lindzen, R. S., Batten, E. S., and Kim, J. W. (1968). Oscillations in atmospheres with tops. *Mon. Weather Rev.* **96,** 133-140.

McIntyre, M. E. (1982). How well do we understand the dynamics of stratospheric warmings? *J. Meteorol. Soc. Jpn.* **60,** 37-65.

Matsuno, T. (1970). Vertical propagation of stationary planetary waves in the winter Northern Hemisphere. *J. Atmos. Sci.* **27,** 871-883.

Mechoso, C. R., Suarez, M. J., Yamazaki, K., Spahr, J. A., and Arakawa, A. (1982). A study of the sensitivity of numerical forecasts to an upper boundary in the lower stratosphere. *Mon. Weather Rev.* **110,** 1984-1993.

Mechoso, C. R., Yamazaki, K., Kitoh, A., and Arakawa, A. (1985). Numerical forecasts of stratospheric warming events during the winter of 1979. *Mon. Weather Rev.* **113,** 1015-1029.

Palmer, T. N. (1981). Diagnostic study of a wavenumber-2 stratospheric sudden warming in a transformed Eulerian-mean formalism. *J. Atmos. Sci.* **38,** 844-855.

Quiroz, R. S. (1979). Tropospheric-stratospheric interaction in the major warming event of January-February 1979. *Geophys. Res. Lett.* **6,** 645-648.

Schlesinger, M. E. (1976). A numerical simulation of the general circulation of atmospheric ozone. Ph. D. thesis, Dept. Atmos. Sci., University of California, Los Angeles.

Schlesinger, M. E., and Mintz, Y. (1979). Numerical simulation of ozone production, transport and distribution with a global atmospheric general circulation model. *J. Atmos. Sci.* **36,** 1325-1361.

Simmons, A. J., and Strüfing, R. (1983). Numerical forecasts of stratospheric warming events using a model with a hybrid vertical coordinate. *Q. J. R. Meteorol. Soc.* **109,** 81-111.

Suarez, M. J., Arakawa, A., and Randall, D. A. (1983). The parameterization of the planetary boundary layer in the UCLA general circulation model: Formulation and results. *Mon. Weather Rev.* **111,** 2224-2243.

MECHANISTIC EXPERIMENTS TO DETERMINE THE ORIGIN OF SHORT-SCALE SOUTHERN HEMISPHERE STATIONARY ROSSBY WAVES

EUGENIA KALNAY AND KINGTSE C. MO*

Laboratory for Atmospheres
National Aeronautics and Space Administration
Goddard Space Flight Center
Greenbelt, Maryland 20771

1. INTRODUCTION

Studies by van Loon and Jenne (1972), van Loon *et al.* (1973), Trenberth (1979), and others indicate that stationary waves in the Southern Hemisphere are of planetary scale, and generally weaker than in the Northern Hemisphere. On the other hand, Kalnay and Halem (1981) reported the presence of unexpectedly large-amplitude, short-wavelength stationary waves over and in the lee of South America during January 1979, the first month of the First GARP Global Experiment (FGGE) Special Observing Period 1 (SOP-1). Even more remarkable was the fact that the waves disappeared during February, the second month of SOP-1.

Kalnay and Paegle (1983) and Kalnay *et al.* (1986) have discussed several possible mechanisms that could have given rise to the waves during January and not during February: orographic forcing by the Andes, sea surface temperature (SST) anomalies, tropical heating, and resonant enhancement of forcing due to convection over the Amazonia region and over the South Pacific Convergence Zone (SPCZ) region.

Briefly, they concluded that the January waves were not forced by the Andes because of their phase: instead of a lee trough their phase corresponds to a lee ridge or poleward flow over the Andes. In an equivalent barotropic fluid this indicates transfer of kinetic energy from stationary waves to the mean flow (Kalnay-Rivas and Merkine, 1981), so that the waves seem to exist *despite* the Andes. Kalnay and Paegle (1983) also pointed out that SST anomalies of the same wavelength as the atmospheric waves were observed in the South Atlantic during SOP-1. However, their phase and time change indicated that the SST anomalies were driven by the atmospheric stationary

---

* Department of Meteorology, University of Maryland, College Park, Maryland 20742.

waves, rather than being their cause. Finally, they suggested that the waves were related to the stronger convective heating over the Amazon and central Pacific during January. This stronger source of stationary forcing was probably the cause of weaker tropical easterlies and stronger tropical westerlies in the upper atmospheric flow during January. They concluded that the presence of stronger tropical westerlies during January allowed freer propagation of energy from the stationary low-latitude forcing into the extratropics.

Simple numerical experiments with a hemispheric two-level quasi-geostrophic model (J. Paegle, personal communication) and with a shallow water barotropic model (F. Semazzi, personal communication), are in general agreement with these arguments. They indicate that placing a source of tropical heating in the Amazon region leads to the development within a few days of a train of short subtropical waves not unlike the ones observed during January. The results also show a strong sensitivity of the stationary waves in the South Atlantic to the precise position and extension of a heat source representing the SPCZ. However, the experiments with simple models are not conclusive because the waves have a phase relationship to the Andes corresponding to a *lee trough* rather than the observed *lee ridge.*

In the present study, we perform experiments with a comprehensive three-dimensional general circulation model (GCM). In the first two experiments, we attempt to reproduce as closely as possible the observed January and February 1979 atmospheric circulation by performing 15-day forecasts from real initial conditions. Because of the well known limited predictability of the atmosphere, we cannot expect *a priori* to simulate well the observed stationary circulation. However, since the GCM succeeds, at least partially, in reproducing the stationary waves in January and their absence in February, we have the possibility of modifying the model and performing other *mechanistic* experiments designed to determine the precise role of orography, tropical heating, etc.

In Section 2, we present the two control experimental forecasts: 15-day integrations with the Goddard Laboratory for Atmospheric Sciences (GLAS) fourth-order GCM, starting from 5 January and 4 February 1979. We compare them with the GLAS analyses for the same periods and discuss the extent to which the observed January waves are reproduced in the forecast. The rest of the paper is devoted to several mechanistic experiments in which we artificially modify the Andean orography (Section 3), tropical heating (Section 4), and regional heating (Section 5) and observe the effect of these changes on the stationary waves. In Section 6 we show that a slight deceleration of the initial westerly flow results in a remarkable amplification of a regular train of zonal wavenumber 5 waves, similar to those observed in the Southern Hemisphere during the FGGE summer (Salby, 1982; Randel and Stanford, 1983). Finally, Section 7 contains a summary of the experiments and conclusions.

2. ANALYSIS AND CONTROL EXPERIMENTS

In this section we compare the time averages of 15-day control forecasts started from 5 January 1979 and from 4 February 1979 with analyses averaged over the same period. For simplicity, in the rest of the paper we will refer to these 15-day averages as "January" and "February," respectively. Since operational forecasts retain useful skill for only about 1 week (e.g., Bengtsson, 1984) and we are trying to reproduce the persistent anomalous waves observed in January 1979, it does not seem warranted to make forecasts longer than 2 weeks.

The forecasts are performed with the GLAS fourth-order GCM as documented in Kalnay *et al.* (1983), which has a resolution of 4° latitude, 5° longitude, and 9 vertical levels. The rather coarse resolution is partially compensated by the use of fourth-order horizontal differences, so that the model has a short-range forecast skill comparable to that of models of finer resolution (e.g., Atlas, 1984; Halem *et al.,* 1982; Baker *et al.,* 1984). The model has a complete parameterization of subgrid-scale physical processes, and when used in a general circulation simulation mode, reproduces rather well the summer and winter atmospheric circulation (Kalnay *et al.,* 1983).

For verification, we use the GLAS Analysis, developed by Baker (1983), which has the same horizontal resolution and has been used for many satellite data impact studies (e.g., Halem *et al.,* 1982; Baker *et al.,* 1984). There is very good agreement between the time-averaged fields of meridional velocity, which we present here, and the corresponding fields of the European Centre for Medium Range Weather Forecasts (ECMWF) analysis.

Figure 1a and b presents the observed (15-day averaged) January and February mean meridional velocity at 200 mbar. A comparison with the same maps averaged over 30 days during January and February (Kalnay and Paegle, 1983) shows the same characteristics almost everywhere except that the stationary waves appear somewhat stronger in the 15-day average than in the 30-day average. The focus of our attention is the very large amplitude (up to 30 m sec^{-1} mean meridional velocity) and short zonal wavenumber (approximately 7) stationary waves centered over and lee of South America during January at about 30°S. These subtropical waves are virtually absent in the February average, which also shows weaker waves in the SPCZ region. As indicated by Kalnay and Paegle (1983), independent satellite observations agree well with this picture; showing the presence of strong convergence zones in the South Pacific and South Atlantic in January, and their virtual absence in February.[1] The corresponding zonal velocity averages at 200

---

[1] It should be noted that the January SPCZ is displaced eastward from its normal position, a phenomenon similar to that observed during "El Niño" years for longer periods.

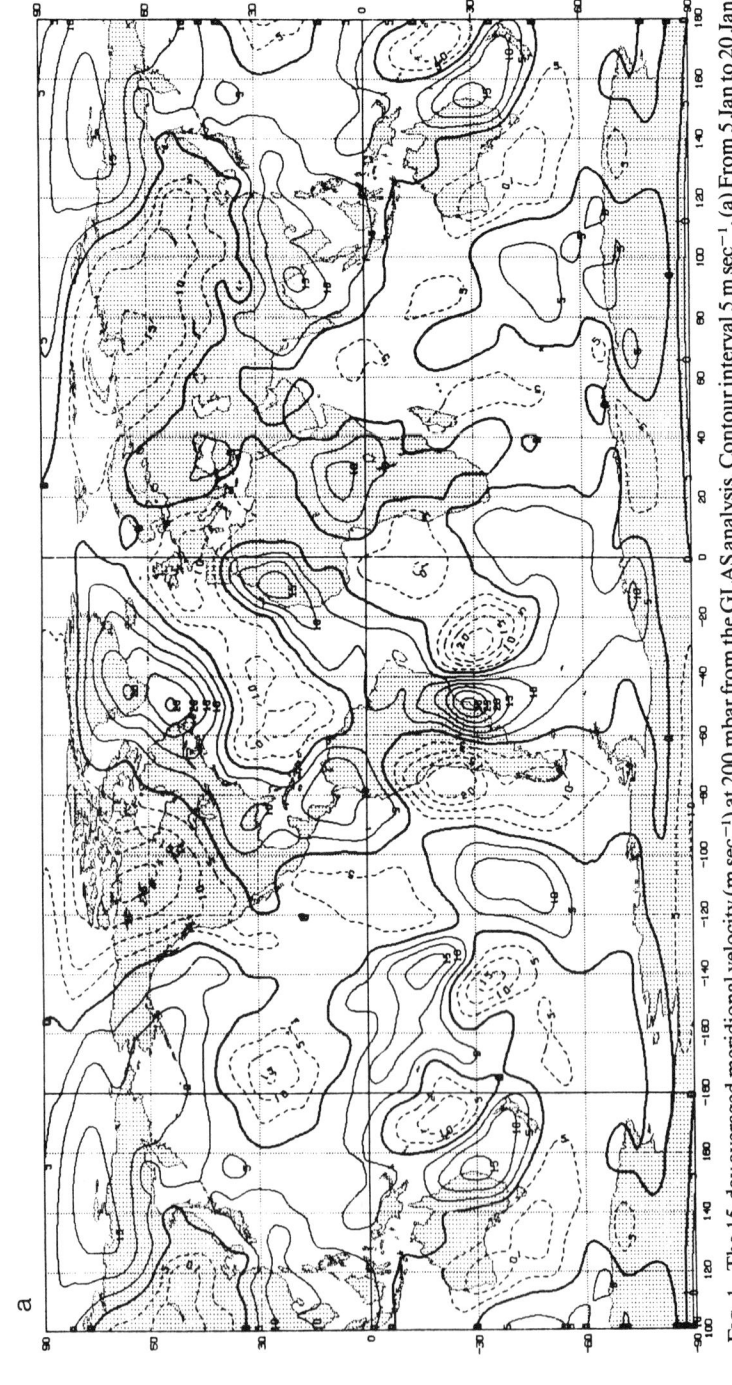

FIG. 1. The 15-day averaged meridional velocity (m sec^{-1}) at 200 mbar from the GLAS analysis. Contour interval 5 m sec^{-1}. (a) From 5 Jan to 20 Jan; (b) from 4 Feb to 20 Feb.

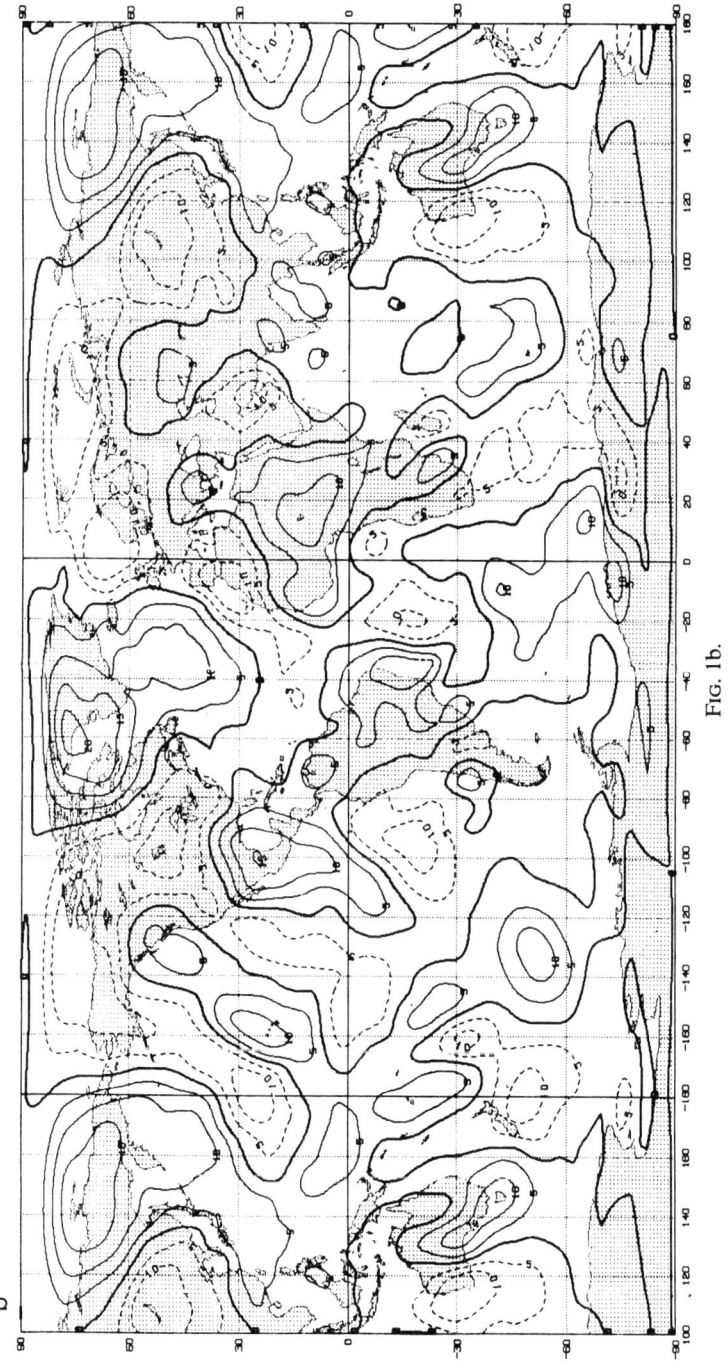

FIG. 1b.

mbar (not shown) indicate stronger upper level tropical westerlies and weaker easterlies in January, in agreement with Kalnay and Paegle (1983). It is also important to notice that during February, the center of maximum cyclonic vorticity ($\partial v/\partial x$ min) is located at about 20°S 30°W, which is very close to its climatological position (Taljaard and van Loon, 1969; Newell *et al.*, 1972). In contrast, during January, the cyclone center is shifted to 30°S and 40°W, and is much more intense.

Figure 2a and b presents Hovmoller diagrams of the meridional velocity at 200 mbar at 30°S for the same January and February periods. The difference in the stationary wave behavior in January and February is truly striking, indicating that during January, stationary waves over South America were not only present, but they dominated the subtropical circulation.

Next we present the 15-day average control forecasts for January and February (Fig. 3a and b). With some notable exceptions (e.g., northwest Africa in February), the model succeeds in predicting the average direction of the meridional flow, and its changes between January and February. In our region of interest, South America, we observe that during January there

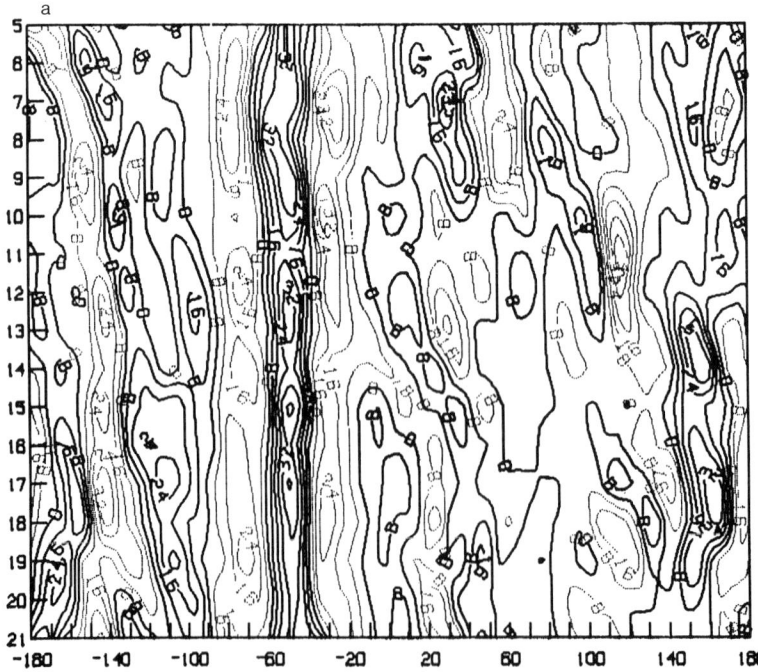

FIG. 2. Hovmoller diagram of the meridional velocity at 200 mbar and 30°S for (a) January and (b) February. Contour interval is 8 m sec^{-1}.

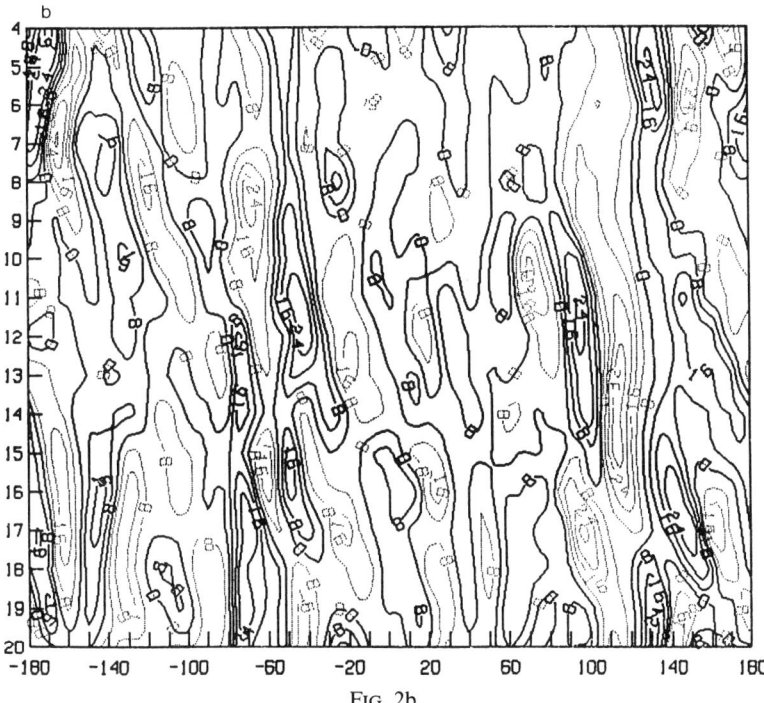

Fig. 2b.

are short stationary waves elongated mainly in the north–south direction which are similar but weaker than the January waves of the analysis.

In the February forecast, the stationary waves over South America are weaker than in January, and, in agreement with the analysis, are tilted in the northwest–southeast direction. The forecast has also succeeded in placing the cyclone center in the correct position, 20°S and 40°W.

Figure 4a presents the Hovmoller diagram for 200 mbar for the January forecast. It is apparent that during the first 10 days, the model is able to predict and even amplify the stationary waves over South America. On 10–12 January, the waves reach an amplitude of 40 m sec^{-1} in the meridional wind, about 10 m sec^{-1} stronger than in the analysis. After that, however, there is a sudden breakdown of the stationary waves, with a partial regeneration at about 16 January. In order to test the possibility that the sudden breakdown of the forecasted stationary waves may be due to barotropic instability, we computed the barotropic stability parameter $\beta - (\partial^2 u/\partial y^2)$. A Hovmoller diagram of this parameter computed at the same latitude of 30°S confirms this hypothesis (Fig. 4b). The stationary waves are associated with a strong minimum in the zonal velocity, which results in negative

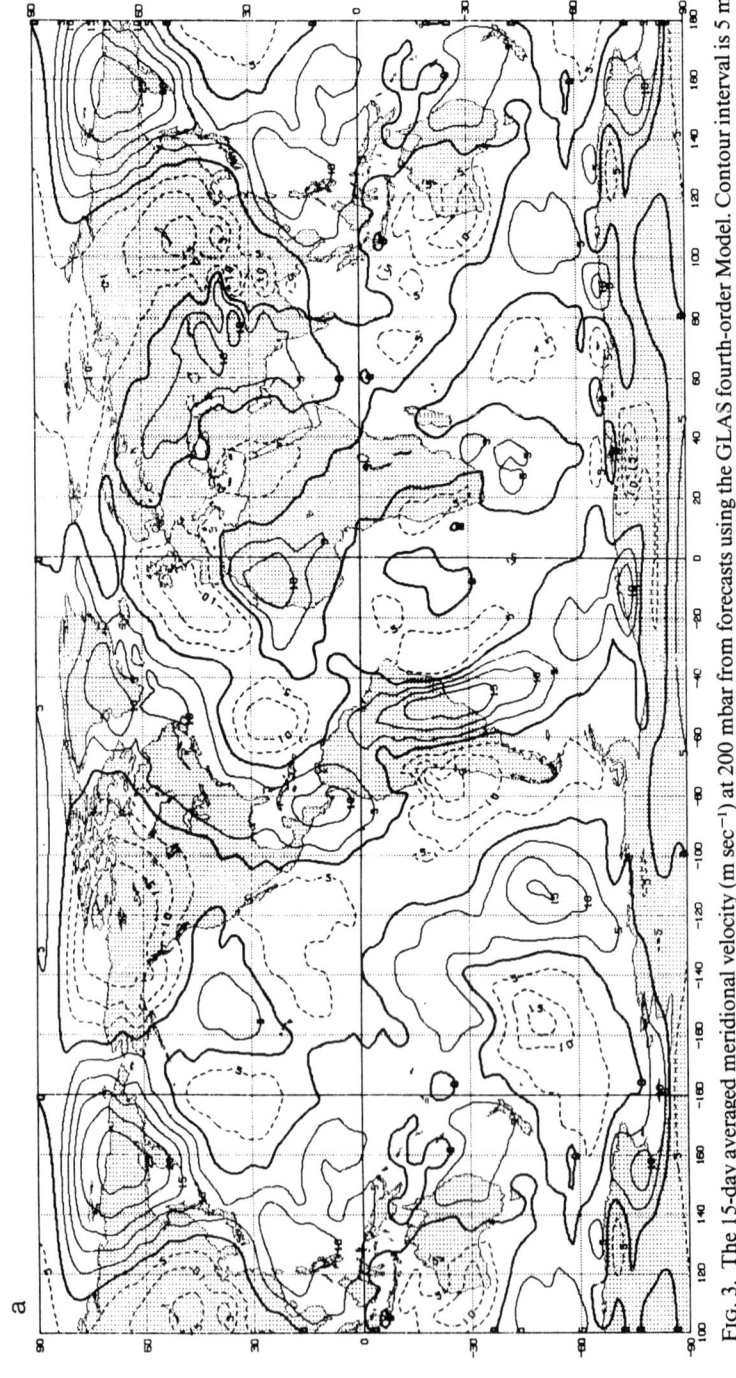

FIG. 3. The 15-day averaged meridional velocity (m sec^{-1}) at 200 mbar from forecasts using the GLAS fourth-order Model. Contour interval is 5 m sec^{-1}. (a) January control forecast; (b) February control forecast.

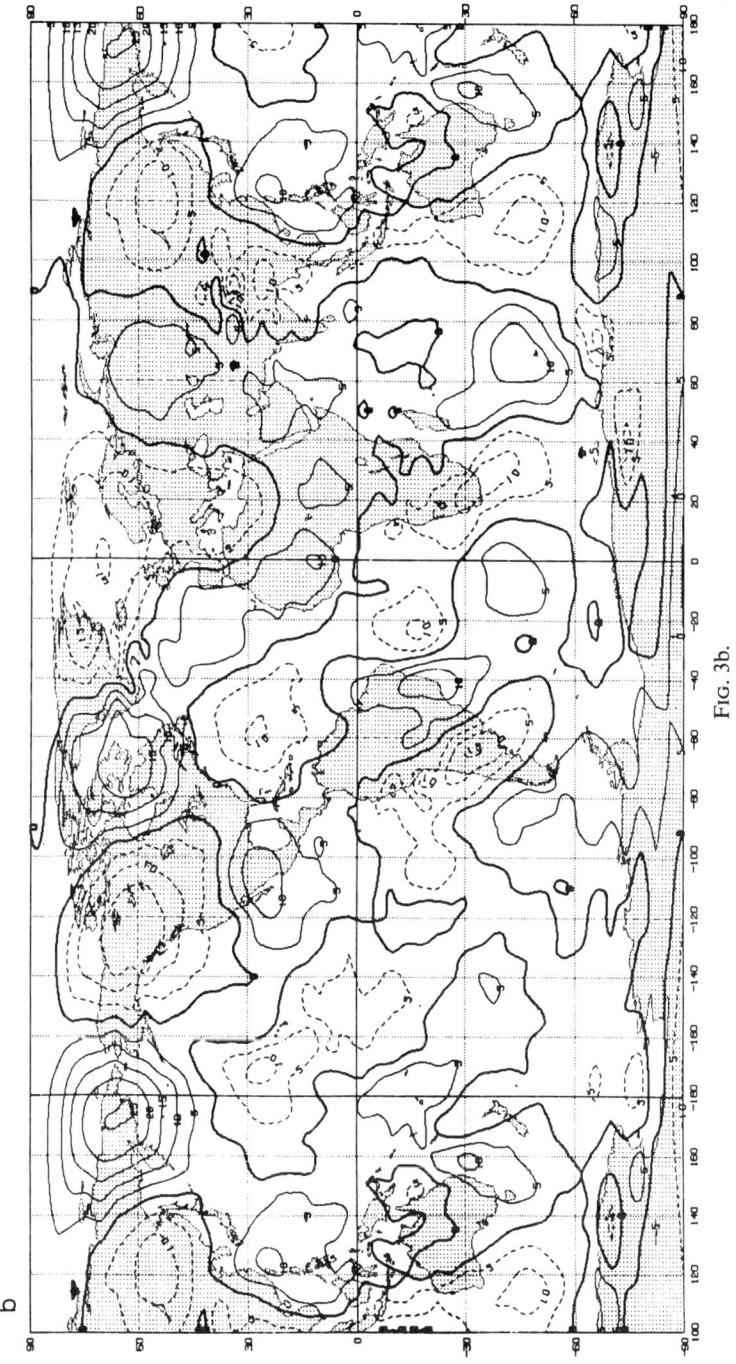

Fig. 3b.

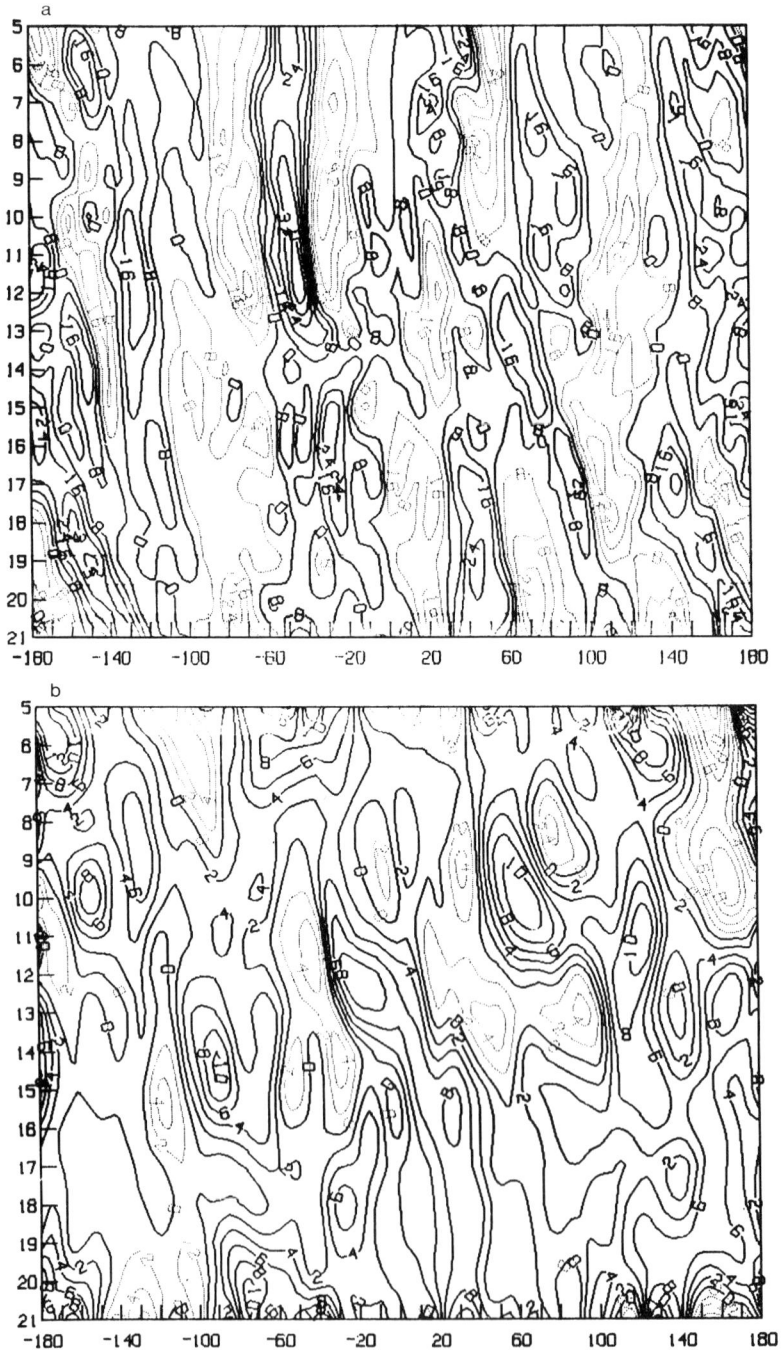

FIG. 4. (a) Hovmoller diagram of the meridional velocity at 200 mbar and 30°S for the January control forecast; contour interval is 8 m sec^{-1}. (b) Barotropic stability parameter, $\beta - (\partial^2 u/\partial y^2)$ at 30°S for the January control forecast.

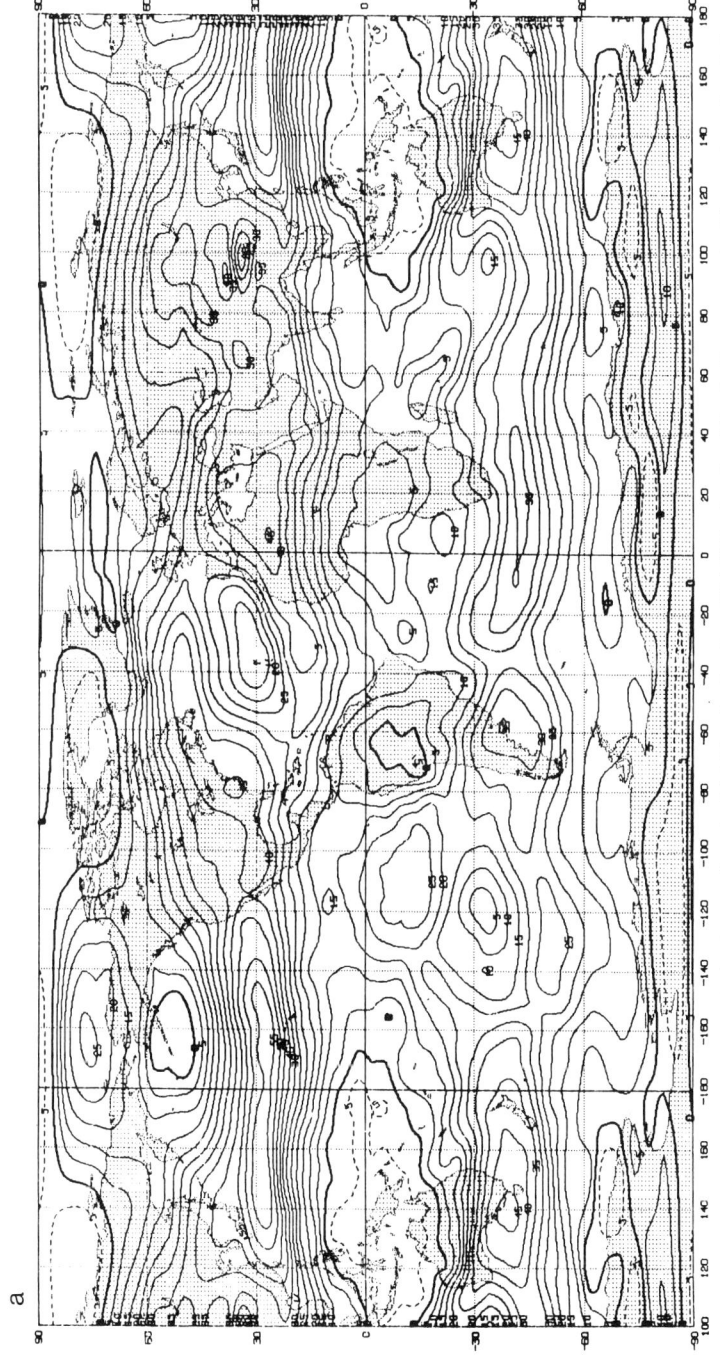

FIG. 5. The 15-day averaged zonal velocity (m sec^{-1}) at the (a) 200-mbar level and (b) 700-mbar level for the January control forecast; contour interval is 5 m sec^{-1} for (a) and 4 m sec^{-1} for (b).

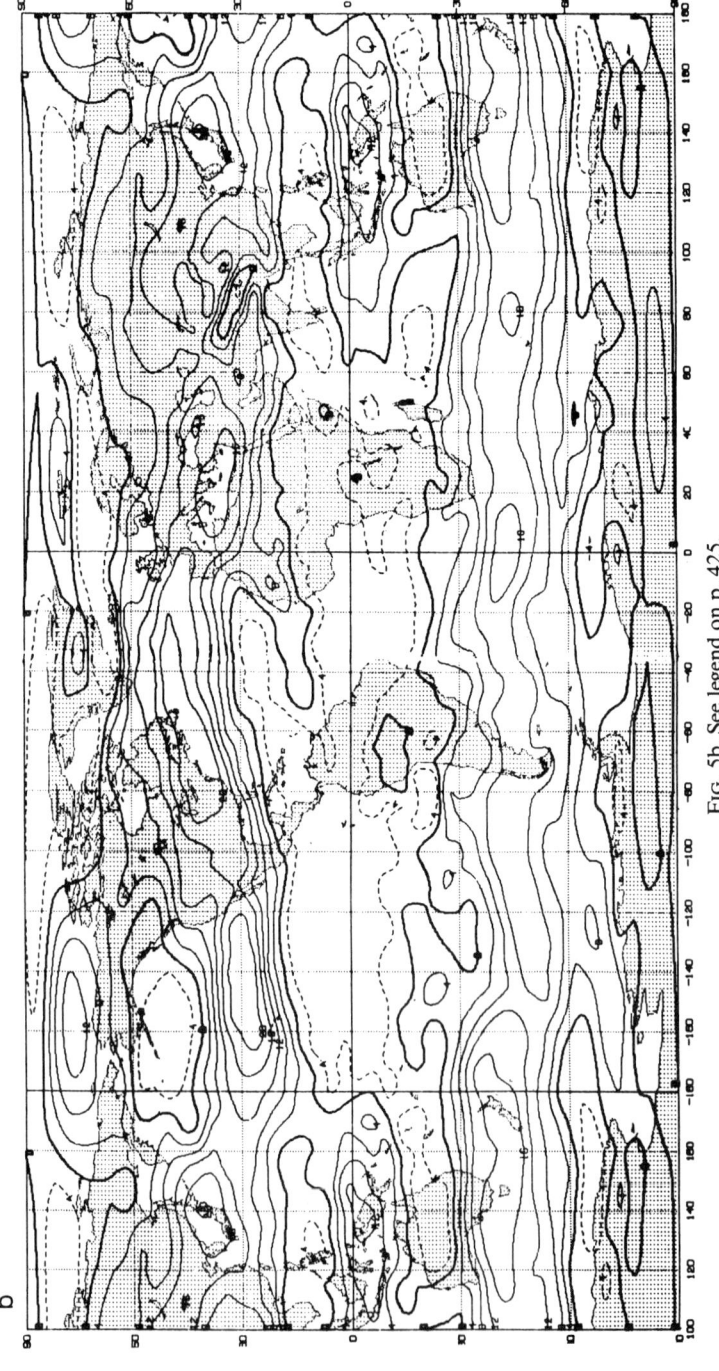

Fig. 5b. See legend on p. 425.

values of $\beta - (\partial^2 u/\partial y^2)$ at about 10-12 January, indicating barotropic instability. A similar computation for the analysis (not shown) indicates that in the real atmosphere the observed stationary waves were barotropically stable.

As a reference for the following sections, we also pesent the 200- and 700-mbar time-averaged forecasts of the zonal wind (Fig. 5a and b). In the upper levels, the January forecast has stronger tropical westerlies than the February forecast (not shown). At 700 mbar, both forecasts have somewhat similar strength in the tropical flow.

Since the model succeeded in predicting the presence of stationary waves over South America in January and their absence in February, it has become a powerful tool to perform experiments designed to test Kalnay and Paegle's (1983) previously formulated hypotheses about the possible origin of these waves. In the following sections we discuss the results of such experiments, which, because of the realism of the model, can provide more precise answers than simple model experiments.

3. "No Andes" Experiment

As indicated in Section 1, Kalnay and Halem (1981) and Kalnay and Paegle (1983) argued that the January waves over South America cannot be due to the Andes because they correspond to a lee ridge rather than a lee trough. However, this argument is valid only to the extent that the atmospheric response to narrow orographic forcing is equivalent barotropic, and it is not possible to demonstrate whether this is strictly true in the presence of vertical and horizontal shear.

In order to test this argument, we performed a forecast from the same initial conditions as the January control forecast but eliminating orography over the Andean region. The 15-day averaged meridional velocity field for the "No Andes" experiment is shown in Fig. 6a, and the corresponding Hovmoller diagram at 30°S is shown in 6b. The results conclusively confirm the validity of the argument that the South American waves occurred during January *despite* the Andes. The forecast of the stationary waves has actually improved, not only over South America, but also in the tropical central Pacific and over northwest Africa. This result is not an indication of a general deleterious effect of the Andes upon the model forecasts. It simply confirms that since the mountain forcing tends to produce a lee trough, its absence enhances the lee ridge, in closer agreement with the observations. Therefore, the absence of the Andes resulted in a better forecast of the stationary-wave structure over South America, which remains accurate during all of the 15

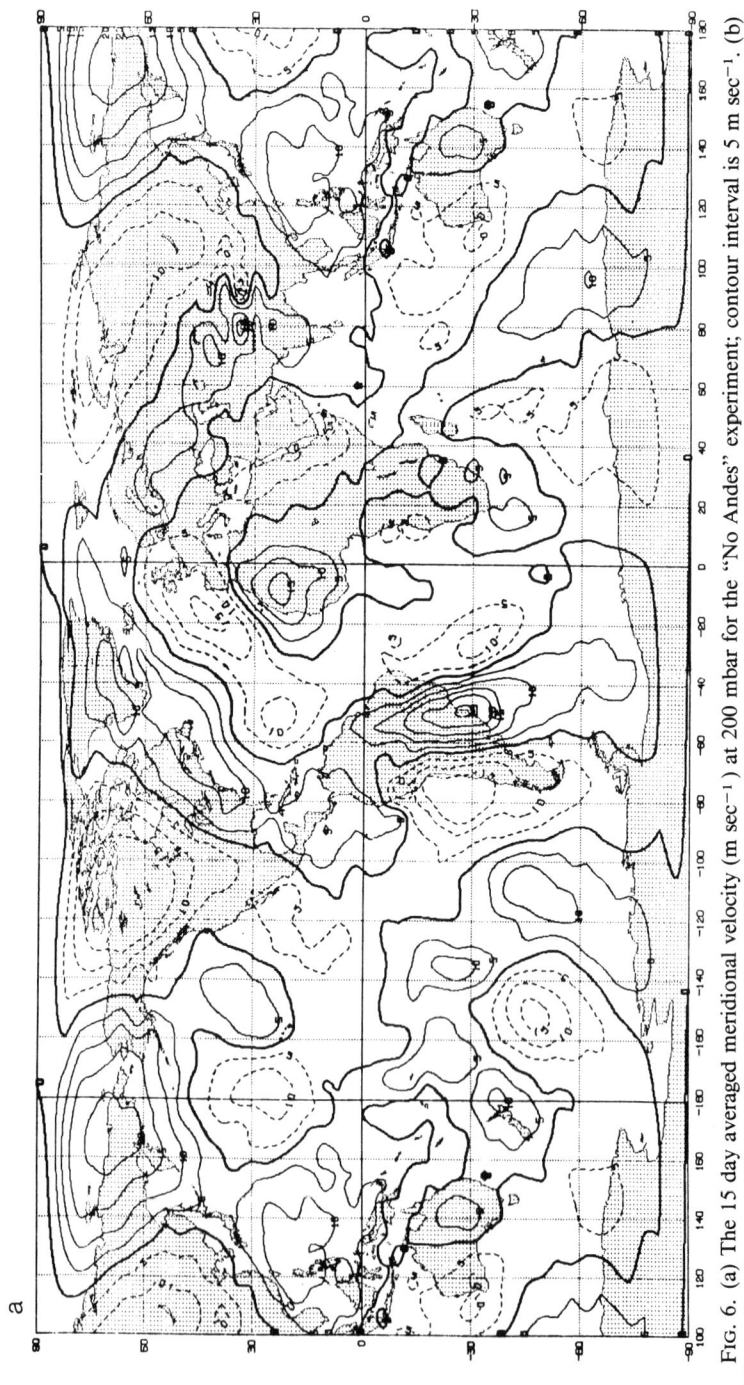

FIG. 6. (a) The 15 day averaged meridional velocity (m sec^{-1}) at 200 mbar for the "No Andes" experiment; contour interval is 5 m sec^{-1}. (b) Hovmoller diagram of the meridional velocity at 200 mbar and 30°S for "No Andes" experiment; contour interval is 8 m sec^{-1}.

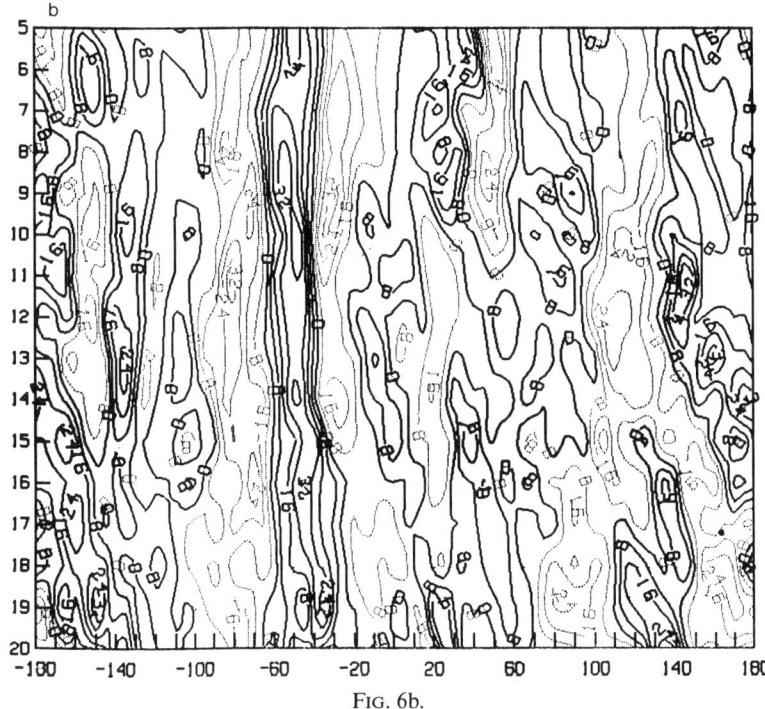

Fig. 6b.

days of integration. As a result, the associated convection is better represented and tropical forcing is closer to the real atmospheric forcing.

4. "Reduced Tropical Heating" Experiment

In this section, we try to assess the role of tropical heating in the maintenance of the Southern Hemisphere stationary waves by reducing tropical heating. Following an idea developed by Paegle and Baker (1983), we simply modify the value of the coefficient of latent heating by multiplying by a factor which is latitudinally dependent and smaller than 0.5 between 30°S and 30°N 1(Fig. 7a).

The 15-day averaged forecast of $\bar{v}_{200\,\text{mbar}}$ with the "reduced tropical heating" model, and the corresponding Hovmoller plot at 30°S, are presented in Fig. 8a and b. It is apparent that all stationary waves have been strongly reduced in the Southern Hemisphere, even in the extratropical regions. In the extratropical Northern Hemisphere the relative change in the stationary waves is much smaller, indicating that in that region tropical heating is less

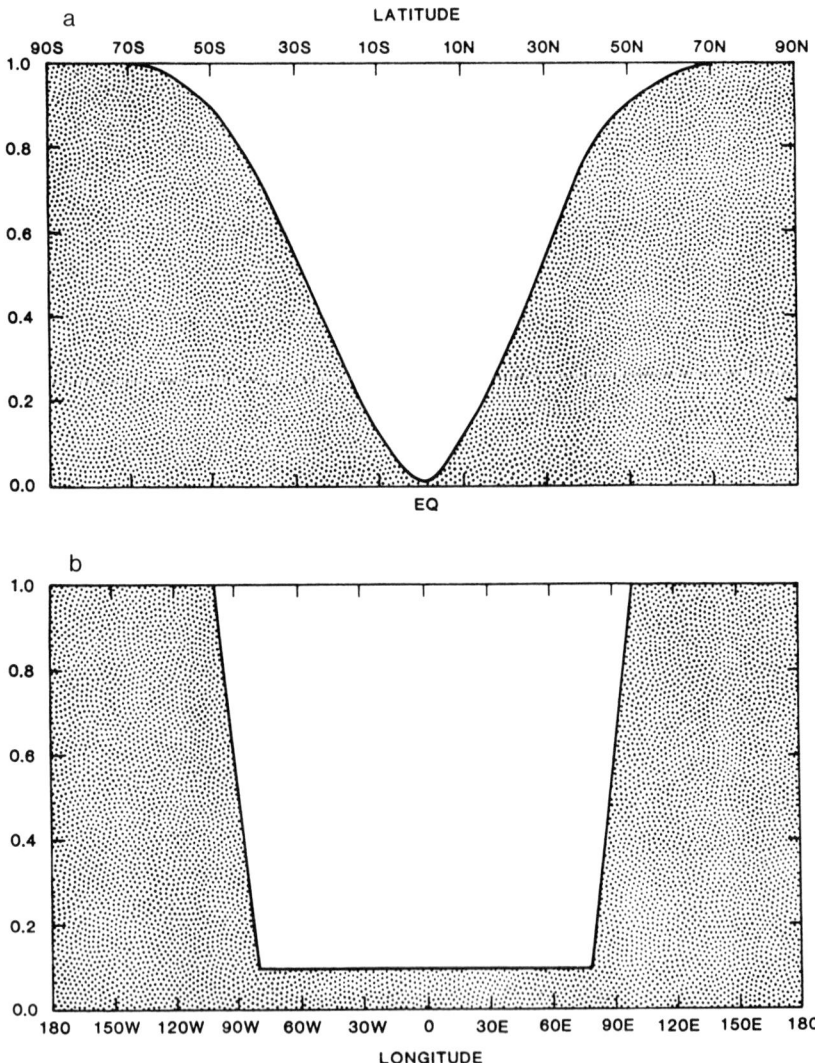

FIG. 7. Factor which modifies the coefficient of latent heating for (a) "reduced tropical heating" and (b) "suppressed Atlantic heating" experiment.

important than orographic forcing. Although there is a remnant of a stationary wave over South America, the Hovmoller plot indicates that it is essentially a residual from the initial conditions.

Figure 9a and b show the difference in the mean zonal velocity at 200 and 700 mbar, respectively, between the "reduced tropical heating" forecast and

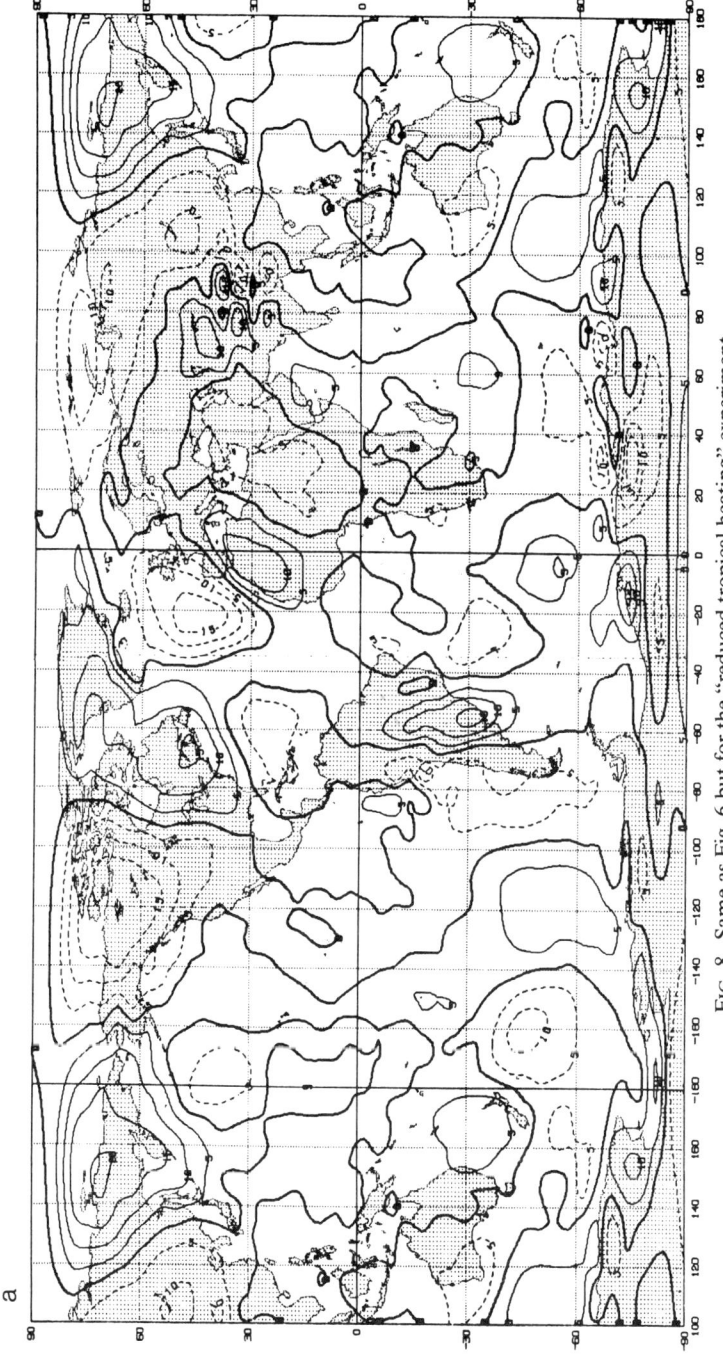

FIG. 8. Same as Fig. 6 but for the "reduced tropical heating" experiment.

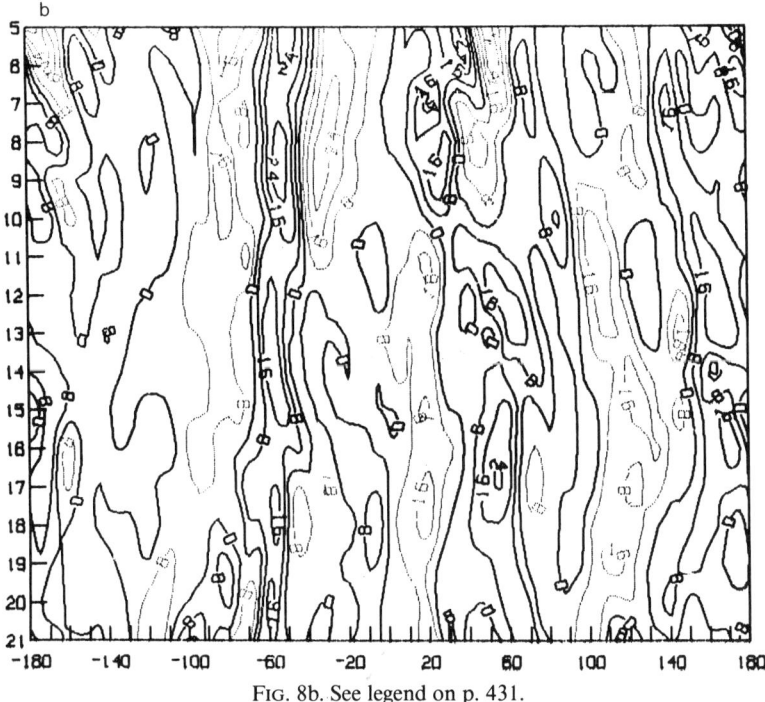

Fig. 8b. See legend on p. 431.

the control forecast at 200 and 700 mbar, respectively. The result is very similar to the forced tropical circulation discussed by Gill (1980), Matsuno (1966), and Webster (1972), but with negative sources of heat in the regions where convection has been strongly suppressed: the Amazon, equatorial Africa, and most especially, Indonesia. Over and slightly westward of these regions, the anomaly corresponding to the reduced convection results in low-level easterlies and upper level westerlies, exactly in opposition to the zonal flow in the control experiment. In the rest of the equatorial belt, there are stronger westerlies at low levels and easterlies at upper levels, presumably because forced Kelvin waves have propagated in these regions (Gill, 1980). Again, as described by Gill, there is a reduction in the subtropical westerly jets directly north and south of the suppressed convection regions. However, although at low levels there is good agreement with Gill's model, there are significant quantitative differences in the upper levels. The reduction in the subtropical jet is about three times *stronger* than the increase in westerly flow over the convective regions. This is in contrast with Gill's linear solution, where the subtropical response was about three times *weaker* than the equatorial change. The change in the subtropical jets is rather symmetric over

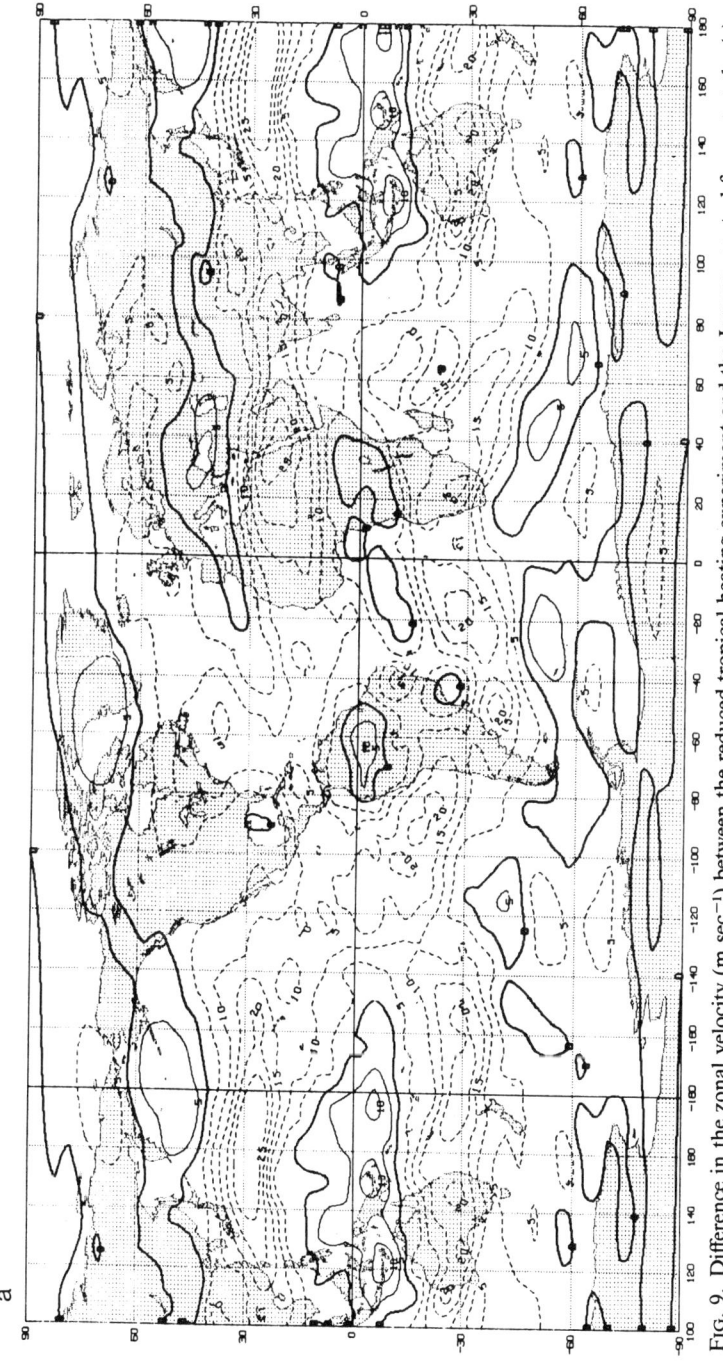

FIG. 9. Difference in the zonal velocity (m sec^{-1}) between the reduced tropical heating experiment and the January control forecast at the (a) 200-mbar level and (b) 700-mbar level; contour interval is 5 m sec^{-1}.

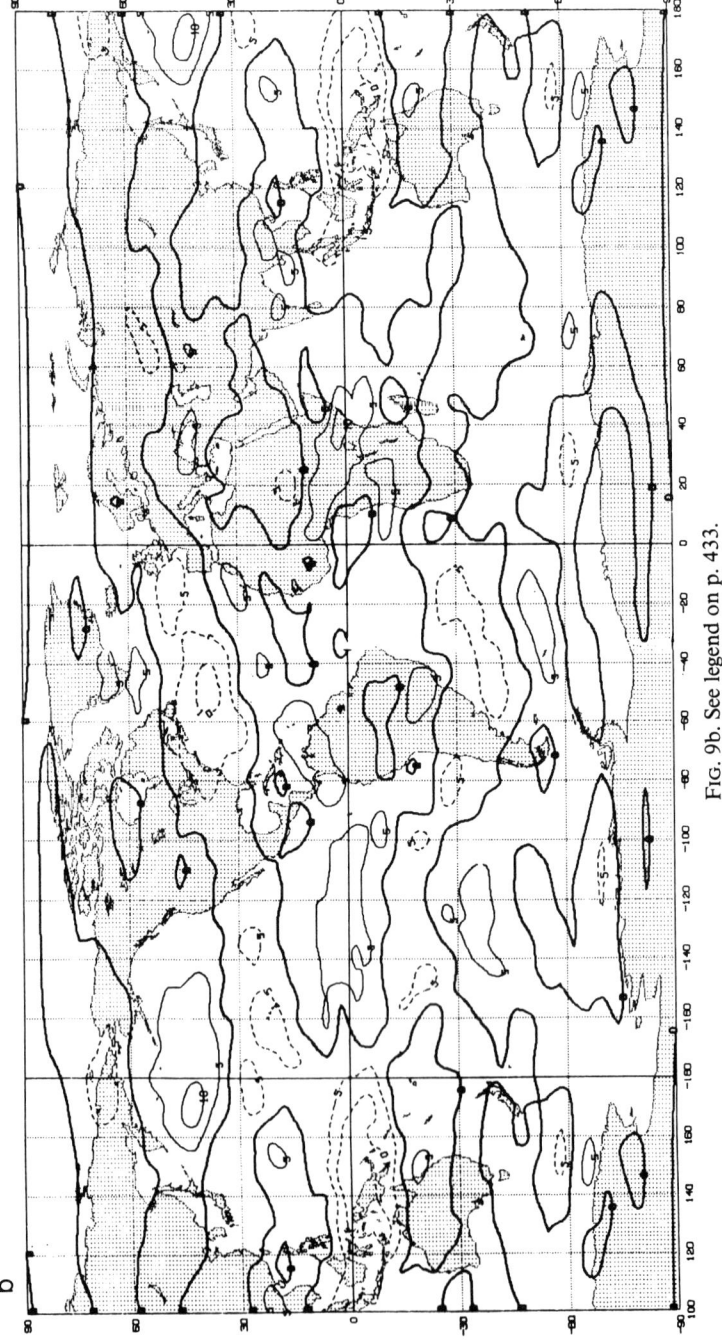

FIG. 9b. See legend on p. 433.

Indonesia and equatorial Africa, and agrees well with the intuitive notion of a nonlinear "local Hadley cell circulation."

5. "Suppressed Regional Heating" Experiments

The experiment discussed in the previous section indicated that tropical heating was an important contributor to the January waves over South America, and, indeed, over all the Southern Hemisphere.

Kalnay et al. (1986) have indicated that instances of a strong SPCZ shifted to the east tend to be associated with strong "South Atlantic Convergence Zones" (SACZs), such as occurred during January 1979. This association is also apparent in the correlation maps of outgoing long-wave radiation with convection at the dateline computed by Lau and Chan (1983).

In order to determine whether local convection, convection originating in the SPCZ, or both, were necessary to maintain the South American January waves (and the corresponding SACZ), we performed two more experiments. In the first one, which we will denote "suppressed Pacific heating," we reduced the latent heat of condensation to one-tenth of its value at all latitudes and over a sector of 180° longitude centered on the dateline (Fig. 7b). In the second experiment, denoted "suppressed Atlantic heating," the latent heat release was similarly suppressed but in the hemisphere centered on the Greenwich meridian.

The 15-day forecasts of $\bar{v}_{200 \text{ mbar}}$ and Hovmoller plots corresponding to the "suppressed Pacific heating" are presented in Fig. 10a and b. The reductions in amplitude and change of the South American waves are even stronger than in the "reduced tropical heating" experiment.

Although the latent heat is suppressed in the complete hemisphere centered on the dateline, there are rather small changes in the Northern Hemisphere, and in the Southern Hemisphere the largest changes are in the SACZ region. In the Hovmoller diagram, it is apparent that in the Pacific, the stationary waves are immediately weakened. In our region of interest, the SACZ waves remain stationary for about 6 days, and then they propagate eastward with a speed of about 5° per day.

When the Atlantic hemisphere heating is suppressed, the corresponding stationary waves and Hovmoller plots are those of Fig. 11a and b. The South American waves weaken and start propagating immediately.

The fact that the weakening of the South American waves in these two experiments is much stronger than in the "reduced tropical heating" experiment suggests that subtropical convection in the SPCZ and SACZ plays a role in their maintenance, and they are not just a tracer of a circulation determined by equatorial convection over South America and Indonesia.

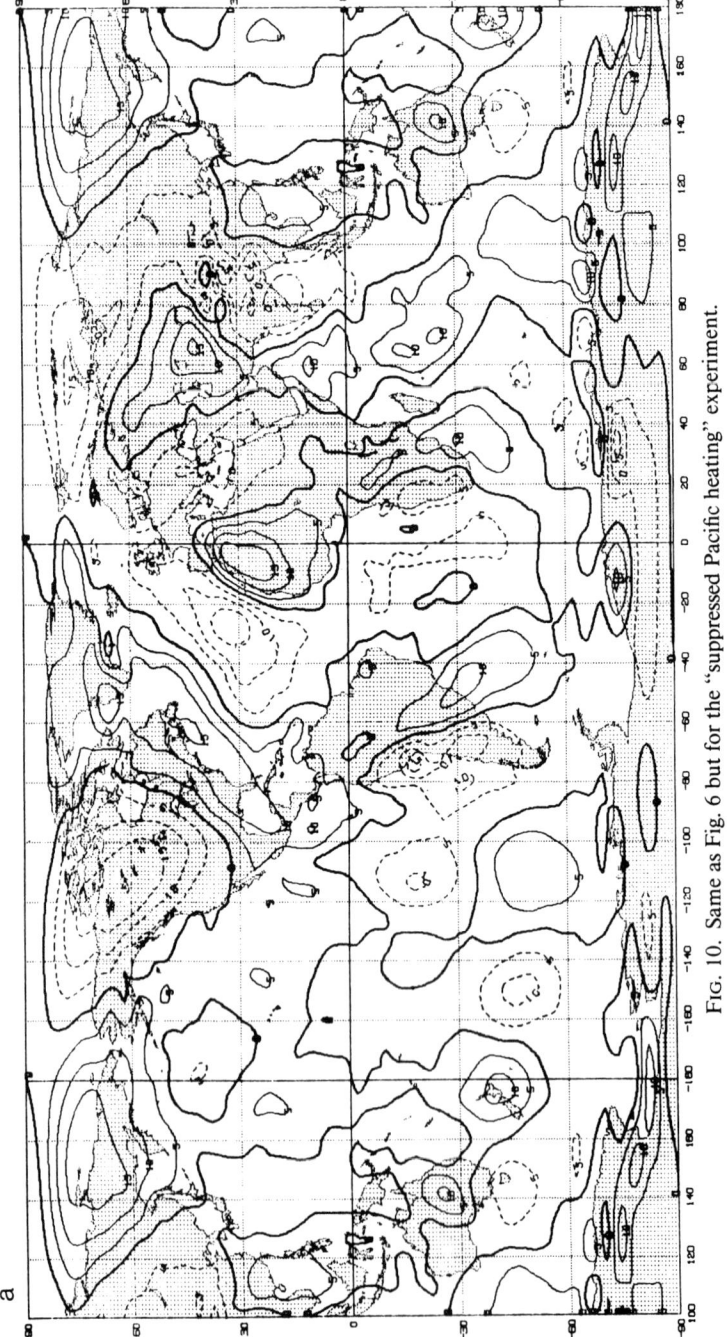

FIG. 10. Same as Fig. 6 but for the "suppressed Pacific heating" experiment.

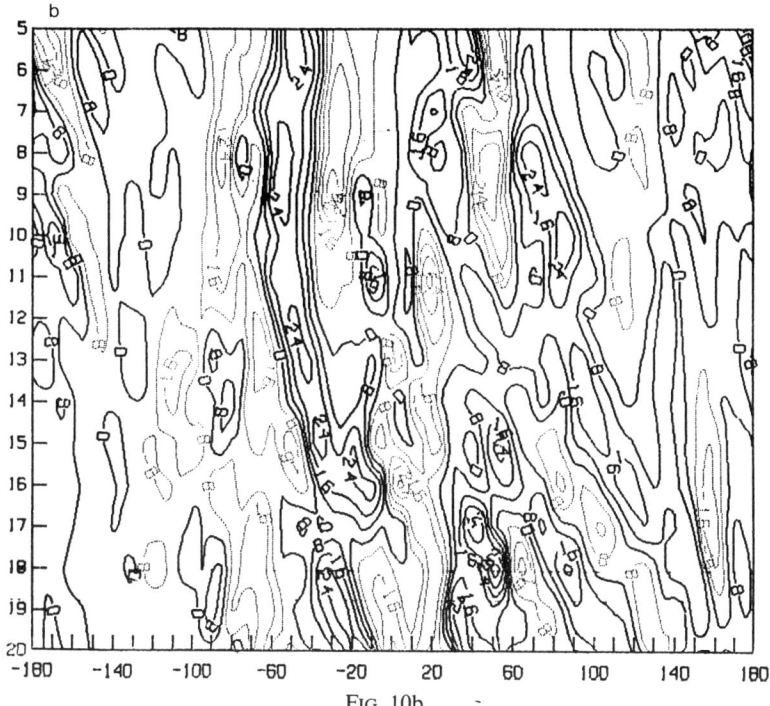

Fig. 10b.

6. "Easterly Deceleration" Experiment

We performed an experiment in which we decelerated the initial zonal flow by adding a solid body rotation velocity corresponding to -5 m sec^{-1} at the equator. The sea level pressure was also modified so as to maintain geostrophic balance with the change in zonal flow.

The purpose of this experiment was to determine whether the South American waves were sensitive to a shift in the position of the critical surface separating easterlies and westerlies, as argued by Kalnay and Paegle (1983).

The result of this experiment was very unexpected, and is apparent in Fig. 12. The forecast of the stationary waves in the Northern Hemisphere is not very affected by the change in zonal wave, whereas the Southern Hemisphere changes dramatically. The intensity of the short subtropical stationary waves centered at 30°S in the control January forecast has indeed decreased. On the other hand, it is clear that the change in zonal flow has resulted in the very abrupt generation of almost stationary waves of zonal wavenumber 5 centered at 45°S. This result is reminiscent of the "ubiquitous" pentagonal wave

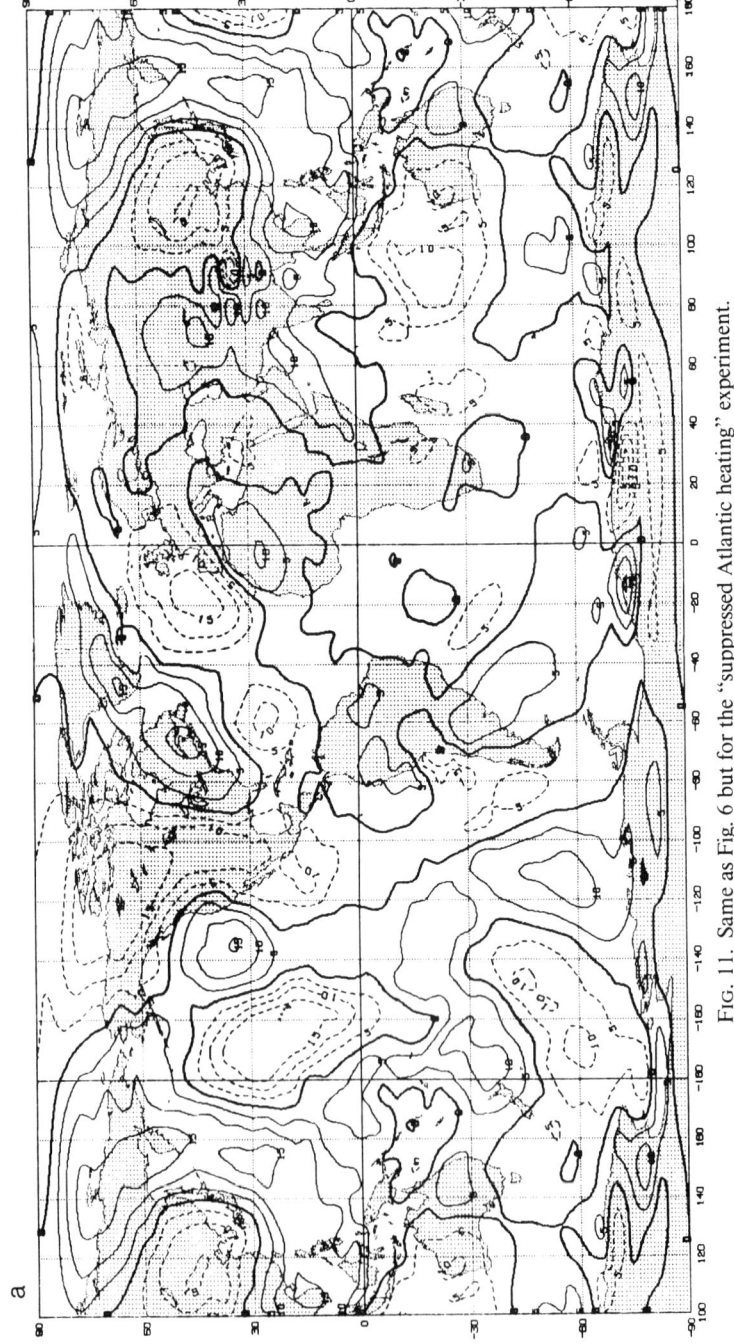

FIG. 11. Same as Fig. 6 but for the "suppressed Atlantic heating" experiment.

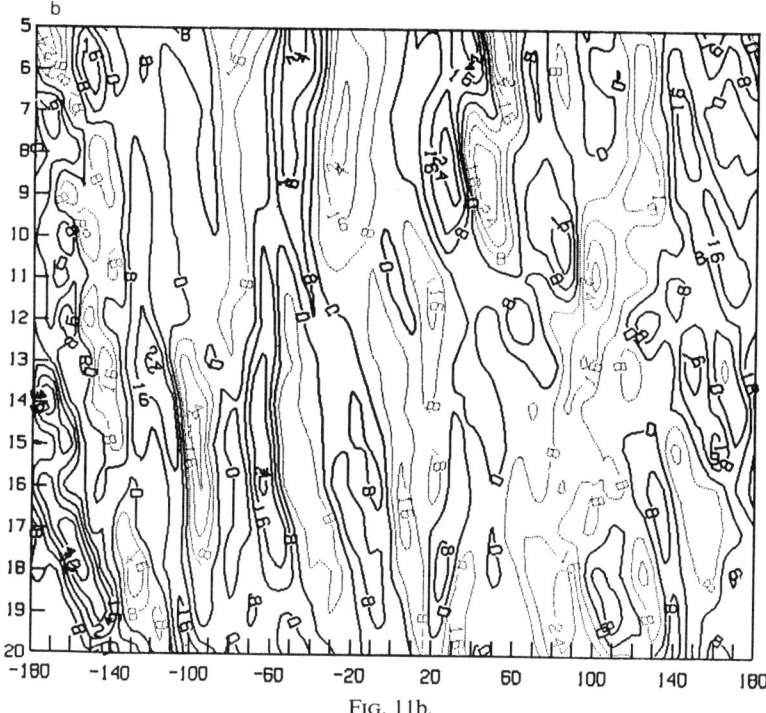

FIG. 11b.

in the Southern Hemisphere observed during certain periods of the southern summer and discussed by Salby (1982), Hamilton (1983), and Randel and Stanford (1983).

7. Summary and Conclusions

Through the use of a comprehensive general circulation model, we have been able to determine the mechanisms that maintain the large-amplitude short-wavelength stationary waves in the Southern Hemisphere. The model control forecast represents well the maintenance of the waves for about 10 days, after which barotropic instability results in their sudden disappearance.

Several mechanistic experiments are performed. The "No Andes" forecast shows that the waves exist independently of the orographic forcing of the Andes. Several experiments modifying the coefficient of latent heat lead to the conclusion that tropical heating is important in the maintenance of the waves. However, the convection in the subtropical waves themselves is im-

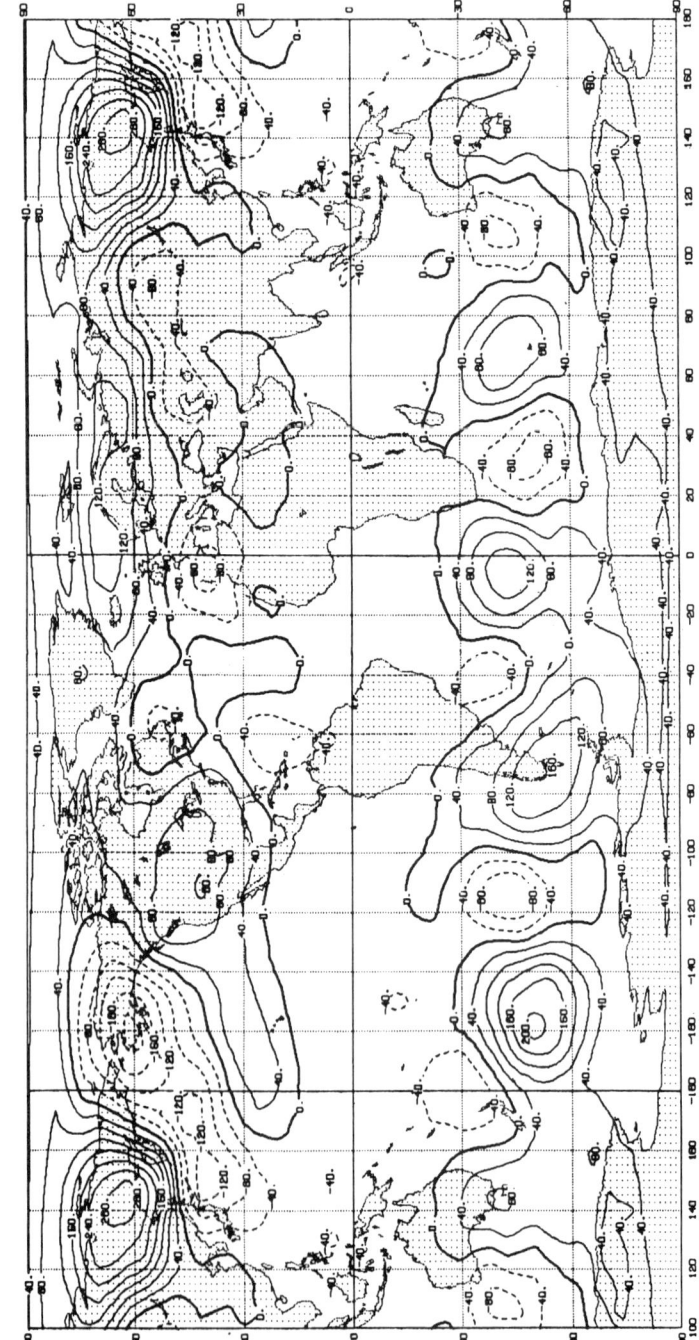

FIG. 12. Same as Fig. 6a but for the "easterly deceleration" experiment. Mean height difference (M) at 500 mbars.

portant in sustaining their amplitude and phase, and the Walker-type circulation associated with the SPCZ is also a contributor to the maintenance of the South American waves. These results confirm the existence of a relationship between the occurrence of a strong SPCZ, somewhat eastward from its climatological position, and the strong "South Atlantic Convergence Zone" observed in outgoing long-wave radiation maps.

A "pentagonal" stationary wave was the unexpected result of an experiment in which the zonal flow was decelerated by a solid rotation of -5 m sec^{-1} at the equator. This phenomenon, probably related to the observed presence of waves with zonal wavenumber 5 in the Southern Hemisphere summer, will be studied in further detail.

ACKNOWLEDGMENTS

This work is a continuation of the research started in collaboration with Jan Paegle, whose interest in the problem is very gratefully acknowledged. We have had useful discussions with R. Lindzen, M. Halem, W. E. Baker and S. Bloom.

The experiments performed in this study were made possible because of the collaboration of J. Pfaendtner, L. L. Takacs, and R. Balgovind, who contributed significantly to the development of the GLAS fourth-order GCM. R. Dlouhy and M. Almeida helped in setting up the experiments. Most of the computer plots were developed and computed by R. Rosenberg and F. Tahmasebi. We are very grateful to all of them for their enthusiastic help.

The manuscript was typed by B. Richardson and L. Wright, and the figures were prepared with the collaboration of L. Rumburg and B. Sherbs.

REFERENCES

Atlas, R. (1984). The effect of physical parameterizations and initial data on the numerical prediction of the President's day cyclone. *Proc. Conf. Weather Forecast. Anal., 10th, Clearwater Beach, Fla., June 25–29.*

Baker, W. E. (1983). Objective analysis and assimilation of observational data from FGGE. *Mon. Weather Rev.* **111**, 328–342.

Baker, W. E., Atlas, R., Halem, M., and Susskind, J. (1984). A case study of forecast sensitivity to data and data analysis technique. *Mon. Weather Rev.* **112**, 1544–1561.

Bengtsson, L. (1984). Operational medium range weather prediction at ECMWF. *Proc. Conf. Weather Forecast. Anal., 10th, Clearwater Beach, Fla., June 25–29.*

Gill, A. E. (1980). Some simple solutions for heat induced tropical circulation. *Q. J. R. Meteorol. Soc.* **106**, 447–462.

Halem, M., Kalnay, E., Baker, W. E., and Atlas, R. (1982). An assessment of the FGGE satellite observing system during SOP-1. *Bull. Am. Meteorol. Soc.* **63**, 407–426.

Hamilton, K. (1983). Aspects of wave behavior in the mid and upper troposphere of the southern hemisphere. *Atmos. Ocean.* **21**, 40–54.

Kalnay, E., and Halem, M. (1981). Large amplitude stationary Rossby waves in the southern hemisphere. *Proc. Int. Conf. Early Results FGGE Large Scale Aspects Monsoon Exp., Tallahasse Fla., Jan. 12–17.*

Kalnay, E., and Merkine, L.-O. (1981). A simple mechanism for blocking. *J. Atmos. Sci.* **38**, 2077-2091.

Kalnay, E., and Paegle, J. (1983). Large amplitude stationary waves in the southern hemisphere observations and theory. *Proc. Int. Conf. Southern Hemisphere Meteorol., 1st.*

Kalnay, E., Balgovind, R., Chao, W., Edelmann, D., Pfaendtner, J., Takacs, L., and Takano, K. (1983). Documentation of the GLAS fourth order general circulation model, Volumes 1, 2, and 3. *NASA Tech. Memo. 86064.*

Kalnay, E., Mo, K. C., and Paegle, J. (1986). Large-amplitude, short-scale stationary Rossby waves in the Southern Hemisphere: Observations and experiments to determine their origin. *J. Atmos. Sci.* **43**, 252-275.

Lau, K. M., and Chan, P. H. (1983). Short term climate variability and atmospheric teleconnection from satellite observed outgoing longwave radiation, Part I: Simultaneous relationships. *J. Atmos. Sci.* **40**, 2735-2750.

Matsuno, T. (1966). Quasi-geostrophic motions in the equatorial area. *J. Meteorol. Soc. Jpn.* **44**, 25-43.

Newell, R. E., Kidson, J. W., Vincent, D. G., and Baer, G. J. (1972). "The General Circulations of the Tropical Atmosphere," Vol. 1. p. 116. MIT Press, Cambridge, Massachusetts.

Paegle, J., and Baker, W. E. (1983). The influence of the tropics on the prediction of ultra long waves, Part II: Latent heating. *Mon. Weather Rev.* **111**, 1356-1371.

Randel, W. J., and Sanford, J. L. (1984). Structure of medium-scale atmospheric waves in the southern hemisphere summer 1983. *J. Atmos. Sci.* **40**, 2312-2318.

Salby, M. L. (1982). A ubiquitous wave 5 anomaly in the southern hemisphere during FGGE. *Mon. Weather Rev.* **110**, 1712-1720.

Trenberth, K. E. (1979). Planetary waves at 500 mb in the Southern Hemisphere. *Mon. Weather Rev.* **108**, 1378-1389.

Taljaard, J. J., van Loon, H., Crutcher, H. L., and Jenne, R. L. (1969). Climate of the upper air: Southern Hemisphere, Vol. 1. Temperatures, dew points and heights at selected pressure levels. NAVAIR 150-1C-55, Chief Naval Oper., Washington, DC.

van Loon, H., and Jenne, R. L. (1972). The zonal harmonic standing waves in the Southern Hemisphere. *J. Geophys. Res.* **77**, 992-1003.

van Loon, H., Jenne, R. L., and Lakitzke, K. (1973). Zonal harmonic standing waves. *J. Geophys. Res.* **78**, 4463-4471.

Webster, P. J. (1972). Response of the tropical atmosphere to local steady forcing. *Mon. Weather Rev.* **100**, 518-541.

SST ANOMALIES AND BLOCKING

J. SHUKLA

Center for Ocean-Land-Atmosphere Interactions
Department of Meteorology
University of Maryland
College Park, Maryland 20742

1. INTRODUCTION

There is a large body of published literature on the possible role of sea surface temperature (SST) anomalies in producing quasi-stationary circulation anomalies in midlatitudes, to be referred to, loosely, as blocking. We present here only a summary of what we consider to be important physical concepts that have appeared during the past decade in connection with the problem of the influence of SST anomalies on extratropical circulation anomalies.

Let us first ask a basic question. How does a SST anomaly influence the atmospheric circulation? The immediate effect of a SST anomaly is to alter the sensible heat flux and evaporation over the region of the SST anomaly. This will alter the temperature and humidity of the overlying air. If that was all the SST anomaly did, there would be little change in the atmospheric circulation. However, if the sensible heating of the air produces gradients of surface pressure, then it can change the convergence of the moist air and precipitation in the region of the SST anomaly. The magnitude and the structure of the convergence pattern will depend upon the magnitude and structure of the SST pattern itself. Warmer and moister surface air might lead to deeper moist convection and heating of the atmospheric column. This might lead to further enhancement of the surface pressure gradients, an increase in moisture convergence in the boundary layer, and an increase in precipitation. This Convective Instability of the Second Kind, or CISK-type positive feedback, among heating of the vertical column, convergence in the boundary layer, and precipitation can transform the initial surface heat flux anomaly into a three-dimensional heating anomaly which can produce changes in large-scale atmospheric circulation. Based on these arguments it is possible to present the following simple conceptual framework to understand the effects of SST anomalies on atmospheric circulation.

$$\partial T \to \partial Q \to \partial C$$

The SST anomaly (∂T) produces a deep heating anomaly (∂Q), which in turn produces a circulation anomaly (∂C). For longer periods, ∂C in turn can affect ∂T, however, the discussion in this paper will be confined to those time scales for which ∂T is quasi-steady (about a month or a season). It is now simple to state, at least in this conceptual framework, that whether a SST anomaly will produce a circulation anomaly will depend upon two things: the SST anomaly should produce a heating anomaly, and the heating anomaly should produce a circulation anomaly. We shall examine the two factors separately.

Let us first examine the factors which will determine whether a given SST anomaly will produce a heating anomaly or not. These factors are as follows:

1. The magnitude and structure of the SST anomaly.
2. The magnitude and structure of the climatological SST on which the anomaly is superimposed.

These two factors together determine the modified SST field due to presence of an anomaly. Since the atmosphere is influenced by the total SST, not just the SST anomaly, the anomaly alone is not of any special significance except as a convenient descriptor of the SST field. The basic SST field is of utmost importance because a change in saturation vapor pressure for a 1° change from 29° to 30°C is much larger than that for a similar change from 20° to 21°C. Moreover, the structure of the actual SST field determines the magnitude and the pattern of convergence which is crucial for the establishment of a deep heat source.

There are additional factors which determine the transformation of a SST anomaly into a deep-heating anomaly and which suggest that the tropical SST anomalies are more effective than the midlatitude SST anomalies in producing deep-heat sources.

3. The latitude of the SST anomaly. Changes in the thermal wind produced by the changes in SST depend upon the rotation rate. Even weaker SST gradients in the tropics can produce large changes in thermal wind and convergence of moisture in the boundary layer.

4. The location of the SST anomaly with respect to the large scale flow. If a warm SST anomaly occurs under the ascending branches of the Hadley and/or Walker cells in the tropics, it can increase the moisture convergence and heating, whereas a similar warm anomaly under the descending branch will not be able to produce moisture convergence and rainfall because even the additional moisture due to enhanced evaporation will be diverged away.

5. The dominant instability mechanism. Warm SST anomalies in the tropics might enhance the moist convection quickly and produce a deep heat source due to moist convective instability. If the effect of the SST anomaly

were only to change the shear of the flow and thereby to change the dynamical instability properties of the flow, that may involve a relatively longer time scale.

A similar consideration of the factors which determine the influence of a given heating anomaly on the circulation anomaly also suggests that there are distinctly different processes that determine the response due to tropical and midlatitude heating anomalies. It is therefore appropriate to divide the remaining discussion into four broad topics covering the influence of tropical and extratropical SST anomalies on the tropics and extratropics.

There is a vast amount of literature covering these topics and we cannot present a comprehensive review of the previous works for each topic. We shall confine our discussion mainly to the topics in Sections 2 and 3, and we shall summarize the results of only a few representative investigations.

2. Influence of Tropical SST Anomalies on Extratropical Circulation

The question of the influence of the tropical heating on midlatitude circulation has been investigated using past atmospheric observations and models of varying degrees of complexity ranging from linear and simple nonlinear models to global general circulation models (GCMs). The most significant difference between the simple models and the GCMs lies in the treatment of the heating anomaly $\partial \dot{Q}$. In GCMs, ∂T is explicitly used to calculate $\partial \dot{Q}$ from model physics and dynamics, whereas in simple models $\partial \dot{Q}$ is prescribed. The GCMs enable us to examine the effects of all the processes involved in $\partial T \rightarrow \partial \dot{Q} \rightarrow \partial C$, whereas the simple models examine the processes involved in $\partial \dot{Q} \rightarrow \partial C$ only. Simple models also prescribe the large-scale mean flow which can be either zonally symmetric or have simple asymmetries. In spite of these limitations of the simple models, they are useful tools for carrying out a large number of experiments and for helping to understand the mechanisms for the remote response to tropical heating anomalies.

The mechanisms suggested so far to explain the midlatitude effects of tropical heating anomalies can be summarized under the following broad categories.

1. *Rossby wave propagation.* Forced (steady) Rossby wave solutions of the linearized barotropic vorticity equation show remarkable similarity to some of the observed patterns of circulation anomalies associated with tropical heating. However, this mechanism is not adequate to explain the amplitudes of the observed circulation anomalies. Moreover, the tropical and the

midlatitude flows are not necessarily steady for the time scales that are required to set up stationary Rossby waves.

2. *Instability of the horizontally and vertically sheared zonal flows.* The horizontal and vertical shears of observed flows can allow a variety of perturbations to grow by barotropic and/or baroclinic instability mechanisms. The tropical heating, depending upon its magnitude and location, can act as a trigger for the growth of some preferred normal modes.

3. *Modification of the Hadley circulation.* Tropical heat sources can modify the intensity of the Hadley circulation, which in turn can change the zonal flow in the midlatitudes. The modified zonal flow, interacting with the preexisting quasi-stationary heat sources (and mountains), can produce quasi-stationary circulation anomalies. Circulation anomalies can also be produced by growth and decay of perturbations unstable with respect to the modified zonal flow. There have been no systematic modeling or diagnostic studies to determine the role of modified zonal flows.

If one examines the results of numerical experiments with multilevel global primitive-equation models, one can find partial support for each of the mechanisms described above. It is found that the differences in the model simulations with and without tropical heating show, especially for the first 5–15 days, well-defined Rossby wave trains with very small amplitudes (~ 10 m) emanating from the heat source. However, after about 2 weeks or less, the differences due to a tropical heat source are indistinguishable from the differences that could arise due to instabilities and nonlinear interactions in the midlatitude flow itself, irrespective of the tropical heat sources.

A spatially coherent pattern of significant correlations between equatorial Pacific SST anomalies and the circulation over North America has been found by several investigators (Horel and Wallace, 1981). This pattern of significant correlations is also reproduced by a long integration (15 years) of a GCM using the observed SST over the equatorial Pacific (Lau and Oort, 1985).

Linear model calculations have shown that the midlatitude response due to tropical heating anomalies depends upon the amplitude and vertical structure of heating, the structure of the prescribed zonal flow (especially the presence/absence of a zero wind line), vertical resolution and the upper boundary of the model, and the parameterization of dissipation. If one chooses a suitable value of dissipation for producing a reasonable tropical response, there is no guarantee that the midlatitude response will also be equally reasonable.

We have chosen to describe the results of Nigam (1985) because he does not make arbitrary assumptions about the heating and the zonally averaged zonal flow. These fields are taken from a GCM integration, and the linear model has been constructed by linearizing the same GCM equations. The horizontal and vertical resolutions as well as the vertical spacing of the model levels are identical for the GCM and the linear model. The GCM solution is used to verify the results of the linear model. It is found that after a suitable tuning of the dissipation term, the stationary solution produced by the linear model is remarkably similar to the GCM simulation. The dissipation term in the linear model can be assumed to mimick, at least partially, the effects of nonlinearity and transients. Separation of the linear solutions forced by tropical heating and orography shows that the stationary-wave amplitude in middle latitudes due to tropical heating is only about 50 m, whereas the orographically forced solution is about 300 m. Results of this and several other linear model studies suggest that the tropical heating is not an important forcing for the midlatitude stationary waves. However, the limitations of the linear models, and especially the prescription of damping, should be thoroughly examined before accepting these results.

Sardeshmukh and Hoskins (1985) have carried out diagnostic studies using atmospheric observations and shown that the regions of deep tropical heating and large upper level divergence occur in conjunction with very small absolute vorticity, and therefore a linear balance between the β term and the divergence term, which describes the dominant balance in the linear models, is not valid. They find that nonlinear advection is quite important to get a reasonable vorticity balance at the upper levels. They also find the transients to be important to describe the observed flow. Schneider (1985) has shown that the essential features of the large-scale tropical flow can be simulated by a steady nonlinear vorticity equation with prescribed divergence sources.

It should be noted, however, that most of these studies are concerned with the explanation of climatological mean, stationary waves and not that of the interannual variability of quasi-stationary anomalies. The role of SST anomalies in producing interannual variability of monthly or seasonal means and the relative importance of the prescribed zonal flow and the prescribed heating have not been fully examined using multilevel linear models.

Sensitivity experiments to determine the influence of tropical SST anomalies on midlatitude circulation using GCMs have also produced a variety of results which are too ambiguous to fit any conceptual framework for the physical mechanisms involved. For example, the transient response (change in circulation for the first few weeks after the SST anomaly is introduced) can be quite different from the equilibrium response (change in circulation after a long-term integration of the model with and without the SST anomaly). In a series of short integrations (~ 75 days) with the Goddard Laboratory for Atmospheric Sciences (GLAS) model, Shukla and Wallace (1983) found that the simulated response due to composite El Niño SST anomalies was very similar to the observed circulation anomaly during the El Niño years. However, the observed midlatitude circulation anomaly during the winter of 1982/1983 could not be simulated well using a similar model with the observed SST anomaly during 1982/1983 (Fennessy et al., 1985).

Several GCMs have been used to simulate the effects of the observed 1982/1983 SST anomaly in the equatorial Pacific, and there are considerable differences among the midlatitude responses simulated by different GCMs. The following impressions emerge from a large number of GCM sensitivity studies:

1. The tropical response (change in location of maximum precipitation and accompanying changes in wind flow, surface pressure, and vertically integrated temperature) is well simulated by most of the models.

2. The initial spin-up time (the time taken by a GCM to produce a well-defined large-scale precipitation anomaly after the SST anomaly is introduced) strongly depends upon the parameterizations of boundary layer and moist convection.

3. The midlatitude response is different for different GCMs. In particular, the transient response depends on the structure of the initial conditions, and the equilibrium response depends upon the model climatology.

Since the transient midlatitude response depends strongly on the initial conditions, it can be conjectured that the interactions between the perturbations forced by the tropical heating and the preexisting stationary and transient waves in the midlatitude could be an important factor in determining the "final" midlatitude response. The location of the tropical heating and the location of the corresponding forced wave trains with respect to the amplitudes and phases of the midlatitude planetary waves could be an important factor in determining the evolution of the midlatitude flow. Direct interactions between the forced wave trains and the mountains can also influence the subsequent evolution of the flow.

A long-term integration (~15 yr) by the Geophysical Fluid Dynamics Laboratory (GFDL) model by Lau and Oort (1985) using the observed SST anomalies in the tropical Pacific has shown that the tropical as well as the midlatitude circulation anomalies were well simulated. In particular, the pattern of correlation between the tropical SST anomaly and midlatitude circulation anomaly is remarkably similar to that of the observations. This suggests that the high-frequency transient disturbances in the tropics are not large enough to alter the large-scale, low frequency tropical variability due to the influence of the boundary conditions.

There have been only a few experiments which have examined the predictability of the midlatitude flow with and without the tropical SST anomaly. Such experiments should provide direct confirmation of the effect of tropical

SST anomalies on midlatitude flow. There is some evidence of an improvement in the midlatitude forecasts for 30-day averages using the observed SST compared to the climatological SST; however, such studies are only few in number, and a large number of such controlled experiments are required to establish the robustness of this result.

3. Influence of Extratropical SST Anomalies on Extratropical Circulation

The earlier discussion on the factors that determine the transformation of a SST anomaly into a deep-heating anomaly ($\partial T \rightarrow \partial \dot{Q}$) suggests that it is relatively more difficult to transform a SST anomaly into a heating anomaly in the extratropics, compared to the tropics. The most important reasons are (1) the mean SST in midlatitudes is considerably lower than that in the tropics; (2) the pressure and wind fields are in quasi-geostrophic balance and therefore it is not possible to produce large convergence or divergence; (3) there are already large asymmetries in surface heating due to land–ocean contrast; and finally (4) the midlatitude flow is highly variable due to dynamical instabilities. However, there have been numerous observational studies [see the collected papers by Namias (1975)] which suggest a significant relationship between the SST anomalies and circulation anomalies in the midlatitudes. Also, there have been several modeling studies (using simple models and complex GCMs) to investigate the effects of midlatitude SST anomalies on midlatitude circulation but the results are inconclusive. Some meteorologists have expressed the opinion that the tropical SST anomalies might be more effective than the mid-latitude SST anomalies in producing quasi-stationary circulation anomalies in the midlatitudes. Most of the early investigations (Charney and Eliassen, 1949; Smagorinsky, 1953) were primarily concerned with explaining the observed climatological stationary waves in the atmosphere as responses to orographic forcing or forcing due to stationary heat sources. It is now generally accepted that the mountains provide the most important forcing for the stationary waves. It is further argued that if the climatological heat sources are not effective in producing stationary anomalies, it is inconceivable that the SST anomalies and the associated weak heat sources can produce changes in the circulation sufficiently large to be distinguished from the natural variability of otherwise active midlatitude flow.

The results of the sensitivity experiments with GCMs are also ambiguous, and at times contradictory. For example, Kutzbach *et al.* (1977) and Chervin *et al.* (1980) reported that for the National Center for Atmospheric Research (NCAR) model, even an unrealistically large SST anomaly (about four times larger than the observed) over the northern Pacific could not produce a statistically significant change in the circulation away from the anomaly. Shukla and Bangaru (1978) found that a SST anomaly about twice as large as the observed SST anomaly could produce large changes in the planetary-scale circulation, giving rise to significant changes away from the SST anomaly. A similar result has also been reported by Pitcher et al. (1984) using a new version of the NCAR model. They found significant changes in circulation away from the anomaly using a SST anomaly twice the observed SST anomaly. However, they did not find any significant changes in circulation for the observed anomaly itself. It appears that these results are largely model dependent. It has been pointed out by Blackmon and Lau (1980) that the version of the NCAR model used by Chervin et al. (1980) had serious deficiencies in simulating the transient variability, and therefore results of sensitivity studies with that model should not be considered a sufficient evidence for the inability of midlatitude SST anomalies to produce circulation anomalies. In fact it is not unconceivable that even the present models, which show a significant response to a SST anomaly twice the observed anomaly, can be further improved, and then it may not be necessary to artificially increase the anomaly.

Heuristically, it can be argued that it would be more difficult to detect the influence of midlatitude SST anomalies on midlatitude flow because the flow already contains preexisting wave patterns due to stationary forcings and dynamical instabilities. The response of the flow is going to be very different depending upon the location of the SST-induced heat source with respect to the structure of the preexisting wave patterns. Moreover, the steady and the transient components are equally important in describing the dynamics of the midlatitude flow, and therefore the effects of SST-induced heat sources have to be significantly larger than the transient variability which is naturally present at the midlatitudes.

For this reason, it will also be difficult to detect the impact of SST anomalies on deterministic prediction using GCMs. If the model physics and boundary-layer parameterizations do not generate a strong heat source within about 2 weeks of the introduction of the SST anomaly, the midlatitude flow can change significantly before being influenced by the SST anomaly.

We do not wish to discuss here the influence of tropical SST anomalies on tropical circulation nor the influence of extratropical SST anomalies on tropical circulation. It would suffice to state that it has been clearly established, from observational as well as modeling studies, that the tropical SST anomalies exert profound influence on the tropical circulation. The most notable examples are the El Niño–Southern Oscillation–monsoon interactions, and the role of SST anomalies in droughts over northeast Brazil and tropical Africa. As regards the influence of extratropical SST anomalies on tropical circulation, the results are as ambiguous as those discussed in Section 3; however, there have been several suggestions that even the small circulation anomalies produced by the midlatitude SST anomalies can trigger changes in convection in the tropical regions which can further amplify to produce large heating anomalies in the tropics.

REFERENCES

(Selected bibliography on the influence of SST anomalies on circulation anomalies)

Bjerknes, J. (1969). Atmospheric teleconnections from the equatorial Pacific. *Mon. Weather Rev.* **97**, 163–172.

Blackmon, M. L., and Lau, N. C. (1980). Regional characteristics of the Northern Hemisphere wintertime circulation: A comparison of the simulation of a GFDL general circulation model with observations. *J. Atmos. Sci.* **37**, 497–514.

Blackmon, M. L., Geisler, J. E., and Pitcher, E. J. (1983). A general circulation model study of January climate anomaly patterns associated with interannual variation of equatorial Pacific sea surface temperatures. *J. Atmos. Sci.* **40**, 1410–1425.

Branstator, G. (1985a). Analysis of general circulation model sea surface temperature anomaly simulations using a linear model. I: Forced solutions. *J. Atmos. Sci.* **42**, 2225–2241.

Branstator, G. (1985b). Analysis of general circulation model sea surface temperature anomaly simulations using a linear model. II: Eigenanalysis. *J. Atmos. Sci.* **42**, 2242–2254.

Charney, J. G., and Eliassen, A. (1949). A numerical method for predicting the perturbations of the middle latitude westerlies. *Tellus* **1**, 38–54.

Charney, J. G., and Shukla, J. (1981). Predictability of monsoons. In "Monsoon Dynamics" (Sir James Lighthill and R. P. Pearce, eds.). Cambridge Univ. Press, London.

Chervin, R. M., Kutzbach, J. E., Houghton, D. D., and Gallimore, R. G. (1980). Response of the NCAR general circulation model to prescribed changes in ocean surface temperature. Part II: Midlatitude and subtropical changes. *J. Atmos. Sci.* **37**, 308–332.

Davis, R. E. (1978). Predictability of sea level pressure anomalies over the North Pacific Ocean. *J. Phys. Oceanogr.* **8**, 233–246.

Fennessy, M. J., Marx, L., and Shukla, J. (1985). General circulation model sensitivity to 1982-83 equatorial Pacific sea surface temperature anomalies. *Mon. Weather Rev.* **113**, 858–864.

Horel, J. D., and Wallace, J. M. (1981). Planetary-scale atmospheric phenomena associated with the Southern Oscillation. *Mon. Weather Rev.* **109**, 813–829.

Keshavamurty, R. N. (1982). Response of the atmosphere to sea surface temperature anomalies over the equatorial Pacific and the teleconnections of the Southern Oscillation. *J. Atmos. Sci.* **39**, 1241–1254.

Kutzbach, J. E., Chervin, R. M., and Houghton, D. D. (1977). Response of the NCAR general circulation model to prescribed changes in ocean surface temperature. Part I: Mid-latitude changes. *J. Atmos. Sci.* **34**, 1200–1213.

Lau, N. C. (1981). A diagnostic study of recurrent meteorological anomalies appearing in a 15-year simulation with a GFDL general circulation model. *Mon. Weather Rev.* **109**, 2287–2311.

Lau, N. C., and Oort, A. H. (1985). Response of a GFDL general circulation model to SST fluctuations observed in the tropical Pacific Ocean during the period 1962-1976. *Elsevier Oceanogr. Ser. Coupled Atmos. Ocean Models* pp. 289–302.

Manabe, S., and Hahn, D. G. (1981). Simulation of atmospheric variability. *Mon. Weather Rev.* **109**, 2260–2286.

Moura, A., and Shukla, J. (1981). On the dynamics of droughts in northeast Brazil: Observations, theory and numerical experiments with a general circulation model. *J. Atmos. Sci.* **38**, 2653–2675.

Namias, J. (1975). "Short Period Climatic Variations, Collected Works of J. Namias 1934 through 1974 (2 Vols.). Univ. of California, San Diego.

Nigam, S. (1983). On the structure and forcing of tropospheric stationary waves. Ph. D. dissertation, Princeton University.

Pitcher, E. J., Blackmon, M. L., Bates, G. T., and Munoz, S. (1984). The effect of midlatitude Pacific sea-surface-temperature anomalies on the January climate of a general circulation model. (Unpublished.)

Rowntree, P. R. (1972). The influence of tropical east Pacific Ocean temperatures on the atmosphere. *Q. J. R. Meteorol. Soc.* **98**, 290–321.

Sardeshmukh, P. D., and Hoskins, B. J. (1984). Vorticity balance in the tropics during the 1982-83 El Niño-Southern Oscillation event. *Q. J. R. Meteorol. Soc.* **111**, 261–278.

Schneider, E. K. (1985). A model of the modified Hadley circulation. Paper presented at the IAMAP/IAPSO Joint Assembly, August 5-16, 1985, Honolulu, Hawaii.

Shukla, J., and Bangaru, B. (1979). Effect of a Pacific SST anomaly on the circulation over North America: A numerical experiment with the GLAS model. *GARP Publ. Ser.* **22**, 501–518.

Shukla, J., and Wallace, J. M. (1983). Numerical simulation of the atmospheric response to equatorial Pacific sea surface temperature anomalies. *J. Atmos. Sci.* **40**, 1613–1630.

Simmons, A. J. (1982). The forcing of stationary wave motion by tropical diabatic heating. *Q. J. R. Meteorol. Soc.* **108**, 503–534.

Simmons, A. J., Wallace, J. M., and Branstator, G. W. (1983). Barotropic wave propagation and instability, and atmospheric teleconnection patterns. *J. Atmos. Sci.* **40**, 1363–1392.

Smagorinsky, J. (1953). The dynamical influence of large-scale heat sources and sinks on the quasi-stationary mean motions of the atmosphere. *Q. J. R. Meteorol. Soc.* **79**, 342–366.

Webster, P. J. (1981). Mechanisms determining the atmospheric response to sea surface temperature anomalies. *J. Atmos. Sci.* **38**, 554–571.

Webster, P. J. (1982). Seasonality in the local and remote atmospheric response to sea surface temperature anomalies. *J. Atmos. Sci.* **39**, 41–52.

Webster, P. J., and Holton, J. R. (1982). Cross equatorial response to middle latitude forcing in a zonally varying basic state. *J. Atmos. Sci.* **39**, 722–733.

INDEX

A

Amplitude distribution, waves with bimodal, 105–113
Anomalies
  developing, nonlinear simulations and, 291–301
  mature, *see* Mature anomalies
  monthly-mean, representation, 329–333
  persistent, *see* Persistent anomalies
  SST, *see* Sea surface temperature anomalies
Anomalous circulation
  blocking, 15–25
  steady states, 8–15
Anticyclones, blocking
  maintenance by eddies, 147–158
  forcing by synoptic-scale eddies, 183–197
    results, 186–197
    stochastically forced planetary modes, 184–186
Atlantic blocking episode, eddy forcing
  data manipulation, 139–142
  eddy straining mechanism, 136–139
  eddy vorticity flux divergence patterns, 145–147
  Ertel potential vorticity analysis, 147–158
  E vectors and momentum forcing, 142–145
  synoptic situation, 139–142
Atmospheric blocking, *see* Blocking
Atmospheric circulation
  effect of SST anomalies, 443
  during Northern Hemisphere winter of 1979, 377–379
Atmospheric energetics
  third power law for kinetic energy distribution, 7–8
  waves with bimodal amplitude distributions, 113–123
Atmospheric flow, probability density distribution
  data, 230–231
  nonparametric estimation, 231–233
  patterns of 500-mbar geopotential height and, 235–243
  results, 233–235
  theoretical background, 228–229
  zonal wind and, 246–247
Available potential energy
  atmospheric energetics
    third power law for kinetic energy distribution, 7–8
    waves with bimodal amplitude distributions, 113–123
  concept of, 4–5
  eddy
    conversion from total zonal, 126
    conversions to eddy kinetic energy, 6
    and zonal, calculation, 5
  in spectral domain, 5–6

B

Baroclinic instability
  linear, strong blocking and, 85
  local
    eddy forcing and circulation, 177–180
    properties, 169–170
Baroclinic model atmosphere
  energetics, 215–219
  resonance bending in, 219–223
Bimodal amplitude distribution, waves with, 105–113
Blended orographies, experiments with, 355–359
Blocking
  anticyclones
    maintenance by eddies, 147–158
    forcing by synoptic-scale eddies, 183–197
      results, 186–197
      stochastically forced planetary modes, 184–186
  Atlantic, eddy forcing during
    data and synoptic situation, 139–142

eddy straining mechanism, 136–139
eddy vorticity flux divergence patterns, 145–147
Ertel potential vorticity analysis, 147–158
E vectors and momentum forcing, 142–145
synoptic situation, 139–142
Charney–DeVore model, 90–91
climatology of, 17
effect of extratropical SST anomalies on extratropical circulation, 449–451
effect of tropical SST anomalies on extratropical circulation, 445–449
  instability of horizontally and vertically sheared zonal flows, 446
  modification of Hadley circulation, 446
  Rossby wave propagation, 445–446
North Pacific
  breakdown, 59–64
  data for 500-mbar analysis, 32–33
  development of 500-mbar height anomaly patterns
    low-pass filtered, 34–41
    Pacific jet region zonal flow and, 46–48
    unfiltered anomaly analyses, 42–46
    zonal Fourier harmonics and, 48–50
  procedure, 33–34
  synoptic characteristics of height and thermal patterns, 54–59
  vertical and thermal evolutions, 50–54
numbers, 77–80
  energy fluxes
    correlation analysis, 85–87
    regression analysis, 87–89
  energy parameters and data, 72–76
  kinetic energy budget and, 80–85
prediction of, 318–323
role of cyclone waves in, 127–131
strong, linear baroclinic instability and, 85
as structural entity, 15–25
types, 89–94
  characteristics, 93
  energy parameters and data, 72–76
as wave–wave interaction, 12
Blocks, growth of
  intermediate modes, 289
  Pacific and Atlantic onset-of-blocking dipole modes, 284
Pacific onset-of-blocking dipole modes, 283–284

C

Charney–DeVore circulation model, 200–209
  wave equation modification, 209–215
Charney–DeVore theory, 8, 10, 229
  bimodal zonal wind and, 246–247
  energetics of flux, counterpart for, 91
Charney–Eliassen model, 229–230
Circulation
  extratropical, effect of
    extratropical SST anomalies, 449–451
    tropical SST anomalies, 443–449
  middle latitude, properties, 199–224
    baroclinic energetics, 215–219
    Charney–DeVore model, 200–209
    Charney–DeVore wave equation modification, 209–215
    resonance bending in baroclinic model atmosphere, 219–223
Climatology of blocking, 17
Convective instability of the Second Kind, 443
Cutoff lows, prediction of, 318–323
Cyclone waves, role in blocking, 127–131

D

Data assimilation system at ECMWF, 306–307
Deep-heating anomaly, SST anomalies and, 444
Density distribution of atmospheric flow
  data, 230–231
  nonparametric probability density estimation, 231–233
  patterns of 500-mbar geopotential height and, 235–243
  results, 233–235
  theoretical background, 228–229
Disturbance streamfunctions
  intermediate modes, 289
  monopole cyclogenesis modes, 281–282
  North Atlantic mature anomaly pattern modes, 287–289

INDEX

Pacific and Atlantic onset-of-blocking bipole modes, 284
Pacific-North American mature anomaly pattern modes, 285–287
Pacific onset-of-blocking dipole modes, 283–284

E

ECMWF, *see* European Centre for Medium Range Weather Forecasts
Eddies
  available potential energy conversion to kinetic energy, 6
  fluxes, *see* Eddy fluxes
  forcing, *see* Eddy forcing
  maintenance of blocking anticyclones by, 147–158
  straining, *see* Eddy straining
Eddy fluxes
  associated with local baroclinic instability, 172–177
  in two-layer model, 167–168
Eddy forcing
  during Atlantic blocking episode
    data manipulation and synoptic situation, 139–142
    Ertel potential vorticity analysis, 147–158
    E vectors and momentum forcing, 142–145
    vorticity flux divergence patterns, 145–147
  blocking anticyclones by, 183–197
    results, 186–197
    stochastically forced planetary modes, 184–186
  and circulation, relation, 177–180
Eddy straining, mechanism, 136–139
Eigenmode structure, and storm tracks, 170–172
El niño, 260–262
Energetics
  atmospheric
    third power law for kinetic energy distribution, 7–8
    waves with bimodal amplitude distributions, 113–123
  baroclinic model atmosphere, 215–219
  of flux analysis, 90–91

Enhanced orography, *see* Envelope orography
Enstrophy budget, waves with bimodal amplitude distributions, 113–123
Envelope orography, 340–343
  effects in tropical regions, 363–370
  mean orography vs., 343–355
  mountain torque and zonal flow, 359–363
Equilibria, multiple, *see* Multiple equilibria theories
Ertel potential vorticity, eddy forcing, 147–158
European Centre for Medium Range Weather Forecasts (ECMWF)
  accuracy of forecasts, 310–318
  assessment methods, 308–310
  data assimilation, 306–307
  developments in predictive skill, 323–328
  forecast model, 307–308
  gridpoint model, orographic forcing and systematic error of, 340–343
  monthly-mean anomalies, 328–333
  prediction of blocking and cutoff laws, 318–323
  systematic model errors, 333–334
E vectors, eddy momentum forcing and, 142–145
Extratropical circulation
  effect of tropical SST anomalies, 445–449
    instability of horizontally and vertically sheared zonal flows, 446
    modification of Hadley circulation, 446
    Rossby wave propagation, 445–446
  effect of extratropical SST anomalies, 449–451

F

Flow, zonal, in envelope orography, 359–363
Forcing
  eddy
    during Atlantic blocking episode, 139–158
    Ertel potential vorticity analysis and, 147–158
    local baroclinic instability and, 177–180
    momentum forcing and E vectors, 142–145
    planetary-scale blocking anticyclones, 183–197

orographic, and systematic error of
ECMWF gridpoint model, 340–343
stationary, linearized response of
stationary waves to, 262–269
Forecasts
accuracy in medium range, 310–318
models
at ECMWF, 307–308
at UCLA, 380–381
numerical, *see* Numerical prediction
Fourier harmonics, zonal, persistent
anomalies and, 48–50
Frequency distribution, planetary-wave
amplitude indicator, 103–105

G

General circulation models, 259, 263
ECMWF, envelope orography and, 339–340
origin of Southern Hemisphere stationary
Rossby waves
analysis and control experiments,
417–427
"easterly deceleration" experiment,
437–439
"no Andes" experiment, 427–429
"reduced tropical heating" experiment,
429–435
"suppressed regional heating"
experiment, 435–436
UCLA, medium-range experimental
forecasts, 375–377
Global circulation
anomalous
blocking, 15–25
steady states, 8–15
theoretical development, 315
Gridpoint model at ECMWF, orographic
forcing and systematic error of,
340–343

H

Hadley circulation, modification, 446
Heating, tropical, Pacific-North American
pattern and, 259–262

I

Instability theory
basic state, 279–280

disturbances, 280–281
disturbance streamfunctions
intermediate modes (class F), 289
monopole cyclogenesis modes (class A),
281–282
North Atlantic mature anomaly pattern
modes (class E), 287–289
Pacific and Atlantic onset-of-blocking
dipole modes, 284
Pacific-North American mature
anomaly modes (class D), 285–287
Pacific onset-of-blocking dipole modes
(class B), 283–284
evolution of mature anomalies, 289–291
growth rates and phase frequencies, 281
model details, 278
nonlinear simulation
disturbance kinetic energy spectra,
293–294
disturbance streamfunctions, 294–301
model details, 292–293

K

Kinetic energy
budget, and blocking numbers, 80–85
eddy, conversion from eddy available
potential energy, 6
Kolmogorov–Smirnov test, 232–235, 246

L

Local baroclinic instability
eddy fluxes and tendencies, 172–177
eigenmode structure, 170–172

M

Mature anomalies
evolution, 289–291
intermediate modes, 289
North Atlantic pattern modes, 287–289
Pacific-North American pattern modes,
285–287
Maximum penalized likelihood, method of,
232
Medium-range forecasts, accuracy of,
310–318
Model atmospheres, baroclinic, *see*
Baroclinic model atmosphere

INDEX 457

Model errors, systematic, *see* Systematic model errors
Momentum forcing, E vectors and eddy, 142–145
Monopole cyclogenesis modes, disturbance streamfunctions, 281–282
Monthly-mean anomalies, representation of, 329–333
Mountain torque, in envelope orography, 359–363
Multiple equilibria theories, objections to, 255–259

N

Nonlinear simulations of baroclinic waves, instability solutions and, 291–301
Nonparametric estimation theory, 231–233
Northern Hemisphere
  extratropical, predictability for three regions, 313
  middle latitude circulation, properties, 199–224
    baroclinic energetics, 215–219
    Charney–DeVore model, 200–209
    Charney–DeVore wave equation modification, 209–215
    resonance binding in baroclinic model atmosphere, 219–223
  tropospheric and stratospheric forecasts for winter of 1979, 381–398
North Pacific, persistent anomalies and blocking
  breakdown, 59–64
  data, 32–33
  development of 500-mbar height anomaly patterns, 34–50
  procedure, 33–34
  synoptic characteristics of height and thermal patterns, 54–59
  vertical and thermal evolutions during development, 50–54
Numerical forecasts, *see* Numerical prediction
Numerical prediction
  assessment methods, 308–310
  blocking and cutoff lows, 318–323
  ECMWF forecasting system
    data assimilation, 306–307
    forecast model, 307–308

medium-range forecast accuracy, 310–318
Northern Hemisphere winter of 1979
  atmospheric circulation during, 377–381
  stratospheric forecasts, 392–398
  tropospheric forecasts, 381–392
skill developments in, 323–328

O

Orographic forcing, and systematic error of ECMWF gridpoint model, 340–343
Orography
  blended, experiments with, 355–359
  envelope
    effects in tropical regions, 363–370
    mean orography vs., 343–355
    mountain torque and zonal flow, 359–363

P

Pacific jet region zonal flow, and North Pacific persistent anomalies, 46–48
Pacific–North American pattern, role of tropical heating in, 259–262
Persistent anomalies
  duration of, 253–255
  free Rossby waves and, 269–271
  multiple equilibria theories and, 256–259
  over North Pacific
    breakdown, 59–64
    data, 32–33
    development of 500-mbar height anomaly patterns
      Fourier harmonics and, 48–50
      low-pass filtered, 34–41
      Pacific jet region zonal flow and, 46–48
      unfiltered analyses, 42–46
    procedure, 33–34
    synoptic characteristics of height and thermal patterns, 54–59
    vertical and thermal evolutions during development, 50–54
  tropical heating role in Pacific–North American pattern, 259–262
  teleconnections and, 259–262
Planetary waves
  with bimodal amplitude distributions
    eddy available to eddy kinetic energy conversion, 126

458   INDEX

energetics and enstrophy budgets for, 113–123
frequency distribution of amplitude indicator, 103–105
structure, 105–113
total zonal to eddy potential energy conversion, 126
stationary, linearized response to stationary forcing, 262–269
Potential vorticity, eddy straining and, 137
Predictability, three regions of extratropical Northern Hemisphere, 313
Prediction, numerical, *see* Numerical prediction
Predictive skill, developments in, 323–328

R

Resonance
existence of, 257–259
bending in baroclinic model atmosphere, 219–223
Rossby waves
free planetary-scale, 253
and persistence, 269–271
propagation, and SST anomalies, 445–446
Southern Hemispheric stationary, origin, 417–439
analysis and control experiments, 417–427
"easterly deceleration" experiment, 437–439
"no Andes" experiment, 427–429
"reduced tropical heating" experiment, 429–435
"suppressed regional heating" experiment, 435–436

S

Sea surface temperature anomalies
extratropical, effect on extratropical circulation, 449–451
heating anomaly production by, 444
transformation into a deep heating anomaly, 444
tropical, effect on extratropical circulation, 445–449

instability of horizontally and vertically sheared zonal flows, 446
modification of Hadley circulation, 446
Rossby wave propagation, 445–446
Southern Hemisphere, origin of stationary Rossby waves
analysis and control experiments, 417–427
"easterly deceleration" experiment, 437–439
"no Andes" experiment, 427–435
"reduced tropical heating" experiment, 429–435
"suppressed regional heating" experiment, 435–436
Stationary waves
linearized response to stationary forcing, 262–269
orographically induced, 228
Steady states, 8–15
Storm tracks, eigenmode structure and, 170–172
Straining, eddy, 136–139
Stratospheric forecasts, Northern Hemisphere winter of 1979, 392–398
Synoptic-scale variability, *see* Persistent anomalies
Systematic model errors, 339
at ECMWF, 333–334
and orographic forcing of ECMWF gridpoint model, 340–343
UCLA general circulation model
lowered upper boundary level impact, 408–411
upper boundary impact on tropospheric forecasts
error distributions at 500 mbar, 402–405
root-mean-square error fields, 398–402
ultralong wave forecasts, 405–407

T

Teleconnections, role of, 259–262
Temperature
sea surface, anomalies, *see* Sea surface temperature anomalies
zonal inhomogeneities, effects of local baroclinic instability, 165–180

circulation induced by eddy forcing, 177–180
eddy fluxes and tendencies, 172–177
eddy fluxes in two-layer model, 167–168
eigenmodes and storm tracks, 170–172
properties of local baroclinic instability, 169–170

Thermal patterns, evolution during development of persistent anomalies, 50–54

Torque, mountain, in envelope orography, 359–363

Tropical heating, Pacific-North American pattern and, 259–262

Tropical regions, effects of envelope orography in, 363–370

Tropospheric forecasts, Northern Hemisphere winter of 1979, 381–392

U

University of California, Los Angeles (UCLA)
general circulation model
description, 380–381
stratospheric forecasts, 392–398
tropospheric forecasts, 381–392
upper boundary impact on tropospheric forecasts, 398–411
error distributions at 500 mbar, 402–405
root-mean-square error fields, 398–402
ultralong wave forecasts, 405–407

V

Variability, synoptic-scale, *see* Persistent anomalies

Vertical structures, evolution during development of persistent anomalies, 50–54

Vorticity
eddy, flux divergence patterns, 145–147
Ertel potential, eddy forcing and, 147–158
potential, eddy straining and, 137
zonal inhomogeneities, effects of local baroclinic instability, 165–180

circulation induced by eddy forcing, 177–180
eddy fluxes and tendencies, 172–177
eddy fluxes in two-layer model, 167–168
eigenmodes and storm tracks, 170–172
properties of local baroclinic instability, 169–170

W

Wave equations, Charney–DeVore, 209–215

Waves
cyclone, role in blocking, 127–131
planetary, with bimodal amplitude distributions
eddy available to eddy kinetic energy conversion, 126
energetics and enstrophy budgets for, 113–123
frequency distribution of amplitude indicator, 103–105
structure, 105–113
total zonal to eddy potential energy conversion, 126

Rossby
free planetary-scale, 253
and persistence, 269–271
propagation, and SST anomalies, 445–446
Southern Hemispheric stationary, origin, 417–439

stationary
linearized response to stationary forcing, 262–269
orographically induced, 228

Wave–wave interaction, blocking as, 12

Z

Zonal flow
in envelope orography, 359–363
horizontally and vertically sheared, instability, 446
Pacific jet region, and North Pacific persistent anomalies, 46–48

Zonal Fourier harmonics, persistent anomalies and, 48–50

Zonal wind, probability density distribution and, 246–247, 446

RAYMOND H. FOGLER LIBRARY

DATE DUE

BOOKS ARE SUBJECT TO
RECALL AFTER TWO WEEKS

OCT 18